Quantum Cosmology and the Laws of Nature

Scientific Perspectives on Divine Action

Second Edition

A Series on "Scientific Perspectives on Divine Action"

First Volume
Quantum Cosmology and the Laws of Nature:
Scientific Perspectives on Divine Action
Edited by Robert John Russell, Nancey Murphy, and C. J. Isham

Second Volume
Chaos and Complexity:
Scientific Perspectives on Divine Action
Edited by Robert John Russell, Nancey Murphy, and Arthur R. Peacocke

Future Scientific Topics
Evolutionary Biology
Neuro-biology and Brain Research
Quantum Physics and Quantum Field Theory

Jointly published by the Vatican Observatory and
The Center for Theology and the Natural Sciences

Robert John Russell, General Editor of the Series

Supported in part by a grant from
the Wayne and Gladys Valley Foundation.

Quantum Cosmology and the Laws of Nature

Scientific Perspectives on Divine Action

Second Edition

Robert John Russell
Nancey Murphy
C. J. Isham

Editors

Vatican Observatory
Publications,
Vatican City State

The Center for Theology
and the Natural Sciences,
Berkeley, California

1999

Robert John Russell (General Editor) is Professor of Theology and Science In Residence, Graduate Theological Union, Founder and Director, The Center for Theology and the Natural Sciences, Berkeley, California.

Nancey Murphy is Associate Professor of Christian Philosophy, Fuller Theological Seminary, Pasadena, California.

C. J. Isham is Professor of Theoretical Physics, The Blackett Laboratory, Imperial College, London, England.

Copyright © 1993, 1996 by the Vatican Observatory Foundation

Reprinted 1999

Jointly published by the Vatican Observatory and
The Center for Theology and the Natural Sciences

Manufactured in the United States of America

Distributed (except in Italy and the Vatican City State) by
    The University of Notre Dame Press
    Notre Dame, Indiana 46556
    USA

Distributed in Italy and the Vatican City State by
    Libreria Editrice Vaticana
    V-00120 Citta del Vaticano
    Vatican City State

ISBN 0-268-03975-5
ISBN 0-268-03976-3 (pbk.)

ACKNOWLEDGEMENTS

The editors wish to express their gratitude to the Vatican Observatory and the Center for Theology and the Natural Sciences for co-sponsoring this research. Particular appreciation goes to George Coyne and Bill Stoeger, whose leadership and vision made this series of conferences possible.

Editing for this volume began with an initial circulation of papers for critical responses before the conference in 1991, and continued with extensive interactions between editors and authors after the conference. The editors want to express their gratitude to participants for their written responses to early drafts and for the enthusiastic discussions during the conference.

Regarding this volume, special thanks go to Karen Cheatham, CTNS Editing Coordinator, who devoted meticulous attention and long hours in preparing the camera-ready manuscript for this volume. Her effort ensured the quality of the final manuscript and made it possible for us to meet our deadline. Thanks also to Robin Ficklin-Alred, Alison Levin, and other staff members for their efforts, and to George Coyne for overseeing printing, jacket design and distribution. One of us (RJR) wishes to thank the Vatican Observatory for its hospitality during the initial editing of the volume.

The Center for Theology and the Natural Sciences wishes to acknowledge the generous support of the Wayne and Gladys Valley Foundation. Their initial grant makes our on-going collaboration with the Observatory possible. A second grant from the Valley Foundation supported the production of this volume.

TABLE OF CONTENTS

Introduction
R. J. Russell 1

I. SCIENTIFIC BACKGROUND: STANDARD AND QUANTUM COSMOLOGIES

Introduction to General Relativity and Cosmology
G. F. R. Ellis and W. R. Stoeger 35

Quantum Theories of the Creation of the Universe
C. J. Isham 51

II. METHODOLOGY: RELATING THEOLOGY AND SCIENCE

On Theological Interpretations of Physical Creation Theories
M. Heller 93

Metaphors and Time Asymmetry: Cosmologies in Physics and
Christian Meanings
S. Happel 105

III. PHILOSOPHICAL ISSUES: TIME AND THE LAWS OF NATURE

The Debate over the Block Universe
C. J. Isham and J. C. Polkinghorne 139

The Intelligibility of Nature
P. C. W. Davies 149

Quantum Cosmology, the Role of the Observer, Quantum Logic
A. A. Grib 165

Divine Action, Human Freedom, and the Laws of Nature
W. Alston 185

Contemporary Physics and the Ontological Status of
the Laws of Nature
W. R. Stoeger 207

IV. THEOLOGICAL IMPLICATIONS 1: TIME AND QUANTUM COSMOLOGY

The Temporality of God
J. R. Lucas 235

God as a Principle of Cosmological Explanation
K. Ward 247

The Trinity In and Beyond Time
T. Peters 263

Finite Creation without a Beginning: The Doctrine of Creation in Relation to Big Bang and Quantum Cosmologies
R. J. Russell 291

A Case Against Temporal Critical Realism? Consequences of Quantum Cosmology for Theology
W. B. Drees 327

V. THEOLOGICAL IMPLICATIONS 2: THE LAWS OF NATURE

The Theology of the Anthropic Principle
G. F. R. Ellis 363

Evidence of Design in the Fine-Tuning of the Universe
N. Murphy 401

The Laws of Nature and the Laws of Physics
J. Polkinghorne 429

LIST OF CONTRIBUTORS 441

INDEX 443

INTRODUCTION

Robert John Russell

It is not at all clear that the notion of Divine action makes any sense, or what sort of sense it makes. If it makes no sense, the Christian faith may for a while cling on to a tenuous and marginal existence as a set of legends outlining an optional policy of life. But it will eventually evaporate to take its place with the great legends of Greek and Roman mythology, its policy of life at last becoming as quaint and archaic as that of ancient Athens. It is therefore a matter of vital importance to examine the idea of Divine action, starting again from first principles, to discover what may be said of it in view of the many difficulties raised in the modern age.[1]

1 *Background to the Conference*

In October, 1991, twenty-one scholars with cross-disciplinary expertise in physics, cosmology, philosophy of religion, philosophy of science, philosophical theology, systematic theology, history of religion, and history of science, met for a week-long research conference at the Vatican Observatory near Rome. The purpose of the conference was to explore the implications of quantum cosmology, as well as the status and meaning of the laws of nature, for theological and philosophical issues surrounding the topic of divine action. The resulting papers form the contents of this volume. The conference was the first of a series of five such research conferences being planned for the decade of the 1990s on theology, philosophy and the natural sciences, jointly sponsored by the Vatican Observatory and the Berkeley-based Center for Theology and the Natural Sciences (CTNS). The overarching goal of these conferences is twofold: to contribute to constructive theological research as it engages current research in the natural sciences and to investigate the philosophical and theological elements in ongoing theoretical research in the natural sciences.

The Vatican Observatory, or "Specola Vaticana," is housed in the Papal Palace in the picturesque town of Castel Gandolfo, poised overlooking Lake Albano thirty miles southeast of Rome. Since 1935 it has been the site of basic research in both observational and theoretical astronomy. It is also here that the current Pope resides during the summer. In former years John Paul II, then Bishop of Cracow, had frequently entered into conversations on cosmology and philosophy with Polish friends and colleagues. On becoming Bishop of Rome, he continued his interest in this kind of dialogue and sought to improve the relationships between the Church and the scientific community. Early in his

---

[1] Keith Ward, *Divine Action* (London: Collins Religious Publishing, 1990),

Papacy in 1979 in an address to commemorate the 100th anniversary of the birth of Albert Einstein he said:

> ... I hope that theologians, scholars, and historians, animated by a spirit of sincere collaboration, will study the Galileo case more deeply and, in loyal recognition of the wrongs from whatever side they come, will dispel the mistrust that still opposes, in many minds, the fruitful concord between science and faith, between the Church and the world. I give my support to this task which will be able to honor the truth of faith and of science and open the door to future collaboration.[2]

As a partial response to this call of the Pope, the Vatican Observatory, together with the Center for Interdisciplinary Studies in Cracow, organized a conference at Cracow in 1984 which resulted in a major publication on the Galileo case.[3]

As a result of these initiatives, the Pope through the Secretariat of State, asked the Vatican Observatory to organize a conference to further the science-faith dialogue on the occasion of the commemoration of the 300th anniversary of the publication of Isaac Newton's *Principia*. During this conference, again held in Cracow,[4] a group of scholars, including George Coyne, Bill Stoeger, Michael Heller, and me, began to consider the possibility of organizing an entire series of such conferences. To test the feasibility of this idea, a first conference, organized by Bill Stoeger and me, was held at the Specola Vaticana in September 1987 with the participation of twenty-one scholars. The resulting publication, *Physics, Philosophy and Theology: A Common Quest for Understanding*,[5] includes a message by the Pope on the relations between the church and the scientific communities. This was the first major Pontifical statement on science and religion in three decades. It was reprinted, together with nineteen responses by scientists and theologians, in *Pope John Paul II on Science and Religion*.[6]

Based on this work, George Coyne, as Director of the Vatican Observatory, proposed a major new initiative: a series of five conferences spaced over a decade. The goal would be to expand upon the research agenda begun in 1987, moving into additional areas in the physical and biological sciences. Coyne convened a meeting at the Specola in June, 1990, to plan the overall direction of the research, out of which a long-term steering committee was formed with Nancey Murphy, Associate Professor of Christian Philosophy, Fuller Theological

---

[2] *Discourses of the Popes from Pius XI to John Paul II to the Pontifical Academy of Sciences* (Vatican City State: Pontificia Accademia Scientiarum, 1986), Scripta Varia 66, 73-84.

[3] *The Galileo Affair: A Meeting of Faith and Science*, ed. G. V. Coyne, M. Heller, and J. Zycinski (Vatican City State: Libreria Editrice Vaticana. 1985).

[4] *Newton and the New Direction in Science*, ed. Coyne, Heller, and Zycinski (Vatican City State: Libreria Editrice Vaticana, 1988).

[5] Robert John Russell, William R. Stoeger, and George V. Coyne, eds. (Vatican City State: Vatican Observatory, 1988; and Notre Dame: University of Notre Dame Press, 1988).

[6] Russell, Stoeger, and Coyne, ed. (Vatican City State: Vatican Observatory Publications, 1990).

Seminary, Stoeger, and me as members. Coyne then invited CTNS to co-sponsor the decade of research. CTNS was able to accept this offer through the generous support of the Wayne and Gladys Valley Foundation.

2 Guiding Theme: God's Action in the World

A major issue in the debate over theology and science regards the role science ought to play. Too often science tends to set the agenda for the theological discussion with little if any initiative taken by theology. From the beginning it was the clear intention of the steering committee that our research expand beyond this format to insure a 'two-way interaction' between the scientific and theological research programs. In order to achieve this goal, we decided on the following two-fold strategy. First, we looked for an overarching topic of a theological nature to thematize all the conferences, so that the sciences did not unilaterally guide the direction of the discussion. The topic of God's action in the world was quickly singled out as a promising candidate, since it seems to permeate the discussions of theology and science in both philosophical and systematic contexts and it allows for a variety of particular issues to be pursued.

Next, given this overarching topic, individual conferences could be developed around specific topics in the natural sciences. These include quantum cosmology; the origin and status of the laws of nature; chaos, complexity, and self-organization; the mind-brain problem; biological evolution; and quantum physics.

Quantum cosmology and the laws of nature were combined to form the scientific agenda of the first conference in the series, held at the Specola in Fall, 1991. Here questions included the following: What are the status of time and the laws of nature in contemporary physics and cosmology? How does our scientific understanding of time relate to divine temporality and divine action in the world? What are the continuing methodological issues which underlie research on theology and science? How should we understand God's action as creator and redeemer in light of the proposals for a quantum interpretation of the origin of the universe and in light of issues regarding the origin and status of the laws of nature?

We also discussed the overall research methodology for the series of conferences. It was decided that all papers for each conference should be circulated at least once in advance of the conference for critical written responses from all participants. Revisions would also be read in advance of the conference to maximize the productivity of conference discussions. Post-conference revisions would be elicited in light of these discussion. We agreed to hold regional, advance conferences to provide an introduction for participants to relevant technical issues in science, philosophy and theology and to foster joint research and collaboration among participants prior to the conference. An organizing committee would guide the preparation for, procedures of, and editorial process following each conference.

Since the topic of God's action in the world was chosen as the guiding theological theme for the Vatican Observatory/CTNS conferences, a brief introduction to the topic is in order. In the following paragraphs I will examine this theme in its traditional context and touch on some of the alternative interpretations given it by contemporary theologians. Then I will present a summary and brief analysis of the papers included in this volume.

3 Overview of "Divine Action": Historical and Contemporary Perspectives

3.1 Biblical and Traditional Perspectives

The notion of God's acting in the world is central to the biblical witness. From the calling of Abraham and the Exodus to the raising of Jesus from the dead, God is represented as making things happen in the world—in nature and in human history—in order to accomplish God's purposes. This God is known as both creator and redeemer, giving rise to an intimate linking of the biblical theologies of creation and of redemption.[7] Rather than seeing divine acts as occasional effects in otherwise entirely natural and historical processes, both the Hebrews and the early Christians conceived of divine action as the basis of all that happens in nature and in history.[8]

The view that God works in and through all the processes of the world was commonly held in Patristic and medieval times, as even a cursory reading of Augustine and Aquinas demonstrates. Here God was understood as the first cause of all events; all natural causes are secondary or instrumental causes through which God works. In addition, God was thought to act immediately in the world through miracles without using, or by surpassing, finite causes. This view of divine action led to such problematic issues as: Can one unified action

---

[7] The Psalms provide evocative instances of the connection between redemption and creation. In Psalm 18, for example, David calls upon the LORD for deliverance from his enemies and God acts in response to David's cries. Compare this with Psalm 139, where our mother's womb is the intimate location of God's action as creator. Another particularly clear instance is found in Psalm 136. The prophets repeat this connection; see for example Isaiah 44:24. This theme is continued in the New Testament; see for example John 1.

The ordering of the relation between redemption and creation theologies in the history of Israel is the subject of diverse opinions. It has been generally held that the Exodus experience served as the basis for Hebrew faith in God the Creator, a position developed by Gerhard von Rad, *Old Testament Theology*, 2 volumes (New York: Harper & Row, 1957-65), 1:138 and incorporated in standard treatments of the doctrine of creation. See for example Langdon Gilkey, *Maker of Heaven and Earth* (Garden City, NY: Doubleday, 1959), 70. Recent scholarship, however, has questioned this view. See for example, R. J. Clifford, "Creation in the Hebrew Bible," in *Physics, Philosophy and Theology*, 151-170. For a recent discussion of the relation between creation, redemption, and natural science, see Ted Peters, "Cosmos as Creation," in *Cosmos as Creation* (Nashville: Abingdon Press, 1989), 45-113.

[8] See Gen. 45:5; Job 38:22-39:30; Ps. 148:8-10; Is. 26:12; Phil. 2:12-13; 1 Cor. 12:6; 2 Cor. 3:5.

issue simultaneously from two free agents (e.g., God and human agents)? How does an infinite agent (God) preserve the finite freedom of a creaturely agent when they act together? And finally, is not God responsible, ultimately, for evil, leading to the problem of theodicy?

The conviction that God acts universally in all events, and that God and free human agency can act together in specific events, was maintained by the Protestant Reformers and the ensuing Protestant orthodoxy. Calvin, for example, argued that God is in absolute control over the world and at the same time maintained that humans are responsible for evil deeds.[9] In general, the problem of divine and human agency was treated as part of the doctrine of providence and formulated in terms of divine preservation, concurrence and government. Questions about human freedom and the reality of evil were seen more as problems to be solved than as reasons for abandoning belief in God's universal agency.

The rise of modern science and modern philosophy in the seventeenth century led to a rejection of the traditional views of divine action, especially of belief in miracles. Although Isaac Newton argued for the essential role of God in relation to the metaphysical underpinnings of his mechanical system, and in this way for the sovereignty of God in relation to nature, Newtonian mechanics seemed to depict a causally closed universe[10] with little, if any, room for God's action in specific events. With the ascendancy of deism in the eighteenth century, the inheritors of the mechanistic tradition abandoned the 'God hypothesis' as part of the overall explanation of nature; the scope of divine agency was limited to an initial act of creation.[11] Moreover, the impact of David Hume and Immanuel

---

[9] See, for example, Calvin's handling of the problem of multiple agents in one event in *The Institutes*, II, iv, 2.

[10] We now know, however, that even simple deterministic equations can give rise to seemingly random results, as studied in the theories of chaos, complexity, and self-organizing systems.

[11] Newton's mechanics and his system of the world led to profound philosophical issues through his introduction of absolute space and absolute time to ground the distinction between uniform and accelerated motion, as well as to important theological reconstructions of the relation of God to nature in terms of the *divine sensorium* and the design of the universe. See E. A. Burtt, *The Metaphysical Foundations of Modern Science* (Garden City, NY: Doubleday, 1954). Michael Buckley has argued that the reliance on Newtonian science as a foundation for theology, and the abandonment of the 'God hypothesis' thereafter, were principle causes of the rise of atheism in the West. See Michael J. Buckley, *At the Origins of Modern Atheism* (New Haven: Yale University Press, 1987). For a briefer historical account see Ian G. Barbour, *Issues in Science and Religion* (New York: Harper Torchbook, 1966), ch. 3. One can also argue that the concept of inertia played an important role in deflecting attention away from the need to view God as acting ubiquitously to sustain nature in being. See Wolfhart Pannenberg, "Theological Questions to Scientists," in *The Sciences and Theology in the Twentieth Century*, ed. A. R. Peacocke (Notre Dame: University of Notre Dame Press, 1981), 3-16; see especially 5-6.

Kant was to undercut natural theology, metaphysical speculation about causality and design, and belief in miracles; to limit the scope of reason; and to relocate religion to the sphere of moral agency.

Given this background and the rise of the historical-critical approach to the Bible, theology in the nineteenth century underwent a fundamental questioning not only of its contents and structure, but even of its method. The response of Friedrich Schleiermacher (1768-1834) was to understand religion as neither a knowing (the activity of pure reason) nor a doing (the activity of practical reason) but as grounded in an entirely separate domain, the "feeling of absolute dependence." Theological assertions can only have as their basis the *immediate* assertions of religious self-consciousness. Schleiermacher understood God's relation to the world in terms of universal divine immanence. By miracle we mean ". . . simply the religious name for event. Every event, even the most natural and usual, becomes a miracle, as soon as the religious view of it can be the dominant."[12]

3.2 *The "Travail" of Divine Action in the Twentieth Century*

Protestant theology in the first half of the twentieth century was largely shaped by Karl Barth. In his rejection of nineteenth century liberal theology, Barth began with the sovereignty of the God who is "wholly other" and stressed God's initiative in the redemptive act of faith. "The Gospel is . . . not an event, nor an experience, nor an emotion—however delicate! . . . [I]t is a communication which presumes faith in the living God, and which creates that which it presumes."[13] But do Barthian Neo-Orthodoxy and the ensuing "biblical theology" movement succeed in producing a credible interpretation of "act of God?"

In a well-known article written in 1961, Langdon Gilkey forcefully argued that they do not.[14] According to Gilkey, Neo-Orthodoxy is an uneven composite of biblical/orthodox language and liberal/modern cosmology. It attempts to distance itself from liberal theology by retaining biblical language about God acting through wondrous events and by viewing revelation as an objective act, not just a subjective inference. Yet, like liberalism, it accepts the modern premise that nature is a closed, causal continuum as suggested by

---

[12] Friedrich Schleiermacher, *On Religion: Speeches to its Cultured Despisers* (New York: Harper Torchbook, 1958), 88. In a long discussion in *The Christian Faith*, he wrote: ". . . As regards the miraculous . . . we should abandon the idea of the absolutely supernatural because no single instance of it can be known by us. . . ." Friedrich Schleiermacher (Edinburgh: T & T Clark, 1968), #47.3, 183. For an excellent analysis of Schleiermacher and other important developments in the nineteenth century Claude Welch, *Protestant Thought in the Nineteenth Century*, 2 volumes (New Haven: Yale University Press, 1972).

[13] Karl Barth, *The Epistle to the Romans*, 6th ed. (London: Oxford University Press, 1968), 28.

[14] Langdon B. Gilkey, "Cosmology, Ontology, and the Travail of Biblical Language," *The Journal of Religion* 41 (1961): 194-205.

classical physics. The result is that, whereas orthodoxy used language univocally, Neo-Orthodoxy uses language at best analogically. Worse, since its language has been emptied of any concrete content, its analogies devolve into equivocations. "Thus the Bible is a book descriptive not of the acts of God but of Hebrew religion. . . . [It] is a book of the acts Hebrews believed God might have done and the words he [*sic*] might have said had he done and said them—but of course we recognize he did not."[15]

Thus a 'two-language' strategy sets in: Neo-Orthodox theologians use biblical language to speak confessionally about God's acts, but secular language when speaking historically or scientifically about 'what actually happened'. Similarly, the insistence that revelation presupposes faith founders when one asks about the initial event in which faith originates. Thus Neo-Orthodoxy, and with it much of contemporary theology, involves a contradiction between orthodox language and liberal cosmology.

In the wake of these problems, several approaches to divine action are being explored in current literature. The following lists a few of the more prominent ones.[16]

3.2.1 *Neo-Thomism* Here traditional Thomistic distinctions between primary and secondary causality, characteristic of both Roman Catholic and Protestant orthodoxy, have been modified in light of Kant. Advocates include Bernard Lonergan, Joseph Mareschal, Jacques Maritain, and Karl Rahner. Related works are those of Austin Farrer and Eric Mascall.

3.2.2 *Process Theology* Process philosophy represents a fundamental shift away from Thomistic metaphysics. Based on the metaphysics of Alfred North Whitehead, it rejects the traditional and widespread conception of causality in terms of primary and secondary causes. Instead, God is seen as acting in all events, though never exclusively determining their character since each such "actual occasion" also includes an irreducible element of genuine novelty as well as the causal efficacy of the past. Representatives include Ian Barbour, Charles Birch, John Cobb, Jr., David Griffin, Charles Hartshorne, and Schubert Ogden.

---

[15] Ibid., 198.

[16] For a helpful overview of the field as it stood a decade ago, see Owen C. Thomas, ed., *God's Activity in the World: The Contemporary Problem* (Chico, CA: Scholars Press, 1983). According to Thomas, only process and Neo-Thomistic theologies offer full-blown metaphysical theories. They are therefore superior to the personal action approach with its limited analogies, and these are all preferable to liberal, uniform action and two perspectives theologies, which offer neither a theory nor an analogy. In a more recent publication, Ian Barbour offers a lucid description and creative comparison of the problem of divine action in classical theism, process theism, and their alternatives, including several types of personal agency models. See Ian Barbour, *Religion in an Age of Science*, The Gifford Lectures, 1989-1991, vol. 1 (San Francisco: HarperSanFrancisco, 1990), ch. 9. See also Arthur Peacocke, *Theology for a Scientific Age: Being and Becoming—Natural and Divine* (Oxford: Basil Blackwell, 1990), ch. 9, for another recent discussion.

3.2.3 *Uniform Action* Here God is thought of as acting uniformly in all events in the world. Distinctions in meaning and significance are due entirely to human interpretation. This view is found in the writings of Gordon Kaufman and Maurice Wiles.

3.2.4 *Personal Agent Models*

3.2.4.1 *Literal Divine Action* Some philosophers of religion, most notably William Alston,[17] have questioned the assumption that causal determinism of all natural events by other natural events prevents us from speaking of God, literally, as a personal agent who brings about particular states of affairs at particular times and places. First, he holds that there is no sufficient reason to adopt a naturalistic causal determinism. Second, he argues that deterministic laws hold only within closed systems, and, for most systems with which we work, we cannot suppose that they are closed to outside influences (including divine ones).

3.2.4.2 *Embodiment Models* The analogy of divine embodiment, namely that God:world::mind:body, has been explored in differing ways by such authors as Arthur Peacocke,[18] Grace Jantzen[19] and Sallie McFague.[20] Peacocke views the universe as dynamic and interconnected, suggesting biological and feminine analogies for divine agency. These analogies counteract the tendency of traditional language which stresses God's externality to creation. Peacocke adopts a panentheistic approach which combines the language of immanence with transcendence in speaking about God's relation to creation.[21] Jantzen proposes that the entire universe is God's body. God is immediately aware of all events in nature, and acts both universally throughout nature and particularly in unique events. Moreover, all of God's acts are basic or direct acts, analogous to the direct acts we perform when we move our own bodies. She recognizes that the embodiment model raises several problems, including the problem of evil, the relation of God's action to the laws of nature, and the significance for the divine life if the universe has a finite past and/or future, but suggests important responses to each of these. McFague argues that the embodiment model need not lead to pantheism since other models are employed which stress "God as mother,

---

[17] William P. Alston, *Divine Nature and Human Language* (Ithaca: Cornell University Press, 1989).

[18] A. R. Peacocke, *Creation and the World of Science* (Oxford: Clarendon Press, 1979), 142 ff., 207; and *idem*, *Intimations of Reality* (Notre Dame: University of Notre Dame Press, 1984), 63 ff., 76.

[19] Grace Jantzen, *God's World, God's Body* (Philadelphia: Westminster Press, 1984).

[20] Sallie McFague, *Models of God: Theology for an Ecological, Nuclear Age* (Philadelphia: Fortress Press, 1987), 69-78; and *idem*, *The Body of God: An Ecological Theology* (Minneapolis: Fortress Press, 1993).

[21] According to panentheism, "the world is regarded as being . . . 'within' God, but the being of God is regarded as not exhausted by, or subsumed within, the world." Peacocke, *Creation and the World of Science*, 207.

lover and friend . . ."²² She admits, however, that the embodiment model suggests that God is at risk, being vulnerable to the sufferings of nature.

3.2.4.3 *Non-Embodiment* Thomas Tracy rejects both the claim that personal agency requires embodiment and that the world is like an organism. Instead he develops a conception of God as a non-embodied agent.²³ Tracy sees himself as combining aspects of classical and process theism. Ian Barbour echoes Tracy's objection to viewing the world as the body of God since it lacks bodily unity and an environment.²⁴ He also believes that embodiment would fail to provide adequately for the independence of God and the world. Instead he opts for a "social or interpersonal analogy."²⁵

3.2.4.4 *Interaction Models* The embodiment approach has been subject to criticism from other positions too. John Polkinghorne²⁶ challenges both the panentheistic implications of, and the scientific conceptions underlying, embodiment. He does not deny that God should be vulnerable to the world; indeed, God suffers with creation, as *kenosis* implies. However, if the universe is God's body then the ultimate fate of the universe ("freeze" or "fry") would surely have "significant consequences for the One embodied in it."²⁷ Moreover, Polkinghorne argues that the universe is simply not like a body—nor is it like a machine.

Instead, Polkinghorne argues that a clue to God's interaction with the world can come from examining how humans interact with the world. He turns to quantum physics and chaos theory to suggest how the combination of lawlike behavior with openness and flexibility in nature makes human and, in some preliminary way, divine agency conceivable.

Owen Thomas, the editor of a major volume on divine action in 1983, returned to the problem in 1991, beginning with this rather caustic comment:

> Theologians continue to talk a great deal about God's activity in the world, and there continue to be only a very few who pause to consider some of the many problems involved in such talk.²⁸

According to Thomas, the question of double agency still remains "the key issue in the general problem. . . ."²⁹ How can we assert coherently that both

---

²² McFague, *Models of God*, 71.

²³ Thomas F. Tracy, *God, Action and Embodiment* (Grand Rapids: Eerdmans, 1984).

²⁴ Response to the CTNS Fellows' Lecture by Arthur Peacocke, Berkeley, 1986 (unpublished).

²⁵ Barbour, *Religion in an Age of Science*, 259.

²⁶ John Polkinghorne, *Science and Providence* (Boston: Shambhala, 1989).

²⁷ Ibid., 20. Polkinghorne is aware of Jantzen's response, that God could change the course of nature when needed. But for Polkinghorne, this possibility would violate the freedom God has given nature, the freedom to be truly other than God and to evolve consistently with its internal laws.

²⁸ Owen Thomas, "Recent Thought on Divine Agency," in *Divine Action*, ed. Brian Hebblethwaite and Edward Henderson (Edinburgh: T & T Clark), 35-50.

divine and creaturely agents are fully active in one unified event? After evaluating the current state of the discussion, Thomas' position is that one must either follow the primary/secondary path of traditional theism or the process theology approach. He asserts that if there *is* another solution to the problem, he has not heard of it and concludes that this question should be a major focus of future discussions.[30]

4 *The Present Publication: Summary and Analysis*

4.1 *Summary*

Much of the literature mentioned above has focused on philosophical issues in divine action with only a secondary concern for the actual results of physical and biological research. Alternatively, many of those who have mined these sciences for their theological implications have not kept clearly in mind the philosophical problems associated with divine action. The following essays are one attempt at remedying that situation. Here the specific scientific foci are on standard and quantum cosmology and the laws of nature, with particular attention to the problem of time. As we shall see, the broad theological and philosophical themes outlined above surface in nuanced and interwoven patterns in these essays.

Section I introduces the basic scientific issues in physics and cosmology. Section II raises methodological issues in the relationship of theology and science. Philosophical problems are explored in Section III. Sections IV and V focus directly on constructive theology in the contexts of cosmology (IV) and the laws of nature (V).

4.1.1 *Scientific Background: Standard and Quantum Cosmology* **George Ellis** and **Bill Stoeger** present the standard model of our universe, the Friedmann-Lemaitre-Robertson-Walker (FLRW), or Big Bang model, including its foundations in special relativity and general relativity, and its observational features. They also discuss inflationary models which solve many of the problems of the standard model. Attention is given to the assumptions which go into cosmology, some of which are technical, others more philosophical in nature. In a key passage, Ellis and Stoeger focus on the initial singularity, t=0, "the boundary to the universe where the laws of physics break down. . . ." Is t=0 a "creation" event, or is it a transitory feature of the standard cosmological model which will disappear when a quantum theory of gravity is achieved? They also touch on questions of the origin of structure, complexity, and life in the universe.

The problem of constructing a quantum theory of gravity and, from it, quantum cosmology is the subject of **Chris Isham**'s paper. Mindful of the problem of t=0 in standard cosmology, Isham stresses the requirements that quantum cosmology be singularity-free and render meaningful the idea of the "beginning" of time. In addition, it is hoped that quantum cosmology will

---

[29] Thomas, "Recent Thought," 46.
[30] Ibid., 50.

produce unique predictions about the universe. Still, Isham stresses that no cosmological theory can answer the philosophical/theological question, "why is there anything at all?" Moreover, he is careful to underscore the tentative scientific status of current quantum theories: they are incomplete and highly speculative. Indeed many physicists "think the whole subject of quantum cosmology is misconceived."

The paper then explores technical details of quantum cosmology. Isham begins by describing two distinctive kinds of theories: those working with a pre-existing spacetime, and those which describe the origin of space and time using general relativity and quantum physics. Opting for the latter approach, Isham begins his discussion by focusing on the concept of time in the four-dimensional framework of general relativity. Here time is not thought of as a fixed, eternal reality in which the three-geometry evolves, but as an "internal" property of the universe defined in terms, say, of the volume of each three-geometry. Internal time continues to be a key concept when moving to a quantum theory of gravity. Such a theory is necessary when dealing with the very early universe and the problem of t=0. However a quantum treatment moves one towards a probabilistic perspective on physical processes and an instrumentalist interpretation of physical theories. This in turn leads to severe conceptual problems when quantum theory is applied to the universe as a whole, most notably the measurement problem (e.g., what constitutes the "observer").

There are several distinct approaches to quantum gravity using general relativity and quantum physics. Having previously described the Hartle and Hawking proposal,[31] Isham chose to focus here on the work of Vilenkin. This approach leads to a plausible *description* of the origination event/region, though it is less successful at *predicting* a unique state function for the universe. The origination can be thought of as the 'emergence' of a real-time, classical universe from an imaginary-time, non-classical universe, where 'emergence' is understood in a symbolic, and not a temporal, sense.

Isham concludes by stressing again the technical as well as the conceptual problems of all such quantum cosmologies—problems which should be borne in mind as these proposals are given philosophical and theological interpretations.

4.1.2 *Methodology: Relating Theology and Science* The two papers in Section II raise methodological issues in the relationship between theology and science.

How should scientific theories be interpreted philosophically and theologically? According to **Michael Heller** there are three possibilities: (a) pseudo-interpretations which contradict physical theory; (b) consistent interpretations which are neutral with respect to the theory's mathematical structure; (c) exegetical interpretations which restate the mathematical structure in everyday language and are in strict agreement with the theory. Heller is highly critical of (a). Moreover, since physical theories as such "say nothing about religious matters" (c) is ruled out. Therefore, to be valid, an interpretation must

---

[31] C. J. Isham, "Creation of the Universe as a Quantum Process," in *Physics, Philosophy and Theology*, 375-408.

be of type (b) and should be taken "seriously but not literally." Heller adds that science may also serve as a source of insight and metaphor for theology, or provide a suitable context for theological reformulation. Still, theology should interact with the overall scientific image of the world and not with a particular theory. He also suggests that physical theories, with the addition of specific premises, can provide grounds for theological conclusions, and that theological and scientific theories might have mutually dependent or even equivalent implications. Finally, the very existence of successful physical theories poses an important philosophical problem.

For **Stephen Happel**, the methodological bridge between theology and science comes through language—in particular, through metaphor. Happel argues that scientists, as well as theologians, use ordinary speech, constructed of metaphors, to originate, process, and communicate their insights. By studying how scientists employ metaphors, theologians can discover an important role for cosmology in religious discourse. Similarly scientists can gain from understanding the hermeneutical framework they employ in their discourse.

Happel begins by studying the general reasons for focusing on metaphor in both fields, namely the conviction that science, like theology, is a hermeneutical venture. Next he focuses on the metaphors that adorn cosmology and the ways they narrate the story of the universe. Happel then argues that there is a basic relationship between the particular metaphors chosen by cosmologists and the actual temporal asymmetry of the universe. To support his claim, Happel critically evaluates four competing theories about the nature of metaphor. In the process he argues that in both science and religion metaphors communicate more than feelings: they indicate a state of affairs. Because of this, Happel is critical of Paul Ricoeur's paradoxical "is/is not" view of metaphor and of theologians who appropriate it, preferring instead the moderate realism advanced by Mary Hesse and Bernard Lonergan.

Next Happel attempts to understand why some scientists hold an atemporal perspective on cosmology while others see the universe in terms of a temporal narrative such as the Big Bang. The difference is due less to physics or mathematics, Happel argues, than to the presuppositions one holds about metaphor. Atemporalists tend to have a Ricouerian view of metaphor as paradoxical, whereas temporalists tend toward a moderately realist theory of metaphor. Lonergan's theory and its relation to emergent probability as the explanation of temporality provide just the needed ontological reference for the metaphors of time asymmetry.

Happel acknowledges that deconstructionists, drawing heavily on the writings of Jacques Derrida, propose an alternate theory in which the surplus of metaphors is sheer play without directionality or finality. He contrasts the teleological approach of scientists like John Barrow and Paul Davies, who treat cosmological metaphors as narrative, with the non-teleological approach of Stephen Hawking, who seems closer to Derrida and Ricoeur. Happel sees these approaches as parallel to the theological typologies of prophecy (narratives which stress the ethical imperative) and mystical communion (non-narrative descriptions of the atemporal identity with the divine). Both narrative and non-narrative languages are integrally intertwined in the doctrine of the Trinity. As a result, theologians might investigate how narrative and non-narrative interpretations of spacetime might be "equiprimordial, requiring a co-implicating dialectic." Scientists, in turn, are encouraged to develop an approach which overcomes the symmetrical-asymmetrical arguments concerning temporality. Ultimately, whether the universe is seen in mystical or prophetic terms, God's involvement is to be trusted, and God's gift of time leads us to combine narrative and non-narrative language.

4.1.3 *Philosophical Issues: Time and the Laws of Nature* Five essays are included in Section III, each focusing on philosophical issues raised by cosmology and the laws of nature but with special attention to the problem of time.

Is ours a world of timeless being (the "block universe") or of flowing time and true becoming? The current debate over the block universe, represented in the essay by **Chris Isham** and **John Polkinghorne**, brings together scientific, philosophical, and theological arguments in a tightly-knit, interwoven pattern.

Proponents of the block universe appeal to special and general relativity to support a timeless view in which all spacetime events have equal ontological status. The finite speed of light, the light cone structure, and the downfall of universal simultaneity and with it the physical status of "flowing time" in special relativity result in a heightened tendency to ontologize spacetime. The additional arbitrariness in the choice of time coordinates in general relativity makes flowing time physically meaningless. Thus no fundamental meaning can be ascribed to the "present" as the moving barrier with the kind of unique and universal significance needed to unequivocally distinguish "past" from "future." Instead the flowing present is a mental construct, and four-dimensional spacetime is an "eternally existing" structure. God may know the temporality of events as experienced subjectively by creatures, but God cannot act temporally, since flowing time has no fundamental meaning in nature. Theologians must accept the Boethian and even Gnostic implications of the block universe.

Opponents of the block universe begin by distinguishing between kinematics and dynamics. Special relativity imposes only kinematic constraints on the structure of spacetime. The dynamics of quantum physics and chaos theory encourages a view of nature as open and temporal, thus allowing for both human and divine agency. The problem of the lack of universal simultaneity is lessened since simultaneity is an *a posteriori* construct. Philosophically disposed to critical realism, opponents are wary of the incipient reductionism of the block view. They resist the Boethian implications of relativity, and argue instead that divine omnipresence must be redefined in terms of a special frame of reference, perhaps one provided by the cosmic background radiation. God's knowledge of spacetime events in terms of this frame of reference will be constrained by both the world's causal sequence and the distinction between past and future. Similarly God's actions will be consistent with relativity theory.

In the end, is the debate merely philosophical or could it actually have scientific consequences? *Proponents* of the block universe challenge their opponents to decide between a mere *reinterpretation* of the existing theories of physics and the much stronger claim that these theories should be *changed*. If forthcoming, such changes ought to be testable empirically and would constitute a major achievement in the debate over time. Proponents also point to additional complexities in the debate, such as the problem of giving a realist interpretation of quantum physics. These problems become even more acute when dealing with quantum cosmology, making an atemporal interpretation almost inevitable. They do not object to positing that God experiences the world through a special frame of reference or that God is aware of the experience of temporality of living creatures. However they find it hard to understand how God's *action* on the world can respect the causal constraints on such action entailed by special relativity.

Paul Davies begins with the claim that our ability to understand nature through the scientific method is a fact which demands an explanation. He proposes that our mind and the cosmos are linked, that consciousness "is a fundamental and integral part of the outworking of the laws of nature." In particular, the laws of nature which make possible the emergence of life must be of a form such that at least some species which arise according to them have the ability to discover them. Thus science can explain the rise of species which can engage in science without appealing to a God who either intervenes in or guides nature. Still the ultimate explanation of the origin of the laws lies outside the scope of science and should be pursued by metaphysics and theology. Whether this leads to God "is for others to decide."

Davies begins with the sociological debate over the origin of science. Although science is clearly a product of Western European culture, he sees no simple relationship between Christian theology and the emergence of science. Whatever its origins, though, the validity of science is transcultural and warrants a realist interpretation.

What is most significant about nature is that the universe is ". . . poised, interestingly, between the twin extremes of boring over-regimented uniformity and random chaos." Accordingly, it achieves an evolution of novel structures through self-organizing complexity. "The laws are therefore doubly special. They encourage physical systems to self-organize to the point where mind emerges from matter, and they are of a form which is apprehensible by the very minds which these laws have enabled nature to produce." Does our ability to "crack the cosmic code" lead to an argument for God? No. Davies prefers an evolutionary interpretation of mind as emergent within the material process of self-organization. The emergence of mind with its ability to pursue science is not just a "biological accident." Instead it is inevitable because of the laws of physics and the initial conditions. Hence life should emerge elsewhere in the universe—a claim which Davies sees as testable.

If mind emerged because of the laws of nature, is it surprising that mind is capable of discovering these laws? Davies first stresses that evolution is a blend of chance and necessity; it is neither teleological nor is it a "cosmic anarchy." The laws "facilitate the evolution of the universe in a purposelike fashion." Still the actual laws of the universe are remarkable. They not only encourage the evolution of life and consciousness but they support the evolution of organisms with the ability for theoretical knowledge. Here the ability to do mathematics is particularly surprising. Davies connects mathematics with the physical structure of the world through "computability" and thus to physics, since computing devices are physical. In this way mathematics and nature are intertwined. Moreover, mathematics is capable of describing the laws of physics which govern the devices which compute them.

The intimate relation of mind and cosmos need not lead to a theological explanation, but Davies is equally critical of a many-universe explanation, opting instead for a form of design argument. This, however, takes us to the limits of science. "The question of the nature of the laws themselves lies outside the scope of the scientific enterprise . . . (and) belongs to the subject of metaphysics . . ."

The central argument in **Andrej Grib**'s paper is that temporal existence, that is, movement through time, allows the human mind to obtain information about the universe governed by quantum physics. The universe is characterized by non-standard logic (non-Boolean logic), whereas we interpret the universe in terms of ordinary (Boolean) logic. Thus we must experience events successively as past, present, and future in order to gain knowledge of the objective but incompatible (non-commuting) character of nature. Using this basic argument, Grib speculates about World Consciousness and suggests how quantum cosmology can provide plausibility arguments for orthodox Christian theology.

Grib begins by describing several key interpretations of quantum physics. According to Niels Bohr, complementary quantum phenomena lack independent reality since the measurement apparatus and the quantum object form an indivisible whole. John von Neumann interpreted the collapse of the wave packet during measurement as a process without a physical cause. Instead it is due to the consciousness of the observer, conceived as an abstract self. For Fritz London and Edmond Bauer the key feature of consciousness is introspection, not abstract ego, giving the problem a more objective character. Eugene Wigner developed this argument further by proposing that any living system could have the capacity to collapse the wave function.

The next step in Grib's account was taken by John Bell whose famous theorem forces us to chose between idealism (in which quantum objects with non-commuting properties only exist when observed) and realism (in which quantum objects with non-commuting properties have a qualified existence independent of observation). Still the latter is far from "naive realism" for, even admitting the qualified existence of quantum objects (objects characterized by non-Boolean logic), the existence of macroscopic objects in nature (objects characterized by Boolean logic) is the direct result of observation.

Grib now turns to the problem of quantum cosmology where, as a quantum theory, one faces the fundamental problem of measurement: who is the observer? Clearly, to speak here about a "self-originating universe" and an "objectively contingent universe" is misleading, since the existence of the universe *per se* now requires an "external observer." Grib proposes that we opt instead for the qualified existence afforded by quantum logic and apply this to cosmology. Thus, given that quantum logic involves conjunctions and disjunctions which do *not* satisfy the distributivity law, in order for our (Boolean) minds to grasp the (non-Boolean) quantum cosmos we must experience the world through *temporal* sequence. Each event in the sequence is a different

Boolean substructure (which *is* accessible to us) of the overall non-Boolean universe (which *is not* accessible). Thus the temporality of the world arises out of our mental processes, which integrate our Boolean experiences with physical, non-Boolean, structures.

In his paper **Bill Alston** studies the philosophical aspects of the problem of divine action in relation to both the laws of nature and the meaning of human freedom. To set the stage, he begins by stating two presuppositions which characterize his general approach: first, he takes "seriously and realistically" the idea of God as a personal agent, and second, God's activity extends beyond creation and conservation to include special acts performed by God in light of knowledge of the world and to achieve a purpose.

By "seriously" Alston means that, at least in some cases, we understand the statement "God acts" literally and not just figuratively. By "realistically" Alston means that religious discourse, along with scientific discourse, aims at "an accurate portrayal of an independently existing reality" with objective characteristics. God's actions result in outcomes, at least on some occasions, which are different from the outcomes that would have been if only natural factors had been at work. God's acts include not only revelation but also such "super-spectacular miracles" as the parting of the Sea of Reeds and the resurrection, as well as daily divine-human interaction, in prayer for example. Thus God acts as a personal agent, "possessed of intellect and will."

With this as background, Alston proceeds to the main burden of the paper, relating his convictions to the topics of natural law and human freedom. He defines "determinism" as the doctrine that "every happening is uniquely determined to be just what it is by natural causes within the universe." Although determinism has a strong hold on contemporary culture, Alston takes quantum mechanics to provide a "definitive refutation" of it. Hence because of quantum indeterminism, God can act without violating physical laws. Moreover, acts such as these which begin on the sub-atomic level can lead to differences in macroscopic states. It is thus possible that God designed the universe in this way to allow for divine action.

Still, Alston's main point does not depend on the indeterminacy of quantum mechanics. According to Alston, even deterministic laws only provide sufficient conditions for predicting the behavior of *closed* systems. But we *never* have reason to believe a system is actually closed, i.e., that we know all the operative forces at work. Any system can be open to outside influences, including the acts of God. Hence, in this more general sense, God's acts do not violate natural law regardless of whether these laws are probabilistic or deterministic.

Next Alston turns to the problem of human freedom. He takes a libertarian view of free action in which "nothing other than my choice itself uniquely *determines* me to choose one way rather than another." Does human freedom pose a problem for conceptualizing God's action? Alston first argues that, with the exception of acts by free creatures, all events which we attribute to God's specific acts could in fact be the unfolding of what God designed initially (whether "initially" means temporally first, as in a universe with a first moment, or first in order of priority, as in a universe with an infinite past). But if we assume that humans, at least, have libertarian free choice, the strategy of initial design cannot work—unless God can be said to possess "middle knowledge," defined as knowledge of what (actual and possible) free agents would choose to do in *any* situation in which they found themselves. Alston then argues that middle knowledge is impossible: God cannot know what a free agent would decide in situations which the agent never actually encounters. Thus if there is libertarian free agency and if middle knowledge is impossible, we must conclude that those of God's acts that appear to take place in time in response to the choices of free agents do indeed take place in precisely that way, and not merely by means of God's initial design.

Alston then turns to physical cosmology and its possible bearing on divine action. The choice between cosmological models such as the Big Bang, the oscillating universe and inflation makes little difference to Alston's position in general, though the status of time in these models might be significant. To pursue this question, Alston distinguishes between a block universe view of time and the process view of time. Does human freedom require the process view, or is the block view sufficient? Alston argues that, although the latter denies the passage of time, it does not imply that all events exist at *all* times, but only that they each exist at *their own* time. Thus the block view does *not* undercut human freedom, since all that is required for an act to be free is that it not be determined by anything prior to it, and this is possible even on the block view. Similarly God knows each event in its own time.

Finally, what about God's existence: is it temporal or atemporal? Alston argues that even if God is atemporal, God's acts can produce temporally ordered consequences in the world. Moreover, relativity theory and quantum cosmology suggest that time should not be viewed as a "metaphysically necessary form of every kind of existence, including the divine existence." Thus physical cosmology, and with it the status of time, has little bearing on how we should best think of divine action in the world and its relation to the laws of nature and human freedom.

How should we think of the laws of nature? **Bill Stoeger** poses this as "an absolutely crucial question" underlying the entire discussion of science, philosophy and theology. In his essay, Stoeger defends the thesis that the laws, although revealing fundamental regularities in nature, are not the source of those regularities, much less of their physical necessity. They are descriptive and not prescriptive and do not exist independently of the reality they describe. Stoeger thus rejects a "Platonic" interpretation of the laws of nature. They have no pre-existence with respect to nature; this means that they do not ultimately explain why nature is as it is. Instead, the regularities which the laws of nature describe stem from the regularities of physical reality itself, a reality whose complexity subverts any attempt at a reductionist approach to science. Thus a "theory of everything" is ruled out, and the possibility in principle of God's acting in the world is strongly affirmed.

The laws of nature are approximate models, idealized constructions which can never be complete and isomorphic descriptions of nature. Prevailing theories are eventually replaced or subsumed, often entailing a radically different concept of nature. Moreover, no theory, no matter how complete, can answer the ultimate question: why nature is as it is and not some other way. Stoeger is thus critical of those realists who make excessive claims about the correspondence between theory and the structures of reality. "The illusion that we are somehow discerning reality as it truly is *in itself* is a pervasive and dangerous one." Stoeger also argues that ontological reductionism and determinism are untenable. The laws of nature are in fact human constructions guided by careful research. The intermediate-level regularities which they model originate "in the relationships of fundamental entities in a multi-layered universe," many of which remain beyond our purview. An understanding of the ultimate origins of these underlying regularities takes us to the limits of what can be known.

Stoeger's account of the status of the laws of nature leads him to argue that the laws neither exist independently of the universe nor are they prescriptive of its behavior. It thus does not make sense to suppose that there may be other sets of actual or potential laws that might describe universes different from our own. This reduces the cogency of "many-worlds" arguments which hypothesize the existence of other universes as a means of explaining away the (supposed) fine-tuning of our own universe.

Stoeger then turns to the problem of divine action in light of his nuanced realism. God can be thought of as acting through the laws of nature. However, the term 'laws' refers here to the underlying relations in nature and not principally to our imperfect and idealized models of them. Moreover, as their ultimate source, God's relationship to these laws will be "from within" and God will not need to formalize it. Our relationship to them will always be "from without" and it will be only partially manifested through our laws. Finally, as imperfect models of the regularities and relationships we find in nature, our laws only deal with general features in nature. They cannot subsume the particular, special, and personal aspects, though these aspects are part of the deeper underlying regularities and relationships of nature. It is through these aspects, as well, that God acts.

4.1.4 *Theological Implications 1: Time and Quantum Cosmology* Section IV begins the exploration of the theological implications of cosmology, with special attention to the problem of time in scientific and theological perspectives.

John Lucas defends the temporality of God against both traditional theism and the difficulties raised by relativity and quantum cosmology. For Lucas, the temporality of God is essential if we are to claim that God is personal and therefore conscious of the passage of time. Against traditional orthodoxy and deism, Lucas cites both Barth and process theology in support of divine temporality. Moreover, the Biblical witness is unalterably to a God who acts in specific ways. Though God may be beyond time in the sense that time was created by God, Lucas insists that God is not timeless.

But how can God experience the world in time if physics undercuts the temporality of the world? Lucas argues that while special relativity on its own provides no absolute temporal reference frame, it is consistent with the possibility that one exists, such as the cosmic background radiation. This in turn might provide a reference-frame by which God has temporal knowledge of the world.

What about the creation of the universe by God? Lucas points out that the proposals by Hartle and Hawking and by Vilenkin explain the origin of the universe not as a result of conditions existing before the Big Bang, but as an instantiation of important rational *desiderata*. This reflects a parallel move in the philosophy of science in which the kind of explanation sought shifts from a deductive-nomological explanation (in which temporally antecedent conditions evolve through covering laws) to a top-down explanation (i.e., the instantiation of *desiderata*). Lucas concludes that time can be thought of modally as a transition from possibility to actuality.

In his paper **Keith Ward** moves "both ways" between theology and cosmology. He begins with a summary of the traditional doctrine of creation: God is a non-spatio-temporal being, transcending all that is created, including spacetime, although immanent to all creation as its omnipresent creator. Divine eternity is thus timeless, for God has neither internal nor external temporal relations. The act of creation is one of non-temporal causation. Whether there was a first moment is irrelevant to the doctrine.

Ward admits that this view of God is congruent with the block universe interpretation of special relativity, but he is highly critical of it. Ward maintains that the doctrine of creation does not entail a timeless God. Although God transcends spacetime as its cause, God is nevertheless temporal, since ". . . by creating spacetime, God creates new temporal relations in the Divine being itself." Allowing God to have temporal relations makes it possible for God to act in new ways, make new decisions and bring into being in time an infinite number of new things. The inclusion of divine contingency along with divine necessity enriches the concept of omnipotence.

Ward distinguishes his view of God from that of process theism. He maintains God's omnipotence and still affirms free will by appealing to divine self-limitation. The advantage over Whitehead is that God's omnipotence will "ensure that all the evil caused by the misuse of creaturely freedom will be ordered to good. . . ."

Ward then relates nomological models, which are dominant in physics and involve general principles and ultimate brute facts, to axiological models, which arise in the social sciences and describe the free realization of ultimate values. A nomological model realizes an aesthetic value, since the laws of nature are elegant and simple. An axiological model is ultimately factual, since values arise out of the natural capacities of sentient beings as described by physics and evolutionary biology. This inter-relationship is central to the Christian claim that ". . . goodness is rooted in the nature of things, and is not some sort of arbitrary decision or purely subjective expression of feeling."

Quantum cosmologists attempt to offer a secular explanation of ultimate brute facts, but this minimizes the importance of freedom, creativity, and the realization of values. Theism can offer a comparable explanation of nature, but its advantage lies in its combination of nomological and axiological explanations. Theism is thus "the best possible intelligible explanation of the universe" and "the completion of that search for intelligibility which characterizes the scientific enterprise." He urges that we reconstruct the doctrine of creation in terms of creative emergence, that is, the novel realization of intrinsic values grounded in the divine nature and emerging through the cooperative acts of rational creatures.

Modern cosmology "sets the notion of Divine action in its broadest and most all-embracing context." The laws of nature realize God's purposes, understood as potentialities in the structure of reality and not interferences from an alien power. Miracles are "transformations of the physical to disclose its spiritual foundation and goal...." Thus theism "can be seen as an implication of the scientific attitude itself, and the pursuit of scientific understanding may be seen as converging upon the religious quest for self-transforming knowledge of God...."

The central concern of **Ted Peters'** paper is how an eternal God can act, and be acted upon, in a temporal universe. Classical theology made the problem particularly difficult by formulating the distinction between time and eternity as a "polar opposition." Peters' fundamental move is to presuppose a trinitarian doctrine of God, thus including relationality and dynamism within the divine. By relating the economic and the immanent Trinity we take the temporality of the world into the divine life of God. To substantiate this move, Peters turns to the understanding of temporality in physics and cosmology. His overall aim is to show that the Trinitarian doctrine of God leads us to expect that the temporality of the world will be taken up eschatologically into God's eternity.

According to Peters, Scripture depicts God in temporal terms. With the theology of Gregory of Nyssa, Augustine, and Boethius, however, divine agency was understood as timeless, a view which came to pervade traditional Christian thought down to the present situation. Might contemporary physics shed any light on this issue? Peters' cites Eleonore Stump, Norman Kretzmann, Ian Barbour, and Holmes Rolston, each of whom suggest ways in which God might relate to the temporality of the cosmos. Still, Peters claims that, to the extent that their proposals conceive of eternity as timeless, they all fail to solve the underlying problem posed by God's eternal experience of a temporal universe. Can we instead conceive of God as "enveloping time," transcending its beginning and its end and taking it up into the divine eternity? According to Peters, Hawking would answer "no" to this question, for Hawking's cosmology has no beginning and challenges the temporality of the universe as such. Indeed Hawking draws "anti-theological" implications from his work: with no initial singularity there is no need whatsoever for God. Peters is critical of Hawking's "anti-religious agenda" and points out that the God whom Hawking attacks is the God of deism, not the God of Christians, Jews, and Muslims. Moreover, an alternate interpretation of the Hawking cosmology has been offered by Isham, who shows how God can be thought of as present to and active in all events of the universe even if there were no initial event.

Peters then returns to the problem of reconceptualizing the divine eternity. He is appreciative and yet critical of the thought of Wolfhart Pannenberg, who draws on holistic principles to interpret eschatology. Such principles have important scientific as well as theological warrant. Proleptic eschatology adds to the whole/part dialectic of science the claim that the whole is present as one part among others. This theme is developed by Robert Jenson, who stresses that Yahweh's eternity is "faithfulness through time," and by Jürgen Moltmann, who turns to Christology and the dynamics of shared suffering to connect eternity and temporality. This results in Moltmann's modification of Rahner's Rule: the identification of the economic and the immanent Trinity will only be achieved eschatologically.

Peters concludes by pointing to new directions for future research. The doctrine of God might be required to explain the temporality of the world, including the arrow of time. Moreover the movement between economic and immanent Trinity, through creation, incarnation, spiration, and consummation, could be seen as bringing the history of creation into the life of God.

My paper (**Robert Russell**) is divided into two sections. In the first section I focus on inflationary Big Bang cosmology and the problem of $t=0$. The theological reaction to $t=0$ has thus far been rather mixed. Some (such as Peters) have welcomed it as evidence of divine creation; others (such as Barbour and Peacocke) have dismissed it as irrelevant to the core of the creation tradition. As I see it, the argument on both sides has been shaped by the work of Gilkey in his *Maker of Heaven and Earth*. Here Gilkey acknowledges that the problem of relating empirical and ontological language is a *fundamental* issue for theologians, reflecting what he later calls the "travail of Biblical language." However, I am critical of Gilkey's resolution of the problem, which begins with his use of the traditional distinction between what can be called "ontological origination" and "historical/empirical origination." Gilkey, citing Aquinas, seems to view these as *strictly dichotomous alternatives*. One then either rejects the latter as theologically irrelevant (Gilkey's position) or elevates the latter into the essential meaning of the former (the position Gilkey rejects). In the first case, science, insofar as $t=0$ is concerned, plays no role in theology; in the second case it plays a normative role.

I criticize both extremes by attempting to undermine Gilkey's assumption that the alternatives should form a strict dichotomy. Instead I believe that historical/empirical origination provides an important corroborative meaning for ontological origination, although it is neither its essential nor even its central meaning, a view, incidentally, which I take to be more in keeping with that of Aquinas. I then argue that an important way of relating historical/empirical origination to ontological origination is through the concept of finitude. This abstract concept, initially closely connected to ontological origination, can take on an important historical/empirical meaning in the context of cosmology, where the past temporal finitude of the universe is represented by the event, $t=0$. Hence I argue that $t=0$ is relevant to the doctrine of creation *ex nihilo* if one interprets arguments about historical origination (such as found in $t=0$) as offering confirming, but neither conclusive nor essential, evidence for ontological origination. In this way science plays a more vigorous role in the doctrine of creation than many scholars today allow but *without* providing its essential meaning. In particular, taking a cue from the writings of Ian Barbour, Nancey Murphy, and Philip Clayton, I frame my approach in terms of a Lakatosian research program in theology. Creation *ex nihilo* as ontological origination will form the core hypothesis of this program, with $t=0$ entering as confirming evidence through the use of a series of auxiliary hypotheses involving the concept of finitude deployed in increasingly empirical contexts of meaning.

In the second section, I discuss the Hartle/Hawking proposal for quantum cosmology. Their startling claim is that the universe, though having a finite age, has no beginning event, t=0; that is, that the universe is finite but unbounded in the past. How should this result affect the arguments in Part I? To answer this, I first critically discuss the positions developed by Isham, Davies, and Wim Drees regarding the theological significance of the Hartle/Hawking proposal. Next, I present Hawking's own theological views and offer a counterargument to them. Finally, my constructive position is that the Hartle/Hawking proposal, even if its scientific status is transitory, *can* teach us a great deal theologically. First, given their work, we should distinguish between the theological claim that creation is temporally finite in the past and the further claim that this past is bounded by the event, t=0. This leads to the important recognition that the first claim by itself is actually quite sufficient for *creatio ex nihilo*. Hence we can set aside arguments specifically over t=0 and yet retain the historical/empirical sense of the past temporal finitude of creation. Moreover, this insight, which I term "finite creation without a beginning," is valid whether or not the Hartle/Hawking proposal stands scientifically; thus it suggests that we can in fact work with "speculative proposals" at the frontiers of science instead of restricting ourselves *necessarily* to well-established results, as most scholars cautiously advise. I view this generalization of the meaning of finitude as an additional auxiliary hypothesis to our research program, and following Lakatos again, look for novel predictions it might entail and without which it would be *ad hoc*.

Thus, I analyze the temporal status of the universe in terms of quantum gravity and general relativity. The variety of ways time functions here (external, internal, phenomenological) and their implications for the temporality of the universe lead to important new directions for understanding God's action as creator and the doctrine of creation. From one perspective, the combination of quantum gravity and general relativity describes the universe as having domains of a temporal, of a timeless, and of a transitional character. Accordingly we must reconsider God's relation as creator to *each* of these domains. Here the generalization of the concept of finitude to include an unbounded finitude might allow us to claim the occurrence of the transition domain as a Lakatosian "novel fact" of our research program. From a different perspective, however, we can take quantum gravity as the fundamental theory replacing general relativity. Here God's relation to the universe as a whole will need to be reinterpreted in terms of the complex role and status of temporality in quantum gravity. In either case, God's activity as creator is not limited to a "first moment" (whether or not one exists) but to the entire domain of nature, returning us to the general problem of divine action in light of science. I close by pointing, then, to the need to rethink the current models of divine agency and of the relation between time and eternity in terms of a more complex understanding of temporality from a Trinitarian perspective informed by quantum physics and quantum cosmology.

It is **Wim Drees'** argument that, unlike other topics in science, Big Bang and quantum cosmology are equally compatible with a timeless, static view of the universe and a temporal, dynamic view. Critical realists must face this ambiguity squarely. If they want to make the ontological claim that nature is temporal they must relinquish the epistemological claim for the hierarchical unity of the sciences by leaving out relativity and cosmology (due to this ambiguity). Yet this is problematic to realists, where both ontological and hierarchical claims are pivotal. Drees himself sees the timeless character of cosmology as more compatible with a Platonizing tendency in theology, in which God is timelessly related to the world rather than temporally related via specific divine acts in the world.

Drees begins by defining "temporal critical realism" as the combination of critical realism with an evolutionary view of the world and a temporal understanding of God. The Big Bang appears at first to offer just such a highly dynamic worldview consistent with temporal critical realism. However, its underlying theories (special relativity and general relativity) undercut this dynamic perspective, challenging universal simultaneity and re-interpreting time as an internal, rather than an external, parameter. Quantum cosmology, and its underlying theory, quantum gravity, further challenge the dynamic view of nature. Although they overcome the problem of the singularity at $t=0$, quantum cosmology and quantum gravity offer an even less temporal view of nature than does relativity, since they move from a four-dimensional spacetime perspective into a three-dimensional, spatial perspective in which time plays a minimal role at best.

Critical realists such as Barbour, Peacocke, and Polkinghorne have been careful to avoid theological speculations about $t=0$, recognizing that its status is controversial and subject to the shift in theories. However, they have not been equally attentive to the challenge to temporality *per se* by special relativity and general relativity, let alone by quantum cosmology and quantum gravity. Moreover, Drees claims the latter ought not be dismissed merely because they are speculative. Such a strategy to insulate temporal critical realism is *ad hoc*, since temporal critical realists are already committed epistemologically to a hierarchical unity of the sciences, and thus changes—even if only potential ones—at the fundamental level of the hierarchy carry enormous epistemic leverage. For its part, the timeless character of physics and cosmology leads us to view God in more Platonic terms. Drees explores this option in some detail, including the problem of divine action, the arguments for viewing God as an explanation of the universe, and the constructivist view of science as myth. He concludes by suggesting that axiology may be a more apt focus for theology than cosmology, and this in turn would lessen the impact science has on theology.

4.1.5 *Theological Implications 2: Laws of Nature* The volume concludes with three essays on the theological implications of the laws of nature.

George Ellis's paper combines reflections on the anthropic principle with the theology of William Temple. He calls this a "Christian Anthropic Principle," which seeks to account for the particular character of the universe in terms of the design of God who intends the evolution of creatures endowed with free will and the ability to worship the creator. Ellis thereby hopes to provide a synthesis of science and theology which will take into account recent work in cosmology and provide a better understanding of how these two fields might be related.

Ellis begins by distinguishing between the patterns of understanding in science and in theology. Still, both religion and science can be relevant when we consider the nature of the universe and its ultimate cause. Five approaches to such a cause are available: random chance, which is unsatisfactory unless one accepts reductionism; high probability as in chaotic cosmology, which is hard to quantify; necessity (only one kind of physics is consistent with the universe), but since the foundations of the sciences are debatable, an argument from the unity of the sciences is far from available; universality (all that is possible happens), but such Many Worlds arguments are controversial and probably untestable; and design of the laws of physics and the choice of boundary conditions. Design requires a transcendent designer.

The anthropic principle (AP) speaks to two questions: "Why do we exist at *this* time and place (weak AP), and why does the universe permit evolution and our existence at *any* time or place (strong AP)?" The strong AP can be linked to quantum mechanics through the role of the "observer," but this is controversial and it leaves unanswered the question of why quantum mechanics is necessary. Thus Ellis looks for ultimate causes beyond the confines of science. Religion can provide just such an approach, since it is capable of dealing with ultimate causation without being incompatible with science.

Ellis provides a Christian setting for the design argument by describing the "essential core" of New Testament teaching based on Temple's theology and his own Quaker perspective. God is understood as creator and sustainer, embodying justice and holiness. God is personal, revealed most perfectly in Jesus, and active in the world today. The Kingdom is characterized by generosity, a forgiving spirit and loving sacrifice. The universe arose as "a voluntary choice on the part of the creator, made because it is the only way to attain the goal of eliciting a free response of love and sacrifice from free individuals."

This interpretation of divine action guides Ellis in his proposal of a Christian anthropic principle (CAP), combining design with divine omnipotence and transcendence. The nature, meaning, and limitations of creation are determined by the fundamental aim of God's loving action, that of making possible in our universe the reality of sacrificial response. God's design, working through the laws of physics and chemistry, allows for the evolution of such modes of life in many places in the universe. "From this viewpoint, fine-tuning is no longer regarded as evidence for a Designer, but rather is seen as a consequence of the complexity of aim of a Designer whose existence we are assuming. . . ."

This entails five implications for the creation process. The universe must be orderly so that free will can function. God attains this goal through creating and sustaining the known physical laws which allow for the evolution of creatures with consciousness and free will. God has also given up the power to intervene directly in nature. The existence of free will makes pain and evil inevitable and requires that God's providence be impartial. Moreover, God must remain hidden from the world, allowing for "epistemic distance." God achieves both an impartial providence and epistemic distance through the impartiality of the laws of nature. Yet revelation must be possible, so that God can disclose to the faithful an ethical basis for life. None of this contradicts the standard scientific understanding of the universe, but adds an "extra layer of explanation" for the universe and its laws that is basically metaphysical. Finally, Ellis turns to quantum indeterminacy to provide a basis for divine inspiration. Other forms of intervention or action are thus excluded, including amplification by chaotic systems; these would "greatly exacerbate the problem of evil."

While Ellis has argued that it is highly probable that life exists throughout the universe, he claims that the number of individuals in the universe must be finite if God is to be able to exercise care for each. If the universe were infinite instead, it would not be possible for God to have the requisite knowledge of the infinite number of individuals and their infinite number of relations to one another. Thus the SETI project is of "tremendous religious significance" in testing the hypothesis of a caring creator.

CAP leads us to the following questions: Is our physical universe the only way to achieve the divine intention? How, more precisely, is the ultimate purpose imbedded in, and manifested by, the laws of physics? What proof can be given for CAP? To the last question Ellis argues that the evidence for CAP is stronger than evidence for inflation or the quantum creation of the universe.

The purpose of **Nancey Murphy**'s paper is to assess the possibility for using the "fine-tuning" of the laws of nature in constructing a new design argument. Her paper is closely linked with that of Ellis but with an important difference: she treats the thesis advanced by Ellis as an argument for the existence of God. Murphy first shows how recent developments in philosophy overcome the traditional Humean objections to design arguments. Carl Hempel's theory that science employs hypothetico-deductive reasoning undercut Hume's assumption that knowledge proceeds by induction. Holist accounts of the structure and justification of knowledge, offered by both W. V. O. Quine and Imre Lakatos, show that a hypothesis is never tested on its own, but rather in conjunction with a network of beliefs into which it fits. Lakatos provides a detailed theory about the structure of this network (or "research program"), in which a core theory is surrounded by a belt of auxiliary hypotheses which, in turn, are both supported and challenged by the data. Lakatos's structure includes theories of instrumentation for relating the data to the auxiliary hypotheses and a positive heuristic for the expansion of the auxiliary hypotheses into new domains of data.

To avoid circularity and relativism, Lakatos provides external criteria for choosing among competing research programs. A progressive program is one in which an additional auxiliary hypothesis must both account for anomalies, predict "novel facts," and occasionally see them corroborated. Moreover, such new hypotheses must fit coherently into the existing program. "The only reasonable way to assess the claim that fine-tuning provides evidence for divine creation is to consider the design hypothesis not as a claim standing alone . . . but rather as an integral part of . . . a theological research program" which can then be assumed as progressive or not. But what should constitute the "data" for theology? Murphy recognizes that this is a central issue for her proposal. Her sources typically include both Scripture (incorporated through an appropriate doctrine of revelation; i.e., a "theory of instrumentation") and experience. Murphy suggests that the church's practice of communal discernment could minimize the subjectivity of religious experience.

Murphy then reconstructs Ellis' paper in terms of the Lakatosian structure. Her aim is to show that theology can be regarded as a science, that cosmological fine-tuning can serve as an auxiliary hypothesis in a theological research program, and that theological theories can be compared directly with scientific theories.

In what way is the Temple-Ellis program confirmed? According to Lakatosian standards, it must produce novel facts, and Murphy claims that it does. Ellis added an auxiliary hypothesis to Temple's theology: in order for there to be the free will required by Temple, God's plan for the world had to include that the world be law-governed *as well as* fine-tuned. The facts supporting the law-like character of the world were irrelevant to Temple's theology, but in the Temple-Ellis program these facts now take on theoretical meaning. The key here is that Ellis did not set out to explain the facts supporting the law-like character of nature but only the presence of free will in nature. Thus, Murphy concludes, these facts are "weakly novel" since they were already known but irrelevant to Temple's theology before Ellis modified it. Finally, Murphy suggests ways in which the Temple-Ellis program might be expanded to include the theological problems of theodicy, moral evil and natural evil, and the scientific discussion of thermodynamics, the arrow of time, and perhaps even consciousness. These could make the program even more progressive.

In his paper, **John Polkinghorne** defends a version of critical realism in which the process of discovering the laws of nature is interpreted "verisimilitudinously as the tightening grasp" on reality. Yet these laws ought not be reduced to those of fundamental physics; instead our experiences of macroscopic nature are to be taken equally seriously. Polkinghorne accepts a "constitutive reductionism" (in that we are composed merely of fundamental particles) but he opposes "conceptual reductionism" (since the laws of biology cannot be reduced to those of physics). Thus constitutive and holistic laws must be combined in some way.

Polkinghorne's proposal is that, to our usual notions of upward emergence (which address the qualitative novelty of mind and life), we must add "downward emergence, in which the laws of physics are but an asymptotic approximation to a more subtle (and more supple) whole." Polkinghorne sees his approach as contextualist: the whole and the environment influence the behavior of the parts. It is guided by the principles of coherence (the need to explain the known laws of physics, given this wider view), historic continuity (the world must permit our experience of free agency), and realism (not only the general claim that the world can be known through science but the explicit claim that "epistemology models ontology").

The reality thus known must include the phenomena of mind-brain. Polkinghorne admits that no solution to the mind-brain problem is forthcoming, but hopes that his is a "suggestive way of beginning." Here the dynamic theory of chaos provides a vital clue. Chaotic systems, though governed by deterministic equations, are highly sensitive to environmental circumstances and initial conditions. They represent a form of "structured randomness" whose intrinsic unpredictability, according to Polkinghorne's form of critical realism, means they are in fact ontologically indeterminate. Thus he concludes that the future is open, involving "genuine novelty, genuine becoming." This in turn allows for human intentionality and divine action.

Polkinghorne then conceives of the operation of agency as the exchange of "'active information', the creation of novel forms carried by a flexible material substrate." Here Polkinghorne is contrasting agency as the transmission of information with agency as causal influence, which would include the "transaction of energy." Information transmission thus becomes a very general characteristic of living processes. Quantum physics provides similar insights to chaos theory, but Polkinghorne is cautious about relying on it. We ought not confuse randomness with freedom, and we need to remember both that the interpretation of quantum theory is still in dispute and that quantum equations may not exhibit chaotic solutions.

With chaos theory as a basis, Polkinghorne returns to a suggestion he has previously considered, that the mind/brain problem leads to a "complementary metaphysics of mind/matter." Here, however, he relates this suggestion to the problem of divine agency. A conception of nature as open allows us to understand God's continuing interaction with nature as "information input into the flexibility of cosmic history." This entails a "free-process" defense in relation to physical evil, a hiddeness to God's action in the world, and a limitation on that for which we can pray. Polkinghorne rejects the criticism that his is a "God of the gaps" strategy, since the open character of chaotic processes are intrinsic gaps in nature revealed by science, not flaws in our knowledge of nature. Likewise, he does not see himself making God into a finite causal agent, since God's interaction with nature is through information, not energy. Finally, in his proposal, God is highly temporal since the world is one of "true becoming." In this "dipolar (time/eternity) theism," eternity and time are bound together in the divine nature. God cannot know the future, since the future is not there to be known. The divine *kenosis* thus includes an emptying of God's omniscience. But God is "ready for the future," being able to bring about the eschatological fulfillment even if by way of contingent processes.

4.2 Analysis

Our overarching topic is divine action. Individual essays, however, range widely in terms of scientific, philosophical, and theological interests.

In Section I, Ellis and Stoeger present standard Big Bang cosmology, while Isham gives us a taste for the changes in store as we move to quantum gravity and quantum cosmology. If anything, the role of time seems curtailed from its arguably questionable status in the former to its more or less secondary status in the latter. How we speak about God's acting ought to be affected in turn.

In Section II, both Heller and Happel acknowledge the role of metaphor when introducing scientific discoveries into theology. Heller's position is that any theory of divine action should respect the integrity and neutrality of the scientific/mathematical account, although science can be used to support a theological view of divine action when augmented by further extra-scientific assumptions. Happel believes that the interpretations scientists give of their work are colored more by the presuppositions conveyed through the metaphors which they bring to their research than by the research itself. One could argue that when the research is later analyzed for its implications for theology, what one obtains is more a retrieval of these presuppositions than a discovery of the intrinsic implications of the research itself.

The status and significance of the laws of nature is the predominant theme running through all the philosophical papers in Section III. Do the laws of nature suggest that the universe is temporal and dynamic, or timeless and static? How do the laws become known to the human mind? Is the mental experience of time a necessary feature in permitting the mind to imagine the laws of nature? These questions set the stage for the theological questions which follow about God's relation to time and the laws of nature as Creator.

The temporality of the universe is debated by Isham and Polkinghorne. The laws of nature seem to allow both a timeless and a temporal interpretation. Until new evidence emerges to decide the case on empirical grounds, the debate remains philosophical, allowing both perspectives equal status, and both figure in the discussions that follow.

Davies and Grib point in distinct ways to a fundamental link between mind and matter. Davies argues that the laws of nature must be such as to allow the evolution of minds which can construct them successfully. Through mathematics we can describe the processes which produce the minds out of which mathematics arises. Grib links mind and matter in a different but related way. Grib relates our experience of moving in time to our ability to understand the world, given that the world obeys a non-ordinary form of logic.

Alston and Stoeger both defend the possibility of divine action by analyzing, each in his own way, the problems raised by the laws of nature. Alston brings new insights to an argument he has previously developed. Even for deterministic systems, we can affirm that God acts in the world in specific events as well as through the lawfulness of nature without invoking an interventionist strategy, since we never have sufficient reasons to claim that we know all the operative factors. Here he adds to this that the block universe does not undercut genuine human freedom or God's ability to produce temporally ordered responses to human action. Stoeger, too, defends the possibility of God's acting in the world. A key argument is that the laws of nature are descriptive, not prescriptive. Moreover, as generalizations which ignore unique features in nature, they cannot be used to support a reductionist and deterministic account of all that eventuates in nature or to overrule God's action in specific cases.

All of the papers in Section IV deal specifically with the problem of God's action in relation to the temporality of the world. Lucas and Ward defend the temporality of eternity as well as God's action in time against traditional, timeless, interpretations of divine eternity and action. Lucas claims that divine temporality is consistent with special relativity; moreover, cosmology can provide a unique, universal temporal framework via the microwave background.

In other writings, Lucas, in distinction from Ward and Polkinghorne, makes the additional claim that divine temporality might require a change in the fundamental laws of physics.

Ward argues for the temporality of God and the possibility of divine action in the world by internalizing the temporal relations in creation within the divine being. Thus God can act in new ways, make new decisions and bring into being an infinite number of new things in time. Yet these actions are consistent with the laws of nature. Like Stoeger, Ward sees God's purposes realized as potentialities in the structure of reality and not as interferences from an alien power.

Peters, like Ward, Lucas, and Polkinghorne, argues that the divine being inherently includes temporal relations. Like Ward and Polkinghorne, he leaves the laws of physics unquestioned. Unlike all of them, however, Peters works within an explicitly *trinitarian* framework throughout his writings. It is via the relation between the economic and the immanent Trinity that the temporality of the world is taken up into the divine life of God. For Peters, however, this happens eschatologically, such that the economic Trinity becomes identified with the immanent Trinity at the end of time. To support this move Peters turns to temporality in physics and cosmology. Though he admits the evidence is mixed, Peters opts for a temporal interpretation against Hawking and other non-temporalists.

Russell and Drees, unlike Lucas, Ward, and Peters, emphasize the ambiguous meaning of temporality in general relativity, inflationary Big Bang cosmology, quantum gravity, and quantum cosmology. They, however, pay close attention to the detailed scientific arguments involved, arguments emphasized earlier by the Isham/Polkinghorne debate. Russell begins by focusing on the problem of $t=0$ in Big Bang cosmology, constructing a position intermediate between those who do not find it significant for theology and those who find it pivotal to the meaning of creation *ex nihilo*. Turning to the Hartle/Hawking proposal, he supports Isham and Drees in arguing that quantum cosmology can be seen in both temporalist and atemporalist perspectives, especially if quantum gravity and general relativity are maintained in tension. He emphasizes that time is subordinate to space in general relativity and even more fully in quantum gravity. Russell opposes Hawking's deistic interpretation of quantum cosmology, however, arguing that God's action should be seen throughout nature. He closes by urging that the theological problem of time and eternity and the philosophical problem of divine action be reconsidered in light of the complex problem of temporality in physics and cosmology.

Drees, on the other hand, argues strenuously for the ambiguous verdict of relativity and cosmology regarding the temporality of nature and the problem their ambiguity poses for critical realists, drawing on much the same material as in the Isham/Polkinghorne debate but pressing the problem facing a critical realist position. In essence, Drees tries to force critical realists to choose between a unified epistemology (including relativity) without a clear ontology, given relativity's ambiguity about time, or a temporalist ontology based on a partial epistemology in which relativity, being ambiguous about time, is left out and the temporality of the other disciplines such as thermodynamics or biology is accepted. If his wedge is successful, it means that a decision about the problem of divine action must be postponed until the philosophical ambiguities underlying a critical realist interpretation of physics is settled—or abandoned.

Section V deals with God's action in relation to the laws of nature, seen first from cosmological and then from macroscopic scales. The former is formulated in terms of the Anthropic Principle by Ellis and Murphy while the latter is discussed in terms of dynamic chaos by Polkinghorne.

According to Ellis' Christian anthropic principle, God acts through the laws of nature to achieve the divine purpose, namely the evolution of creatures capable of sacrificial response. Apart from revelation, for which quantum indeterminacy (and *not* dynamic chaos) provides a possible basis, God never intervenes in these laws. Consequently God is hidden from the world, and since providence is impartial, pain and evil are both real, inevitable, and impartial as well. All this leads Ellis to make an empirically testable (!) claim: that the universe is finite in size and contains a multitude of life-forms.

Murphy interprets Ellis' work in terms of a Lakatosian research program which extends Temple's theology into the domain of cosmology. Unlike Ellis, however, Murphy treats fine-tuning as confirmation for the Temple-Ellis proposal and, thus, as evidence for the existence of God. That is, she argues that the Temple-Ellis analysis comprises a progressive theological research program, since fine-tuning, introduced by Ellis, provides additional, non *ad hoc* confirmation of the already-existing arguments for God advanced by Temple. To the extent that she works within the Temple-Ellis framework, it can be assumed that Murphy accepts Ellis' claims that God's action is consistent the laws of physics, and that divine revelation may occur through events at the quantum level.

Polkinghorne, on the other hand, explicitly moves beyond the quantum domain to find avenues for God's direct action in the world. He begins by arguing that the world in intrinsically open and that it is consistent with human *and* divine agency. Although the laws of physics are deterministic, they are only a part of the complete laws of nature. Macroscopic nature, with its newly studied phenomena of dynamic chaos, provides additional laws which are indeterministic in character, leading to the view of nature as intrinsically open.

The key to Polkinghorne's thesis is the assumption that "epistemology models ontology," that is, that the epistemic characteristic of the laws of nature arises from the intrinsic character of nature. This view is part of Polkinghorne's particular version of generic critical realism. Given this view, together with the theory of macroscopic dynamic chaos, Polkinghorne asserts that nature is open: the future is not determined uniquely from the present; indeed, the future does not exist as an actuality. If this is so, human agency, as well as divine agency, can be operative in the world. Agency in general, however, is best understood in terms of the exchange of information rather than the effect of energetic causes (forces). Thus, God is not a secondary cause, and our understanding of God's action is based not on ignorance but on knowledge of the ontology of the world.

These seventeen essays comprise the first of five planned volumes on divine action in the perspective of the natural sciences. Clearly the problem of divine action underlies the program of systematic theology, even when it is not explicitly manifest. The editors hope that this discussion of quantum cosmology and the laws of nature provides a first sortie into the general problem which will be useful to other scholars in the field. We await further research with great expectation.

I
SCIENTIFIC BACKGROUND: STANDARD AND QUANTUM COSMOLOGIES

INTRODUCTION TO GENERAL RELATIVITY AND COSMOLOGY

G. F. R. Ellis and W. R. Stoeger

1 *Introduction*

In this brief essay we shall present the basic ideas of contemporary classical cosmology, including its foundations of special relativity and general relativity (Einstein's theory of gravity), the standard Friedmann-Lemaitre-Robertson-Walker (FLRW) cosmological models upon which much of cosmology relies, and the observational features which tend to confirm that these models describe the large-scale structure and evolution of our universe to a fairly good approximation. In our treatment we shall not deal at all with either the cosmology of the very early universe, where quantum and unified field theory considerations become important, or with quantum cosmology. These topics are being treated in other essays. We shall, however, briefly discuss the inflationary universe idea, since it is so important for resolving certain problems, especially the horizon problem, in classical cosmological models.

2 *Special Relativity*

Before discussing cosmology itself, it is important to talk about the special and general theories of relativity.

Special relativity deals with the physics of particles and light rays in flat spacetime. It ignores from the complications of gravitational fields and gives an accurate account of situations where gravitation is weak enough to be neglected for all practical purposes. For particles or other bodies moving slowly—much less than the speed of light—the description provided by special relativity is well-approximated by the classical, pre-relativistic physics of Galileo and Newton; for example, they describe the motions of cars, aircraft, and the planets in the solar system extremely well. For particles and bodies traveling at velocities approaching that of light, the modifications provided by special relativity become essential for describing the velocities, masses, and time intervals actually measured by observers, and for correctly giving the laws of conservation of energy and momentum, without which physics cannot proceed.

Special relativity can be easily understood and compared with classical physics by speaking of "reference frames." Two automobiles traveling along the same road—one at 25 mph and the other at 60 mph—provide a good example. There are three natural reference frames, or coordinate systems, in this situation: the frames of each of the two drivers, and that of someone standing by the side of the road, stationary with respect to his or her surroundings. In classical physics, if the two automobiles are traveling in the same direction, then, although the static observer by the side of the road measures their velocities to be 60 mph and 25 mph respectively in that direction, in the frame of reference of the slower driver the faster driver is traveling at only 35 mph in that direction, and the observer by the side of the road is moving at 25 mph in the opposite direction. Similarly, in the reference frame—from the point of view—of the faster driver, the slower driver is traveling at 35 mph in the opposite direction, and the static observer at 60 mph, also in the opposite direction. And so we speak naturally of the velocity one driver has *relative* to the other. This becomes more complicated to determine if the drivers are traveling not along the same straight road, but on roads at angles to each other. A trivial but important point is that *any observation or measurement is always made with respect to such a reference frame*—and the result of the measurement, for example, velocity, depends on the characteristics of that reference frame—on its velocity with respect to the object being measured (obviously!), and on whether it is accelerating or not.

Thus, observers in different reference frames will obtain different measurements of the same body. In special relativity, not only will the velocities they measure be different, but also the mass of the object, its length, and the rate at which a clock traveling with it will tick. For example, the simple additive composition of velocities in classical physics we illustrated above is replaced by a more complicated rule, such that the velocity of an object in any reference frame whatsoever will not exceed the velocity of light. This relativistic rule for velocity addition is like the usual Newtonian law (discussed above) at slow speeds, but quite unlike it, and indeed very counter-intuitive, when high speeds are involved. Furthermore, and perhaps most difficult to comprehend, the measurement of simultaneity depends on the motion of the observer. As we have already mentioned, these differences will only become significant for objects moving at velocities approaching that of light with respect to the observers' frames.

Now special relativity deals only with frames of reference which are non-accelerating, that is, traveling at constant velocity relative to each other. These are often called *inertial frames*. There are two postulates which provide the fundamental basis of special relativity. The first is that *the laws of physics must not depend in their formulation on the frame of reference—they are invariant under a change of inertial frame*. This means that there will be a way of expressing each law, for example, the conservation of energy, so that its form in one inertial reference system will be the same as its form in any other inertial reference system. This is usually expressed by saying that the laws themselves are invariant, and their expressions in different coordinate systems and reference frames (and the transformations between one coordinate system and reference frame and another) are "covariant." The general class of transformations which take us from one inertial frame of reference to another, and under which the laws must be invariant, are called *Lorenz transformations*. This first postulate has the important implication that *absolute velocity has no invariant—physical— meaning*, for an absolute velocity cannot be experimentally assigned to any object or observer. Only relative velocities determine the results of physical measurements. Consequently only relative velocities are measurable.

The second postulate is that the *velocity of light in a vacuum, c, is finite and has the same value in every frame of reference*. No matter how quickly or slowly you are moving, you will always measure the speed of light (and indeed of all electromagnetic radiation, e.g., radio waves or X-rays) to be c—approximately 186,000 miles per second. When combined with the first postulate, this leads to the theoretical prediction that the speed of light is a limiting speed for motion of all massive particles and for all signals that can convey information from one place to another. This is confirmed by experimental observation, and is one of the fundamental limitations of modern physics: we know of no circumstances where it is violated.

It turns out that these two postulates can be most easily incorporated into the spacetime geometry, which is the arena for physics, by treating the speed of light multiplied by time, ct, as a fourth dimension, almost on an equal basis with the three spatial dimensions. It is, however, distinguished from them as the "time dimension" by attaching a minus sign to $(ct)^2$ in calculating the square of "spacetime distances." This is a generalization to spacetime of the Pythagorean result that $d^2 = x^2 + y^2 + z^2$ gives the distance d of the point with Cartesian coordinates (x,y,z) from the origin of coordinates. For four-dimensional spacetime the analogous expression for the *invariant interval* S^2 of special relativity is:

$$S^2 = -c^2t^2 + x^2 + y^2 + z^2 \quad (1)$$

S^2 determines the spacetime "distance" between different events, determining both how rigid rulers measure distances and perfect clocks measure times.

Here we shall not go into further detail about special relativity, except to point out that these two postulates have the consequences we have alluded to above: (a) the mass of a particle moving relative to our inertial frame will increase with its velocity, and no particle can be accelerated to a velocity greater than c; (b) the length of an object we measure will decrease with velocity in the direction of its motion; (c) the rate at which a clock carried by the moving particle ticks will appear to run more slowly with increasing velocity; (d) the simple additive composition of velocities in classical physics is modified so that no object can be accelerated to a velocity greater than that of light in any reference frame. In all these cases, these measurements are *relative* to our frame of reference; they cannot be taken as absolute in any sense. Thus, in special relativity, we must take the *idea of relativity* as applied to mass, and spatial and temporal intervals, very seriously. Nevertheless, not all is relative; for all observers agree on the value assigned to the invariant interval S^2. Finally, another consequence of special relativity is (e) the equivalence of mass and energy, the famous formula $E = mc^2$, which tells us how much energy E can be obtained from a given amount of mass m, or how much mass can be created from a certain amount of energy. In light of this, the law of the conservation of energy must be generalized into the law of the conservation of mass-energy.

Special relativity has been extraordinarily successful. Its description of physical phenomena has been confirmed again and again in experimental and observational situations, including the behavior of particles in accelerators, the decay of cosmic rays penetrating the atmosphere, the functioning of nuclear reactors, and the evolution and light-producing capacity of stars.

3 *General Relativity*

Special relativity describes physics in flat spacetime, but does not include a adequate account of gravitational fields, the way they are produced and the effects they have on particles and light rays. General relativity answers this need. In so doing it gives us a way of dealing with accelerating reference frames, not just inertial ones as in special relativity, demonstrating a profound equivalence between acceleration effects and gravitational forces.

The basic feature of general relativity is the relationship it expresses between the geometry of spacetime and the distribution of mass-energy within it. The presence of mass-energy curves spacetime. The spacetime of special relativity can be flat precisely because the gravitational effect of matter is ignored. Once matter—mass-energy—is introduced, spacetime will no longer be flat, but rather curved. Equivalently, the *geodesics*, the paths of freely falling particles and light rays in these curved spacetimes will, in general, be curved. A helpful picture to keep in our imagination is that of rubberized membrane, a section of a tire tube, for instance, stretched over the top of an open cylinder to form a drumhead. The circular section of tire tube over the end of the cylinder is a two-dimensional representation of space. As long as nothing heavy is placed on it, it remains flat, and a light marble will role in a straight line right across it, without any deviation. Now place a heavy shot-put in the middle of the drumhead. It will obviously distend and distort the surface of the membrane, curving its surface from the edge of the cylinder to the place where the shot-put rests on it. And the heavier the shot-put, the greater the distention or curvature. This is exactly what a local distribution of mass-energy does to the space around it. Now the light marble will no longer roll straight across the drumhead; its path will be curved. If it has high enough initial velocity, it will follow a curved path—influenced by the shot-put—from one side of the cylinder rim to the other. But if it doesn't have enough speed, it will be trapped in the middle, and roll in orbit around the shot-put, similar to the way the planets orbit the sun. So the presence of mass-energy distorts space, and particles and light rays now travel on apparently curved paths in it. Gravitation is represented by this spacetime curvature. In Newton's theory of gravity, light rays are not affected by gravitational fields; in Einstein's general relativity, they are. They are bent and, additionally, light either loses or gains energy (it is redshifted or blueshifted) as it climbs out of or falls into a gravitational field. Both of these effects have now been observed and confirmed in a variety of situations—and are an accepted part of contemporary physics.

Thus, in general relativity, there is no pre-defined spacetime with an invariant geometry. Spacetime itself, and its geometry, are determined by the distribution of mass-energy within it. But in turn, the mass-energy—the particles of matter and electromagnetic radiation—is affected by this geometry, which determines the paths they follow—and the way signals are propagated in space and time. The invariant interval of special relativity is generalized in terms of quantities g_{ab}, which are the components of the spacetime *metric tensor*. The notation using subscripts is *tensor notation*, where a and b each can represent time or any of the three spatial dimensions. Such notation is required in order that our equations can be true no matter what coordinate system is used (and so ensuring this even when the observer is in accelerated motion). These ten quantities[1] determine a local distance function on the spacetime manifold, generalizing relation (1) to a curved spacetime; without them there is no way of specifying the distance or time from one spacetime point to another.

This close link between the distributed mass-energy and geometry is expressed primarily in the Einstein field equations, which are written:

$$R_{ab} - 1/2 R g_{ab} = 8p T_{ab} \quad (2)$$

where the left-hand side of the equations[2] represents the geometry (R_{ab} and R are curvatures, the first being the Ricci curvature and the latter being the scalar curvature), and the right-hand side gives the mass-energy and stress (pressure, viscous stress, etc.) distribution. So stress-energy, according to these field equations, determines geometry. If there is no stress-energy, $T_{ab} = 0$; thus we can see clearly that stress-energy *causes* the curvature of the spacetime.[3]

Thus, when we solve the Einstein field equations, we are solving for the components of the metric tensor g_{ab}, the distance functions we have just mentioned. Given g_{ab} the geometry of spacetime is completely specified. There is, however, one further catch. In order to solve the field equations, we need to have adequate *initial data* specified either on a space-like surface (that is, throughout three-space at one given time), or on our past *light cone*, which is the surface determined in spacetime by all the light rays we receive here and now from all directions in the sky, extended as far as they can go. These initial data must be determined by observations or assumed on the basis of considerations outside the theory of general relativity itself. Once we have the initial data, general relativity tells us how they evolve forward and backward in time to determine the structure of spacetime. Without the data we cannot determine a solution to the field equations. It is precisely here that observational issues enter cosmology and general relativity most forcefully.

There is also the issue of the experimental verification of general relativity itself. Essentially we can say that it has been rather well verified for

---

[1] There are ten quantities because g_{ab} is a symmetric tensor, i.e., $g_{ab} = g_{ba}$, where a and b take the values 0, 1, 2, or 3.

[2] Here again each tensor component in these equations is symmetric. These then are ten field equations in (2). For example one relation is obtained by setting $a = 0$, $b = 0$ in these equations, another by setting $a = 0$, $b = 1$, and so on.

[3] For those who are familiar with calculus, the curvatures are just combinations of first and second derivatives of the metric tensor components.

weak gravitational fields (such those found in the solar system), and on length scales up to and including the scale of the solar system. We have already mentioned the observational verification of *light bending* in gravitational fields. This effect was first observed by Sir Arthur Eddington and his colleagues F. W. Dyson and C. Davidson who measured light rays from a distant star as they just grazed the rim of the sun during the solar eclipse of May 29, 1919. This is one of the classical tests of general relativity. And now very precise astronomical observations have revealed that the images of a number of distant quasars are gravitationally lensed by intervening galaxies to form two, three or more images of the single quasar. These are clear cases of effects resulting from the bending of light by gravitational fields. The other two classical tests of general relativity are the *gravitational redshift* of light as it climbs out of a gravitational field (as it loses energy, its velocity of course remains the same but its frequency decreases—moves farther towards the red end of the electromagnetic spectrum), and what is known as the *precession of the perihelion* of the planet Mercury (the gradual shift in time of the orbital position of the point where the planet is closest to the Sun). General relativity has passed these classic, weak-field tests with flying colors. Since the early days of research on general relativity, other types of experiments have been developed—involving such things as radar echo delays of signals shot past the sun and bounced off a planet or the moon. General relativity has been confirmed time and time again through these experiments as well.

When it comes to strong gravitational field experiments, and those for gravitational fields operating over large distances—larger than the scale of the solar system—there is much less verification of general relativity.[4] We are not yet able to make precise enough measurements in those rather inaccessible regimes to test general relativity adequately. Verifying the existence of gravitational radiation also fits into this category—although we now have indirect evidence for it from the decay of the orbits of binary pulsars. The existence of black holes would also be a strong-field test of general relativity— they are predicted by it. We *do* have abundant confirmation that very large, extremely compact objects with very strong gravitational fields exist, on scales from the size of stars 10 times as massive as our sun to scales of a billion times the mass of our sun in the centers of anomalously bright galaxies and in quasars. But whether or not these compact objects are the black holes of general relativity has not been confirmed. At present, although we have strong indications that they may exist, we do not yet have direct confirmation that these objects are surrounded by event horizons, as the theory describes them (see below). At any rate we do not yet have any evidence from the strong field or cosmological distance scale regimes which clearly contradicts or disconfirms general relativity. And that is an important point to emphasize.

The prediction of the existence of black holes follows directly from one of the simplest and most important solutions to the Einstein field equations—the *Schwarzschild solution*. If we impose spherical symmetry on the geometry and then solve the Einstein equations for the vacuum around a perfectly spherical

---

[4] Over such distances not even Newton's theory of gravity—the inverse-square law —is well verified.

object of mass M, we obtain a very simple solution for the metric tensor g_{ab}, which gives the general relativistic description of spacetime around any such object (such as a spherical planet, star, or galaxy), and thus determines the detailed behavior of particles and light rays in that environment. At radii relatively far from the spherical object (relative to its mass), and for test particles moving in the gravitational field at relatively low velocities (therefore *not* light), Newton's theory of gravity is adequate, and therefore is still used in most astronomical calculations, such as those for interplanetary space probes. But for very compact—very dense—objects, which have surfaces at very small radial distance from their centers relative to their overall mass, the behavior of particles and light rays near these surfaces will be very different from what is predicted by Newtonian theory. This would be the case, for example, if the entire mass of the sun were packed into a sphere with a radius of approximately two kilometers and we tried to describe the behavior of particles and light rays between its surface and, say, 100 kilometers from its center. Or, similarly, if the entire mass of the earth were packed into a sphere 3 centimeters in radius, and we studied the behavior of particle and rays within a meter of its surface.

One of the most extreme of these relativistic effects is that for objects compact enough, there is a radius inside which all light rays, radio waves—all electromagnetic radiation—is trapped by the gravitational field. At that radius the effective escape velocity from the object is equal to c, the speed of light. Inside that radius, which is called the *event horizon*, the escape velocity will be greater than c, and hence even light cannot escape from it. This means than anything inside the event horizon will not be seen by an observer outside it—that no communication is possible from inside the event horizon to the world outside. It also means, of course, that anything or anyone falling through the event horizon has irretrievably done so: there is no possibility for any material object to escape from within that radius. An object with such an event horizon is called a *black hole*.

It can be shown, furthermore, that general relativity predicts that the spherical or quasi-spherical collapse of any object with a mass larger than about 2 or 3 solar masses will result in a black hole. It will be too massive to become a neutron star, in which the collapse is eventually brought to a halt and stabilized by neutron degeneracy pressure—when the strong gravitational field forces electrons into protons to create neutrons. As we mentioned above, we are not sure that black holes—objects shrouded by event horizons—really exist. But we are sure than neutron stars (pulsars) exist—and that there are highly collapsed objects much more massive than neutron stars, which, if general relativity is correct in describing very strong gravitational fields, must be black holes. Demonstrating that such objects really do possess event horizons would be one way of confirming the correctness of general relativity in these extreme situations. But that is very, very difficult to do!

The Schwarzschild solution exhibits another feature characteristic of many solutions to the general relativistic field equations. This is the existence of a *singularity*—also called *an essential singularity*. A singularity is a point, line or surface on the boundary of spacetime where at least one of certain physical parameters are predicted to become infinite—the density, the temperature, the curvature, etc. In the Schwarzschild solution this happens at the center of the black hole, where the radius $r=0$. Of course, it is unlikely that the parameters predicted to diverge actually become infinite. It is more likely that the model given by general relativity breaks down, and gravitational quantum effects set in under those very extreme conditions, even though this is long before an infinity in density, temperature or spatial curvature is actually reached. Very shortly we shall see another important case of a singularity in the standard FLRW models—the initial singularity, or Big Bang, at time $t=0$, in which this interpretation in terms of the break-down of the non-quantum gravity models of general relativity is the standard one.

Before going on to discuss cosmology itself, it is important to stress again that general relativity is a generalization of the Newtonian theory of gravity—in the sense that Newtonian theory is an approximation to general relativity that is adequate for weak gravitational fields and slowly moving objects. For these cases general relativity must give the Newtonian results. Similarly, too, special relativity fits in with general relativity—there will be a small neighborhood of every point in spacetime in which special relativity holds very accurately. In that small neighborhood spacetime can be considered flat and is called a Minkowski spacetime (as described by equation (1)).

Now we go on to discuss one of the principal applications of general relativity—cosmology.

4 *Classical Relativistic Cosmology*

What is the scientific subject of cosmology? Essentially it is *the physics of the observable universe, treated as a single object of study*. Cosmology investigates the structure and evolution of the universe, and to some extent its origin and destiny. In doing this it must determine ways in which its conclusions about the universe can be observationally confirmed, and then it must carry out those observations. Much of that can only be done in close collaboration with astronomy and astrophysics.

There are a number of key assumptions[5] that must be made in order to do cosmology. One obvious assumption is that *the laws of physics which we know hold locally also hold throughout the rest of the observable universe, and are valid on scales much larger than those on which we have been able to test them*. Recall our discussion above concerning the verification of theories of gravity on length scales larger than that of our solar system. There is some observational justification for this assumption, but not yet enough to confirm it adequately.

---

[5] Cf. W. R. Stoeger, "Contemporary Cosmology and Its Implications for the Science-Religion Dialogue," in *Physics, Philosophy and Theology: A Common Quest for Understanding*, ed. Robert John Russell, William R. Stoeger, and George V. Coyne (Vatican City State: Vatican Observatory, 1990), 219-247. See also R. J. Russell, "Cosmology," in *A New Handbook of Christian Theology*, ed. Donald W. Musser and Joseph L. Price (Nashville: Abingdon Press, 1992), 101-105.

The *cosmological principle*—that there are no privileged points in space, not even our own—is also invoked as an assumption. In other words we assume that we are not at the center of the universe, whatever that might mean, but rather at a point like any other one. Therefore, what the universe looks like from our point of view as we study it is what it would look like from any other point.

Finally, we must assume *a theory of gravity* in cosmology—we will assume general relativity, with its Newtonian gravity limit—and something about the matter which makes up the universe. It is almost always assumed that (like the molecules in the air around us) the matter is so distributed throughout the observable universe that it can be modeled by a fluid—usually what is known as a *perfect fluid*. This fluid approximation, to be a valid description, requires that on some very large length scale—say, 20 million light years—the matter is distributed homogeneously enough that a meaningful average density can be assigned to it (galaxies or even clusters of galaxies corresponding to molecules in the fluid). Actually, what is important, recalling the equivalence of mass and energy in relativity, is the density of mass-energy. We also need a relationship between the pressure (p) and density of this fluid. This relationship is called an *equation of state*. If a fluid description of the matter in the universe turns out to be unjustified, then something more complicated will have to be used. We usually assume that the fluid constituted by the clusters of galaxies in the recent cosmological past is *pressureless*. So the equation of state is simply $p = 0$. Both the use of general relativity and the reliance on the fluid approximation for the mass-energy distribution in the universe are partially justified—on the basis of what we know from local physics and from what little we can ascertain about the physics and the state of matter over astronomical and cosmological distances. There are puzzles, yes, but no glaring inconsistencies which would indicate that we should abandon either of these assumptions.

If we use Einstein's field equations, impose spherical symmetry and homogeneity (uniformity) on space at each moment of time—that is, on each time slice—and use a perfect-fluid equation of state, we can easily solve the equations to obtain what are referred to as the Friedmann-Lemaitre-Robertson-Walker (FLRW) models,[6] mentioned at the very beginning of this essay. These are considered the standard models of the universe in contemporary classical cosmology. They can also be constructed without assuming specific dynamics or solving the field equations by simply imposing isotropy and spatial homogeneity on spacetime as a rigorous imposition of the cosmological principle (but then their time evolution is undetermined).

What are the characteristics of these models? And how do they fare when confronted with observations? First of all, it is important to recognize that FLRW represents a whole class of cosmological models. It takes an equation of state plus two other parameters to determine one member of this FLRW class and the time at which it is observed.[7] Observationally, these two specifying

---

[6] Also referred to as the Friedmann-Robertson-Walker (FRW).

[7] We assume here that the "cosmological constant" Λ is zero; whether this is so or not has been a subject of continual debate in cosmology.

parameters are the present value of the Hubble parameter, written H_0, which gives the rate of expansion of the spatial sections of the universe, and the present value of the deceleration parameter q_0, which indicates how rapidly that expansion is speeding up or slowing down. If it is slowing down, the universe may eventually stop expanding, and recollapse under its own gravitational attraction, or it may expand forever. If, on the other hand, it is speeding up, or not changing at all, then the universe will necessarily expand forever. Since this cannot happen for ordinary matter, we will not consider it further, thereby excluding the case where a cosmological constant dominates the present-day universe.

Thus, there are three types of FLRW models: open universe models, with $q_0 < 1/2$, which expand forever; flat universe models, with $q_0 = 1/2$ which expand forever also, but just barely do so—a slight increase in q_0, or, equivalently, in the total mass-energy these models contain, and they would eventually collapse: and closed models, with $q_0 > 1/2$, in which the expansion will eventually cease, to be followed by collapse. The open and flat universe models with their usual topology have infinite spatial sections, and so contain an infinite amount of matter (i.e., there is always more matter beyond any particular particle or galaxy, without end), whereas the closed universe models (and some of the flat and open ones, with unusual topologies) have finite spatial sections and contain a finite amount of matter.[8] Finite spatial sections are like the surface of the earth: they are finite in size even though they have no edge or boundary. If you set off in any direction and continue without deviating from your course, you eventually end up where you began.

---

[8] In a number of contexts within general relativity and cosmology we can return to a space-and-time (3 + 1) view of spacetime. This is very natural in FLRW cosmologies, because the homogeneous spatial sections (spatial surfaces on which the density and the pressure are constant) define a unique time slicing of four-dimensional spacetime, and a unique cosmic time parameter. Thus, for FLRW models the temporal part of the metric tensor g_{ab} can be completely split from the three-dimensional spatial part. This cannot be done uniquely for other cosmological models, even though they are very close to FLRW.

As just implied, whether our universe is open, flat, or closed depends on how much matter it contains—the more matter, the more gravitational force there is to slow the expansion. The mass density necessary in an FLRW model to make it flat is called the *critical density*; if the density of the universe is below the critical density, it will be open—if it is above the critical density, it will be closed. At present, we do not know whether our universe is open or closed. The ratio of the actual mass density of the universe to the critical density, written as W, is usually employed as the parameter representing the amount of matter present. If the density of the universe is nearly equal to the critical density, W = 1, in which case it is nearly flat, on the border between being open and closed. The amount of luminous matter in the universe, matter that we actually see, is certainly less than the critical mass—leading to an estimate of no more than, say, W = 0.06 or 0.07. However, from the motion of stars and gas in galaxies, and of galaxies in clusters of galaxies, it is clear that a great deal of matter is not luminous, but is dark or hidden. Is there enough of it to close the universe? We do not know yet. Nor do we know of what the hidden matter is composed. Some of it could be *cold baryonic matter*—like the matter we are familiar with, composed of protons and neutrons—in various combinations to make the familiar elements. But a good deal of it—up to 95 percent of it—might be *non-baryonic*—composed of particles which have mass but which only interact very weakly with electromagnetic and nuclear fields and are virtually undetectable. This is one of the important outstanding questions of contemporary cosmology.

Two characteristics of the FLRW models which we have already seen but which must be emphasized are their isotropy and spatial homogeneity. In fact, they were constructed precisely that way. *Isotropy* is another word for spherical symmetry—they look the same in all directions; there are no angular variations whatsoever. That means, for instance, that if we as observers are *fundamental observers*—moving with the overall expansion of the fluid in the universe—and the universe is isotropic about our spatial position, we will detect no cosmological proper motions, that is, motions of distant galaxies across our line of sight. *Spatial homogeneity* means that at each moment the three-dimensional space constituting the universe at that time is perfectly smooth and uniform—it is not lumpy! Because of these two characteristics, a FLRW universe will look exactly the same from all points in it at a given time, to all fundamental observers.

But of course the universe as we know it *is* lumpy! There are planets, stars, galaxies, clusters of galaxies, and so forth. So how could an FLRW model be helpful in describing our universe? The presumption is, as we have already hinted, that the model applies *on the average*: that is, when averaged on some very large length scale, say, larger than 300 million light years, the real universe is almost perfectly smooth—even though it is not so on smaller scales. The smoothness of the microwave background radiation provides an indication that this is indeed so. However, recent developments in astronomical cosmology[9]

---

[9] These include the observation of very large-scale structures—filaments, chains, and walls made of galaxies, surrounding huge empty voids, known as the

have raised the possibility that the universe may not, in fact, be smooth on very large scales. This is another key question which must be resolved.

The fact that the FLRW models *are expanding* is probably the key characteristic they possess, and deserves some discussion. This implies that they are *evolving*. This can be seen more forcefully by realizing that as an FLRW model expands, the density and temperature of the matter and radiation in it decreases. We can measure the current value of the Hubble parameter within a factor of two—this gives an "age" to our universe of between 10 and 20 billion years. We can also measure the temperature of the blackbody microwave background radiation (see below); we find that it is about 2.7 degrees Kelvin (that is, 2.7 degrees above absolute zero). And the (large-scale) average density of our universe now is between 10^{-29} and 10^{-31} grams per cubic centimeter—the universe is almost empty, but not quite! But as we go back to earlier times in the universe we find that the universe was much hotter and much denser. In fact, in the FLRW models there is a very simple relationship between the time, density, and temperature. So the earlier we look at the universe, the hotter and denser we find it to be. The limit to these hotter denser phases is at time $t=0$, when we encounter a singularity, the Big Bang where the model says that both the temperature and the density are infinite.

But if we leave that initial singularity aside for a moment, we can examine briefly some of these temperature-density phases. For instance, when the temperature was above 4,000 degrees, all the matter was ionized (neutral atoms had not yet formed from their constituent negatively charged electrons and positively charged nuclei). And, because it was ionized, the matter was strongly coupled to the radiation—and so (like the interior of the Sun) very opaque to electromagnetic radiation. Looking out in space, and therefore back in time, we can see back to the time when this ionization ended as atoms first formed, known as the epoch of recombination, but we can see back no further. Our line of sight encounters a "fog bank," known as "the surface of last scattering."

From that time on into the future the universe is transparent to light and other electromagnetic radiation. Only lumps within it are still opaque—stars and parts of galaxies. It is this surface of last scattering (at about 4000 degrees) which we believe to be the origin of the cosmic microwave background (see below). After this epoch, beginning at about 300,000 years after the Big Bang, hydrogen was able to enter its neutral, non-ionized state (nuclei and electrons combined to form neutral atoms); thereafter, stars and galaxies began to form and develop. There were undoubtedly some slight density perturbations—regions of underdensity or overdensity with respect to the average density of the almost homogeneous matter—which would eventually grow and collapse to form clusters of galaxies, individual galaxies, and stars. But with the matter tied to the radiation when the universe was still ionized, very little growth would have been possible before decoupling. This is an example of an important point for evolutionary considerations: the temperature and density of the matter in the

---

"soap-bubble" structure of the universe—and discovery of very large speeds of motion of galaxies.

universe at a given time can severely limit what form it can take—and the objects into which it can evolve.

If we go back to times earlier than the surface of last scattering, we find the matter and the radiation in even hotter and denser states. Roughly three minutes after the Big Bang, we find the temperature at about 6 billion degrees, which is just the temperature at which hydrogen nuclei can fuse to helium and a small percentage of other light elements, like deuterium, lithium, and beryllium, in nuclear reactions (like those that occur in a hydrogen bomb). This is the epoch of *nucleosynthesis*; at earlier times nuclei could not exist because as fast as they tried to form, they would have been destroyed by high energy radiation or particle collisions. As a matter of fact, we do find a surprisingly large percentage of helium in our universe, about 24 percent by weight, too much to be explained by stellar nucleosynthesis, and a significant trace of deuterium which also cannot be produced in stars. These observed abundances are taken as strong confirmation that our universe was once in such a very hot dense phase, for detailed calculations using the standard FLRW models give just the sort of abundances we measure. This is taken as strong support, too, for these models on cosmological length scales.

At even earlier times, much less than a second after the Big Bang, the temperature and density would have been above other key transition thresholds. There was a point, for instance, when it was too hot for there to be single protons and neutrons: all that existed was a sea of quarks—components of protons and neutrons. As the universe expanded and the temperature dropped to lower values, hadrons (specifically protons and neutrons) condensed out of the quark sea. This is referred to as "the quark-hadron transition." Quantum field specialists believe that at higher temperatures some of our physical laws were different. There was a time and a temperature at which the electro-magnetic interaction and the weak nuclear interaction were unified into a single undifferentiated "electro-weak" force. And even earlier, there was a transition, referred to as "the GUTs transition," prior to which there were just two interactions, gravity and the GUTs (Grand Unified Theory) force. At this transition, the latter separated into the strong nuclear force and the electro-weak interaction. At an even earlier time and higher temperature (energy), it is speculated that there was just one superforce, including gravity. At this point the model we have been using breaks down— general relativity is no longer valid in this regime. Nor is the FLRW model we have based upon it. In fact, it is difficult to even identify space and time as we know them at this point. A detailed quantum theory of gravity, and of spacetime itself, is necessary for describing this epoch just after the Big Bang.

This brings us to an interpretation of the *initial singularity* at $t=0$ in the FLRW models. Classically, this is a boundary to the universe where the laws of physics break down, matter ceases to exist, and even the concepts of space and time cease to have meaning. It is the "creation" event in the sense that it is where all these things come into being, not having existed before.[10] However before we get to that time looking back into the past from the present, conditions will be

---

[10] This sentence shows how difficult it is to discuss this concept, for even the concept of *before* has no temporal meaning on this boundary.

such that the classical model we have been using for gravitational physics will have broken down. To treat this regime a more sophisticated, quantum model of gravity will have to be called into service. Presumably, in the more accurate description it will provide, essential singularities such as $t=0$ will disappear. At least that is the hope.

At this point we should briefly turn our attention to the *observational aspects* of cosmology. In the course of our discussion we have already referred to these. But it is good to take an opportunity to summarize the observational situation in a focused way. There are really three major pieces of observational data which lead to the conclusion that the universe is expanding, and possessed hotter, denser phases in the past. The first—and one we have not yet directly mentioned—is the *systematic redshifts* of distant galaxies and quasars. This indicates that they are moving away from us—and that the farther away they are, the faster they are receding from us. There are no systematic blueshifts observed—or even a class of highly blueshifted objects—which would indicate objects moving towards us. It should be pointed out here that we interpret this expansion to be, not the movement of the galaxies away from us within a static space, but rather the expansion of the space itself, with the galaxies essentially fixed in space and moving with it.

The second key observation is the *cosmic microwave background radiation*: both its isotropy (sameness in all directions), and blackbody character. We have already seen that it can be interpreted as originating at the last scattering surface during the epoch of recombination, when matter in the universe was freed from radiation. If today we measure very precisely the microwaves coming from all directions in the sky, we find that there is a constant noise that comes uniformly from all directions—no discrete objects can be resolved as responsible for its origin. It turns out to be equilibrium, or thermalized (blackbody) radiation, and thus can be assigned a specific temperature. Practically all of the possibilities of a local origin for this radiation have been ruled out, and it seems most likely that it is indeed an "afterglow" of the Big Bang, originating from the surface of the last scattering we described above. A great deal of effort is being expended to try to detect small anisotropies, or variations, in this microwave background temperature T which would indicate the presence of the underdense and overdense perturbations we mentioned above as precursors of the lumpiness of the later universe. Such perturbations should be there, and detectable at some level—galaxies do exist, now, and they must have formed from something! In fact, in April 1992 it was announced by a large group of researchers, based in Berkeley, that such fluctuations have been detected by instruments aboard the satellite COBE. The amplitude of the variations is $DT/T \_ 6 \times 10^{-6}$ on angular scales greater than 7 degrees in the sky. Furthermore, their scale-invariant character is consistent with an origin during an inflationary epoch (discussed below).

Finally, the third key cosmological observation is the *abundances of helium, deuterium, and lithium*, as we have discussed above. These certainly show that there was an epoch when the universe was at a temperature of over a billion degrees, and had a density and a neutron-proton ratio suitable for the formation a large amount of helium (as happens in the early expansion of an

FLRW universe). Additionally, number counts of distant objects have given evidence that evolution is taking place in the universe. There are other observations, too, which could be interpreted as cosmological, but those above are the primary ones to remember.

Before concluding, it is worth saying something about *inflation* and the *horizon problem* in cosmology, since we hear so much about them. One of the puzzles of the standard Big Bang model as given by FLRW cosmologies is: At very early times, how could so many billions of different regions of the universe which were out of causal contact—since the velocity of light and the age of the universe are finite—have the same density and temperature? How could initial conditions be so accurately imposed, or coordinated? A way of explaining this so-called horizon problem is by means of *inflation*—in a period very soon after the Big Bang, the universe is considered to have expanded extremely rapidly, indeed exponentially, through very many orders of magnitude. Because of this, what was a very, very small region, already in causal self-contact immediately before inflation, became bigger than our observable universe. Thus, there is no problem about the causal contact of the regions ancestral to the observable universe. In other words, our entire observable universe originated in an extremely small, causally connected region shortly after the Big Bang. Furthermore, inflation provides a mechanism that can smooth the universe greatly, thereby helping explain what is otherwise a second very puzzling feature: that the very large-scale structure of the universe, described by the FRW geometry, is highly improbable because it is so smooth.

In order to drive such exponential expansion, there would have had to be a tremendous amount of energy suddenly dumped into the universe, and then some reheating of the universe—or at least our part of it, after the inflationary process came to a halt. Many[11] have supported the idea that this energy came from a scalar field, possibly associated with the GUTs phase transition. That may be. We should note that this suggestion is separate from the idea and possibility of inflation itself. Finally, it is often maintained[12] that to have an inflationary epoch we need to have a universe with $W = 1$ today, that is, with critical density. Although it is the case that there must have been a time (including the epoch between nucleosynthesis and decoupling) when W was close to unity, it is our opinion that it is by no means necessary for the inflationary picture that the density of the universe should be critical today.

An outstanding problem is the question of the *origin of structure*—how galaxies, clusters of galaxies, voids, walls, and so on, evolved from an almost exactly uniform early stage of the universe. The inflationary universe idea gives a tantalizing glimpse as to how this might have happened, but has not fully solved the problem, which remains one of the major open issues of cosmology. A further—highly controversial—issue is the question of *the origin of intelligent life in the universe*. Is this highly probable, or improbable? Is the universe teeming with intelligent life, or are we a highly exceptional outcome of the action

---

[11] Cf. E. W. Kolb and M. S. Turner, *The Early Universe* (Reading, MA: Addison-Wesley, 1990), ch. 8, 270-282.

[12] Ibid., 309-313.

of physical laws? Answering this question is very difficult, because we do not know how probable the evolution of planetary systems is, let alone how probable the evolution of life is, given a suitable habitat in the form of a hospitable planet. One should treat with great caution highly dogmatic statements about the existence of extra-terrestrial life. Scientifically speaking, we simply do not know the answer at present: life may be plentiful in the universe, or it may be extremely rare.

There is much more that could be said about all these topics.[13] This essay has been written to sketch the broad features of contemporary cosmology and its foundations in special and general relativity.

---

[13] For more details on spacetime structure, see G. F. R. Ellis and R. M. Williams, *Flat and Curved Space-times* (New York: Oxford University Press, 1988), particularly chs. 1, 5 (on spacetime structure), 6 (on black holes), and 7 (on cosmology); for more details specifically on cosmology, see S. Weinberg, *The First Three Minutes* (New York: Basic Books, 1977; reprinted in *Bubbles, Voids and Bumps in Time: The New Cosmology*, ed. J. Cornell [New York: Cambridge University Press, 1989]); or at a more technical level, ch. 6 of J. D. Barrow and F. J. Tipler, *The Anthropic Cosmological Principle* (New York: Oxford University Press, 1986). An extremely good general discussion at an accessible level of all the topics presented here, except inflation, is contained in E. R. Harrison, *Cosmology: The Science of the Universe* (Cambridge: Cambridge University Press, 1981).

QUANTUM THEORIES OF THE CREATION[1]
OF THE UNIVERSE[2]

C. J. Isham

1 *Introduction*

1.1 *Topics To Be Discussed*

The aim of this article is to provide a short introduction to recent scientific attempts to describe the origin of the universe as some type of quantum-mechanical event. The main topics to be discussed are:

(1)   The aims and aspirations of those who work on quantum theories of the origination of the universe.

(2)   The scientific status of such theories.

(3)   Specific examples, including (and distinguishing between):

• theories that describe the origination of the material content of the universe within a pre-existing spacetime structure;

• theories that describe the origination of space and time themselves.

(4)   The concept of *time* in these theories and, in particular, the notion of the *beginning* of time.

(5)   The severe conceptual problems that arise whenever quantum theory is applied to the universe as a whole.

The appendix contains a brief discussion of some of the basic ideas of quantum theory and, in particular, those conceptual issues that are of particular relevance to quantum cosmology.

---

[1] The term 'creation' raises problems at many levels, not the least of which is a psychological tendency to assume the existence of a creator who is responsible for the act of creation. This is certainly not what is intended when the term is used in a scientific context where it is essentially synonymous with the less emotion-laden word 'origination'. For this reason, I shall mainly employ the latter term when discussing the scientific theories.

[2] A preliminary version of this paper appeared in *Interpreting the Universe as Creation*, ed. Vincent Brümmer (The Netherlands: Kok Pharos, 1991).

1.2 *The Big Bang*

A major motivation for trying to construct a theory of the quantum origination of the universe is the substantial observational evidence that we live in an expanding universe in which each piece of matter is moving steadily away from every other. When extrapolated backwards, this motion suggests that at about 15 thousand million years ago the material content of the universe was compressed into a tiny region from which it expanded at great speed against the force of gravity. Such a behavior is consistent with the classical[3] theory of general relativity. However, the same theory implies that the origination event must have been in a regime where the mathematical structure of general relativity breaks down, which removes any possibility of extrapolating the behavior of the system to a time before this event. Thus it is "as if" the universe began at some finite time in the past—a situation that raises intriguing questions of both a scientific and a philosophical nature.

A variety of reactions is generated by the idea that the universe may be temporally finite and, in particular, by this striking confluence of theory breakdown with the point at which the universe appears to have "come into being." One extreme school of thought maintains that it is, and always will be, impossible to describe the origination of the universe in scientific terms, and that the singularity in the mathematical framework of general relativity is merely one reflection of this basic truth. It then becomes pertinent to ask whether such a fundamental limitation on the applicability of physical science has any philosophical or theological significance. There is a great range of possible positions that can be taken on this question. For example, a rather naïve reaction is to posit a God who performs a creation of the universe at the precise point where the theory breaks down but who is such that the subsequent development of the universe is described exactly by the existing theoretical structure. The invocation of such a deistic creator is psychologically understandable even if it cannot be justified logically.

The second school of thought is less reactionary and accepts that there is no fundamental objection to describing the origination in a strictly scientific way. From a theological perspective, this is in accord with the idea of seeking a *consonance* between the world-views of science and religion.[4] However, a common scientific caveat is that our current observational information about the early universe is so limited that, in practice, it is impossible to produce anything other than highly speculative theories that cannot be tested. This is often augmented with the claim that radical theoretical advances will be needed before any genuine theory can be constructed.

---

[3] The appellation *classical* means that quantum-theoretical ideas are not included.

[4] For example, see the papers by I. Barbour, E. McMullin, T. Peters, and R. J. Russell in *Physics, Philosophy and Theology: A Common Quest for Understanding*, ed. Robert John Russell, William R. Stoeger, and George V. Coyne (Vatican City State: Vatican Observatory, 1990), and the articles in Sections IV and V of this volume.

A third view is that it *is* possible to construct a coherent origination theory using existing scientific ideas. Furthermore, the requirement of internal self-consistency can replace the acknowledged lack of observational information about the very early universe.

The present paper is concerned with this third position and its claim that the origination event can be described using moderately conventional scientific ideas. Again, it is reasonable to ask whether such a striking claim has any extra-scientific significance. Stephen Hawking certainly thinks so. In his famous book he implies that the construction of such a theory would form a substantial step towards "knowing the mind of God."[5]

One approach to the singularity in classical general relativity might be to modify the theory so that it possesses a solution in which the universe collapses to a small size (but not a point) and then re-expands to give what we now regard as the Big Bang. Such a picture of a bouncing universe poses no major conceptual problems and could be described using techniques drawn from conventional theoretical physics. However, theories of the quantum origination of the universe are *not* of this type but instead take seriously the idea that there is some real sense in which the universe "began" around 15 thousand million years ago. The challenge is to ascribe scientific meaning to such an event and, in particular, to the associated concept of the *beginning of time*.

The central question therefore is whether this coming into being of the universe can be explained, or at least described,[6] using the methods of theoretical physics. For many years following the initial development of the Big Bang model of cosmology, the answer to this leading question was thought to be "no," since the construction of an origination theory is extremely difficult, if not impossible, within the frameworks of Newtonian physics, special relativity, classical general relativity, or standard quantum theory. However, within the last 10 to 15 years, the situation has changed dramatically and much effort has been directed towards the goal of constructing a genuine theory of the origination of the universe in which the addition of quantum ideas to general relativity plays a central role. It is with this research program that the present paper is concerned.

1.3 *Aims of a Quantum Origination Theory*

The aims of research in this area can be summarized as follows. The minimal requirement is to construct a theory that affords a singularity-free description of the origination event[7] and that gives a satisfactory meaning to the "beginning" of

---

[5] Stephen Hawking, *A Brief History of Time* (London: Bantam Press, 1989).

[6] The distinction between *explaining* and *describing* opens up a Pandora's box of questions concerning the status and role of mathematical theories of physics. Nevertheless, the distinction is one with which most practicing physicists are familiar, even if it *is* somewhat ambiguous.

[7] One must be careful not to be prejudiced in advance about the meaning of the word 'event'. The term is convenient for referring to the processes at work in the very early universe, but it must not be taken too literally, particularly in regard to any temporal connotations.

time. A far stronger requirement of some who work in this area is to construct a theory that predicts a *unique* universe. In particular, this means:

- what *type* of matter (particles, fields or whatever) is present in the universe;
- how *much* of it there is;
- what it is *doing* at a given time.

Note that the one question that even a very ambitious creation theorist cannot (or, perhaps, should not) address is "Why is there anything at all?" That is strictly a job for philosophers and theologians!

In constructing any origination theory, a key question is how to handle the basic dualism that exists in most branches of theoretical physics between the equations of motion—which describe all *possible* motions of the system—and the boundary conditions—which determine the specific motion of any particular system. For example, if I throw a piece of chalk across a lecture room it could move along a number of possible trajectories, each of which satisfies Newton's second law of motion. However, the actual path followed by the chalk is determined by the position of my hand when I release it and the velocity I impart to it (and, if it hits one of them, the position of the walls in the room).

This analysis extends to a Newtonian universe as a whole and shows that, for a given set of forces, the positions and velocities[8] of the particles at any time t_2 are uniquely determined[9] by their values at any earlier (or later time) t_1. Thus the equations of motion can provide only *conditional* histories: they give the state at any time t_2 in terms of the state at some other time t_1, but that is all. To produce an origination theory in this framework it would be necessary to argue that there is some special time (perhaps the big bang itself?) at which the state of the universe can be specified in some natural way and which then leads to the complex world we see around us today.

Similarly, Einstein's field equations in general relativity possess many solutions that describe a universe emerging from an initial singularity. All of these are *possible* histories for the universe, that is, they are compatible with the theory of general relativity, but the theory itself cannot select any particular one. The deistic solution to this problem is to ascribe to God the role of setting the initial conditions which, in the case of the piece of chalk, is played by my hand and the will that moves it. The scientific solution is to overcome the dualism of laws and initial conditions by bringing together general relativity and quantum theory to form a unified quantum gravity structure that admits a "natural" set of some appropriate analogue of initial conditions. We shall see later how this might be done.

---

[8] The exact values of the complete set of these variables specifies a possible *state* of the system. This idea is developed further in the appendix.

[9] Modern ideas of chaotic motion show that this statement needs to be expanded somewhat. The ensuing subtlety is not directly relevant here but nevertheless it is something that should be kept in mind when talking about determinism in classical physics.

1.4 *Approach to a Quantum Origination Theory*

When talking about "predicting a unique state of the universe," one crucial question is: "Predict in terms of *what*?" That is, what *assumptions* are fed into the theory from which the desired results are to be obtained? It must be emphasized that, in practice, the primary input to quantum gravity research is not any particular experimental or observational results—not even the existence of the Big Bang—but rather a desire to satisfy certain general theoretical principles. In particular, much emphasis is placed on the requirement of mathematical consistency. For example, in the United Kingdom, most research in this field is funded by the *Mathematics* Committee of the Science and Engineering Research Council: an applicant only seeks assistance from the Physics Committee if his or her proposal concerns events well after the origination regime or at length scales well beyond the Planck length! This predilection for mathematical formalism affects both the style of research and the way in which results are interpreted, particularly the decisions by the scientific community as to what constitutes a genuine advance.

The starting point for all recent quantum origination theories is some view on how to construct a consistent unification of quantum theory and general relativity. The minimum requirement for such a theory is that it be compatible with classical general relativity and quantum theory in their appropriate domains. An important role here is played by the *Planck length* L_p defined to be $\sqrt{(G\hbar/c^3)}$ where G is Newton's constant (which sets the scale or "size" of gravitational effects, and is therefore a key constant in general relativity), $\hbar$ is Planck's constant (which sets the scale of quantum effects) and c is the speed of light. The quantity L_p has the dimensions of a length, and a value of approximately 10^{-35} meters; the associated *Planck time* has the value 10^{-42} seconds and is defined as the time $t_p = L_p/c$ it takes light to travel a Planck length. These minute quantities set the scale at which quantum gravity effects are expected to become highly significant. Thus the basic requirement for a theory of quantum gravity is that at length scales significantly larger than L_p (and at time scales much bigger than t_p) it reproduces classical general relativity and quantum theory.

It is very difficult to construct theories of this type; indeed, no universally agreed-upon unification of general relativity and quantum theory has yet been achieved. Therefore one should not be too pedantic about how the next move—from quantum gravity to quantum cosmology—is to be made. But, broadly speaking, the situation is as follows. Any consistent quantum theory of gravity should be able to provide a description of the origination event, but many different such pictures are expected, corresponding to the set of all possible universes (and their time evolutions) that are compatible with the particular theory. The prediction of a *unique* universe requires the addition of some natural initial, or boundary, conditions[10] that select a unique solution from this plethora

---

[10] Care must be taken in interpreting the phrase 'boundary conditions'. It does not necessarily refer to a boundary in space or time. For example, in the Vilenkin

of possibilities. Note that in such a theory it would not be possible to find a solution that describes, say, only the gross structure of the universe. One might find various approximations to the unique solution, but they would not themselves be solutions. This contrasts strongly with the use of simple model cosmologies in classical general relativity that *are* exact solutions to the Einstein field equations. In a full origination theory this is not possible—there is just one (presumably very complicated) solution to the basic equations and boundary conditions.

1.5 *Status of Quantum Origination Theories*

Before embarking on a detailed description of specific theories, it might be prudent to make some general comments about the scientific status of this rather esoteric branch of theoretical physics. In particular, the following qualifying remarks should be kept in mind when assessing the recent scientific developments.

- More than one scheme has been proposed. Thus it is misleading to talk about science predicting a "unique" universe. At best, the uniqueness refers to the predictions of a particular theoretical structure, and different structures may lead to different "unique" universes.

- None of the schemes proposed so far are in any sense rigorous theories. This stems partly from the lack of any proper unification of general relativity and quantum theory. However, even setting this aside, the extant proposals are incomplete; in particular, it is by no means clear that they do in fact lead to a unique quantum state.

- Major conceptual problems arise when trying to apply quantum theory to the universe as a whole. This problem is so severe that many highly respectable theoretical physicists think the whole subject of quantum cosmology is misconceived.[11]

It follows from the above that theories of the quantum origination of the universe are highly speculative and do not have anything like the scientific status of, say, even the more exotic branches of modern elementary particle physics. On the other hand, quantum gravity itself has been the subject of intense study for over 25 years and some of the ideas that have emerged during this period are at least vaguely plausible. In particular, I would be prepared to defend the thesis that some of the new ideas about the nature of time have a validity that transcends any particular origination theory of the universe that is currently in vogue.

The rest of the paper is concerned with developing an insight into the more technical ideas that lie behind modern quantum origination theories. The

---

proposal discussed later, the boundary is the edge of the mathematical space of all possible universes.

[11] Private communications from a variety of friends and colleagues!

treatment is not mathematical, but nonetheless it is complicated in places, and it may try the patience of the unenthusiastic reader. I would recommend any such to jump first to the concluding section where the main objections and weaknesses of quantum cosmology are summarized in a non-technical way. He or she can then decide whether it is worth making the effort to read the intervening sections. Some of the basic ideas about quantum theory itself can be found in the appendix which can be read independently of the rest of the paper.

2 Quantum Origination in a Fixed Spacetime

The instinctive picture of space possessed by most people is of a big box that contains the matter in the universe. Thus it is understandable that the first serious theories of the origination of the universe used a mathematical framework in which space and time form a fixed background and where matter is created at some specific point in this background. This is illustrated in *Figure 2.1* where the lines moving away from the origination point are the "world-lines" of the created particles. A point on the world-line specifies the position in space of the particle at a certain time. The spatial position is obtained by projecting the point onto the *x-y* plane, and the corresponding time is the value along the vertical axis. Thus each world-line represents the entire history of the particle from its moment of production. Note that I have had to drop the third spatial direction in order to draw the (four-dimensional) diagram on a piece of paper.

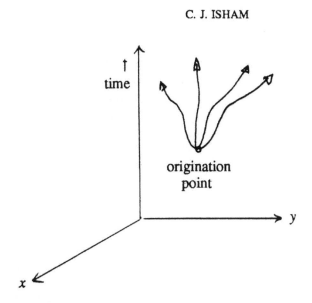

Figure 2.1: Origination in a fixed spacetime (the third spatial dimension z is not shown).

There is a sense in which theories of this type can be said to involve the idea of "creation from nothing," but that which is created is only the material world.

Space and time themselves are "eternally pre-existent" and the nothing from which the world emerges is empty space, namely "no thing."

The obvious question to ask of such a theory is, what happens to the conservation of energy? The initial prognosis for satisfying this universal requirement is not promising. The rest-mass energy $(E = mc^2)$ and kinetic energy of the created matter are both positive, whereas it can plausibly be argued that the energy of empty space ("no thing") is zero (what else could it be?). However, as E. P. Tryon pointed out, every piece of matter in the universe attracts every other piece gravitationally, and the potential energy of this attraction is always negative (because energy has to be expended in separating matter against the force of gravity).[12] Thus the possibility arises that the total negative gravitational energy might cancel the positive rest-mass energy of all the matter in the universe, and various detailed arguments (with various degrees of plausibility) have been proposed purporting to show that this is indeed so. The picture that emerges is thus one in which empty space is potentially unstable to the production of matter, since the combined total energy of both is zero!

It is not possible to construct such an origination theory within the rigidly deterministic framework of classical Newtonian physics, but the introduction of quantum ideas changes the situation. In practice, theories of this type have involved some sort of nucleation process in which a minute particle of matter is produced at some time. This generates a gravitational field which, by a standard quantum-mechanical process, gives rise to the production of particles, which produce more gravitational fields . . . and so on. Thus, there is a sort of zero-energy conserving, fire-ball explosion away from the initial nucleation, and this is what can be thought of as a model for the Big Bang.

This picture is interesting, if for no other reason than it points to the crucial role that might be played by the *gravitational* field in an origination theory. However, it is subject to several serious objections. For example, how does one explain the initial nucleation event? The scientific papers tend to be distinctly coy on this rather important question! But, without doubt, the most crucial problem is to decide what determines the time (or the place) of the initial nucleation.

The difficulty is that, within an infinite, homogeneous timeline, there is simply no way whereby the mathematics can select one particular time at which the origination of matter occurs. Similarly, there is no way of selecting a particular spatial point for the origination. Instead, theories of this type are prone to predict an origination of matter at every time or, to be more precise, as quantum theories they predict a non-zero *probability* of origination within any finite time interval, and with the origination points being evenly distributed throughout space. This leads at once to an infinite number of origination points within the background spacetime. One difficulty with this picture is that there is no reason why the material emerging from one origination point could not cross that associated with another, and we do not actually see anything like this happening in the universe.

---

[12] E. P. Tryon, "Is the Universe a Vacuum Fluctuation?" *Nature* 246 (1973): 396.

For these reasons, plus a general reluctance to employ a theoretical structure in which a background spacetime is an essential ingredient, physicists have turned to more radical types of origination theories in which time and space are themselves created in some appropriate sense. The arguments in support of such a step have interesting historical precedents. The impossibility of determining the actual instant of origination was invoked by Kant[13] in proving the first half of his antinomy that (i) the world began a finite time ago, and (ii) it did not do so. The same problem had been considered much earlier by Augustine[14] who concluded that space and time should themselves come into being at the moment of creation.

Note that since quantum mechanics deals only with the *probability* of events occurring, prior to a creation of the Tryon type the theory could at best be said to describe the potentiality or latency of the event. This accords nicely with the old Greek idea of chaos as the yawning abyss of infinite empty space that contains only the potentiality of being.[15] Thus the scientific move away from Tryonic theories and towards the idea that space and time are also created, mirrors the construction by early Christian theologians (such as Augustine) of the *creatio ex nihilo* doctrine in which a demiurgic creation from potential being is replaced by God's creation from total non-being. From a scientific perspective, this requirement leads naturally to general relativity and, ultimately, to the modern quantum theories of origination.[16]

3 *Space and Time in General Relativity*

3.1 *Curved Spaces*

The picture of space and time which is most relevant for origination theories is that provided by the general theory of relativity. In particular, and unlike the situation with Newtonian physics or special relativity, it is now possible to begin to talk about the creation of time, and hence the temporal origin of the universe.

To speak of the "creation" of time sounds like a contradiction in terms. After all, "create" is a verb and hence its use would appear to be meaningless if there was no time at which the act of creation was in the present, not to mention earlier times in which it lay in the as yet un-realized future. However, it is

---

[13] I. M. Kant, *Critique of Pure Reason*, trans. N. Kemp Smith (London: Macmillan Publishing Co., 1929), book II, ch. II, particularly 396-402.

[14] Augustine, *Confessions*, trans. F. J. Sheed (London: Sheed & Ward, 1960), book 11.

[15] Ernan McMullin, ed., *The Concept of Matter in Greek and Medieval Philosophy* (Notre Dame: University of Notre Dame Press, 1965). See also the discussion of the difference between *ouk on* and *me on* in Paul Tillich, *Systematic Theology*, vol. I (London: James Nisbet, 1960), part II/I.

[16] Christopher J. Isham, "Creation of the Universe as a Quantum Tunneling Process" in *Physics, Philosophy and Theology*, 375-408. See also the article by Robert J. Russell in this volume.

important to realize that the Newtonian/Kantian picture of a transcendent space and/or time has been largely abandoned by twentieth century physics and replaced with a more fluid image in which the structure of spacetime is as much fair game for the investigations of experimental physicists (or the fantasies of theoreticians) as is the matter contained therein. Of particular relevance in this respect is the theory of general relativity, to which we now turn.

The basic idea of general relativity is that the three-dimensional space in which we live is *curved*. The deviation from flatness at any particular point is a measure of the strength of the gravitational field at that point, and this can change in time. In the context of Big Bang cosmology, the reason why the galaxies appear to be moving apart from each other is not because they all originated at some specific point in a fixed space at some time in the past from which they are receding (like shrapnel from an explosion, as in *Figure 2.1*) but rather because they are embedded in a closed, curved space whose size is increasing with time. A useful two-dimensional model is provided by the surface of a balloon covered with dots that represent the galaxies. If air is blown into the balloon, it expands in size and each point moves away uniformly from all its neighbors.

Spacetime diagrams play an important pedagogic role in general relativity. To draw such things on a piece of paper it is useful to drop two dimensions and suppose (as an illustrative model only) that physical space has one, rather than three, dimensions. The simplest example is where space is a circle that does not change with time. The resulting spacetime is static and has the structure of the (two-dimensional) infinite cylinder that is sketched in *Figure 3.1*.

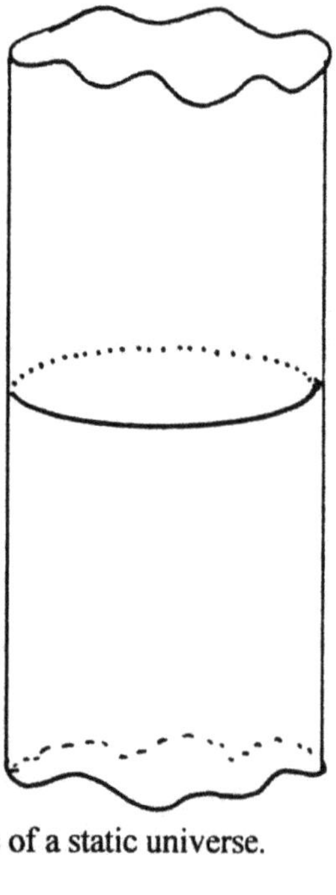

Figure 3.1: The spacetime of a static universe.

However, the actual solutions to Einstein's equations are not this simple and, in general, the curvature of space changes in time, leading to a dynamic spacetime of the sort shown in *Figure 3.2*, where the spatial structure is time-dependent. A special (and very important) example is when the radius increases steadily with time leading to a shape which is something like the surface of an ice-cream cone (*Figure 3.3*).

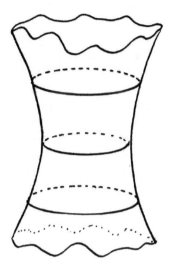

Figure 3.2: A spacetime in which the curvature of space changes with time.

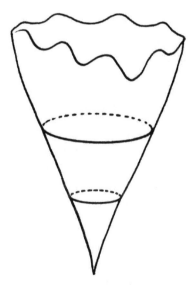

Figure 3.3: The spacetime of the Big Bang.

This model of a cone is used frequently in cosmology because, extrapolating backwards in time, we come to a region where the radius of the circle becomes very small and then, at the tip of the cone, vanishes completely. In the four-dimensional analogue of this two-dimensional picture, the circles are replaced by three-dimensional spheres[17] (models of the space in which we actually live) and the tip of the cone (a limiting three-dimensional sphere of zero radius and infinite curvature) is identified with the Big Bang. The cone itself provides the basic general relativity model of the spacetime structure of the evolving universe.

It is important to emphasize that, as remarked earlier, there exist many solutions to Einstein's equations which exhibit this type of behavior and, within the framework of the theory itself, there is no way of determining which of these possible histories of the universe is actually realized. On the other hand, a full origination theory would predict a *single* history (i.e., spacetime). Thus, as with all classical deterministic systems, the most that general relativity can do is to provide a prediction of the state of the universe (that is, its curvature and, if appropriate, matter content)[18] at some time t_2 in terms of its state at some earlier time t_1. Looking at *Figure 3.3*, there is clearly some sense in which space and time "begin" at the tip of the cone (there is certainly nothing "before" then in this model!) and so perhaps a natural choice for the time t_1 would be its value at that point. However, there is a severe problem in specifying such initial conditions since the mathematics becomes singular at this point and any conventional analysis is likely to break down.[19] Thus, although the introduction of general relativity does give some insight into the sense in which time (and space) might be said to begin, it does not naturally lead to a genuine origination theory since such a theory should include a description of the initial conditions and singularity too.

3.2 *Time in General Relativity*

A key issue in general relativity is the position it takes on the general question of the absolute versus the relational nature of space and time. At a first glance, the theory appears to resemble Newtonian physics in granting spacetime a positive ontological standing. Thus spacetime pictures like *Figure 3.1-3.3* have the same status in general relativity as does a picture like *Figure 2.1* in the Newtonian

---

[17] A sphere is defined to be the surface of a ball. In particular, a three-dimensional sphere is the surface of a four-dimensional ball. However, it is only the surface that is meant to have any physical reality, and the ball itself must not be confused with the four-dimensional spacetime. Note that a one-dimensional sphere is the same thing as a circle regarded as the boundary of a two-dimensional ball (which is a disc).

[18] The equations of the gravitational part of general relativity need to be supplemented with additional terms to describe the matter found in the universe. These days, one would probably look to a grand unified theory to provide this additional information.

[19] But see, for example, Roger Penrose, *The Emperor's New Mind* (Oxford: Oxford University Press, 1989); and Frank Tipler, "The Omega Point as Eschaton: Answers to Pannenberg's Questions for Scientists," *Zygon: Journal of Religion & Science* 24 (1989): 217.

world view. However, there is another aspect to the concept of time in general relativity that contrasts with this view and that is of the greatest importance in the construction of quantum theories of origination. In looking at a diagram like *Figure 3.1* it is essential to appreciate that it is the cylinder in *itself* that represents the spacetime—the fact that it has been drawn in a space of one higher dimension has no intrinsic significance.[20] Furthermore, and unlike the analogous situation in Newtonian physics, there is no fundamental way to choose which variable we call time. Although I have shown time as going up the page, this is not to be construed as implying that there is any sort of *universal* time that points in this direction, and least of all one that is related in some way to the higher-dimensional space in which the spacetime cylinder happens to have been displayed.

So what then is time in general relativity? We seem to be in the peculiar position of knowing what is meant by spacetime but not by space or time considered separately. A complete answer to this question requires a deep excursion into the technical structure of the theory, but the crucial idea can be gleaned from the observation that time is best thought of as an *internal* property of the system. For example, consider a particle moving in a one-dimensional circle that is expanding in time. Then we could say that to specify the configuration of this system at a particular time consists in giving both the radius of the circle, and the position of the particle on the circle, at that time. However, we can avoid reference to this undefined time by specifying instead the position of the particle at each value of the radius of the circle. In other words, we can *define* time to be the radius of the circle. Of course, the statement that the universe is expanding in time then becomes essentially tautological[21] but, as far as the matter content of the universe is concerned, this affords a perfectly workable definition.

In the real situation where space has three[22] rather than one, dimensions, one natural definition of an internal time is the *volume* of the universe. Thus, in this approach, to specify the dynamical evolution of the universe is to say how the gravitational field and the matter variables are correlated with the volume of space.

However, there are many other ways of defining an internal time, none of which has any preferred status. In fact, given a four-dimensional spacetime M, a choice of time is associated with *any* way of slicing M into a collection of

---

[20] This is just as well for the full theory since there are many four-dimensional spacetimes that cannot be represented at all as curved surfaces in a flat five-dimensional space!

[21] Any further discussion of this question leads to the well-known problem of the arrow of time. The notion that the universe is expanding (rather than contracting) still has an absolute meaning but it involves correlating the global cosmological time we are discussing with the entropy of thermodynamically complex systems. The problem is too subtle and complicated to be developed further here.

[22] Some of the more speculative "theories of everything" suggest that the dimension of space may be bigger than three at very small distances, for example the Planck length. However, this does not affect the basic ideas being developed here.

three-dimensional spacelike surfaces in which the elements of the collection are labeled by a single real number which increases steadily from one slice to the next (*Figure 3.4*). All the points on a single slice are simultaneous with each other with respect to the time defined as the value of this labeling parameter. This idea that, rather than being a fixed, external concept, time is an internal property of the system is of considerable significance in and beyond physics and plays a central role in quantum theories of origination.

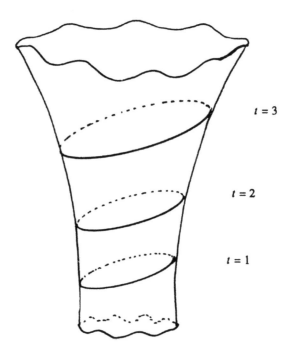

Figure 3.4: Time as a slicing of spacetime.

4 Quantum General Relativity

4.1 Quantum Theory

The crucial step in the scientific discussions of the origin of the universe is the introduction of quantum theory into the picture given by general relativity. To illustrate the basic ideas of quantum theory let us consider a simple Newtonian model of a particle moving in a two-dimensional plane with coordinates (x,y). The classical state of the system at a given time t_1 is specified by the position (x,y) of the particle and its velocity; the causality structure in the model is reflected in Newton's equations of motion which lead to a unique state at any later (or earlier) time t_2 as a function of these initial data.

The quantum version of this situation is radically different. The state of the system[23] at time t_1 is now specified by a *function* $\Psi(x,y)$ that leads only to the *probability* that the particle will be found at (x,y) if its position is measured.[24] This introduction of a probabilistic structure corresponds to an irreducible property of Nature in the sense that, even *in principle*, the theory cannot say what the position of the particle actually *is*; indeed, the concept of the particle *having* a position (independently of the measurement process) is very difficult to sustain without the introduction of hidden variables[25] that reduce the structure to that of conventional statistical physics.[26] These references to measurement and finding the particle somewhere (as opposed to the particle *being* somewhere) introduce an instrumentalist/operationalist element into quantum theory that is difficult to remove. This poses a particular problem in the context of quantum cosmology since it is now the state of the entire universe that is being considered and there is no place where an external observer can sit to make his, her or its measurements! We shall return later to some of these conceptual issues.

It is important to realize that, notwithstanding the emphasis placed on probability, dynamical evolution of a system in quantum theory is as causal as it is in classical physics. The only difference is that it is the *probabilities* of finding various configurations that evolve deterministically, rather than the configurations themselves. The mathematical representation of this situation is that (i) the Ψ-function possesses a time label t, that is, there is a separate wave function Ψ_t for each value of time, and (ii) the change of $\Psi_t(x,y)$ in time is determined by a differential equation that yields a unique state function $\Psi_{t_2}(x,y)$ at any time t_2 given the value of the state function $\Psi_{t_1}(x,y)$ at some earlier (or later) time t_1. One consequence of this deterministic picture is that the problem of constructing an origination theory is no easier in the quantum world than it is in the classical one. A full quantum origination theory would yield, for each time t a *unique* state function Ψ_t from which the probabilities of the results of all possible measurements of the entities in the universe could be computed. However, all that can be extracted from normal quantum theory is the possibility of computing

---

[23] See the appendix for a more detailed description of the structure of quantum theory. In particular, the idea is developed there that the function ψ is best thought of as a vector in an infinite-dimensional vector space.

[24] More precisely, $\Psi(x,y)$ is a complex number, and it is $|\Psi(x,y)|^2$ that gives this probability.

[25] Hidden variables introduce peculiar features of their own, in particular a striking non-locality that is related to the violation of the Bell inequalities by quantum theory. For a discussion of Bell's important work see the article by Andrej Grib in this volume.

[26] In classical statistical physics, the use of probability is primarily *epistemological*, that is, it is a measure of our lack of knowledge of the details of the system which, it is assumed, could always be improved by making more accurate measurements. On the other hand, in conventional interpretations of quantum theory, probability tends to be thought of in a more *ontological* sense as referring to an intrinsic property of reality itself, rather than to our knowledge of it.

the state function at some given time in terms of the state at any earlier (or later) time, which does not qualify as an origination theory.

4.2 *Quantum Gravity*

To escape this impasse we must employ a formalism in which it is no longer appropriate to attach a conventional time label to the quantum states. This entails moving away from the Newtonian (and, indeed, special relativistic) concept of time, and by far the most-studied framework for doing this is general relativity. Thus we are led to contemplate the construction of an origination theory that is based on some combination of quantum theory and general relativity. This will not be easy; indeed, as remarked already, a complete reconciliation or unification of these disparate conceptual frameworks remains one of the major goals of theoretical physics. The current favorite is superstring theory but it is not known how to apply this to the type of quantum cosmological problem in which we are interested or, indeed, if superstrings really do provide a consistent quantum theory of gravity at all. However, certain general properties are expected to hold in *any* quantum gravity theory, and these will suffice for our present purposes.

In the quantum theory of a particle moving in the *x-y* plane, the state of a system at time t is represented by a function $\Psi_t(x,y)$ that yields the probability of finding the particle at position *(x,y)* at time t. In a quantum theory of general relativity, the analogous object is expected to be a function $\Psi(curv)$ where *curv* is some numerical measure of the curvature of space. By analogy with the quantum theory of a particle, this function $\Psi(curv)$ should tell us about the probability of finding a particular three-dimensional space *curv* if we measure the curvature of space (i.e., the gravitational field) everywhere.[27] If matter is present (in the form of fields or particles) the appropriate function has the form $\Psi(curv, matter)$ (where *matter* denotes a numerical measure of the matter that is present) and gives the probability distribution of the matter variables and the curvature of space.

A key property of this quantum gravity state function is that it does *not* carry a time label. This reflects the highly significant property of general relativity alluded to earlier whereby "time" is not an external entity but is rather to be regarded as an internal function of the system's variables (such as, for example, the volume of the curved three-space).[28]

At first sight it is not easy to see how any idea of dynamical evolution (and especially origination!) can be extracted from the function $\Psi(curv,$

---

[27] It must be emphasized that this view of quantum gravity has been much debated over the years and it is by no means universally accepted. This particular issue is one facet of the general problem of understanding the nature of time in quantum gravity. Unfortunately, the issues involved are too complex to be discussed here, but they are of crucial importance in studying quantum cosmology at a deeper level. For a recent review see Christopher J. Isham, *Canonical Quantum Gravity and the Problem of Time*, to appear in the Proceedings of the NATO Advanced Summer Institute "Recent Problems in Mathematical Physics, Salamanca, 1992."

[28] The phrase 'three-space' is short for 'three-dimensional space'.

matter)—since it carries no time label, surely nothing can change in time? This so-called "frozen formalism" caused much confusion when it was first discovered and it took some time for the paradox to be unraveled. But the basic idea is simple enough: one must take seriously the idea of an internal time and use *it* to discuss how the quantum states of the remaining variables evolve. To be more specific, let us choose volume as internal time, and let *curv'* denote all those numerical curvature variables that remain after removing the volume variable. Then our state function can be written as Ψ(*curv'*, *matter*, *vol*), and the crucial idea is to interpret this as the quantum state for the *curv'* and matter variables at the value *vol* of the internal time. Alternatively, some of the matter variables could be used to define the passage of time,[29] giving rise to the idea of quantum clocks. In either case we may say that time is *internal* to the system and is determined by the contents (gravitational or material) of the universe rather than, as in the case of Newtonian physics, being a fixed, external measure. There are many possible choices for such a clock and a recurring problem in quantum gravity is to understand if, and how, the results depend on the choice made.[30]

A major technical step is to show that, if they are re-expressed in terms of these new variables, the quantum equations satisfied by the state function Ψ(*curv*, *matter*) look like standard dynamical evolution equations. This almost works—"almost" in the sense that, for the various choices of internal time tested so far, the ensuing equations resemble, but are not exactly the same as, the time evolution equations of conventional quantum theory. But this "almost" is just what we want. If the reconstructed evolution equations were exactly the usual ones, the same rigid causality problem would be encountered and it would not be possible to construct an origination theory. However, the deviation from "normality" is precisely what is needed to avoid this causal problem and hence to build a genuine origination theory. Of course, this also entails a subtle change in the meaning ascribed to time around the critical region, and I shall explain later what happens in the context of one specific theory.

But we are jumping ahead of ourselves since, even if an internal time can be introduced in this way, we are still far from having an origination theory. Such a theory would yield a *unique* function Ψ(*curv*, *matter*) which could be interpreted in the way indicated above. Note that such a function would predict (i) *how much* matter and curvature is present, within the general probabilistic

---

[29] I say 'define' rather than 'measure' because, unlike the situation in Newtonian physics, there is no absolute background time to which the matter clock is responding. Thus, the position of the hands of a clock does not measure time—it *is* time. Of course, this changes somewhat the meaning of the question of whether a particular clock is, or is not, a good timekeeper!

[30] Note that the approach to quantum gravity we are discussing here starts with a three-dimensional space, and hence there is a sense in which space is more fundamental than time. On the other hand, to actually *locate* a point in this space also requires internal measurements using, for example, the values of various matter variables: there is no more of an absolute spatial reference grid in general relativity than there is a universal time. Thus, one must also study the dependence of the quantum results on the choice of spatial coordinates.

limitations of quantum theory; and (ii) what they are *doing* as a function of the internal time. However, the quantum equations satisfied by this function are expected to admit *many* solutions, and so we are back once more to the problem encountered already in the contexts of Newtonian physics, classical general relativity and conventional quantum theory. To proceed further we must look more carefully at the state function $\Psi(curv, matter)$ and ways in which a unique one might be selected.

5 Quantum Theories of Origination

5.1 Superspace

There are several ways in which one might address the key problem of how to select one particular solution from the plethora of possible probabilistic distributions for the universe. One well-known example is the work of Hartle and Hawking.[31] However, since I have written about this elsewhere[32] I shall concentrate here on Alex Vilenkin's method, which is also representative of several other schemes.

A key concept in quantum gravity is *superspace*, defined to be the mathematical space S of all curved, three-dimensional spaces. If matter is present, the definition of superspace is extended to be the set of all pairs (*curv, matter*) of curved three-spaces and matter configurations on such spaces. Thus superspace is the space of all *possible* universes, where "possible" means within the framework of a specific choice for the extension of general relativity to include matter.

Superspace is important because it is the *domain* space of the quantum gravity state function Ψ, that is, Ψ assigns a complex number $\Psi(curv, matter)$ to each point (*curv, matter*) in S. The mathematical space S is very large (it is infinite dimensional) and is difficult to picture. However, as we shall see shortly, it plays a central role in the construction of origination theories.

It is particularly important to understand what a classical spacetime looks like in this context. To this end, consider a path in S that joins together the points ($curv_1$, $matter_1$) and ($curv_2$, $matter_2$) where $matter_1$ and $matter_2$ denote configurations of matter variables on the curved spaces $curv_1$ and $curv_2$ respectively. Each point on the path represents a matter configuration on a particular curved three-space, and the entire family of such points corresponds to a four-dimensional spacetime (plus matter) which has the three-spaces $curv_1$ and $curv_2$ as its edges. Thus the path corresponds to a preferred slicing of this spacetime with each point on the path representing the corresponding slice (*Figure 5.1*).

---

[31] James B. Hartle and Stephen W. Hawking, "Wave Function of the Universe," *Physics Review* D28 (1983): 2960-2975.

[32] Isham, "Creation of the Universe as A Quantum Tunneling Process."

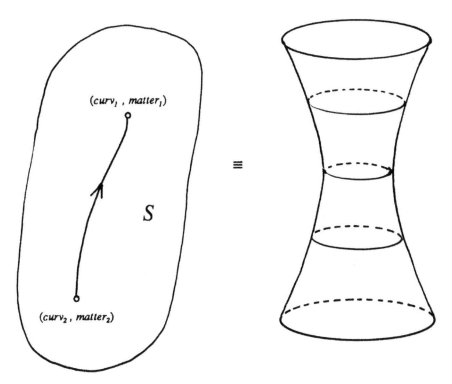

Figure 5.1: A curve in S corresponds to a four-dimensional spacetime plus matter.

Any parametrization of the path provides an admissible measure of time on the spacetime provided only that the parameter value increases steadily along the path; in this sense, each path is a possible *history* of the universe. Conversely, given a four-dimensional spacetime with a pair of boundaries $curv_1$ and $curv_2$ (on which the matter variables have values $matter_1$ and $matter_2$), one can choose any way of slicing it into a one-parameter family of spacelike three-surfaces (plus a matter configuration on each slice) such that the two end slices are $curv_1$ and $curv_2$ and with the matter values $matter_1$ and $matter_2$ respectively. Then this one-parameter family defines a path in S which joins together the points ($curv_1$, $matter_1$) and ($curv_2$, $matter_2$).[33]

To arrive at the quantum-theoretical analogue of the above one might consider first a state function Ψ which vanishes except on those curved three-spaces that are some spacelike slice of a particular four-dimensional spacetime M. Then the probability of finding a three-space is zero unless it is such a slice, and in this sense Ψ could be said to be associated with the classical spacetime M. In practice, the equations of quantum gravity are such that a typical state function

---

[33] But note that this slicing can be performed in many different ways (corresponding to the many different choices of time), and each generates a different path in S. Thus a more accurate statement is that a four-dimensional spacetime plus matter corresponds to a collection of paths in S, each of which joins together the two points representing the boundary three-spaces and their matter configurations.

is expected to be non-vanishing on a set of three-spaces that is *larger* than the set of spacelike slices of any one four-space. The most that can be hoped for is that Ψ is sharply peaked around such a set, thus giving an *approximate* spacetime picture.

Although this description is fairly easy to understand it does not correspond to the sort of state function produced by a typical quantum origination theory. Instead, such a state oscillates rapidly in certain directions in superspace, and in such a way as to lead to a high quantum-mechanical correlation between the curvature and matter variables and their velocities. In turn, these correlations generate a whole *family* of classical histories for the universe and its constituents. Thus, not only is the spacetime picture afforded by the quantum theory an approximate one, it is also not unique since the single state function Ψ is associated with more than one classical spacetime. This is the ultimate example of the Schrödinger cat paradox!

5.2 *The Boundary of Superspace*

A key role is played in quantum origination theories by the natural *boundary* of superspace S. A point in the boundary is associated with a sequence of points in S (i.e., a sequence of possible universes) that appears to converge to something that does not itself belong to S. In some respects this is analogous to the way in which a sequence of rational numbers can converge to an irrational number (for example, the sequence 1/1, 14/10, 141/100, 1414/1000, 14142/10000 . . . converges to the irrational number $\sqrt{2}$).

A boundary point represents a configuration of the universe that is singular in some way. Examples are when an infinite value is possessed by the curvature or volume of the three-space, or by one or more of the matter variables in the theory. Some of these points lie on curves in S for which the associated four-dimensional spacetime is also singular (i.e., as measured by the properties of the *four*-dimensional geometry)—for example, the tip of the cone in *Figure 3.3*—and are said to belong to the *singular* part of the boundary. Thus a curve in S that represents a classical universe starting with a big bang would emerge from this singular boundary (see *Figure 5.2*).

It is important to appreciate that not all points in the boundary are of this form. For example, if one imagines a sequence of three-spaces obtained by slicing a four-dimensional sphere then the limit at the pole does not describe a singular point on the sphere itself (since the curvature computed using the four-dimensional geometry is finite). This is in sharp contrast to the sequence of slices on a conical Big Bang spacetime which converges to the singular tip. Those singular three-geometries which do not correspond to a singularity in spacetime are said to belong to the *regular* boundary of superspace.

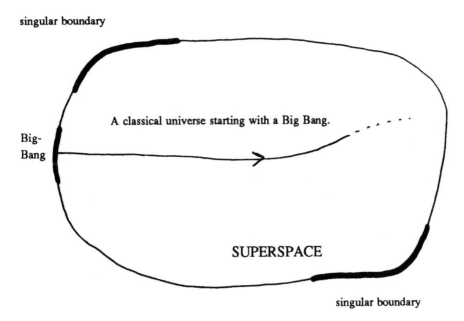

Figure 5.2: The boundary of superspace.

5.3 *The Construction of Origination Theories from Boundary Conditions on Superspace*

The key idea is to try to select a unique function Ψ by specifying its behavior in some natural way on the boundary of superspace.[34] The first attempt to construct a quantum origination theory of this type was made by Bryce DeWitt who, in a seminal paper, suggested that a unique function might be obtained by requiring Ψ to vanish on the boundary of superspace.[35] In particular, this implies that the probability of finding a singular configuration is zero, which appears to achieve our goal of avoiding the singularity that arises in the classical theory.[36] This

---

[34] There are a number of well-known precedents in mathematical physics for fixing the value of a function by its value on a boundary. For example, in a charge-free region of normal physical space, the electric field inside the region is uniquely determined by Maxwell's field equations once the value of the field on the boundary of the region is specified. However, the situation in our case is considerably more complicated since the region concerned—the whole of superspace—and its boundary are both infinite dimensional. It is also *not* the case that the quantum gravity equations for Ψ are of the type that automatically admits such a procedure.

[35] Bryce DeWitt, "Quantum Theory of Gravity I. The Canonical Formalism," *Physical Review* 160 (1967): 1113-1148.

[36] This could also be achieved by requiring that Ψ vanish only on the *singular* boundary of superspace.

particular proposal does not seem to work very well but a more promising suggestion was made about eight years ago by Alex Vilenkin,[37] and this is the scheme on which I shall concentrate.

5.4 *The Vilenkin Scheme*

Vilenkin suggested that a particular solution Ψ to the quantum gravity equations should be selected by requiring that, at the singular boundary, the flux associated with Ψ always points out of superspace. The idea of a flux can be given a precise mathematical definition but the following rough idea will suffice for our purposes. First, suppose the function Ψ is such that it has the oscillatory behavior mentioned earlier that is associated with a family of classical solutions to the Einstein equations. In particular, this guarantees that the world that emerges from the originating quantum regime is classical—an important requirement for any origination theory that is to agree with our actual universe.

If this condition is satisfied, Vilenkin's boundary condition is essentially that, at the singular boundary of S, the paths in S corresponding to the classical spacetimes associated with Ψ should all be so oriented that they appear to be moving out of S. This allows classical spacetimes that end at a spacetime singularity or that expand forever (so that the three-slices tend to a three-space with infinite volume) but, for example, it excludes an expanding universe that *begins* at a spacetime singularity.

The crucial question is the extent to which this scheme generates a genuine origination theory. From our introductory discussion we recall that there are two aspects to this. The first is whether the scheme yields a *description* of the origination event, and the second is whether it yields a *unique* quantum state Ψ. Note that if the latter is false the best to be hoped for vis-à-vis the former is that the various quantum states that satisfy the Vilenkin condition all lead to the same general picture of the origination.

Various approximate calculations have been performed which do indeed predict a unique state function. However, these approximations involve ignoring all but a small number of the infinite possible modes of the universe, and it is by no means clear that the uniqueness will be preserved in the full theory.[38] The problem is compounded by the fact that the equations for Ψ that we are trying to solve are mathematically ill-defined, which makes it very difficult to assess the status of a scheme like that of Vilenkin. There is a good case for arguing that, at length scales much greater than the Planck length, a proper quantum theory of gravity will reduce to the approach employed above—with the quantum states appearing as functions of the curvature of three-space and the matter variables—and therefore that a scheme like Vilenkin's might have some approximate validity. But the situation is complicated and any proper resolution

---

[37] Summarized in Alex Vilenkin, "Quantum Cosmology and the Initial State of the Universe," *Physical Review* D37 (1988): 888 f.

[38] Proponents of the competing Hartle-Hawking scheme now believe that their own approach definitely does *not* predict a unique state function.

of this issue must await the discovery of a fully consistent unification of general relativity and quantum theory.

5.5 *Where Does the Universe Come From?*

The situation vis-à-vis the more modest goal of producing a *description* of the origination event/region is a little more promising, and a plausible answer can be given to the question of what is meant by the "beginning" of time. A crucial observation is that the classical spacetimes associated with the state function Ψ have the property that Ψ oscillates rapidly along the path in S corresponding to the spacetime. However, there are regions in S (that depend on Ψ and the precise choice of matter and its interactions) in which no such oscillatory behavior is possible and which cannot therefore correspond in any way to a classical spacetime.

Thus there is an internal boundary in superspace (albeit, rather fuzzily defined) that separates the regions in which Ψ oscillates—where there is some sort of underlying classical picture—from regions where there are no such oscillations and which are therefore purely quantum-mechanical. This is illustrated in *Figure 5.3* which also shows the paths corresponding to the classical universes that we are presuming to be associated with the wave function Ψ.

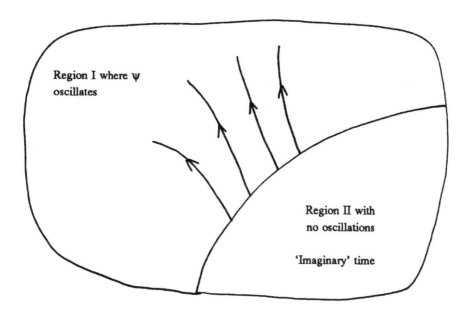

Figure 5.3: The boundary in superspace which separates the regions where oscillations of Ψ occur from those where there are no such oscillations.

A path in the region (I) of superspace where oscillations occur corresponds to a genuine spacetime (plus matter)—typically a standard, expanding universe except that it "starts" (at the internal boundary) with a finite radius, not with the zero radius of the big bang. However, and this is the crucial point, the wave function Ψ "leaks" into the region (II) where it has a strictly non-oscillatory behavior.[39] Although this region is purely quantum-mechanical, one can nevertheless ask whether it corresponds to anything geometrical. The answer is that there is a sense whereby the wave-function in region II can be associated with a spacetime in which time is a purely *imaginary* number (in the mathematical sense of complex numbers, not in the philosophical sense of non-existing). The precise meaning of imaginary is that the time dimension becomes indistinguishable from the space dimensions. Thus space and time have exactly the same geometrical status, in sharp contrast to the situation in normal general relativity. In particular, there is no special *ordering* of events that is characteristic of the real-time situation in general relativity. A typical exact solution of Einstein's equations of this type is a four-dimensional sphere, in contrast to the cone of real-time general relativity.

The net effect is that, just as the state function Ψ approximately corresponds to a genuine spacetime(s) in region I, so there is a sense in which it also describes an imaginary-time spacetime in region II which (very roughly!) can be thought of as being joined to the region I spacetime at the point in superspace where the path meets the internal boundary. Thus the final picture of the origination of the universe is of an imaginary-time spacetime (which is totally non-classical) from which the real-time (and classical) universe emerges with some finite radius that is determined by the parameters in the equation describing the material content of the universe. This process is sketched in *Figure 5.4*.

Of course, the words 'emerge' and 'process' must be understood in a symbolic sense since their usual temporal implications are not appropriate in the present situation: time as we normally understand the word is applicable only in the region of *Figure 5.4* that is well away from the originating four-sphere, and then of course it is an internal time in the sense we have described already, not a fundamental time of the type envisaged by Newton. It is important to emphasize that there are *no* real-time solutions to Einstein's equations of the form in *Figure 5.4*. The nearest one can get is the cone-like spacetime depicted in *Figure 3.3*. And, of course, unlike the four-space in *Figure 5.4*, this has an initial singularity.

5.6 *Other Origination Schemes*

Several other schemes have been suggested for selecting a unique quantum state from among the possible solutions to the equations of quantum gravity. Some of these are similar to Vilenkin's in that they involve setting conditions on the boundary of superspace. The Hartle-Hawking approach is somewhat different and consists of a specific algorithm for computing a state Ψ.

---

[39] This is similar to the quantum tunneling phenomenon that plays an important role in many practical applications of conventional quantum mechanics.

It is not known if this is equivalent to imposing a certain type of boundary condition on superspace.

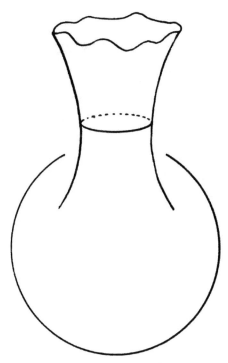

Figure 5.4: The real-time universe emerges from a primordial, imaginary-time, four-dimensional sphere.

Although these schemes differ in their details they all agree on the idea that space and time emerge in some way from a purely quantum-mechanical region which can be described in some respects as if it were a classical, imaginary-time four-space. In particular, the Hartle-Hawking picture of origination is very similar to that of *Figure 5.4*. However, there is another peculiar phenomenon that can arise in the Hartle-Hawking approach, and which also stems from the use of imaginary time—namely the possibility of spacetime *bifurcating* into multiple universes as in *Figure 5.5*. A spacetime of this sort can appear only if an imaginary time is employed near the bifurcation region. However, if an internal time is employed in the two final (disconnected) three-spaces, we do arrive at a picture in which "real" time seems to be bifurcated. Of course, the use of internal time is only appropriate well away from the bifurcation point itself. Near it, the imaginary aspect of time plays the major role.

Note that, unlike the origination in a *fixed* spacetime discussed in section 2, a production of multiple universes does not necessarily mean that the individual universes will interfere with each other. This is because space and time are themselves originated, not just the material content of the universe. This explains why I have used the phrase "multiple universes" rather than always referring to a single universe defined to be "all that there is." In fact, one does not talk about the universe as being "all that there is"—rather one means "all that there is at a given *time*." But there is no special way of correlating the internal times in the two branches of the universe in *Figure 5.5*, which justifies the adjective "multiple."

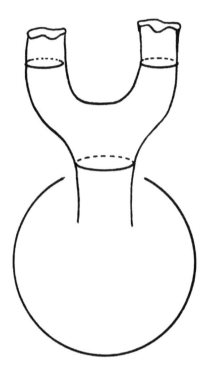

Figure 5.5: A bifurcating universe.

6 Conclusions

We have discussed one approach to constructing an origination theory that involves the imposition of certain conditions on the boundary of the superspace of all possible universes. If successful, such a theory would meet the requirements discussed in the introduction of "how much" and "what it is doing," subject only to the intrinsic (probabilistic) nature of the quantum-mechanical predictions. Furthermore, the scheme provides a description of the origination event in which what we now call space and time emerge from a four-dimensional spacelike region (i.e., imaginary time). A key role in understanding these ideas is played by the concept of an *internal* time which is defined using the curvature (gravitational field) or matter variables of the theory.

Unfortunately, many problems arise in this construction, and it is important to be clear what they are, especially if significant philosophical or theological conclusions are going to be drawn from the world view afforded by quantum cosmology. The issues faced by quantum cosmology are of two types: (i) internal technical/mathematical problems; and (ii) conceptual difficulties. The latter in turn divide into two categories: those that arise from quantum theory in general, and the more specific problems that stem from the clash between the existing quantum-theoretical formalism and relativistic notions of space and time. Since it is arguable whether the two types of problem (i) and (ii) can be so neatly separated, it is particularly unfortunate that we still have no truly coherent quantum theory of gravity. This means that all existing approaches to quantum cosmology are grounded in pious hopes that the techniques employed give a reasonable approximation to the "correct" (but unknown) theory. But they may all be quite wrong.

From a conceptual perspective, the main problem that arises from the lack of any proper theory is the need to fall back on general considerations of what such a theory *might* look like. And, in the absence of anything better, this necessarily focuses on the conceptual problems thrown up by the currently accepted theories of quantum mechanics and general relativity. This may be totally misguided in so far as a real theory of quantum gravity could differ drastically from both and in such a way as precisely to negate the type of discussion in which we are engaged.

6.1 Technical Problems with Quantum Origination Theories

I shall list first some of the technical problems that are characteristic of all current quantum origination theories.

(1) Schemes for describing the quantum origination of the universe are grafted onto what is, in fact, a non-existing theory. In particular, they presuppose that general relativity can be quantized in the way we have described. But this is far from being obvious and a more realistic hope is only that any proper theory of quantum gravity will reduce in some appropriate limit to the heuristic theory sketched in the previous sections. Thus, all the current quantum origination theories are likely to give only an approximate picture, particularly when applied to physical phenomena at scales of the Planck length or time.

(2) The existing quantum origination theories all assume that a unification of general relativity and quantum theory can be achieved without invoking radical conceptual changes in either. But a number of workers in quantum gravity (including the author) feel strongly that, in the region of the Planck length, a profound shift is needed in our concepts of space, time, and matter. If this suspicion is justified, the present pictures of the origination event could be extremely misleading.

(3) For any of the extant theories of the quantum origin of the universe, the evidence in support of the thesis that a *unique* state function Ψ is predicted is rather weak. Vilenkin's scheme gives a unique answer in certain simple models but it is unclear if this result would carry across to the full theory, even if it could be defined properly.

(4) The different quantum origination theories are not equivalent to each other, and each author understandably claims that his or her particular scheme is the "natural" one. Currently, there is no strong reason for preferring any one over the others. In particular, it has not been possible so far to refute any of these theories using genuine observational evidence. This is mainly because of the great difficulty experienced in going beyond very simple calculations that are too approximate to be of any real physical use—a problem that is compounded by the non-existence of a proper quantum gravity theory to which these calculations could be regarded as genuine approximations. Thus it would be difficult to over-emphasize how *extremely* speculative these schemes are.

(5) In the present quantum origination theories there is no way of relating the *type* of matter present to the origination event. This may or may not be a deficiency.

(6) The idea that the universe emerged from a primordial, pleromatic four-sphere is truly mythical and would have delighted all idealists from Plato to Jung.[40] However, one must be careful that spacetime pictures like *Figures 5.3* to *5.5* are not taken too literally, especially near the origination region. These theories are quantum theories and, as such, predict only a probabilistic distribution of three-curvature and matter variables. The classical spacetimes appearing in these diagrams are only an approximate way of depicting the content of the wave-function Ψ, and the regions where something "funny" seems to happen to space and time are especially quantum mechanical.

(7) Just as the concept of time is only an approximate one, so also the use of probabilistic ideas may be only approximately valid. In particular, there is no guarantee that, in the deep quantum regime of the origination event, the probabilities ascribed by the theory add up to one. There are general problems of a conceptual nature associated with the use of probability in quantum cosmology (see below) but a set of probabilities that do not sum to one is also logically inconsistent. It is almost as if quantum theory itself does not apply in the region that is supposed to be especially quantum mechanical—a somewhat paradoxical situation!

(8) Even if there is a classical spacetime picture as in *Figures 5.3* to *5.5*, it is important to realize that a typical quantum origination wave-function is not a wave-packet of the type that is used in conventional quantum physics to approximate a classical situation. Instead, the function

---

[40] Wagner would have appreciated the notion too: the womb-like, archetypal four-sphere resembles the bottom of the Rhine before the Rhinegold was stolen; see for example R. Donnington, *Wagner's Ring and its Symbols* (London: Faber & Faber, 1987).

oscillates in such a way as to correspond to a whole family of classical spacetimes, and the question of how a particular classical world is singled out is one facet of the difficult (and unsolved) conceptual problems discussed below.

6.2 Conceptual Problems with Theories Describing the Quantum Origination of the Universe

A number of general conceptual problems arise when attempts are made to apply quantum theory to the cosmos as a whole. These are so severe that a number of professional physicists believe that the entire quantum cosmology program may be fundamentally misguided. It is clearly important to have some understanding of this situation, but the issues are complex and I can only summarize them here.

The heart of the problem is the thoroughly instrumentalist tone of the conventional interpretation of quantum theory with its references to what happens if a *measurement* is made on a system, rather than to properties of the system itself.[41] In the context of cosmology this raises the obvious question: "Who is the cosmic observer?" Clearly what is needed is a version (or development?) of quantum theory that does not invoke the sharp subject-object dualism of observer and system, and which therefore has a chance of being applied to the universe in its entirety. Similarly, the notion of "state reduction" (which is conventionally associated with certain types of measurement) is an anathema to most people who work in quantum cosmology.

The meaning of state reduction in normal quantum theory has been much discussed in recent years. One approach to understanding a non-reduced state function is to say that all the possibilities to which it refers are realized in some genuine ontological sense—the infamous *many worlds* interpretation. Other solutions involve a change in the technical framework of quantum theory. For example, the idea has been advanced several times that an objective state reduction can be derived from a non-linear version of the theory.[42] Another recurrent idea is that human *consciousness* may somehow play a key role in the reduction process.[43] Any such changes to standard quantum theory would have dramatic implications for quantum cosmology.

The next serious problem for quantum cosmology is the meaning of the probabilistic language used to express the predictions. For example, if the theory predicts that, with probability 0.87, I shall be typing the final version of this article on August 16, 1992, is this to be viewed as a great success? And what does the 0.87 mean anyway? The relative frequency interpretation of probability

---

[41] Indeed, some physicists feel that the world view of quantum theory is sufficiently non-realist that even the concept of the "system itself" is not one that can be sustained.

[42] In particular, Roger Penrose has long advocated the existence of a non-linearity in quantum theory that has its origin in the theory of general relativity and the universal coupling of the gravitational field to all matter. A discussion can be found in Penrose, *The Emperor's New Mind*.

[43] For example, see the article by A. Grib in this volume.

(the form normally used in physics) says merely that if the experiment is repeated many times, then in 87% of the cases I shall indeed be found typing the article. But how is the experiment to be repeated, especially if it involves the origin of the universe?

Opinions are divided sharply on how a prediction of this type should be interpreted. We are faced here with the old problem of what meaning can be ascribed to a probabilistic statement that is intended to apply to a single event. One possibility is to employ a subjective interpretation involving a "degree of belief." One might also try the "latency," or "propensity" interpretations supported by Werner Heisenberg, Henri Margenau and Karl Popper.[44] However, in practice, the tendency is almost always towards an understanding based on some implementation of the idea of relative frequencies, mainly because of the desire to give some direct meaning to the *numerical* value of a probability.

In the case of quantum cosmology, one school of thought maintains that, in some sense, there really is a large ensemble of universes to which the usual statistical interpretation can apply. Various origins have been suggested for such an ensemble. For example:

(1) The universe may continually expand and contract giving rise to an endless series of big bangs and big crunches. Each cycle is to be thought of as a member of the statistical ensemble. Hence we get a version of an "endless return" mythology (but not one of cyclic repetition). This view seems to be incompatible with the spacetime pictures of all existing theories of the quantum origination of the universe.

(2) The initial Big Bang may have been such that many parts of the physical universe emerged in a causally disconnected way (although still part of a single, spatially-connected universe). The probabilistic statements of quantum cosmology are then deemed to apply to the statistical distributions of properties in these separate regions. Whether or not the universe is in fact of this type is something that should be predictable from the theory. More precisely, we might be able to invoke this interpretation if the quantum origination theory predicted that, with probability one, the universe did indeed emerge from the origination region with such a causal disconnectedness.

(3) The physical universe may be spatially *disconnected* and the statistical predictions apply to these separate "universes." The multi-universe production picture of *Figure 5.5* could perhaps be interpreted in this way.

Another approach is to say that the only predictions that have any real meaning are those that affirm something with probability one (or zero). In ordinary quantum theory, the only time it is unequivocally meaningful to assert

---

[44] See also the article by John Lucas in this volume.

that an individual system *possesses* a value for a quantity is when a measurement of the corresponding observable is guaranteed to yield that value with probability one. However, this is the exception rather than the rule, and to stick only to probability-one statements is to ignore most of the theory. On the other hand, there are some intriguing meta-theorems that purport to prove that if the actual universe is sufficiently complex to admit many physical copies of some system, then one of the probability-one statements is that measurements made on this physical ensemble will reproduce the statistical distributions of conventional quantum physics. This is comforting in a way but it does little to help in understanding the content of quantum theory when applied to the universe as a whole.

By now it should be apparent why many physicists believe that it is naïve to expect quantum theory to apply to the universe in its entirety and that, in particular, the subject of quantum cosmology is distinctly dubious. This should not be forgotten when contemplating the papers presented in this volume.

A *Appendix: The Formalism of Quantum Theory*

A.1 *Preliminary Remarks*

The main aim of this appendix is to provide a short introduction to the basic ideas of quantum theory with particular emphasis on conceptual issues. The central points can be approached from a variety of perspectives, but perhaps the most obvious is the way in which conventional expositions of quantum theory focus on the role of *measurement* as a primary, and frequently undefined, concept.

Any exposition of physics will inevitably contain terms which, although undefined, form part of the general scientific background of the age and culture within which they are employed. The meaningfulness and applicability of such terms is usually deemed to be obvious and not worthy of further explication (at least, by the scientists involved; philosophers of science see things otherwise). But, from time to time, new concepts arise that challenge this pre-established order of *a priori* truths and necessitate a radical reappraisal of the foundations of the subject. In twentieth century physics, the two major examples of such a paradigm shift are the theory of relativity and quantum theory. The former caused a major reassessment of the concepts of space and time; the latter challenges the notion of existence itself.

Some examples of such seemingly innocuous terms that arise in many discussions of physics are *system*, *observable*, *property*, *measurement*, *state*, and *causality*. In classical physics, the significance of each of these concepts has, after three centuries of use, indeed become relatively "obvious" and uncontentious. However, in quantum theory, which is still quite new and, in many ways, further removed from ordinary experience, the situation is radically altered and the cumulative shift in meaning of these basic ideas has produced a view of reality that is profoundly different from that associated with the classical world.

Let us start with the word *system* which, in all branches of physics, is understood to refer to the object or objects under study. But what is an "object?" Is the significance of this term really so obvious? We are confronted here with the question whose seeming innocence is so singularly deceptive:

> What is a thing?

This splendid question launched Western philosophy two and a half thousand years ago and, together with the complementary query "what is the nature of our *knowledge* of things?" continues to challenge us today.[45] It is certainly not trivial. In particular, it does not admit a simple answer of the form "A thing is a . . .," since that little word "is" is part of the problem. To address the question "what is a thing" is also to ask "what does it mean 'to be'?"

It is striking that, of all the modern sciences, only quantum physics seems to have been obliged to face this issue directly. However, in truth, no branch of scientific study can avoid this central question entirely; indeed, it should lie at the heart of *any* attempt to grasp the nature of reality, be it scientific or otherwise. For our present purposes it is important to start with a short analysis of the answer given to this question by classical physics.

In short, the answer is that a "thing" is a bundle of properties, or attributes, that adhere to the thing and make it what it is. Thus we talk about the mass of an electron, its position, and its velocity; or the color of my pencil, its position on my desk, its length, and the direction in which it points. Note that mass or color are examples of *internal* properties—attributes referring to the constitution of the thing itself—whereas position, direction, and velocity are *external* properties—attributes that deal with the way the thing appears in, or relates to, the framework of space and time.

This notion of an object as the focus of determinative properties lies at the heart of classical realism, which is one of the main Western, "common sense," philosophical positions. It is encoded in the structure of our language itself: the subject-predicate form of simple sentences reflects the idea of properties adhering to things, and plays a crucial role in determining how we understand the word "truth"; in particular, what is meant by a true statement.

The epistemological question of how we can have knowledge of the properties of an object, is answered in physics by the notion of *observables*. True, quantitative knowledge of a physical system is obtained by an act of *measurement*, by which I mean any physical operation performed from outside the system whereby the numerical value of an observable can be determined and recorded. Such a picture is in accord with standard scientific methodology

---

[45] For example, Martin Heidegger, *What is a Thing?*, trans. W. B. Barton and V. Deutsch (Indiana: Regnery, 1967). A healthy counterbalance to Heidegger's perspective is the famous work by Ludwig Wittgenstein, *Philosophical Investigations*, trans. G. E. M. Anscombe (Oxford: Basil Blackwell, 1984).

whereby part of the natural world is deliberately isolated from its environment so that theoretical and experimental investigations may proceed unhampered by any influence from the rest of the universe.

From the perspective of cosmology this separation of observer and system may already seem rather undesirable. However, generally speaking, such a split has no fundamental significance in classical physics since both observer and observed are deemed to be parts of a single, objectively existing, external world in which, ontologically speaking, both have equal status and are potentially describable by the same physical laws. Similarly, there is nothing special about the concepts of *measurement* or *observable*. The reason why a measurement of an observable yields one value rather than another is simply that the system *possesses* that value at the time the measurement is made. Thus properties or attributes are intrinsically attached to the object as it exists in the external world, and measurement is nothing more than a particular type of physical interaction designed to display the value of a specific quantity.

This general view of the nature of things also determines how we commonly understand the significance of time. We say that things "change in time," by which we mean that the internal or external properties of an object change. Furthermore, in classical physics it is assumed that these changes are *deterministic*, so that a knowledge of the properties of an object at one time suffices to predict with certainty the values of its properties at any later (or earlier) time.

The concept of properties, and the way they change, is coded scientifically into the notion of a *state* of a system: an idea that fits in well with the philosophical position of simple realism. In classical physics, the basic requirements of a state are:

(1) Knowledge of the state at any time suffices to determine the values of all properties of the system. Or, in more operational terms, it suffices to predict with certainty the results of all possible measurements that could be made at that time.

(2) The state at any time t_2 is uniquely determined by the state at any earlier (or later) time t_1. This *principle of causality* is how strict *determinism* finds its way into physics.[46]

A typical question that can therefore meaningfully be posed (and answered), and one that is of direct physical interest, is: "If, at time $t=t_1$, the state of the system is known and an observable A has the value a in this state, what will be the value of A at some other time $t=t_2$?"

To illustrate the idea of states and their mathematical representation, let us return to the example of a particle moving in one dimension under the influence of a force. The motion of the system is described by Newton's second

---

[46] Recent developments in the theory of chaotic motion have enforced important qualifications in the way classical determinism is viewed. However, this does not affect the line of argument being developed here.

law *force=mass* x *acceleration* which can be thought of as a second-order differential equation for the position of the particle as a function of time. This has a unique solution (for a given value of the mass) at any time once the position x and velocity v have been specified at any initial time $t=t_1$. Thus the principle of causality is satisfied if the state space (the mathematical space whose points represent states of the system) is identified with the two-dimensional space of all pairs of values (x,v). An observable is then defined to be any function of x and v, which means that the first criterion for the concept of a state is also satisfied.

The picture sketched above may seem to be in satisfactory accord with common sense but, in fact, under the impact of quantum ideas, every facet of it has come under review. In the conventional presentation of quantum theory a sharp distinction is made between the system and the observer (or observing equipment). The primary emphasis is now on the act of *measurement* and the results obtained for the values of the observable measured. The concept of properties being objectively possessed by an individual system becomes difficult to sustain; in particular, it is *not* in general meaningful to say that the reason why a measurement yields a particular value is simply because that is the value which the corresponding property happens to have at that time. In so far as quantum theory applies at all to single objects (and this is debatable) a "thing" is now best understood as a bundle of *latent*, or *potential*, properties that are only brought into being (in the sense of classical physics) by the act of measurement. This failure of classical realism is reflected at a mathematical level in the *non-distributive* nature of the set of quantum propositions: a feature that is discussed at length in the article by Andrej Grib in this volume.

The concept of a *state* still plays a vital role, but the first condition for a state can now be stated only in the operationalist version of yielding predictions for the results of possible measurements. And, since it is quantum theory we are discussing, it is the *probabilities* of yielding certain values that are predicted, not the values themselves. The notion of the state as the bearer of the causal structure of the theory still stands and, in particular, a specification of the state at any one time is required to determine uniquely the state at any other time. But note that it is the probabilities of measurement results that evolve deterministically (as computed from the evolving state), not the actual values themselves.

These remarks need to be unpacked carefully if the full impact of quantum theory is to be appreciated. But first we need at least some minimal idea of the general mathematical formalism of quantum theory upon which later developments, be they technical or conceptual, can be based.

A.2 *The Mathematical Formalism*

The space of states in a quantum theory is a *vector space H*. For expository purposes it suffices to think of this as an analogue of the familiar space of undergraduate vector calculus. That is, a vector is something that has length and direction. The main differences from the easily visualizable, three-dimensional space of Newtonian mechanics are that (i) the space may have an infinite dimension, and (ii) the components of a vector are complex rather than real numbers. The most important properties of any vector space (Newtonian or quantum mechanical) are that any vector can be multiplied by a number to yield a new vector, and any pair of vectors can be added together to give a third one. This leads to the important idea that quantum mechanical states can be *superimposed* to give new states.

As in classical mechanics, the state changes in time, and in such a way as to satisfy the requirements of determinism. Thus, through each point in the state space H there passes a unique path that represents one particular possible history of the quantum-mechanical system (or, in the statistical interpretation, an ensemble of quantum-mechanical systems). The uniqueness is secured by requiring the states to satisfy a differential equation (the time-dependent Schrödinger equation) that is first-order in time and which therefore has the mathematical property that a solution is uniquely given once the state at any reference time is specified.

Each observable A in the quantum theory is represented mathematically by a *linear operator* $\hat{A}$ that transforms each vector in H into another vector which is usually (but not necessarily) different from the original one. This will in general change both the length and the direction of a vector. However, for each operator $\hat{A}$ there are special vectors known as *eigenvectors* which have the property that only their length is changed. Thus the new vector is simply the old one multiplied by some number, known as the *eigenvalue* of that eigenvector. The operators representing physical observables are always chosen to be of a special type, one of whose properties is that such numbers are *real*. It is then a famous mathematical theorem that (i) the eigenvectors corresponding to different eigenvalues are *orthogonal* to each other (i.e., at 90^0), and (ii) any vector in H can be written as a unique linear combination of the eigenvectors.

The basic interpretive rules of quantum theory using this formalism are as follows:

(1) If a measurement is made of the observable A, the only value that will ever be found is one of the eigenvalues of the corresponding operator $\hat{A}$.

(2) If the state of the system is a vector y, the probability that any particular eigenvalue a will be found is

$$prob(A = a; \Psi) = \left| \cos\theta_{\psi u_a} \right|^2$$

where $\theta_{\psi u_a}$ is the angle between the vector Ψ and the eigenvector u_a associated with the eigenvalue a.

The second result is consistent since it can be shown that the sum of the cosine-squareds of the angles between y and all of the eigenvectors is equal to one. It is of course a fundamental requirement of any theory involving probabilities that the sum of all probabilities is one.

This formalism may seem very abstract since the connection between the mathematical symbols and the physical entities they represent is far more remote than is the case in classical physics. Nevertheless, it has been outstandingly successful in describing atomic and nuclear physics. It is also the basic mathematical framework within which all attempts to construct a quantum theory of cosmology have been made.

A.3 The Basic Conceptual Problems

Discussions of conceptual problems in quantum theory can usefully be centered around four main topics. These are:

(1) The meaning of probability.
(2) The role of measurement.
(3) The reduction of the state vector.
(4) Quantum entanglement.

These can be visualized as the four vertices of a tetrahedron, each vertex being closely linked to the other three. One could start the discussion anywhere: let us choose the notion of *probability*.

Philosophers and probability theorists have argued for centuries about the meaning of this concept, and some of this debate has spilt over into theoretical physics. The use of probability in physics is not peculiar to quantum theory; indeed, the ideas of classical statistical physics were well established in the nineteenth century. The most common meaning attached to probability in classical physics is an epistemic one. Thus it is assumed that probabilities measure the extent to which an (idealized) observer *knows* the precise details of the system. Our ignorance of these details arises from the great complexity of the system (for example, a box containing gas) but the underlying assumption is that the uncertainty in the actual behavior can be made arbitrarily small (at least in principle) with the aid of more precise measurements (or a more detailed theoretical description) of the *possessed* properties of the constituents of the system.

However, unless hidden variables are posited (and these bring problems of their own), the situation in quantum theory is very different. It is generally no longer consistent to talk of individual entities possessing a value for an observable: it is more as if the property concerned is *latent* in the system until such time as an observation brings it into being. As John Bell has put it, "be-ables" are replaced by "observables," and the verb "to be" becomes "to be measured."[47]

This lack in quantum theory of any analogue of the microstates of classical statistical physics has a number of important implications for the way in which the theory is interpreted. For example:

(1) It gives a fundamental role to the notion of *measurement* and places a gap between the system being observed and that which is making the

---

[47] John Bell's papers are conveniently collected together in *Speakable and Unspeakable in Quantum Mechanics* (Cambridge: Cambridge University Press, 1987).

observation. This contrasts sharply with the situation in classical physics where all aspects of the physical world are deemed to be describable by the same set of physical laws; in particular this is so for the interaction between the piece of apparatus and the system it is measuring. This poses the obvious problems of (i) when is an interaction between two systems to count as a measurement by one system of a property of the other? and (ii) what happens if an attempt is made to restore a degree of unity by describing the measurement process in quantum-mechanical terms rather than the language of classical physics that is normally used? There is no universally accepted answer to either of these questions.

(2) The inappropriateness of an epistemic interpretation means that the probabilistic predictions of the theory are usually understood in strict statistical terms. That is, they say what will happen on the *average* if the observable concerned is measured on a large number of identically prepared systems, but they usually say nothing about the behavior of an individual system. Thus a state is not a state of an individual system but refers only to an idealized ensemble of systems. This is closely related to the denial that an individual system can generally be said to possess values for its attributes.

(3) This "minimal" statistical interpretation provides the framework within which quantum theory is most commonly taught these days. It can be viewed as a pragmatic, fall-back position that does not necessarily rule out more adventurous developments in either the technical or the conceptual content of the theory. However, if it is taken as fundamental then the underlying philosophical position is clearly instrumentalist and anti-realist. Quantum theory is seen as a black-box that churns out useful results, but which gives no direct picture of (or assigns any meaning to) the reality that is assumed by most scientists to lie beneath their observations.

The probabilistic interpretation of quantum theory has another striking effect—the so-called reduction of the state vector—which has been the subject of much debate. Let us suppose first that the state of a system (or, perhaps better, the state of an ensemble of systems) has been prepared in some way to be an eigenstate u_a of a particular observable A. Then if a measurement of A is made, the value obtained must be a with probability one (the state vector u_a is orthogonal to every other eigenvector and therefore has a vanishing $\cos \theta$, whereas the cosine of the angle (zero) between a vector and itself is equal to one). This is the only occasion in conventional quantum theory in which it is deemed to be meaningful to say that the system *possesses* a value for the observable A.

Now suppose that the state Ψ is not an eigenvector and that a measurement is made of A. We do not know what value will be obtained, only that it will be one of the eigenvalues of $\hat{A}$, and that the probability of it being any particular a is $|\cos\theta_{\Psi u_a}|^2$. But what happens if the measurement is repeated immediately? If the first measurement is an "ideal" one, that is, it does not destroy the system or otherwise drastically disturb it, then the second measurement must find the same value again. But the only way of guaranteeing this from within the formalism is if the state vector immediately after the first measurement is the eigenvector u_a. Thus the first measurement has caused a reduction of the state vector from Ψ to u_a. Hence the overall time development of a state consists of sharp jumps produced by the act of measurement—jumps that are *not* governed by the laws of quantum mechanics—followed by periods of deterministic evolution governed by the Schrödinger equation.

These non-law-abiding jumps have been the subject of intense debate by those interested in the interpretation of quantum theory. They are not particularly problematical within the statistical interpretation since they correspond merely to the observer deciding to restrict his or her attention to a particular subensemble each time a measurement is made. However, such a view requires a strict enforcement of the system-observer distinction and the related primary role of the concept of *measurement*. As mentioned already, many physicists are unhappy with this situation and think that the observer, the system, and the act of measurement itself should all be capable of being described by a common set of, presumably quantum-mechanical, physical laws. Such reservations are often connected with a move towards trying to associate states with individual systems—a step that raises its own profound conceptual difficulties.

Finally, any naïve attempt to describe the act of measurement in quantum-mechanical terms quickly encounters the phenomenon of quantum entanglement—the fourth in our list of basic problems and one that lies at the heart of some of the most peculiar features of quantum theory. It would take us too far afield to describe this phenomenon in detail but suffice it to say that it refers to the peculiar feature whereby two quantum-mechanical systems that have once interacted remain "linked" by the state vector which describes them. The classical example is the correlations between two particles which have previously interacted but now lie far apart and seemingly isolated from each other (the heart of the Einstein-Podolsky-Rosen paradox). In particular, this applies to a quantum-mechanical description of a system plus a piece of measuring apparatus. It implies that unique, objective properties cannot be ascribed to the apparatus (even though it is a macroscopic object) but that, instead, the state vector gives an equal status to all possible outcomes of the measurement. The infamous Schrödinger's cat paradox belongs to the same stable.

The phenomenon of quantum entanglement imparts to the physical world a degree of holistic interdependence that contrasts strongly with the predominantly atomistic concepts of Western philosophy. It also seems at variance with our actual experience of how large objects behave. As a consequence there have been a number of attempts to change the mathematical structure of quantum theory in such a way that, for example, a macroscopic object automatically disentangles itself from anything with which it has interacted in the past. Or, to put it another way, a reduction of the state vector takes place for such systems as part of their internal dynamical evolution. Theories of this sort usually involve a departure from the linear Schrödinger equation and entail a radical change in the way the theory is interpreted. In particular, they tend to support a more realist view in which states (and properties) can be ascribed to individual systems. It is clear that issues of this type are of great importance in understanding quantum cosmology.

II

METHODOLOGY:
RELATING THEOLOGY AND SCIENCE

ON THEOLOGICAL INTERPRETATIONS OF PHYSICAL CREATION THEORIES

Michael Heller

1 *Introduction*

In recent years many books and articles have appeared dealing with theological interpretations of physical theories. The proceedings of a previous conference can serve as a notable example.[1] Very often it is claimed that the so-called new physics offers new horizons for theological reflection. On the one hand, such a situation should be welcomed as potentially opening ways for rapprochement between science and theology, which for many decades remained blocked by positivistic prejudices. On the other hand, it almost inevitably leads to dangers of pseudo-scientific explanations and compromised "God of the gaps" strategies in theology. A responsible methodological analysis of theological interpretations of scientific theories seems needed more than ever. The aim of the present study is to contribute to this goal.

In section 2 the structure of physical theories is briefly reviewed to prepare the background for discussing, in section 3, their different types of interpretations. In section 4, the results of this analysis are applied to some cosmological models or theories (especially those implying a creation of the universe) and their theological interpretations. Finally, section 5 touches briefly on the theological significance of the comprehensibility of the world as a precondition of any scientific theory.

Sometimes it is difficult to draw a sharp borderline between philosophical and theological interpretations of scientific theories. Although I shall be interested mainly in theological interpretations, the majority of the following analysis remains valid, *mutatis mutandis*, for philosophical interpretations.

2 *The Structure of Physical Theories*

Any physical theory consists of:

(A)   a mathematical structure;

(B)   the domain of the theory—roughly speaking a part or an aspect of the world to which the theory refers; and

---

[1] *Physics, Philosophy and Theology: A Common Quest for Understanding*, ed. Robert John Russell, William R. Stoeger, and George V. Coyne (Vatican City State: Vatican Observatory, 1990), 21-48.

(C)   bridge rules between (A) and (B); they are also called coordinating definitions or interpretation (in a narrower sense) of (A). One usually says that (A) models (B).[2]

This is, of course, a very idealized scheme of a physical theory. Usually, (A) is not given in a ready made state from the beginning. Very often, the mathematical structure of a given theory is only very roughly sketched by its creator as, for example, the mathematical structure of classical mechanics in Newton's work (in fact, at that time the fundamental notions of calculus were not yet clearly elaborated). Sometimes, only certain equations are proposed describing some physical phenomena in terms of a few of their solutions, and the mathematical environment of these equations, the field of their solutions, etc., are discovered much later. This was the case with the general theory of relativity. Initially proposed by Einstein in the form of his field equations together with a few physical postulates, it gradually developed into the very elaborate mathematical structure associated with the rich physical interpretation. Until now only rather "small" regions of the space of all solutions to Einstein's fields equations were known. It is by no means an exception in the history of science that the solutions found in the beginning turn out to belong to a zero-measure subset in the set of all solutions. Einstein's static world model can serve as an example. It was discovered in 1917 as the first relativistic cosmological model, and now we know that it is only one, and in a sense very exceptional, world model in the set of all possible solutions.

When a mathematical structure is more or less *a priori* chosen to model some domain of reality, it is usually a sign of a crisis in physics. This happens when theoreticians nervously search for a theory without having adequate empirical guidelines. In the history of both classical and contemporary physics it is usually an interplay between experimental results and some partial mathematical structures that gives rise to valuable new physical theories, as illustrated, for instance, by the early history of quantum mechanics.

Usually, the full mathematical structure of a physical theory is known only when this theory has been transcended by a new and more general theory, that is, when the limits of applicability of the old theory are already known. For instance, the full mathematical (geometrical) structure of classical mechanics was

---

[2] A word of warning seems necessary. The structure of physical theory was originally analyzed by neopositivists in a logically rigorous manner. I take the liberty of presenting it in a more colloquial style frequently used by physicists. For instance, strictly speaking, coordinating definitions can only relate theoretical statements with observational statements. Therefore, the domain of theory (our element (B)) should be defined as a set of utterances about a part (or aspect) of the world. However, physicists prefer to speak directly of the world, presupposing that the experiments they perform establish a certain type of contact with reality. In such a context, the term 'bridge rules' is often used. This straining a point in physicists' favor does not influence the outcomes of our discussion.

discovered only after elaborating geometric details of the generally relativistic structure of spacetime (by Cartan and others).

The most delicate element in the above scheme is the domain of a physical theory, that is, an aspect of physical reality to which the theory semantically refers. This domain can sometimes be given *a priori* (i.e., before the theory is created or constructed) in terms of commonsense cognition and using ordinary language description; but in other domains this is impossible, as in the case of quarks, black holes, and quantum gravity. However, strictly speaking the only authentic sources of our knowledge about the domain of a theory are the results of experiments which are not independent of the mathematical structure of the theory (without this structure we would not know what to measure, how to measure, and how to interpret what has been measured), such results only sample aspects of reality, that is, give us information about the tip of an iceberg which remains totally hidden from us. We can only believe that the mathematical structure of the theory reflects or discloses (in some approximate way) the structure of the iceberg itself. For this reason some versions of the structuralist theory of explanation are not uncommon among working scientists; usually, this type of explanation is presupposed unconsciously by them.[3]

Bridge rules connecting elements (A) and (B) of physical theories, can be almost straightforward, as for example in classical mechanics where our everyday sense perception is a reasonably reliable guide. In such a case, we can say that the theory in question has a natural interpretation. In more advanced theories of recent physics, bridge rules can be very complicated. They are certainly not one-to-one coordinating definitions, but are entangled in nuances of the mathematical structure and their references are sometimes ambiguous. All discussions concerning various interpretations of quantum mechanics have their origin here.

To sum up: (1) The structure of a physical theory is an organic totality. Elements (A)-(C), enumerated above, are *non-linearly* mixed with each other. (2) The structure of a physical theory is a *dynamic* entity; it evolves, and has a history. (3) The italicized words suggest that physical theories should be treated as *dynamical systems* and perhaps investigated with the help of the dynamical system method. I think this suggestion should be taken seriously. Notice that in this approach Kuhnian normal science and scientific revolutions are naturally

---

[3] Roughly speaking, by the structuralist type of explanation I mean a standpoint according to which the world's structure (very often hidden from our direct perception) is approximated and disclosed by the mathematical structures of physical theories. According to structuralism, there is a certain resonance between the structure of the world and the structures of physical theories; this justifies the agreement, so often achieved, of theoretical predictions with the actual empirical data. For details see, E. McMullin, "Structural Explanation," *American Philosophical Quarterly*, 15 (1978): 139-147.

interpreted as stable states and bifurcation points of a dynamical system, respectively.[4]

3 *Interpretation Strategies*

Now, a few remarks concerning theological (or philosophical) interpretations of physical theories. By an interpretation of this kind I shall understand *any comment on a particular physical theory*. My understanding is, therefore, very broad; in fact, it presupposes only one limitation: the comment in question must concern a single, well-identifiable, physical theory. Usually such comments refer—even if this is not directly noticed by their authors—to the mathematical structure of a given theory since it is these structures that are supposed to disclose the structure of physical reality. The relationship of such comments to the mathematical structure of a given theory can be threefold:

(a) A comment can be inconsistent with or even contradict the mathematical structure of a theory. Such comments include those interpretations of the theory of relativity which claim that it presupposes an "ontologically" unique or absolute time, such as Bergson's[5] or Maritain's[6] "philosophical interpretations" of special relativity; Mach's original interpretation of Newtonian mechanics which claimed to incorporate into its conceptual framework the idea of the relativity of mass; etc.[7] Such interpretations are usually regarded by physicists as "hand waving arguments," and should be treated as inadmissible.

(b) A comment can be neutral with respect to the mathematical structure of a physical theory. For instance, the spacetime of the special theory of relativity can be interpreted either as describing the block universe, with the past and the future on an equal footing, or as a process of becoming with the "now" of a given inertial reference frame as its only real state.[8]

---

[4] I have developed this point of view in the paper, "Nonlinear Evolution of Science," in *Annals of Philosophy (Catholic University of Lublin)*, fasc. 3, 32 (1984), 103-125 (in Polish, with English summary).

[5] H. Bergson, *Durée et simultanéité (a propos de la théorie d'Einstein)*, (Paris: Alcan, 1922); the second edition in 1923 contains three new appendices.

[6] J. Maritain, *Réflections sur l'intelligence et sur sa vie propre*, (Paris: Nouvelle Libraire Nationale, 1926), especially ch. 7.

[7] See my analysis of Mach's views regarding this question: "Between Newton and Einstein: Mach's Reform of the Newtonian Mechanics," in *Newton and the New Direction in Science*, Proceedings of the Cracow Conference, May 25-28, 1987, ed. G. V. Coyne, M. Heller, and J. Zycinski (Città del Vaticano: Specola Vaticana, 1988), 155-173.

[8] See R. Penrose, "Singularities and Time-Asymmetry," in *General Relativity: An Einstein Centenary Survey*, ed. S. W. Hawking and W. Israel

A good symptom indicating that the interpretation of a given theory is neutral with respect to its mathematical structure is the following. Suppose you are able to show that your interpretation does not produce inconsistencies with the mathematical structure of the theory you interpret. Try to say something contradictory with what your interpretation asserts. If this does not require changes in the actual mathematical structure of the theory, this means that both your interpretation and its negation can be referred to the given theory, and consequently your interpretation (and its negation as well) is neutral with respect to its mathematical structure. For example, suppose you interpret the spacetime of special relativity as the block universe. This produces no inconsistencies with the mathematical structure of the theory, but you can also see that if you change to the "flowing time" interpretation you need change nothing in this structure.[9] This means that both interpretations are neutral with respect to it.

(c) A comment can be in strict agreement with the mathematical structure of a given physical theory. Such a comment can be regarded as an exegesis of the mathematical structure of the theory. It so closely follows this structure that any deviation from it would immediately lead to inconsistencies with the structure of the theory. An example (at least in the intention of the authors) of such an exegesis is the book by D. J. Raine and M. Heller, *The Science of Space-Time*[10] in which interpretations (of type (c)) of different spacetime theories (from Aristotle to general relativity) are studied. Such an exegesis can be a very difficult task for many reasons, for instance: (i) mathematical structures of many recent theories still evolve; (ii) everyday language, in which such comments are usually made, is often inadequate to the "internal logic" of more advanced mathematical structures. In this view, it is not true that only those things which can be clearly expressed in everyday language can be made understandable. For instance, everyday language cannot "clearly" describe non-linear situations, but modern physical theories make such situations understandable by modeling them with the help of non-linear mathematics.

Philosophical or theological interpretations of physical theories (such as, for example, creationistic interpretations of cosmology, see below Section 3) assume the form of comments on physical theories. If they are of type (a), they are pseudo-interpretations, and render a poor service to theology, making it ridiculous in the eyes of scientists. They cannot be of type (c) since physical

---

(Cambridge: Cambridge University Press, 1979), 581-638, 883-886, section 12.2.5. See also the article by J. Polkinghorne and C. J. Isham in this volume.

[9] These two interpretations can be regarded as contradictory with each other in the sense that they cannot be true together.

[10] D. J. Raine and M. Heller, *The Science of Space-Time* (Tucson: Pachart, 1981).

theories, from their very nature, say nothing about religious matters (as more fully discussed below). Therefore, they can only be of type (b); that is, neutral with respect to the mathematical structure of a given theory. Interpretations of this type should be taken "seriously but not literally."[11] They can show more than the non-contradiction between a given scientific theory and a given philosophical (or theological) doctrine; if they are successful, they can be regarded as demonstrating a certain consistency or consonance between them.

A warning following from these considerations: Our analysis of type (c) interpretations stressed the inadequate character of everyday language for interpreting the mathematical structures of physical theories. Careful analysis of many contemporary physical theories shows that the structure of the world (as it is approximated and disclosed by the structures of our theories) transcends the possibilities of our language and our imagination. In light of this fact, we must be aware of a sense in which type (b) interpretations are inadequate. In this case our imagination and our linguistic resources are more weakly guided by mathematical structures than is the case with interpretations of type (c). We must be ready to admit that what we believe to be a good comment, consonant with a given theory, could easily turn out to be a metaphor or an image adapted to our limited possibilities rather than truly approaching the nature of reality.

4 *Creation Interpretations in Cosmology*

Great confusion reigns in the philosophical literature dealing with cosmological matters. In the following, I shall illustrate my analysis with examples picked up more or less at random from the manifold instances of such confusion.

It is more or less evident that authors will avoid interpretations which would be in open contradiction with the mathematical structure of a given physical model or theory. An example of such an open contradiction would be the claim that Einstein's static world model supports the idea of a temporal beginning of the universe. Of course, it is possible to argue that God could create the static Einstein universe in a ready-made state at any instant of its existence, but this would be an interpretation superimposed, so to speak, on the model, requiring additional assumptions of an extra-scientific character necessary to enforce this interpretation on the model's structure. Moreover, these assumptions automatically qualify the interpretation as belonging to category (b) rather than category (a).

Usually, inconsistencies (or even contradictions) between a proposed interpretation and a given model or theory are more cleverly disguised. When they are brought to light they are often unmasked as falling under category (a); this shows simultaneously that they cannot aspire to belong to category (c). There is a host of instances. The subject-matter often inspires an eloquent rhetoric. A good example is Robert Jastrow in his once best-selling book *God and the Astronomers*; commenting on the Big Bang theory he writes: "For the

---

[11] See I. G. Barbour, *Myths, Models, and Paradigms* (New York: Harper & Row, 1974), 7.

scientist who has lived by his faith in the power of reason, the story ends like a bad dream. He has scaled the mountains of ignorance; he is about to conquer the highest peak; as he pulls himself over the final rock, he is greeted by a band of theologians who have been sitting there for centuries."[12] The point is that both scientists and theologians are aiming at the *same* peak, that is to say, Big Bang theory and the biblical phrase "In the beginning God created heaven and earth"[13] are supposed to speak about the same event—the creation of the universe. This is clearly stated by John A. O'Keefe in the afterward to Jastrow's book:

> We see then, that the resemblance between our cosmology today and that of the theologians of the past is not merely accidental. What they saw dimly, we see more clearly, with the advantage of better physics and astronomy. But we are looking at the same God, the Creator.[14]

More examples of the same interpretative fallacy are described by Adolf Grünbaum. Some of them are "instructively articulate" as instances of the confusion of concepts.[15] The illegitimacy of such interpretations can be demonstrated in two ways: first, one could show that the authentically theological doctrine of creation says something different from what can be deduced from the Big Bang theory; second, one could appeal to a methodological analysis in order to show that theological questions transcend the limitations inherent in the very nature of the scientific method. In fact, each of these two ways cannot avoid using elements of the other one; they differ from each other in the point of departure rather than in the essence of the argument.

The best approach of the first sort is a historical analysis of the development of the creation doctrine in theology. This method shows persuasively that the theological idea of creation is immensely richer than anything physics or cosmology is able to say. It is not difficult to demonstrate that a proponent of such a theological interpretation says something different than what can be read out of a given scientific model or theory, and in this sense the interpretation in question transcends the structure of a given model or theory. To pursue this analysis further would exceed the limits of the present paper; the reader is referred to the extant literature.[16]

It is interesting to notice that many creation interpretations of type (a) consist in imputing to theologians specific views never (or seldom) shared by them, the most common error being the identification of creation (in the theological sense) with the initiation in time of the existence of the universe.

---

[12] Jastrow, *God and the Astronomers* (New York: Warner Books, 1980), 125.

[13] Jastrow quotes this biblical verse in the context of his interpretation. Ibid., 124.

[14] Ibid., 158.

[15] A. Grünbaum, "The Pseudo-Problem of Creation in Physical Cosmology," *Philosophy of Science* 56 (1989): 373-394.

[16] I especially recommend a short account by Ernan McMullin of the history of the creation and evolution doctrines presented in his introduction to the volume: *Evolution and Creation*, ed. E. McMullin (Notre Dame: University of Notre Dame Press, 1985), 1-56. See also the paper by the same author "Nature Science and Belief in a Creator: Historical Notes" in *Physics, Philosophy and Theology*, 49-79.

There are so many instances of this error that there is no need to quote any of them.

The second way is based on the well founded achievements of the modern philosophy of science, and consists in applying its results concerning the nature and scope of scientific theories to the particular case of cosmological theories. One could repeat here Grünbaum's analysis of this problem[17] with a slight change of emphasis: his results should not be understood as following from an *a priori* positivistic or "over-empiricist" idea of science (which, from the phraseology he uses, leads me to suspect he falls into), but as derived from sound methodological premises (as I believe he actually intended). The way he uses the term "pseudo-problem" with regard to the philosophical or theological doctrine of creation suggests an *a priori* assumption that the limits of the scientific method coincide with the limits of rationality.[18] Such an assumption constitutes an epistemological standpoint which necessarily follows neither from science nor from its philosophy. In fact, the present analyses are based on a conviction that this assumption is not true.

Grünbaum distinguishes two questions: first, "Does the physical universe have a temporal *origin*, and—if so—what does physical cosmology tell us about it?" and second, "Was there a creation of the universe, and—if so—what light can science throw on it, if any?" He argues that the first question is a legitimate physical problem whereas in the second question "the genuine problem of the origin of the universe or of the matter in it has been illicitly transmuted into the pseudo-problem of the 'creation' of the universe or of its matter by an external cause." The above distinction of the two kinds of questions and their evaluation is well known[19] and there is no need to repeat all the arguments usually quoted on its behalf.

If creation interpretations of cosmology are inconsistent (or even contradictory) with this physical theory (as claimed by Grünbaum), that is to say, if they belong to category (a), they cannot be regarded as legitimate exegeses of its mathematical structure, that is, they cannot by of type (c). The third possibility (of their being of type (b)) remains to be discussed.

In the above-analyzed creation interpretations falling under category (a), the error consists in identifying certain physical statements with some theological assertions (physicists and theologians "met at the top of the same peak"). If, in such a contest, identity is replaced by a kind of consonance, the interpretation switches from category (a) to category (b), and no methodological objections can be raised against it. However, in the case of the Big Bang theory and the theological doctrine of creation, to simply say that God created the

---

[17] Grünbaum, "The Pseudo-Problem of Creation."

[18] This is also apparent in Grünbaum's attempt to reduce all questions transcending the scientific method to mere psychological discomforts. In his opinion, a question concerning the creation of the universe "cannot be regarded as a well posed challenge merely because the questioner finds it psychologically insistent, experiences a strong feeling of puzzlement, and desires to answer it." Ibid., 821.

[19] See, for instance, W. B. Drees, *Beyond the Big Bang: Quantum Cosmologies and God* (La Salle, IL: Open Court, 1990).

universe in the Big Bang singularity and therefore these two ideas are "consonant" would be rather a trivial statement. To avoid triviality one should provide a comment showing a compatibility of these two doctrines (but not their identity). Usually, theologians point out that to create in the theological sense means something much more than to bring a thing into existence, the thing which *a moment ago* was just nothing. Even Grünbaum is ready to admit that "the view of *timeless* causation set forth by Augustine" evades his criticism, and he disregards it only because he finds it "either unintelligible or incoherent." However, these two reasons are not a matter of scientific methodology but rather of psychological temperament or of *a priori* assumptions.

Within interpretations of type (b) different strategies are possible, varying from claiming mutual independence of physical theories and theological doctrines, through dialogue or some kind of integrations, to a constructive consonance between them. Some of these strategies have been analyzed by Ian Barbour[20] and Willem Drees (the latter with special reference to cosmological problems).[21]

A good example of an interpretation falling under category (c) is Chris Isham's paper[22] in which he makes a "step by step exegesis" of the Hartle-Hawking model of the quantum creation of the universe. He clearly shows that the theological doctrine of creation and "creation" as a quantum process have hardly anything in common besides the name "creation." He warns, however, that "some attention should surely be paid to the shifting forms in which the archetypes of space and time are impinging on the scientific world." Evidently, as soon as we change from studying the structure of the Hartle-Hawking model to contemplate "the archetypes of space and time" in their role of shaping our intuitions concerning the creation of the world, we immediately leave the secure land of interpretations (c) and enter the territory of those belonging to category (b).

5 *Philosophical and Theological Significance of Physical Theories*

The philosophical and theological significance of physical theories is not limited to categories (a), (b), and (c), with which we have dealt in the previous sections. Theology has always interacted—and, I think, will always interact—with the sciences in manifold ways. For instance, scientific ideas may be for theologians a source of inspiration to reach new theological insights or to invent more appealing metaphors; they might help to evoke the feeling of mystery ("if in physics things go so far beyond our imagination, what can we say about God?") or to create a suitable context for reconsidering or reinterpreting a traditional

---

[20] Barbour, "Ways of Relating Science and Theology," in *Physics, Philosophy and Theology*, 21-48; and *idem, Religion in an Age of Science*, The Gifford Lectures, 1989-1991, vol. 1 (San Francisco: HarperSanFrancisco, 1990), 3-28.

[21] Drees, *Beyond the Big Bang*.

[22] C. J. Isham, "Creation of the Universe as a Quantum Process," in *Physics, Philosophy and Theology*, 375-408.

religious doctrine, and so on. Such interactions are unavoidable since our thinking (in theology and elsewhere) is unable to operate over long periods of time in isolation from the evolving stream of social knowledge. Therefore, such interactions are indispensable if theology is to fulfill its mission with respect to each generation. Usually, theology interacts, in one of these ways, with no particular theory but rather with an overall image of the world as it is drawn by science as a whole in each epoch. Consequently, the above analysis does not refer to this kind of interaction (since I have considered exclusively *interpretations of particular* physical theories).

Sometimes, however, a particular physical theory can enter into a fruitful interaction with a theological (or philosophical) doctrine. Let us consider two instances of such interactions.[23] First, it can happen that some part of a physical theory plus specific additional premises can provide reasonable grounds for a theological conclusion. Second, a theological theory and a scientific theory might have implications that turn out to be mutually dependent or even equivalent. Neither of these cases is an interpretation of a scientific theory in the sense described at the beginning section 3. Of course, one could interpret the theory in question "in the light" of additional premises or of the fact that its conclusions are somehow related to some theological conclusions. If this happens, interpretations fall into category (b), and are neutral with respect to the content of the theory. In both cases some external elements (with respect to the original structure of the theory) are needed (additional premises or theological implications) to make a comment possible.

Let us, finally, mention yet another way in which science can be relevant, and very much so, for philosophical or theological discourse. In my opinion, the very existence of physical theories and their effectiveness in processing our knowledge of the world is of much greater importance for theology than any theological interpretation of a particular theory. As it is well known, the question of why the method of using mathematical structures to model some aspects of the world works, and works so well, is a non-trivial philosophical problem.[24] This question is often expressed in the form: "Why is the world mathematical?" and perhaps it should be regarded as a special instance of a more general question: "Why is the world comprehensible?"[25] Having no possibility to enter the discussion of the problem in a more exhaustive way, I should refer the reader to the huge literature on this problem.[26]

---

[23] They were suggested to me by Nancey Murphy.

[24] Only a very few positivistically minded thinkers, usually non-physicists, would dismiss it as evoking purely emotional reactions.

[25] I say "perhaps" on behalf of those who are inclined to identify the comprehensibility of the world with its mathematical character.

[26] My preferred readings on this subject are: A. Einstein, "Physics and Reality," in *Ideas and Opinions* (New York: Dell, 1978), 283-315, especially the beginning of this paper; E. Wigner, "The Unreasonable Effectiveness of Mathematics in the Natural Sciences," *Communications in Pure and Applied Mathematics* 13 (1960): 1-14; Roger Penrose, *The Emperor's New Mind* (Oxford: Oxford University Press, 1989), especially chs. 3 and 4; and J. D. Barrow, *The World Within the World*

In philosophy one can at most try to demonstrate the non-triviality of this problem, and to conclude with Einstein that it is "the eternal mystery . . . which we shall never understand."[27] In theology one could go a step further and try to offer a theological interpretation of this result. For instance, one could say that the comprehensibility of the world and its existence are but two aspects of the *creation*. Owing to the act of creation the world exists, and through the act of creation the world is comprehensible. The rationality of the Creator is reflected in the created world. To use the old Platonic principle, "God always geometrizes" and consequently every result of God's creative action always has a geometric (or more generally—mathematical) character.

To sum up, I think that legitimate theological interpretations of cosmological theories should be limited to type (b). Such interpretations are useful insofar as they help the Christian "to make his theology and his cosmology consonant in the contribution they make to his world-view."[28] Moreover, I strongly believe that Christians who want to look to the sciences for a deeper understanding of the creation should not take into consideration any particular cosmological model or theory, but turn instead to the most fundamental assumptions presupposed by every scientific endeavor.

Acknowledgments: I gratefully thank Nancey Murphy and Bob Russell for their patient criticism. It has allowed me to substantially improve the text of the paper. My thanks go also to Chris Isham who carefully read the manuscript.

(Oxford: Clarendon Press, 1988), especially ch. 5. See also the article by Paul Davies in this volume.

[27] Einstein, "Physics and Reality," 285.

[28] E. McMullin, "How Should Cosmology Relate to Theology?" in *The Sciences and Theology in the Twentieth Century*, ed. A. R. Peacocke (Notre Dame: University of Notre Dame Press, 1981), 52.

METAPHORS AND TIME ASYMMETRY: COSMOLOGIES IN PHYSICS AND CHRISTIAN MEANINGS

Stephen Happel

There is a grain of curiosity
At the base of some new thing, that unrolls
Its question like a new wave on the shore.

John Ashbery, "Blue Sonata"

1 *Introduction*

Don Cupitt has recently written the following:
> To this day there are theologians who try to link the doctrine of creation to the Big Bang of standard model cosmological theory. But, if I am right, they are making a Big Mistake. For one thing, religious thought is concerned with language and with the middle of life, and has no need to defer to highly specialized and marginalized forms of technical expertise.[1]

Cupitt is correct about religion—that its language involves the rhetoric of ordinary speech. He misinforms himself, however, when he assumes that scientists do not use ordinary speech to originate, process, and communicate their insights.

Current philosophy of science consistently makes the point that experiments in the natural sciences, their languages of analysis and communication are mutually implicated. One need only mention in this context certain major twentieth century thinkers to support this claim. Jürgen Habermas has maintained that "the choice of research strategies, the construction of theories and the methods of their verification" are "bound inexorably" to natural languages and the colloquial communication of scientists.[2] Despite scientists' claims to mathematical or experimental "objectivity," part of the work of science occurs in ordinary rhetoric. Habermas goes on, of course, to make the political point that the interests of the scientific community intrinsically affect (and even effect) the analyses, experiments, and conclusions of science. Ludwig Wittgenstein, in the course of analyzing the processes of mathematics (and in consequence his own earlier interpretation of precise languages), explores the linguistic issues that emerge among the activities of mathematics, art, and natural speech. He is convinced of the relationship between the heuristic dimension of

---

[1] Don Cupitt, *Creation Out of Nothing* (Philadelphia: Trinity Press, 1990), 4.
[2] Jürgen Habermas, "On Hermeneutics' Claim to Universality," in *The Hermeneutics Reader*, ed. Kurt Mueller-Vollmer (New York: Crossroad, 1985), 299.

pictures and proofs. "Without this picture I should not have been able to say how it will be, but when I see it I seize on it with a view to prediction."[3] As he says: "the working of the mathematical machine is only the 'picture' of the working of a machine."[4] The rule-governed language of mathematics is connected to the rules of other language games. According to Hans-Georg Gadamer, what appears as popularization or communication to the masses is actually part of the application of scientific activity.[5] The natural sciences were accustomed to assume a pristine, "objective" isolation from their origins and communication to which they can no longer pretend.

The audiences for the following discussion are therefore two-fold. On the one hand, it is necessary to speak to those who need to understand the processes and hermeneutical frameworks that are operative in the natural sciences, especially physics and astrophysics. On the other hand, however, theologians are addressed in that they need to understand that the cosmological data and interpretive language of the sciences have an important role to play within, or at least as a condition for, religious discourse. The hinge for this interdisciplinary dialogue will be an examination of the process and referentiality of metaphor.

This essay, therefore, queries both publics. What are some of the springs that trigger the understanding of temporality in the natural sciences? Does it make a difference whether there is an open or a closed universe? What is the characteristic way in which science now describes the temporality of the universe? Does Christianity require as a condition for its understanding of its divinity a temporally asymmetrical universe in which future transcends present and past? Must there be a direction (any particular positive or negative direction?) to time for Christian theology to make sense? Though I do not believe there is any simple programmatic way in which Christian theology can predict cosmological consequences or that it has a blueprint for the future (any more than it has a ready-made template for interpreting the past), I remain

---

[3] Ludwig Wittgenstein, *Remarks on the Foundations of Mathematics*, ed. G. H. von Wright, R. Rhees, and G. E. M. Anscombe, trans. G. E. M. Anscombe (Oxford: Blackwell, 1964), III-33, 123e; cf. also V-4, 161³-163e.

[4] Ibid., IV-49, 127e.

[5] Hans-Georg Gadamer, *Truth and Method*, trans. Joel Weinsheimer and Donald G. Marshall, rev. ed. (New York: Crossroad, 1989), 307-341. I am aware that the remarks in this paragraph align differing schools of philosophy and conflicting methods of investigation. My concern in this introductory section is to make clear that philosophers of science have argued that "science" itself is not a single language and that even its most precise languages have connections to the rhetoric of ordinary speech. One of the conditions for comparing the integral role of metaphor in scientific and religious cosmologies is to *assume*, but not presume this interrelationship. Even if Wittgenstein is incorrect about mathematics and picture (*Bild*), mathematics may yet have a metaphoric shape that is entailed as narrative in its logical sequencing of equations. (In this, I am indebted to a conversation with Paul Danove.)

convinced that theology, as a public discourse, must respond to the exigencies of reasonable conversation with the natural sciences (or other disciplines).[6]

When Christianity claims that redemption and resurrection are not "more of the same," are there conditions for this assertion in the universe as a whole? I am aware that I have asked far more questions that I can answer in this essay. By making the methodological choice of examining the language operative in both disciplines with regard to the origination of the universe, I believe that a methodological framework for a solution to these questions may appear. I am proposing as a modest beginning to analyze one of the linguistic forms (metaphor/narrative) that shape science and religion. In this sense, the project is another frame for a multi-layered research program. Metaphor may tell us more than we think we know about the nature of time, religious claims, and scientific cosmology.[7]

Much of the work presented here will be analytic, studying what is going forward in contemporary astrophysical cosmologies and theology. As I have said, the hinge is metaphor (sections 4 and 5), the language that these disciplines have used. The paper will treat (section 2) the general hermeneutical reasons for focusing upon metaphoric language; (section 3) the metaphors that become narratives for time in the natural sciences; (section 4) the structure of metaphors and how this affects the supposed nature of temporality; (section 5) the referential nature of metaphors and their implicit temporal character; and (section 6) the way alternate cosmological metaphors parallel Christian understandings of religious narrative. It will be clear to the knowledgeable that I can claim little or no expertise in mathematics and natural science. On the other hand, theologians will note that I take images, symbols, and metaphors seriously as a source for the nature of philosophical and theological rhetorics.[8]

---

[6] David Tracy, *The Analogical Imagination: Christian Theology and the Culture of Pluralism* (New York: Crossroad, 1981), 9-49.

[7] There are many other conditions (both subjective and objective) for the meaning, meaningfulness, and reference of Christian discourse. But when Christianity makes the religious claim that God has brought something new into the world, that event presumably has some effect upon the cosmos and consequent human history. Are there cosmological, physical conditions for the possible meaning and truth of Christian claims? The questions about the nature of temporality in the universe that I am examining in this essay are one arena in which the issue of methods may be joined. A choice for the study of language, however, does not decide various doctrinal issues about nature and grace, eschatology or protology; but it does, in fact, invite different answers to these traditional problems. To study those questions means to develop an entire systematics—an important project, but one that exceeds the bounds of this essay. I hope that the discussion here is the engagement of a conversation, that is, one in which both partners recognize that through language they have something to say to each other.

[8] Ian Barbour, *Myths, Models & Paradigms: A Comparative Study in Science and Religion* (San Francisco: Harper & Row, 1976), 7. Though I will modify Barbour's analysis of reference in metaphors, I obviously owe a considerable debt to his work. For a discussion of some of the relationships between early modern science and

2 Natural Science as Hermeneutical

Had Gaston Bachelard or Georges Poulet written this paper, they would first have completed an exhaustive phenomenology of metaphors for time in modern literature to show how scientific and religious notions have a matrix in literary genres, tropes, and symbols.[9] Paul Ricoeur has completed a lengthy literary-critical and philosophical recovery of metaphors and narrative in light of the various suspicions about cause, effect, sequence, and connection in history and fiction since the early modern period.[10] Even this is a meager lunch at the unlimited buffet of contemporary literary criticism and philosophy of time.

The natural sciences themselves now recognize their own temporal character. Data are theory-laden; understandings of "things" have a history.[11] Judgments about nature lead to and are informed by values and interests. And paradigms shift, sometimes and partially always for nonconceptual, non-mathematical reasons. It therefore becomes possible to speak of the hermeneutics operative within the praxis of science and within the theories of interpretation that guide scientific activities. "Facts" are standard interpretations. Interpretations imply perceptual screening, conceptual organization, and conditions for judgment. The language about nature no longer naively reflects the "already-out-there-now-real"; it is a language mediated by meaning, cultures, and histories.

Christian practices and theories, partially at the insistence of the early scientific methods, discovered their own evolutionary histories and interpretive contexts. Though the nineteenth century was a long battle over the status, meaning, and reference of religious actions and words (a war that is not yet concluded), theologies in the Protestant traditions have largely given up pretensions to non-contextual, a-historical proclamatory biblical words as the

---

language, see Andrew E. Benjamin, Geoffrey N. Cantor, and John R. R. Christie, *The Figural and the Literal: Problems of Language in the History of Science and Philosophy, 1630-1800* (Oxford: Manchester University Press, 1987); and especially Frederic L. Holmes, "Argument and Narrative in Scientific Writing," in *The Literary Structure of Scientific Argument: Historical Studies*, ed. Peter Dear (Philadelphia: University of Pennsylvania Press, 1991), 165-181.

[9] Gaston Bachelard, *The Poetics of Space*, trans. Maria Jolas (Boston: Beacon Press, 1964); and *idem, The Poetics of Reverie: Childhood, Language, and the Cosmos*, trans. Daniel Russell (Boston: Beacon Press, 1969). Georges Poulet, *Studies in Human Time*, trans. Elliott Coleman (Baltimore: Johns Hopkins University Press, 1956); and *idem, The Interior Distance*, trans. Elliott Coleman (Ann Arbor, MI: University of Michigan Press, 1959). See the general bibliography on time under art, language, and literature in J. T. Fraser, N. Lawrence, and D. Park, *The Study of Time IV*, Papers from the Fourth Conference of the International Society for the Study of Time, Alpbach-Austria (New York: Springer-Verlag, 1981), 242-243. Though incomplete, the entire bibliography is useful.

[10] Paul Ricoeur, *Time and Narrative*, trans. Kathleen McLaughlin and David Pellauer, 3 vols. (Chicago: Chicago University Press, 1984-88).

[11] Barbour, *Myths, Models & Paradigms*, 9, 44.

fundament of faith.[12] The bible and traditions of interpretation are co-implicated, if not co-originated in religious experience, speech, and theologies. In Catholic theologies, the eternal dogmas of tradition, the pretensions to absolutist judgments about texts, historical events, and people have ceded to historical-critical biblical scholarship, the cultural and intellectual development of doctrines, a hierarchy of truths, and the (not so successful) distinctions between unchanging substance (or core) and changing customs (or periphery).[13]

In the nineteenth century, it was often possible to create sharp contrasts between the natural and human sciences, between physics or geology and religion. Now the sciences and religion must discuss their hermeneutical procedures, some of which they hold in common. I have chosen to focus this paper upon the most important common frame of interpretation: language. Whether in science or religion, we live in language. It is not possible to exit the house of language whether through transcendent leaps into an imageless void or through appeals to interior feelings or to private cognitional operations.

Language about art, the hermeneutics of language, and theological discourse have (sometimes) subterranean channels to the hermeneutics of mathematics and the natural sciences. If we are to relate the cosmologists' analyses of the universe's origins to religious stories and theological theories about creation, we must link them through their correlative patterns of language. The language that connects Jerusalem with particle acceleration may be the hermeneutics of art. Such a hypothesis is worth exploring.

3 Metaphors for Time Asymmetry in Physics—The Universe as Story

This section points to the commonly acknowledged fact that metaphors fill the standard model of the universe and that these metaphors structure themselves into a narrative form.[14] The natural sciences have chosen to tell a story about the universe using metaphors and the rhetoric of ordinary language. Stories have beginnings, middles, and ends. The middle evolves through a plot complication

---

[12] See Edward Farley, *Ecclesial Reflection: An Anatomy of Theological Method* (Philadelphia: Fortress Press, 1982), especially 64-105; and Sallie McFague, *Metaphorical Theology: Models of God in Religious Language* (Philadelphia: Fortress Press, 1982).

[13] See, for example, Edward Schillebeeckx, *The Understanding of Faith: Interpretation and Criticism*, trans. N. D. Smith (London: Sheed & Ward, 1974), esp. 20-77. For the various permutations of these issues in Catholic, Reformed, and secular traditions, see Werner G. Jeanrond, *Theological Hermeneutics: Development and Significance* (New York: Crossroad, 1991). It is clear that the issues of what counts as *normative* discourse, how it originates, what weight it has in a historical culture, etc., are not yet resolved.

[14] "Only someone completely unacquainted with the language of the natural sciences could believe that it contains no metaphors at all. . . ." Janet Soskice, *Metaphor and Religious Language* (Oxford: Clarendon Press, 1985), 99.

in the initial phases that functions as a narrative node—a trigger or switch, if you will, that generates the development of the tale. In a coherent story, the plot complication and the narrative evolution lead to an appropriate conclusion.[15]

Eighteenth century metaphors for time focused upon the clock; nineteenth century metaphors thought of the universe as an engine running down. Twentieth century metaphors take as the beginning of the standard model the furnace, explosive devices, or more recently decompression and ice (see below). The issues inherent in current images are first surplus, then imbalance, and ultimately open-ended directionality.[16] The possibility of a temporal orientation of the universe that appears in the current metaphors, however, does not always cohere well with contemporary views on time in mathematical physics.

Efforts have been made to think through the specifics of time asymmetry, though some would argue that it has "never been a well-defined subject."[17] With Paul Davies, we can define time asymmetry in physics as "the basic fact of nature that the contents of the world possess a structural distinction between past and future facing orientations." Note, however, it is the *contents of the world*, the "collective quality of complex systems" that have asymmetry, not *time* itself. For Davies, to think otherwise would be to turn an explanation of time into part of the data. Though it is common to distinguish among different dimensions of time, to recognize that human psychological temporality and cosmological time are not identical, the active character of human temporality is often projected upon the passive spacetime of physics.[18] The asymmetry of macrosystems is due to the fact that "all systems left to themselves (isolated) tend to approach thermal equilibrium and not to leave it again."[19] The *origin* of this entropic macroscopic asymmetry is problematic, perhaps due to the *mismatch* between "the boundary conditions on the global dynamical motion as determined

---

[15] For a recent investigation and categorization, see Didier Coste, *Narrative as Communication* (Minneapolis: University of Minnesota Press, 1989).

[16] The notion of *surplus* denotes something over and above a simple sum of two or more items. To begin with, one could take the common sense economic meaning in the dictionary: an excess of receipts over disbursements. It is the *extra* that remains or appears beyond what seems to be needed. Although mathematics may continue to determine precise meanings for this *surplus*, in this essay, it is the meanings in the metaphors of science and religion that I will analyze.

[17] Paul Davies, *The Physics of Time Asymmetry* (Berkeley: University of California Press, 1974), 1.

[18] Ibid., 3. I note that there are philosophical issues that underlie Davies' remarks that human psychology is acting upon an objectively inert world.

[19] Ibid., 30. The discussion of far-from-equilibrium systems, though crucial, must be left to later work. See Ilya Prigogine, *From Being to Becoming: Time and Complexity in the Physical Sciences* (New York: W.H. Freeman & Company, 1980), 131-150; and Prigogine and Isabelle Stengers, *Order Out of Chaos: Man's New Dialogue with Nature* (Toronto: Bantam, 1988), 177-209. Some of Prigogine's "story" of the universe will be included later in this essay.

by gravity, and the microscopic particle motions of the cosmological material," the "large and small scale motions of the universe."[20]

Language by physicists about the universe as a whole is highly metaphoric. In fact, the possibility of speaking about the universe as an entire unit is itself problematic, even metaphoric, in physics.[21] Physical discussions of beginnings and endings give the cosmos a narrative structure with its own implicit plot, agents, complication, and denouement. Physics and astrophysics, despite their explanatory mathematics, emerge from a rhetoric and use metaphor-driven models to communicate their insights to appropriate audiences.[22] The plot of the standard model about the universe begins in the Big Bang, complicates itself by what causes its expansion (e.g., the above *mismatch*), and, depending upon the complication, concludes in greater order or in utterly random disorder—a Big Crunch or Heat Death. Not surprisingly, it has been easier to tell nature's story in its past from the standpoint of the present than to predict the future.[23] In what follows, I shall examine the rhetoric of cosmological physics to see where it leads. The images of an originating and originated universe have assumed a narrative structure that in itself is metaphoric.

---

[20] Ibid., 7, 197-200. Note that J. T. Fraser argues that the various levels of time (cosmological, biological, psychological) nest inside each other in a hierarchical fashion that leads *narratively* from one to another. He argues that "time itself has evolved along a path corresponding to the evolutionary complexification of matter." J. T. Fraser, *The Genesis and Evolution of Time: A Critique of Interpretation in Physics* (Amherst, MA: University of Massachusetts Press, 1982), 35.

[21] See Chris J. Isham, "Creation as a Quantum Process," in *Physics, Philosophy and Theology: A Common Quest for Understanding*, ed. Robert John Russell, William R. Stoeger, and George V. Coyne (Vatican City State: Vatican Observatory, 1990), 383. In conversation, Isham has pointed out that the term *origination* would be preferable to creation in scientific speech.

[22] This is, of course, a controverted opinion on the nature of mathematical models; on the relationship of such models to the rhetoric of ordinary discourse. (One must recognize that all distensions of speech, whether for the purposes of clear and distinct language [as in science] or for creative extension [as in poetry], move from the rhetorical mid-point of ordinary language. These remarks have their origin in Maurice Merleau-Ponty's interpretation of language; for a systematic overview, see H. L. Dreyfus and S. J. Todes, "The Three Worlds of Merleau-Ponty," *Philosophy and Phenomenological Research* XXII (1961-62): 559-565. For remarks about the atemporality or narrative reversibility of quantum mathematics and the phenomenological appearance of temporal sequence and narrative, see Barbour, *Myths, Models, & Paradigms*, 30; and Wittgenstein's discussion above in n. 3. Ricoeur's remarks concerning Augustine's *Confessions* are useful here since he argues for the intrinsic metaphoric reflexivity of human time. See Ricoeur, *Time and Narrative*, vol. 1, 7-30. See sections 4 and 5 below for a discussion of the operation and referentiality of metaphor.

[23] See Steven Weinberg, *The First Three Minutes: A Modern View of the Origin of the Universe* (New York: Basic Books, 1988); and Penrose, *The Emperor's New Mind*, 322-345.

The inception of the narrative is described as a fireball, blasting asunder any known connectives.[24] This "searing violence"[25] bursts and swells into a constantly expanding, perhaps inflating, universe. The force of the violence impels galaxies to "rush away" from one another.[26] There is a veritable "swarm" of elements. The elemental bits have been "thrown apart." At the "origins," the language of paradox dominates discourse. In the beginning, ordinary mathematical formulae, according to many, do not cohere; they only appear or evolve as initial chaos gives way to the incipient regularities of spacetime. Mathematics and the laws of nature evolve within history (perhaps even *make* history?). Under such "unusual and improbable" conditions,[27] little can be described, let alone explained. At best, we might say that it was an "undifferentiated soup."[28]

A complication in the plot is required to explain the present—whether it is the predominance of matter over anti-matter or the presence of observers.[29] The "excess" of matter,[30] the "abundance" of helium,[31] the "graininess" of the primordial soup,[32] and "statistical" imbalances[33] invite scientists to investigate the critical density of matter and the critical rate of expansion.[34]

It is the unexplained "gap"[35] in the primal chaos, the "bubble" in the soup that generates distance and asymmetry between "one" and an "other." It is a "surplus" that establishes the causal network of before and after.[36] Thus the

---

[24] Stephen Hawking, *A Brief History of Time: From the Big Bang to Black Holes* (New York: Bantam, 1988), 35-51; Richard Morris, *Time's Arrows: Scientific Attitudes Toward Time* (New York: Simon & Schuster, 1986), 186; Roger S. Jones, *Physics as Metaphor* (Minneapolis: University of Minnesota Press, 1982), 99ff; Paul Davies, *Superforce: The Search for a Grand Unified Theory of Nature* (New York: Simon & Schuster, 1984), 183; and Weinberg, *The First Three Minutes*, 12.

[25] Davies, *God and the New Physics* (New York: Simon & Schuster, 1983), 20.

[26] Weinberg, *The First Three Minutes*, 21.

[27] John D. Barrow and Frank J. Tipler, *The Anthropic Cosmological Principle* (Oxford: Oxford University Press, 1988), 411.

[28] Weinberg, *The First Three Minutes*, 102.

[29] Hawking, *A Brief History of Time*, 76; Barrow and Tipler, *The Anthropic Cosmological Principle*, 259; and Morris, *Time's Arrows*, 173. "But something in this primordial state, an asymmetry not yet recognized or perhaps pure chance, has led to the initial predominance of matter. Once assured, antimatter became metastable, if for no other reason than for the presence of an overwhelming amount of matter." (in Fraser, *The Genesis and Evolution of Time*, 125)

[30] Barrow and Tipler, *The Anthropic Cosmological Principle*, 377; Davies, *God and the New Physics*, 30.

[31] Davies, *God and the New Physics*, 22.

[32] Barrow and Tipler, *The Anthropic Principle*, 415.

[33] Ibid., 370.

[34] Hawking, *A Brief History of Time*, 76, 121; Morris, *Time's Arrow*, 173.

[35] Davies, *God and the New Physics*, 51.

[36] Hans Reichenbach, *The Direction of Time*, ed. Maria Reichenbach (Berkeley: University of California Press, 1956), 38-39.

"lopsidedness" of unbalanced matter residue[37] creates a "dis-equilibrium"[38] which establishes the temporal asymmetry of the cosmos, whether it is seen to be "running down" due to entropy[39] or generating order.[40] However this "event" is explained, whether through fluctuations, wave disturbances, etc., it is the semantic overlap in the metaphors of surplus of "getting more than you bargain for" in non-linear expansion[41] that establishes a cosmic sequence we call a story.

Recently, Ilya Prigogine and Isabelle Stengers have told a somewhat different version of the story of surplus, using the same metaphors of excess to describe an irreversible power bringing order out of chaos at all levels.[42] David Layzer has maintained that the disequilibrium that generates time asymmetry is more like a gentle decompression than an explosive fireball, a cosmos that emerges from "critical-point opalescence" like a gas.[43] This occurs through gravitational clustering, a "genuinely historical process."[44] For Layzer, this explains the "clumpiness" of the cosmic medium and the eventual emergence of autonomous self-gravitating clusters.[45] This expansion of space is "responsible for chemical and structural order."[46]

The final act of the drama thus alternates, depending upon the interpretation of *what is generated* by the excess, between an infinite dead equilibrium,[47] a contracting gravitational compression, or infinitely expansive order.[48]

The emphasis that I have placed upon the appearance of a universal narrative out of an originating singularity is not, however, the only interpretation of the excess. Other readings maintain random simultaneity or non-narrative sequences. In the version I have stressed, the excess becomes the "operator" that generates the sequence we call the universe. The cosmic "arrow of time," the temporality of the universe—that the cosmos has an asymmetry between past and future—appears in the metaphors of surplus. In one denouement for this version, the surplus leads through entropy to heat death or cold soup; or in

---

[37] Davies, *God and the New Physics*, 30.

[38] Prigogine and Stengers, *Order Out of Chaos*, 46.

[39] Don N. Page, "Hawking's Wave Function for the Universe," in *Quantum Concepts in Space and Time*, ed. R. Penrose and C. J. Isham (Oxford: Clarendon Press, 1986), 275.

[40] David Layzer, *Cosmogenesis: The Growth of Order in the Universe* (Oxford: Clarendon Press, 1990), 133-170; Prigogine, *From Being to Becoming*, 46.

[41] Peter Coveney and Roger Highfield, *The Arrow of Time: A Voyage Through Science to Solve Time's Greatest Mystery* (New York: Fawcett, 1990), 184.

[42] Prigogine and Stengers, *Order Out of Chaos*, 192. Note that this is dependent upon their understanding of far-from-equilibrium systems.

[43] Layzer, *Cosmogenesis*, 144, 159.

[44] Ibid., 164.

[45] Ibid., 163.

[46] Ibid., 170.

[47] Davies, *God and the New Physics*, 199-213; Reichenbach, *The Direction of Time*, 3.

[48] Prigogine and Stengers, *Order Out of Chaos*, 177-209.

another conclusion, the universe develops into patterns of order the farther it moves away from equilibrium. In an alternate non-narrative interpretation, however, there is a focus upon randomness and the paradox of gaps that seems to preclude directionality altogether. Not only are there "simultaneous" worlds due to problems of observation,[49] but surplus can lead to multiple, unrelated universes.[50] We could have the development of cyclical world generation, based upon expansion and contraction of spacetime or "parent-offspring" universes in which structures emerge from one another out of an inflating bubble of spacetime. Here, "'our' universe is only part of an infinite assemblage of universes, although it is self-contained now."[51] Another possibility is that the three spatial dimensions of our universe are simply those that observers in this world are able to experience.[52] The complexities of alternate interpretations to that of a narratively sequential universe focus upon sheer multiple possibilities.

4 *The Nature of Metaphors and Time Asymmetry*

Having observed the kinds of language that scientists use to describe the origins of the universe, the argument now shifts to the language of excess as it occurs in the hermeneutics of metaphor. It is my contention that the sheer form of metaphor chosen by scientists to identify the cosmic origins generates the notions about the asymmetry of the temporal process. The complex of issues surrounding time is intrinsic to the nature of metaphor itself. There is, therefore, a curious circularity about the discussion of the asymmetry of time in physics. The use of metaphors about the origins of the universe is not simply a device to communicate something about a temporality that has been discovered or explained through prior mathematical probabilities. Metaphoric process and the excess it delivers promote the discussion of temporality.

To understand how physicists are users of metaphor and how those metaphors affect their performance as scientists, it will be necessary to outline current theories about the metaphorizing process. Rather than a complete dialectic of historical positions, I shall outline a typology, using four schemes: (4.1) metaphor as ornament or emotional outlet; (4.2) metaphor and Mary Hesse's analogy; (4.3) Ricoeur's notion of metaphor as confrontational paradox; and (4.4) my own modification of Ricoeur's and Lonergan's understanding on metaphor.

---

[49] Barrow, *The World Within the World* (Oxford: Oxford University Press, 1990), 151-157.

[50] Penrose, *The Emperor's New Mind*, 295-96.

[51] Davies, *The Mind of God: The Scientific Basis for a Rational World* (New York: Simon & Schuster, 1992), 70-72.

[52] Barrow and Tipler, *The Anthropic Cosmological Principle*, 273-275.

4.1 *Metaphors as Ornament*

Scientists, no less nor more than philosophers, can dismiss metaphors as a primarily decorative value in discourse. Reichenbach, for example, maintains that images are a sure sign of "emotional dis-satisfaction." Scientists "make use of metaphors invented to appease the desire to escape the flow of time and to allay the fear of death. They cannot be brought into a logically consistent form."[53] According to Abner Shimony, metaphors more often lead us astray, since the quantized and/or mathematical world is fundamentally unimaginable.[54]

These views fit into an intellectual tradition that originates in theological suspicions about imagination dating at least from Augustine. More positive notions appear in the renaissance and in the early modern period with Ignatius of Loyola, but more suspicious attitudes return in continental philosophy with Kant. The classicist view of the imaginative trope is that it fundamentally produces no new knowledge; its function is decorative and emotional, pleasurable rather than instructive. Were one able to substitute the proper term for the metaphor, one would have plain speech, but true speech. It is this view of imagination that placed art in a marginal position in western societies and made artists alternately messiahs and pariahs in an avant-garde.[55] If Janet Soskice is correct, these attitudes emerged at the same moment that there developed the ideologies of scientific empiricism.[56] It is little wonder that modern and contemporary science hoped to excise imaginative devices from its discourse—especially in language about the cosmos.

4.2 *Analogy*

Hesse initially distinguished analogies from metaphors in her writing, as did medieval philosophers;[57] and the basis for the distinction seem to locate her in the first camp (ornament) mentioned above. To speak *metaphorically* of an angry sky is not to assume that there is any discernible identity between the human subject and the heavens; but when speaking of an angry dog, one can estimate *analogically* a definable similarity between canine and human anger that can become the basis of predictions for future activity on the part of both the human and the animal. In the biological sciences she argues that mathematical understandings of nature themselves are analogical.

---

[53] Reichenbach, *The Direction of Time*, 5.

[54] Abner Shimony, "Events and Processes in the Quantum World," in *Quantum Concepts of Space and Time*, ed. Penrose and Isham, 201-202.

[55] Paul Ricoeur, *The Rule of Metaphor: Multidisciplinary Studies of the Creation of Meaning in Language*, trans. Robert Czerny, et al. (Toronto: University of Toronto Press, 1977), 9-64.

[56] Janet Soskice, *Metaphor and Religious Language* (Oxford: Clarendon Press, 1989), 67ff.

[57] Mary Hesse, *Science and the Human Imagination: Aspects of the History and Logic of Physical Science* (London: SCM Press, 1954), 144-45.

In later work, she does not distinguish so strongly the exploratory process operative in metaphors from analogies. Analogies have heuristic value, not reducible to modern logic's distinction between identity and difference. The internal similarity between an analogy and its comparison is that it can be made the basis of predictions. Such analogies guide by their "surplus meaning."[58] The model or *Vorstellung* (in Hegel's sense of representation) provides ways of looking at the consistency of the measured process and the possibility of extending the theory to other data or situations.[59] And, if I understand her early thought correctly, this is possible because of a pre-predicative unity in difference (a "direct, non-analogical relation") at the level of the world as it is. The observer and the observed share a common world prior to their differentiation.[60]

Hesse notes Black's interactive theory of metaphor[61] and connects it to her prior interpretation of the logic of analogy. If metaphors are not able to be reduced to a single, literal proper meaning, if they gain their meaning from their interaction such that each term modifies the other, then perhaps metaphors and analogies are primary speech, and univocal formulae are rather a highly specialized language. Both sides of the metaphorical equation are affected by juxtaposition of terms; the meaning emerges in their interaction not as a substitution for some priorly determined concept or as a conclusion from antecedently held premises.[62]

"What is disclosed" by this process (the reference) is a "redescription of the domain of the explanandum."[63] "The referent seems to be the primary system, which we choose to describe in metaphoric rather than literal terms."[64] The domain of the explanandum is now described in terms of the analogue: as a result (although it is unclear how), the meaning of both is affected. The original observation language can be extended and shifted so that "predictions in the strong sense will become possible."[65]

Hesse's, Soskice's, and Gerhart and Russell's interpretations of metaphor argue that the surplus of meaning has a referential status in scientific (and other) discussions.[66] It tells us something about the world. If we use the

---

[58] Ibid., 141.

[59] Hesse, *Models and Analogies in Science* (South Bend, IN: University of Notre Dame Press, 1966), 19.

[60] The logic of this process is later developed in Hesse, *The Structure of Scientific Inference* (London: Macmillan, 1974), esp. 197-222.

[61] Hesse, *Models and Analogies*, 152.

[62] Ibid., 154.

[63] Ibid. In this sentence and the following, Hesse seems to step back from the mutually interactive form of metaphor that Black (and I) would espouse. It is not clear that there is a *primary* and a *secondary* system at stake, unless a specific *prior* question is engaged. The metaphor itself, though it may begin temporally with the subject, modifies both subject and predicate by its juxtaposition.

[64] Ibid., 166.

[65] Ibid., 176.

[66] Mary Gerhart and Allan Russell, *The Metaphoric Process: The Creation of Scientific and Religious Understanding* (Fort Worth, TX: TCU Press, 1984). It will be

notion of a Bang or a gentle decompression as metaphors for the origins of the universe, we are attempting, not to express our feelings about the universe, not to communicate what we already know otherwise, but to indicate a state of affairs that we discover through the metaphor.

4.3 *Metaphor as Confrontational Paradox*

Ricoeur's hermeneutical phenomenology, although more welcoming to contemporary scientific explanation than either Heidegger's or Gadamer's, focuses upon a-logical dimensions of discourse.[67] His work on images and their function in discourse began with the comparative study of symbols for evil.[68] Later he distinguished symbols and metaphors and studied the structure, process, and referential claims of metaphors.[69] Part of the difference between symbols and metaphors is that symbols are not totally transparent to speech; they have their roots in the *bios*, the hurly-burly of the lifeworld of desire through which they gain their meaning. As liminal experiences, symbols negotiate between desire with its biological force and culture. Symbols always contain, as a result, a knot of experience that is indecipherable. Metaphors, on the other hand, are primarily semantic, a creation of *logos*, the "free invention of discourse," containing a "surplus of meaning."[70]

Making use of Black's interactive theory of metaphor, Ricoeur argues that the juxtaposed terms of a metaphor produce the reference. They refer by indirection, not as scientific, univocal languages (mathematics, symbolic logic) do. Ordinary referentiality is suspended, abolished, exploded; and a deeper paradoxical reference emerges.[71] So to speak of an angry sky (or an angry dog for that matter) is a metaphor in which the two juxtaposed terms (the emotion of anger and the rainy, thundering sky) construct no logical, coherent sentence. It attributes to an inanimate, material object dimensions of human feeling. For

---

clear by the conclusion of this section of the article that I am convinced that metaphors in use have referential power and contribute to our knowledge of the world. In this, the moderate realism of Hesse, Lonergan, and Soskice is the basis for my analysis of the temporal character of metaphors in section 4.4. Each of the thinkers involved has distinct, if interrelated, theories of the process through which metaphors refer. Soskice stresses the social and casual contexts of language that corroborate experience; Hesse argues the exploratory character of analogical knowledge; Gerhart and McFague depend perhaps too much upon the position of Ricoeur with which I will disagree in section 4.3. The process I will explain in section 4.4 should show philosophically *why* metaphors guide and direct knowledge.

[67] Paul Ricoeur, *Interpretation Theory: Discourse and the Surplus of Meaning* (Waco, TX: TCU Press, 1976), 45-69.

[68] Ricoeur, *The Symbolism of Evil*, trans. Emerson Buchanan (Boston: Beacon Press, 1967), esp. 3-24, 347-67.

[69] Ricoeur, *Interpretation Theory*, 57, 59, 61; and idem, *The Rule of Metaphor*, 51.

[70] Ricoeur, *Interpretation Theory*, 59, 64.

[71] Ricoeur, *Rule of Metaphor*, 245-55.

Ricoeur, the absurdity of this interaction implodes and discloses some deeper reference that affects both human feelings and the nature of the skies. If it is a living metaphor, it will translate us (by our choice) into a world in which we know the inner human world as roiling as a thunderstorm and the outer atmosphere as churned, upset humanity. Metaphors are deliberate category mistakes, disclosing a *virtual, possible* world which the reader must appropriate to bring into reality. Metaphors challenge the reader to appropriate the possibility. They invite existential witness.

Ricoeur's intellectual project for many years has been to establish the possibilities for human freedom in a finite, conditioned world. The surplus meaning that metaphors open is the space for human choice. One can choose to act "as if" the skies were angry. Metaphoric language is the evidence for creativity and the way in which the desire for freedom exceeds its exercise.[72] The vehicle that readers must choose to transport themselves into the world redescribed by the living metaphor is "feeling" (i.e., *le sentiment*).[73] The vehicle is driven or "operated" by existential choice, the witness to what one experiences as possible through the metaphor. The ordinary "logical and established frontiers of language" are obliterated so that readers are continually disoriented. The surplus meaning in metaphors (the excess of paradox) is the gap in which human freedom can operate.[74] Assuming responsibility for the path opened by metaphors, readers can become what they read. But note that readers are constantly challenged by the paradox of the metaphor; what makes the metaphor alive is this confrontational mode. Continually disoriented by the non-accustomed usage (the universe, a Big Bang; Harold, the duck-billed platypus; God, the unjust judge), we must choose to live in a new world offered by the metaphors. Indeed, we will not know the world until and unless we choose to enter it.

I find the previous approaches to the surplus of metaphor unsatisfactory with regard to reference. While Ricoeur's is clearly the most elaborate and successful proposal, problems become particularly evident when his notions are

---

[72] Ricoeur, "Metaphor and the Central Problem of Hermeneutics," in *Hermeneutics and the Human Sciences: Essays on Language, Action, and Interpretation*, ed. and trans. John B. Thompson (Cambridge: Cambridge University Press, 1981), 165-181.

[73] Ricoeur, *Rule of Metaphor*, 309. It is difficult sometimes to determine exactly *what* dimension of the metaphoric process leads toward reference in Ricoeur. But once the interactive nature of subject and predicate occur, ordinary reference is destroyed; and the only guiding access speakers have to the world is through *le sentiment*.

[74] Note that this is similar, but not identical, to Soskice's remarks about the role of the undefinable or indecipherable aspects of metaphor. Metaphors refer without defining, leaving an openness. But unlike Ricoeur, Soskice is convinced that there is a cognitive *lead* that metaphors offer. In section 4.4, I hope to show some of the mechanisms of that guidance toward the future. See Soskice, *Metaphor and Religious Language*, 125-133.

applied to religious language, since claims for religion include not just a *potential* way of being in the world but an *actualization*—at least in the case of the religious founder (e.g., Jesus of Nazareth).[75] There is an underlying assertion that this is the way the world is, rather than this is the way it might be in a continually deferred future. As Soskice maintains: "Christian theism has been undeniably realist about these models, whether it has a right to be or not."[76] The problem is how to understand the kind of reference metaphors intend (whether they are used in physics or in religion).

4.4 *Images, Explanations, and Reference*

In this section we will examine a version of the metaphoric process and its referential capabilities based on Bernard Lonergan's understanding of cognitional judgment and art. For him, metaphors tell us the way the world is and the way it can be if certain conditions are fulfilled. I will argue that there is a judgment (s is p) implicit in the work of art. The so-called "shock of recognition" that is often mentioned in art criticism is not simply existential confrontation or emotional surprise. It is an agreement that this is the way the world is . . . under certain conditions.

For example, we can examine an apocalyptic set of images (the film *Blade Runner*, the novel *A Canticle for Liebowitz*, or Anselm Kiefer's paintings). Under certain conditions, some of which may be realized, this is the way the world is and will be. The images extend the conditions of the present toward an as yet unrealized future. This is the case not simply for utopial or dystopial art, but also for so-called realisms. Under certain conditions, this is the way the world is and will be. The fulfillment of the "angry sky" of our prior example depends upon the conditions that surround the metaphor—personal or cosmic apocalypse.

Ricoeur is correct about artistic experience in that if human beings choose to live certain choices, the world will be the way the metaphor potentially describe the situation. He is incorrect in that he sees no clear rational relation between the "is" of the metaphoric interactions and the "is not yet" of the future whose conditions have yet to be fulfilled. In my reading of Ricoeur, it seems that the reader or speaker of metaphors simply leaps beyond the absurdity or paradox of the present towards the future without guidance, without direction or conditions. It is a philosophical version of fideism to preserve a form of human freedom. That is no doubt part of the reason that it is highly appealing to some

---

[75] One could trace this discussion of virtual worlds as evidenced in Sallie McFague, *Models of God*, 192, n. 37, where language about God "projects a possibility"; or in Tracy, *Analogical Imagination*, 354, and the complex n. 34, p. 362, in which no clear vocabulary emerges for treating the finite actualization of possibility; or in sacramental theology and ritual studies with David N. Power, *Unsearchable Riches: The Symbolic Nature of Liturgy* (New York: Pueblo, 1984), 130-139. In each theological position, the use of Ricoeur's interpretation of metaphor makes the reference of religious language the *existential world* of the believer, actualizing the possibility that metaphors awaken.

[76] Soskice, *Metaphor and Religious Language*, 107.

theologians (e.g., McFague) who emphasize the non-mediation between the divine and the human. For her, there can be no concrete conditions that "constrain" divine freedom.[77]

The relationship therefore between the "is" and the "is not yet" of experiences and judgments of art requires a theoretic vocabulary that will articulate a relationship between present and future that accounts for the continuities as well as the discontinuities, the similarities and the differences. If the conditions in the present are fulfilled, then the sequential entailment follows. Temporal asymmetry *may* be inscribed into the very nature of art as metaphoric. Let us look at Lonergan's understanding of art.

Lonergan's philosophy and theology recognize the importance of images and symbols. Images function in his thought in many ways: 1) as part of the perceptual, sensitive flow of consciousness from which we can never be totally divorced; 2) as factors in discerning some known unknown such as the models and analogies in science; and 3) as abstractly designated signals (as in symbolic logic) constructed to indicate the import of some particular experience or observation. As part of affective life, images function in human self-communication—in dreams, art, and love.[78]

It is in his study of art that Lonergan interprets the role of symbolic discourse, though he applies it to the role of models in science and to the construction and reconstruction of human societies. For Lonergan, symbols and art are more than a *virtual* discourse. For Ricoeur, metaphor is always a deferred reality, making room for human freedom by always proposing possibility. In Lonergan's thought, the artwork draws observers into sharing its vision of the world, sometimes without the participants reflectively knowing what is occurring. Hence the "dangerous" character of images; they "trick" us into entering the world they disclose. We are in a common world they define before we "know" it.

The referential world of metaphors is not univocity, nor is it the paradoxical proclamation of is/is not that Ricoeur maintains. Metaphors tell us the way the world is and the way it can be if certain conditions are fulfilled. The shock of recognition that occurs in a work of art is the agreement by the participant that the conditions for an aesthetic judgment have been fulfilled. This artwork tells the truth. It is a virtually unconditioned judgment in that the interpreter recognizes that unless further conditions appear, all the relevant existential, intellectual, and evaluative questions have been satisfied.

Such an aesthetic judgment is exploratory; it does announce what the world might be like were further appropriate conditions fulfilled. Yet some of those conditions have in fact been fulfilled. The teasing character of art is that it directs the participant toward testing the conditions that are fulfilled and exploring whether others have or have not been completed. The metaphor leads

---

[77] McFague, *Models of God*, 22, 39, 192-93 fn. 37.

[78] In a prior essay, I have elaborated this proposal in relationship to the nature of the Christian sacraments; at the same time, I have criticized other interpretations of metaphor. See Happel, "Worship as a Grammar of Social Transformation," *CTSA Proceedings* (Philadelphia, 1987) 42 (1987): 123-142.

participants through the foreground of a work by means of an intelligent, though non-conceptual affect and sensibility into a horizon that may go beyond the artifact itself. In this sense, a work of art discloses a world that both is and is not yet. As Ernst Bloch has said: "The self-identity of a work of art 'is' not yet manifest."[79] Indeed, part of the truth of any metaphor is its continuing ability to establish a world according to the conditions set forth in the work.[80] When models as metaphors no longer offer guidance toward predictable consequences, they are discarded. In a postmodern world, they become ornament and style.

This asymmetrical pattern of truth-telling in aesthetic judgment is operative in all knowing, except (perhaps) in some univocal forms of logic and mathematics where the conditions for judgment have been severely limited for the sake of precision. The subject both is and is not yet identical with its predicate. Metaphors describe a real world which both is and will be under specific conditions. If the conditions are fulfilled (*one* of which may be the choosing subject), then the world as redescribed by the metaphor exists in a (seeming) identical relationship with its participants. Classic, living metaphors continue to reinvent the world, redescribing, and predicting new consequences. The future can be different from the past, but on the basis of conditions in the present. Notice that by this standard the asymmetry of time is built into the copula of all ordinary sentences.

Hesse and Lonergan, as moderate realists, see their use of metaphor and analogy as delivering not just an external observer's opinion, but a view, however partial, of the way the world is and will be. This view of the world can be judged, tested, reexamined, and overturned or confirmed. Analogies and metaphors have predictive capabilities; under certain conditions, such and such is or will be the case. Ricoeur waffles between a referentiality comprising the existential witness of the observer-participant who actualizes a potential, virtual world and the avowal that this existential actualization can never quite take place. For Ricoeur, a living metaphor is always overturning the state of affairs that the

---

[79] Ernst Bloch, *A Philosophy of the Future*, trans. John Cumming (New York: Herder, 1970), 96.

[80] Gadamer, *Truth and Method*, 285-90. Nancey Murphy remarks that Karl Popper states that all predicates are predictive ("This is a cup"), since they maintain that something is the case about the world, open to falsification if they do not continue to act in the same manner. See Popper, *The Logic of Scientific Discovery* (New York: Basic Books, 1961), 32-33, 313-314; and *idem*, *Objective Knowledge: An Evolutionary Approach* (Oxford: Clarendon Press, 1972), 352-357; and *Realism and the Aim of Science*, ed. W. W. Bartley, II (London: Hutchinson, 1983), 206-209, 211-214. I would note that, although the metaphoric process has a hypothetical and heuristic nature (predictive in its own way), it also contains an *implicit judgment* about the way the world is. This judgment continues to be true if the conditions remain fulfilled. Though there is some similarity between what I am arguing and what Popper maintains, I am making a stronger historical and (simultaneously) ontological claim than I understand him to be arguing. It is not simply a matter of whether the projected state of affairs continues to behave in the same fashion. Popper's point is methodological; mine is initially epistemological, then ontological.

observer is trying to actualize. For the earlier and now largely discredited ornamentalists, metaphors reflect human feelings toward the inexorable, entropic fatedness of the universe or its utopially expanding order.

5 *Metaphoric Reference: Emergent Probability or The Deconstruction of Time*

In previous sections, I have noted that scientists "metaphorize" to articulate their understandings of the universe and that metaphors operate in a particular way that is related to temporal directionality. Now I will outline the generic qualities of the worlds to which metaphors point. The first is a historical ontology; the second a radical ahistorical process.

The answers to the question: "To what do metaphors refer?" are important for both science and religion. The reference is, however, shaped by the metaphoric process itself. Hence Ricoeur's position shapes reference by its emphasis upon paradox. The position I have taken in section 4.4 argues that the metaphoric process has an intrinsic temporal asymmetry. This suggests a referentiality that focuses upon narrative. I have asserted that metaphors operate in this fashion in science or in any other discipline (including theology).

Now it is also clear from the descriptive catalogue in Section 2 that not all physicists and cosmologists see the *surplus* in the Big Bang as generating temporality. I would argue that this is due less to physics or pure mathematics that to particular ways of "holding metaphors" in language. Those scientists who use metaphoric surplus to generate atemporality or multiple worlds hypotheses are linked to a theory of metaphor that maintains their paradoxicality. Those who claim that the cosmic big bang generates a narrative to the universe are indebted in a theory of metaphor that supports a moderately realist historical ontology.[81]

In this section, we will look at the following dimensions of the problem: 5.1 a historical issue—the nature of temporality and art in Kant; 5.2 the referentiality of narrative time; and 5.3 the alternative reference of temporal deconstruction.

Two major explanatory languages discuss the referential surplus of metaphors. The first articulates a set of terms and relations that sort out the emergence of the future from the past under specified conditions. The second maintains that there is no direction, no emergence. There is only differentiating seriality without narrative—except in so far as human beings impose order,

---

[81] To be able to "prove" in science that such metaphors generate predictions or conclusions of time asymmetry would require a search for particular examples in which this has been the case. One would develop experiments that might permit the appearance of temporal asymmetry in the (usually understood) temporal symmetry of the micro-world. One would need to study other solutions to the temporal reversibility of the formulae and "realities" of the quantum world. This would permit linking the signals of the micro-world to the macro-world of entropic time, free energy, and the growing organization and self-organization in biology and human history. Such an investigative program suggests links between analysts of language, method, and experimental scientists.

necessarily oppressive hierarchies of before and after, prior and posterior, superior and inferior. Here the surplus of metaphors discloses not history, but the ambiguous interplay of traces. Lonergan's views on emergent probability coordinate his understanding of cosmic and human time. Deconstructionist attitudes toward time, especially in Jacques Derrida, carry the confrontational and paradoxical role of metaphor to its limit.

5.1 *Kant's "Critique of Judgment"*

It is no accident that Kant discusses the notions of teleology and purposefulness in the context of aesthetics. Kant's correlation of art, purpose, and teleology shows how the philosophical tradition has already linked freedom, metaphor, and directionality in nature. Kant hoped to provide an overarching intellectual unity to theory and practice (his two earlier critiques). This required integrating the examination of the conditions for the laws of nature (which involve a theory of pure reason) and the conditions of particular empirical laws (which involved practical reason). The first conditions are not only the foundations for the necessary truth of Newtonian physics (if it is to reach any level of certainty) but also the basis for all conceptual univocity. Experience (and its basis in pure reason) and practice (practical reason) have a different kind of necessity. The "as if" necessity of practical reason (one must act as if one is free, without absolute speculative justification) "gives a law only to itself, and not to nature."[82] In other words, one cannot postulate either intention or concrete temporality to the processes of nature. But by studying the work of art as an instance, one can learn something about how purposefulness operates in a constructed object, and perhaps understand what it means to call nature beautiful.[83] Investigating the ways in which unity, necessity, and universality function in regard to the truth of the particular and practical, Kant turns to art as a primary instance of a particular object taken to have universal significance. Art opens up a world, but what world?

Subjective judgment attributes to nature a "principle of purposiveness."[84] The cosmos does not have a (speculatively) provable goal-directedness; hence, it becomes imperative to study human making which does seem to have a purpose or goal. Since "we cannot ascribe to natural products anything like . . . purposes," we must study purposiveness as attributed to an constructed object—works of art. By prescribing to its own cognitional faculties a "law for its reflection upon nature," human knowing gains a certain pleasure in

---

[82] Immanuel Kant, *Critique of Judgment*, trans. J. H. Bernard (New York: Hafner, 1968), 17.

[83] In what follows "purposefulness" in Kant must be understood to include both willed intention and teleology. If one understands the goal-directedness of an art object, can one then understand and ascribe a *telos* to nature with any legitimacy? Kant, of course, eventually describes nature as sublime, not beautiful. It *escapes*, as a thing in itself, separated from our intelligence, our ability to know if it is has an intention or a goal.

[84] Kant, *Critique of Judgment*, 20.

bringing order to the variety in things.[85] The aesthetic representation of this ordering brings disinterested, that is, non-possessive, pleasure to the observer. For Kant, it is the subjective agreement of sensibility, imagination, and understanding in a single moment that brings the satisfaction of aesthetic judgment.

The purposiveness of judgment can appear as the harmony of the form of an object with the cognitive faculties or the harmony of the form with the (ideal) possibility of the thing in itself.[86] But purpose is not directly attributable to nature (any more than to a work of art). Aesthetic judgment "alone" provides a "principle which the judgment places quite a priori at the basis of its reflection upon nature, viz., the principle of a formal purposiveness of nature . . . without which the understanding could not find itself in nature."[87] The critique of aesthetic judgment therefore becomes a condition for examining the truth of the laws of nature. Aesthetics provides the "analogy with the causality of purpose, without any pretense" to explain nature in itself.[88]

Metaphors, therefore, play an important role in the discussion of nature's direction and temporality. Kant sees the function of the attribution of purpose to nature as "enlarging for the mind."[89] By the elegance of explanation, cognition is "strengthened."[90] The possibility of a self-organizing purpose in nature is a regulative idea,[91] a heuristic ideal that guides human investigation (ultimately attributable only to reasoning humanity itself as the final purpose of nature). The emergence of this free space is precisely for Kant where human responsibility for the universe appears. Since the intrinsic purpose of nature cannot be determined except as a subjective ideal, it is the moral responsibility of human beings to supply "that in which natural knowledge is deficient."[92]

In the history of modern philosophy, therefore, the topics of temporality, metaphors, and human choice have already been interrelated. The referential nature of art for Kant involves attributing to things a certain purposiveness, a subjective attribution of direction to art and by extension to nature. Where Kant finds in aesthetic and teleological judgment the shock of recognition that awakens and broadens cognitive activity and Ricoeur focuses the reality of human freedom, Hesse, Lonergan, and Soskice maintain that a partial, conditioned, though truthful judgment about the world can be made. The conditioned realism of these latter figures indicates that reference is not merely directed to the free human subject nor to subjectively attributed purpose, but to a temporal (cosmic, historical) movement of which human beings are a part. For

---

[85] Ibid., 23.
[86] Ibid., 29.
[87] Ibid., 30.
[88] Ibid., 206.
[89] Ibid., 211. Note that the subject alone determines the purpose in nature. Just as Kant places the "enlargement" of art in the mind of the viewer, so Ricoeur places the "open space" of metaphor in the freedom of the speaker.
[90] Ibid., 212.
[91] Ibid., 217-18.
[92] Ibid., 298.

Lonergan the term to understand this temporal process is called "emergent probability." It provides the basic terms and relations for understanding the temporal asymmetry of the universe.

5.2 *Emergent Probability*

Emergent probability is the term Lonergan uses to discuss "the combination of the conditioned series of schemes [of recurrence] with their respective probabilities of emergence and survival."[93] It is not progress in the nineteenth century Whiggish sense of the term, nor is it a necessary, antecedently determined scroll from which the universe unrolls. "The intelligibility immanent in Space and in Time is identical with the intelligibility reached by physicists investigating objects as involved in spatial and temporal relations."[94] This means that the emergent probability of the universe is a matter of the cooperative intellectual activity of individuals inquiring, imagining, understanding, judging, and deciding for values in the course of history. It is simply the struggle for and results of intelligent inquiry about the universe. In so far as it is temporal, it is the "ordered totality of concrete durations."[95] It hopes to work out the statistical probabilities of emergence and survival for conditioned series to see if and how there is an intelligibility to our experience. "To work out the answers pertains to the natural sciences."[96]

To recite terms and relations in such a global investigation is incantation without explanation; but it may be helpful to outline the process. The general categories for such investigations include the nature of events or occurrences; the spatio-temporal manifold of events; schemes of recurrence for events; the interlocking conditions under which schemes recur; probabilities, non-systematic divergence (chance), and actual frequencies. Though Lonergan assumes these categories are heuristic, to be filled by the data from the various sciences, the inquiry operates under various hypotheses. (i) All world situations are not simultaneous; there is a succession and the world process is open, a succession of probable realizations of possibilities. (ii) Nonetheless, the world process becomes increasingly systematic since later schemes are conditioned by earlier ones. Sufficiently long periods of time and sufficiently high populations

---

[93] Bernard Lonergan, *Insight: A Study of Human Understanding* (New York: Longmans, 1967), 122; Kenneth R. Melchin, *History, Ethics and Emergent Probability: Ethics, Society, and History in the Work of Bernard Lonergan* (Lanham, MD: University Press of America, 1987), 97-121.

[94] Lonergan, *Insight*, 151.

[95] Ibid., 157.

[96] Ibid., 124. Though in this essay, we are interested in the notion of emergent probability as it applies to cosmology, Lonergan's theory is an explanatory framework for all history as it emerges from the Big Bang through the evolution of matter to the emergence of self-conscious life. In this presentation, I will sketch the entire process as he understands it to provide a basis for arguing that this historical ontology, open-ended and dynamic, is the appropriate reference world for metaphor as narrative.

will assure the continuing emergence of certain schemes. (iii) The determination of the initial, basic world situation (boundary conditions) must contain elementary schemes in sufficient numbers to "sustain the subsequent structure." This process admits enormous differentiation, breakdowns in the probability of survival, and blind alleys. (iv) The intelligibility of the world process as emergent probability involves the complementarily of classical and statistical laws. (v) This intelligibility also leaves room for the appearance of human freedom as a condition for the emergence of later schemes of recurrence.

In the interpretation of emergent probability as the explanatory term for temporality, Lonergan offers an ontological reference for the metaphors of the asymmetry of time. The interlocking levels of the microworld provide the initial patterns and conditions for the later appearance of macroworld differentiations. The temporality appropriate to physics and chemistry provides the matrix from which biological and human time emerge. Time is neither a uniform clock imposed upon all objects, nor a sheer occasional contiguity without connection. Rather the metaphors for asymmetry refer to time as emergent probability.

"Concrete extensions and concrete durations are the field or matter or potency in which emergent probability is the imminent form or intelligibility."[97] Emergent probability, therefore, must be to some extent descriptive since it involves the success of probable schemes of recurrence whose outcome one does not yet know. Moreover, though measurement can move the inquirer from description (metaphors) to explanations (formulae, theories of terms and relations), inquiry itself participates in the process of emergence itself.[98] It assumes a determinate, unidirectional ability. Measurement is not temporally neutral. Its intervention becomes part of the conditioned schemes of recurrence, including the microworld. The purposiveness of the universe, therefore, is open, heuristic, dependent upon the concrete recurrence of some schemes of recurrence. To affirm finality is "to deny that this universe is inert, static, finished, complete. It is to affirm movement, fluidity, tension, approximativeness, incompleteness."[99] To affirm this finality is not to offer any opinion on the ultimate fate of the universe, but it is to maintain that there is a directed dynamism in the universe's operation. This process is directed to

---

[97] Ibid., 172. Though Lonergan in other places argues for the complementarity of classical and statistical laws, here time is clearly in the realm of statistical probabilities. Neither purely random, nor absolutely necessary, time is the descriptive term we use for the emerging schemes of recurrence and the conditions under which they appear. "Emergent probability is the successive realization of the possibilities of concrete situations in accord with their probabilities." See ibid., 170; also 105-115.

[98] Ibid., 445. In other words, measurement (a macroscopic entropic process) affects the concrete succession of durations and extensions.

[99] Ibid., 446. Note that the cosmos itself has temporality, but that this is *not* unrelated to the successive levels of temporality that ultimately include human time. Nonetheless, the conditions for the emergence of the schematics of human time must be present in prior levels of cosmic and biological duration. This position is not dissimilar to that found in Fraser, *The Genesis and Evolution of Time*. See n. 20 above.

"whatever becomes determinate through the process itself in its effectively probable realization of its possibilities."[100]

To the notions of probable schemes of recurrence, of emergence, and finality, Lonergan adds the notion of development:

> . . . a flexible, linked sequence of dynamic and increasingly differentiated higher integrations that meet the tension of successively transformed underlying manifolds through successive applications of the principles of correspondence and emergence.[101]

Lonergan proceeds to outline the way in which purpose and development will not be seen as assured ever hopeful progress that seems to justify the present status quo. He later discusses the movement, differentiation, and integration from chemical compounds and molecules to the emergence of higher forms of organization we call "life" that are self-organizing and self-replicating.[102] Evolutionary development moves from lower manifolds through differentiation to higher integrations and schemes of recurrence.[103] To understand development is to "proceed from the correlations and regularities of one stage to those of the next."[104]

Yet once development reaches the level of self-organization and self-reflection, it can encounter human refusals to take responsibility for change. It faces the fact of evil. There appear in human history ignorance, bad will, and/or ineffectual self-control.[105]

For Lonergan, the possibility of overcoming the personal and social surd, where there is no intelligibility, is dependent upon many conditions. But a crucial one is a symbol he names *cosmopolis*.[106] The decline in human culture and history emerges from the various blindspots of human choice that can only be overcome by the non-violent rhetorical appeal of an interlocking set of critically appropriated *images* that appear in art and literature, theater, broadcasting, journalism and history, school and university. These appear in our common sense world to correct the biases that skew development and make the emergent probability of the universe a dead end or a blind alley. Such metaphors drawing humanity into the future assist individuals and communities to take responsibility for the history of the universe.

Though this enumeration of important concepts in Lonergan can be only a sketch, it should indicate how at least one philosopher is convinced that it is possible to think through the intelligibility of the historical universe from quantum states to its ultimate finality. For Lonergan, the open-ended human desire to know always exceeds asymmetrically its embodiment in questions,

---

[100] Ibid., 450.

[101] Ibid., 454.

[102] See Melchin, *History, Ethics, and Emergent Probability*, 108-117.

[103] Note the similar discussions in Prigogine and Stengers, *Order Out of Chaos*, 177-212; and Layzer, *Cosmogenesis*, 177-230.

[104] Lonergan, *Insight*, 461.

[105] Ibid., 689.

[106] Ibid., 238-42, 690.

images, insights, concepts, judgments, and values; it is the self-conscious reflection of the emergent probability of the universe itself.

Both individuals and nature have histories. Lonergan thinks through the intelligibility of that history and, in effect, argues that without a cosmic history, there can be no human histories. The metaphor that guides the process as a teleologically open goal is *cosmopolis*. The higher schemes of recurrence depend upon and integrate lower manifolds at levels of statistical probability. Human self-organizing self-consciousness enters the story as measurer and reflective agent of change. Lonergan explains the excess of human knowing (and loving) as the propelling power within the telos of narrative. It is this "surpassing" (self-transcendence) that marks one's present from one's future. Moreover, this self-conscious intelligence makes judgments about things, events, people and values that can be revised should new conditions emerge. For Lonergan, humans must take responsibility for the "is not yet" of what is emerging.

In this lengthy section, Lonergan has offered an explanatory language for the temporal indications of metaphor. The pattern of "is-is not yet" found in metaphor asks for a theoretic, historical ontology that can be named emergent probability. This name for the concrete extensions and durations of the physical world, enveloped by biological and human time, provides an abstract language for the temporal asymmetry of the universe. It emerges from the surplus, the excess, that is named in the metaphoric narratives about the universe. But this theoretic language is not the sole interpreter of the referentiality of metaphor. There is another that claims to express theoretically the excess that is operative in metaphors. To that language we must now turn.

5.3 *Deconstruction and Time*

Deconstruction views the excess, the surplus of metaphors, as sheer bi-polar play, without finality or purposiveness. As we shall see, the predilection for a present presence (any simultaneity) and an underlying being (Ontology) is "broken up" into temporal meaning that is at once always passing and always deferred into a becoming whose speech is intertextual and multiple. Deconstruction rejects any narrative as an imposition upon intrinsically chaotic data.

The figure who is most aligned with the positions of deconstruction is Jacques Derrida.[107] To be able to locate his writing on temporality requires some contextualization through Freud's suspicion of the unity of the self, Heidegger's rejection of classical metaphysics in western thought, and Nietzsche's philosophical style.

---

[107] For a useful introduction to Derrida's work, see Christopher Norris, *Derrida* (Cambridge, MA: MIT Press, 1987); for a positive reading of Derrida's work in theology, see Mark C. Taylor, *Erring: A Postmodern A/theology* (Chicago: University of Chicago Press, 1984); for an ambivalent reading, see David Tracy, *Plurality and Ambiguity: Hermeneutics, Religion, Hope* (San Francisco: Harper & Row, 1987), esp. 54-62.

Freud describes the ways in which human subjects are never available to themselves completely. Not only is the subject "broken up" in itself as an id, an ego, and superego, but the unconscious, following Lacan's reading of Freud, can never be objectified. Human beings are forever involved in an otherness over which they have no control. Psychoanalytic analysis is a method for reading all experience. The signs that emerge from the unconscious can never be anything but suspect; they will never be deciphered completely. They split apart in fissures of radical ambiguity. All that human beings have are the signs that appear; there is no permanent personal background (a "subject") present that generates the signs.

Following upon this assertion of the subject's radical plurality, Heidegger de-structs western metaphysics. The age of ontology extending from Plato to (at least) Hegel was a tradition that believed that a fully constituted presence or positive ground supported phenomena. Being was the ultimate foundation of all that is, including that which is not. For Heidegger, Being is not a being among other beings, such as plants, trees, chairs, or even people. Being is both absence and presence, and thus should be written "under erasure". To put the western ontological tradition under erasure makes the thinker concentrate upon the ways in which particular realities are present through Being which cannot be spoken.

Derrida points to the absence, the lack that is the condition of life and thinking. In classical terms, this absence is neither non-being nor nothing. One cannot ask "absence of what?" There is no "what" of which "it" is a lack. Any theoretic attempt to understand resists our attempts to make action, thought, and things coincident. Derrida marks the present absence as a "trace," the way in which the [other] appears in the midst of human attempts to signify anything. Trace points to the ineradicable non-identity that conditions the motion of discourse. There can be no full presence or simultaneity.[108] Hence any attempt to control the meaning of things is impossible. No identities can be articulated in discourse, at best only similarities that are always slipping out of the conversation. One can never say "I am x" or "The world is y" without having to postulate paradoxically and immediately its opposite. The "I" or the "world" are, if they "are" at all, *in the act* between the polarities. Yet even this positive statement falls automatically under erasure.

With Nietzsche, Derrida exercises a philosophical style. The perpetual creativity of the artist continually invents metaphors, tropes that ironically subvert the received pieties of philosophy. For Nietzsche, metaphors, not concepts, are the process of truth-telling. Indeed, concepts, jargon, even precise languages are simply disguised metaphors. The fragmentation of the philosopher's prose, the rapid reconfiguration of his texts from aphorisms to fables to lengthy conceptual commentary reveal the multi-faceted play of the writing that is the creative force of the author. For just as authors are present and absent only in the language they risk writing, so also the reader's access to the

---

[108] David Wood, *The Deconstruction of Time* (Atlantic Highlands, NJ: Humanities Press, 1989), 122.

thinker is only through the pathways of prose. The subjects' freedom is the power to write, to undo a position by espousing another, in unending fashion.

"Remaining oneself," therefore, (an impossibility, of course) requires playing in between the differences, not attempting to master one metaphoric appearance by another. The movement has neither origin (and hence no "original" sign) nor a specific ending (and hence no probable, preferable, or improbable denouement). Eternally creative subjects, by noting the otherness within their own psyches, allow the signs to appear.[109]

Différance marks the reality of absent otherness in what is/is not; it is spelled with an "a" to show its relationship in French to both differentiation and deferral of meaning. *Différance* appears in writing, both the usual marks on paper and a psychic space in which subjects are always distanced from themselves through words. Words are always rifled by their own ambiguity, energizing simultaneous labyrinthine wormholes to other terms, not back to a primordial presence or to an absolute future. The subject knows itself only in the play of different signs, the marks that it traces. Hence any linear continuity of spacetime is recognized as illusory nostalgia, abolished in a stream of differences. De-construction, therefore, as a skeptical critique, notes the sutures that try to sew the spaces together between works, the proposals of presence to hide the absences, the gaps in the psyche. It unravels any proposals of positive meaning.

History, therefore, whether of individuals, peoples, or nature is a construction. To suppose that there is a non-contingent relation or connection between one sign and another is to impose upon the flow of disparate differences a unity from the outside. This is the "logocentrism" of western thought. Since this presumption promotes archeology (the search for origins) or teleology (the search for a potential in the present that intends completeness), it is a fundamental mistake. Narratives are a "reaction formation against the discovery of the 'seriality' of existence."[110]

Chronicles are the appropriate trope or style for events. Annals make no attempt to place a set of events ordered vertically as annual markers into a linear or horizontal process. No explanation is offered for interconnection and no central subject or voice is assumed to be speaking through history. Chronicles appear to extend both backward and forward without end. They are records without relation or motivation.

Narratives, on the other hand, seem to plot the connections between "events." They structure data so that they form a coherent pattern in which one moment follows another in progression. Scattered overlaps are taken for cause and effect; characters emerge from backgrounds; and intelligible wholes appear with privileged moments. The interplay of memory and anticipation marks the filled present for the agents in the story. The historian and the novelist have

---

[109] Ibid., 29.

[110] Hayden White, *Tropics of Discourse: Essays in Cultural Criticism* (Baltimore, MD: Johns Hopkins University Press, 1978), 234. Please note the correspondence between this philosophical position and the theological one taken by Cupitt in the opening quotation of this essay.

much in common; they both create plots, granting tensive relations and meaning to the dashes between the dates.[111]

The construction of history reflects the attempt to master the uncontrollable resistance of time. Repression and domination of one moment by another, of the whole by a part, of the whole by one writer/speaker/narrator, of one gender by another—all these are the internal rhetoric of stories. Unwilling to face the "irreducibility of absence and the inevitability of death," tellers of tales colonize the unknown past and the intangible future. History itself, a narrative whether for humanity, nature, or the universe, is a trope, a metaphor that deconstructs itself in the telling.[112] It subverts all attempts to point directions, tell tales, or name names.

Despite deconstruction's rejection of narratives, it claims that it does not replace history with nihilism or respond to experience with a sense of fatedness. That would simply replace presence with absence and transubstantiate it into a reversed presence. One would be entering the polarity of discourse from the opposing side, so to speak; and then no-thing or the void would become a new foundationalism for thinking. For the deconstructionist who is suspended between, time moves in multiple simultaneities, overlapping and deferring identity, leaving traces. Humanity is polyhorizonal, a "tissue of times."[113] Critical skill is required to keep the moment open without assuming closure and the conviction of meaning. The stance of the subject in such a world has been described as a sojourner, an exile, the undomesticated drifter, the anonymous saunterer who is always transgressing boundaries.[114] Players within time have no purpose other than the game; play risks meaning nothing at all. The need to use others, to consume them for one's own benefit, is shed for the sake of frivolity. It lives in the polysemy of carnival.

The opacity of deconstructionist prose is deliberate; it gives evidence for the very metaphoricity of discourse that it proposes. For scientists or systematic theologians, whose interest is in making words coherent or precise, the whole project must seem bizarre. But, aside from its extraordinary political influence in graduate schools in the United States, deconstructionist positions must be entertained seriously. By maintaining the utterly metaphorical character of all language, by arguing that the experience of metaphors is fundamentally paradoxical, and that metaphors are constantly subverting positive discourse, such interpretations push discourse toward thinking as an act and toward an overturning of any status quo. Yet at the same time, it maintains a quite distinct view on the nature of time and narrative, one which prohibits any attempt to create a coherent story out of nature or history.

The two attempts (Lonergan and Derrida) to provide a language for that to which metaphoric language of surplus refers are distinctly different. Under Lonergan's proposal, one can be responsible for time and in time for a future becoming different than the blindspots of the present. This possibility for human

---

[111] Taylor, *Erring*, 52-73.
[112] Wood, *Deconstruction of Time*, 132-33.
[113] Ibid., 334.
[114] Taylor, *Erring*, 156.

freedom is based upon the emergent probabilities of other schemes of recurrence (cosmological, chemical, biological, etc.). One chooses and establishes the conditions under which the vital, social, personal, and transcendent goods might emerge. In the second interpretation (Derrida), one never exits from metaphors and the slippage of their meanings. One could try to control them, but illegitimately. If one accepts them, then one lives in carnival and tropological play. One values the passing, undefinable "moment" without any assured sense that another trace will cohere, unless one creatively allows it to occur.

6 *Excess, Time Asymmetry, and Christian Meanings*

The previous sections have argued that scientists use metaphors of excess or surplus and that with them they create asymmetrical narratives for the universe, that this story-writing for the cosmos may be due to the nature of metaphors, and that current philosophies understand metaphoric reference in quite differing ways. One of the values of this characterization is that it shows how the hermeneutics of the natural sciences have begun to parallel those of the human sciences. In this final section, I will indicate that a similar parallelism occurs in Christian thinking and that, from the point of view of Christian systematic theology, trinitarian language is a primary locus for examining both scientific and religious versions of the intelligibility of the universe. Both teleological narratives and aimless play parallel notions about the human relationship to the triune God.

Carnival must seem a long journey from the mathematics of time and quantum origination of the universe. However, my argument is that the metaphoric language of excess, surplus, and gaps invite physicists and philosophers into curious questions. Excess and surplus are experienced as a reality to be explained. Some make use of the surplus to investigate and analyze directional arrows of time, so that the surplus generates an asymmetry in which the future exceeds the present in the cosmos. Others hope to avoid origins and endings altogether. Direction is oppression. For one, the intelligibility of the universe discloses a story; for the others, it is ahistorical, contingent playfulness.

In the first group, the narrativists, we might place Barrow, Davies, Morris, and Tipler (and in a differing fashion, Layzer, Prigogine, and Stengers) for whom the standard model is a story. Hesse, Lonergan, and parts of Ricoeur attempt to provide explanatory intelligibility for this mode of discourse. In the second group, the non-narrativists, we should not be surprised to find Reichenbach and Hawking with Derridean deconstruction and Ricoeurian notions of confrontational metaphor as philosophical companions.

Further work would indicate that these starkly drawn alternatives may be interpreted as part of western philosophy's search for the intelligibility of identity and difference,[115] their mutual implication, and their temporality. The past and the future differ from the present, yet coinhere within it; they can be

---

[115] See Martin Heidegger, *Identity and Difference*, trans. Joan Stambaugh (New York: Harper and Row, 1969).

experienced as same or other, identical or different. Past and future can be threatening, dangerous, comfortable, or alien. For Derrideans, time can be described as a polyhorizonal dimensionality of traces in which recollection and anticipation are the vanishing excess. One must simply accept the enjoyment of the slippage and commit without pretending to know the coherence of the past or the possibilities of the future. Or with Lonergan, time can be understood as a passage in which the surplus overtones are collecting the past and hinting the future in a melody that unwinds only in linear playing.[116] One learns remorse for the past and anticipation of the future on legitimate conditions.

For the narrativists, metaphors and scientific analyses truly (however critically) categorize what occurred and predict what could happen under specific conditions. The non-teleologists tend to assume that sameness is being imposed upon the differents. The first group understands the differences to be moving into an identity (whether of disorder or order). For the latter group, linear time becomes imaginary, even a burdensome and domineering construction.

These notions closely parallel theological typologies of mystical communion and prophetic difference.[117] According to this typology, the mystic aims for atemporal identity with the divine; the prophet emphasizes the radical otherness of God.[118] Prophecy stresses ethical imperatives through narrative; mysticism lives the unspeakable no-thingness of an absenting-itself present with its sublations of memory and anticipation. Mystics detach themselves from unrolling temporality and sometimes claim extra-temporal experience of simultaneity. Prophets urge repentance (regret for the differences from identity) and command that the failures of the past must be replaced by a new story for the future. Tracy has summarized these concrete patterns in Christianity's rhetoric as manifestation (sacramentality and mysticism) and proclamation (preaching and prophecy).[119]

These experiential patterns of Christian life have their doctrinal correlates in language about the immanent Trinity. Christianity can be interpreted as a religion in which mysticism and prophecy, manifestation and proclamation are equiprimordial.[120] In the classical systematic theology of the immanent (as opposed to the economic) Trinity, sameness and otherness, identity and difference are differentiated only by opposition of relations among the persons.[121] Moreover, this "opposition" should not be understood as negative

---

[116] Michael Theunissen, *The Other: Studies in the Social Ontology of Husserl, Heidegger, Sartre, and Buber*, trans. Christopher Macann (Cambridge, MA: MIT Press, 1984), 53-54, 308-29.

[117] David Tracy, *Dialogue with the Other: The Inter-religious Dialogue* (Louvain: Peeters Press, 1990), 17-26.

[118] James Robertson Price, "Transcendence and Images: The Apophatic and Kataphatic Reconsidered," *Studies in Formative Spirituality* XI, 2 (May, 1990): 197.

[119] Tracy, *Analogical Imagination*, 405-19.

[120] Raimundo Panikker, "Le Temps Circulaire: Temporisation et Temporalité ," in *Temporalité et Alienation*, ed. E. Castelli (Paris: Aubier, 1975), especially 231-38.

[121] Lonergan, *De Deo Trino* (Rome: Gregorian University Press, 1964), II, 115-51; and Quentin Quesnell, "Three Persons—One God," in *The Desires of the*

conflict, but as co-originating cooperation. In the economic Trinity (God's action in history), creation, Christology and the sacraments are Christianity's announcement that the identity and difference of the divine and the human continue to be co-implied.[122] What should be noted, therefore, is that in classical theological language, narrative and non-narrative dimensions of the ultimate reference for Christian discourse are integrally intertwined. There is a narrative economy of immanence in which identity and difference are operative; and there is a mystical immanence in the economy of divine action to others.

What do these parallels of Christian experience and doctrinal language introduce in our discussion of narrative, deconstruction, and cosmic history? First, they encourage *theologians* to investigate how the two paradigmatic interpretations of spacetime (narrative and non-narrative) might be equiprimordial, requiring a co-implicating dialectic. Despite the fact that this essay has emphasized the temporal dimensions of metaphor, it would be important to continue to explore the non-temporal paradoxes that are often claimed for the surplus awakened by metaphors. Secondly, it encourages *philosophers* to analyze how sameness and otherness, identity and difference co-originate the experience of temporality as narrative and as trans-historical or non-teleological. Furthermore, the relationship of the seemingly symmetrical temporality of the microworld still needs to be correlated with the asymmetry of the macroworld. These reflections further encourage philosophers to discuss the genealogy that links metaphoric excess to narrative and non-narrative modes of expression. Thirdly, it encourages *mathematicians* and *physical cosmologists* to develop experiments and formulae that might overcome the symmetrical-asymmetrical impasse (micro-macro world relationships).

For *theologians*, there are many avenues for investigation: First, if the triune identity co-implies both teleological narratives and mystical identity, can theologians examine how they are mutually (self-)mediating? In other words, does mysticism also generate narrative temporality; does narrative teleology mediate identity? Secondly, if the Christian doctrine of the Trinity includes a narrative relationship between identity and difference, does this imply any particular cosmic narrative as a condition for Christian beliefs? Is not this inclusion of cosmic narratives as a condition part of the claim to public discourse that Christianity makes? Thirdly, would the rethinking of the relationship between prophetic narrative and mystical identity not require reexamination of the traditional relationship between action and contemplation? Fourth, what sort of ecclesial and world politics does the mutual mediation of mysticism and prophecy require?

Finally, the specifically religious dimension for Christianity is its announcement that this experience of temporality (whether narrative or trans-historical) is a gift. It claims that the experience of time is neither neutral nor

---

Human Heart: An Introduction to the Theology of Bernard Lonergan, ed. Vernon Gregson (New York: Paulist Press, 1988), 150-67.

[122] If this begins to sound suspiciously like Hegel's reflections, I imagine there are good historical reasons for this. See G. W. F. Hegel, *The Phenomenology of Mind*, trans. J. G. Baillie (New York: Harper & Row, 1967), 104, 763-68, 806-7.

threatening, but cooperation and mutual understanding. The triune interpenetration of relative otherness, of distinction within unity, of communion among differences, of temporality as both teleological and simultaneous announces that the universe is to be experienced as generosity, rather than as a struggle, competition, stalemate, intrusion, or invasion.[123] As it has been said, the universe may be the "ultimate free lunch."[124]

Christianity claims that whether time is experienced as polyhorizonal simultaneities in mysticism or as a prophetic, linear, narrative challenge toward the good, God's involvement is to be trusted. This is based in the very nature of the Godhead. The divine is the ultimate partner, an other who is never destructive. The dialogue immanent within time of identity and difference is not pernicious or vicious. According to this inclusive view, the history of the universe, whether it expands into infinity or contracts into implosion, remains a tragi-comedy, a combination of narrative and non-teleological forms involving both identity and difference, presence and absence. At the center of the debates about the metaphors for time, whether mathematical, physical, or philosophical, are questions about whether time is a threat or a gift. Christian thinkers maintain that they would investigate the excess as gift.

---

[123] This language appears in the physicists examined here. See Davies, *God and the New Physics*, 228; and Prigogine and Stengers, *Order Out of Chaos*, 189-90.

[124] See John Polkinghorne, *Science and Providence: God's Interaction with the World* (Boston: Shambhala, 1989), 12. I am aware that this *metaphor* has a more precise scientific meaning, but as I have said (perhaps too many times) metaphors awaken a surplus of meanings over which we do not always have control.

I am indebted to critical conversations at the conference at Castel Gandolfo in September, 1991, to the editors for their comments, and to my colleagues at the Catholic University of America and at Gonzaga University where I was privileged to hold the Flannery Chair of Theology, 1992-93.

III

PHILOSOPHICAL ISSUES:
TIME AND THE LAWS OF NATURE

THE DEBATE OVER THE BLOCK UNIVERSE[1]

Christopher J. Isham and John C. Polkinghorne

1 *The Debate*

The central thesis of those who support the idea of a block universe is that (i) the notion of a set of spacetime points is meaningful; and (ii) all such points have an equal ontological status. In particular, no fundamental meaning is to be ascribed to the concepts of "past," "present," and "future." Thus the notion of an ephemeral "now"—a moving barrier that continuously travels into the future, leaving behind an unchangeable past—is regarded as being a construct of the human mind that has no reference to reality as understood in the framework of modern physics. A secondary implication is that our perception of a genuine openness of future events is essentially illusory or, at the very least, needs to be understood within this framework of an "eternally existing" spacetime.

Such a view of spatial and temporal relationships can raise serious difficulties for any theological or philosophical position in which the concept of human freedom plays a significant role. There are also severe problems for process theology and other contemporary theological movements in which God's own future is said to be open.

The proponents of a block universe are usually theoretical physicists who base their case on the picture of space and time afforded by the special or general theory of relativity. The basic mathematical entity in these theories is a single four-dimensional *differentiable manifold*: a set of points, each of which represents a possible spacetime location for an event, along with a structure that enables one to talk about properties varying smoothly from one point to another. In both the special and the general theory of relativity, the spacetime manifold carries an additional structure that reflects the possible causal relationships between events. This causal structure is based on the idea that information (or energy) can be sent only at speeds equal to, or less than, the speed of light. This

---

[1] During the conference, the notion of a "block universe" was mentioned a number of times and the organizers invited the two authors to initiate a debate on the subject. The lively discussion that followed our presentation revealed a considerable disagreement among the participants about the validity and the implications of this concept, and we hope to capture some of the key issues in this short paper. Astute redaction critics may detect which author defended which position. However, we have emphasized the differences between us to highlight the key points in the debate: our personal beliefs on the matter are not as far apart as they may appear.

means that the light cone[2] associated with any spacetime event E divides the set of all spacetime points into three regions:

(1) The *future* of E: defined to be all those spacetime points to which energy could be sent from E at, or less than, the speed of light.

(2) The *past* of E: defined to be all those spacetime points from which energy could be sent to E.

(3) The *elsewhere* of E: defined as all those points that are not in the future or the past of E.

This is illustrated in *Figure 1*. Note that the faster the speed of light the thinner is the wedge-shaped "elsewhere" region.

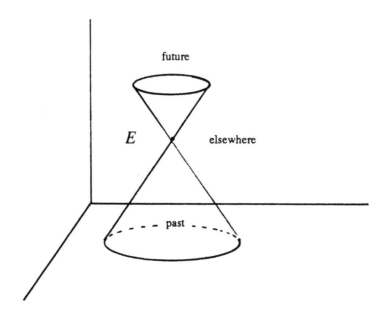

Figure 1: The "past," "future," and "elsewhere" of an event E.

This causal structure is the single most important feature of spacetime. The crucial point is that, although it is meaningful to talk about the past and future of any event, there is no unique way of identifying those events in the "elsewhere" of an event E that can be regarded as contemporaneous with E and

---

[2] The *lightcone* associated with an event E is the set of all paths in the spacetime that could be followed by beams of light focusing onto, or leaving, E. In quantum-mechanical terms, it is the set of all world-lines of photons (the quanta of the electromagnetic field) that could arrive at, or leave, E.

with each other. In common parlance, within the theory of relativity there is no unequivocal meaning to the simultaneity of events, and thus no unequivocal concept of "time." Consequently, no meaning can be ascribed to the notion of *the* future or past: these concepts are viable only in the local sense described above. The most that can be affirmed in special relativity is the existence of an infinite family of possible definitions of time that are related to inertial reference frames. The situation is even worse in general relativity since, in that theory, the role of time can be played by any slicing of the spacetime into a one-parameter family of spacelike surfaces.[3]

In Newtonian physics, which still forms the basis of most of our common-sense views of reality, the situation is somewhat different. In effect, the speed of light is taken to be infinite and, consequently, the "elsewhere" of an event E collapses to a single three-dimensional hypersurface that passes through E. In this case, it *is* possible to give a unique definition of the subset of spacetime points that are "simultaneous" with a given event E: it is all those points that lie on this special hypersurface. Note that the basic mathematical entity in the theory is still the manifold of all spacetime points but, since it is sliced into a one-parameter family in a natural way, it now becomes possible to invoke the "moving-barrier" image of perceived time. The mathematical structure of Newtonian physics does not *require* this psychological addition to the theory, but at least it is consistent with it.

When we come to special relativity, the situation is quite different. The moving-barrier picture can now be employed only by selecting some special reference frame from the infinite set admitted by the theory. In general relativity we have to choose some particular slicing of spacetime from among the even larger set of possibilities afforded by this theory.

The proponents of the block-universe view believe there is no justification for making such a choice and claim that this entails a denial of any "moving now" with its fundamental division into a global past and future. Instead, we must accept the single spacetime picture of relativity theory in which all points in spacetime are equally, and truly, real. If the speed of light had been much smaller, reality would presumably have seemed very different to us and we would never have fallen into the error of assuming that, at each moment of our experience, the whole universe divides into events that have not yet happened, and those that have.

An opponent of the block universe might respond as follows (points 1 to 4 are scientific, 5 to 7 are philosophical, and 8 and 9 are theological, in character):

(1) One should not confuse kinematics with dynamics. The Galilean transformations[4] of Newtonian physics, the Lorentz transformations[5] of special

---

[3] See the articles by W. B. Drees, C. J. Isham, and R. J. Russell in this volume.

[4] These are the transformations from one Newtonian inertial reference frame to another.

[5] These are the transformations from one relativistic inertial reference frame to another.

relativity and the general coordinate transformations of general relativity, each impose certain kinematic constraints on the structure of spacetime in the related theory. It is a separate question to ask what is the nature of the dynamics taking place in that spacetime. We can have Galilean theories with deterministic dynamics (classical mechanics), or with indeterministic dynamics (non-relativistic, conventionally-interpreted quantum theory); Lorentzian theories with determinism (classical electromagnetism), or indeterminism (quantum electrodynamics). Although general relativity partially relates kinematics and dynamics through the way in which matter determines spacetime geometry, there remains the possibility of classical or quantum versions. We have to go beyond kinematical questions, such as the speed at which energy and information are transmitted, if we are to answer the question of the dynamical and temporal character of the universe.

(2) The concept of a block universe has surely arisen from the spacetime picture of classical relativity. These diagrams are really "frozen chunks of history" in a way that is possible because in this case the underlying dynamics (classical mechanics) is indeed deterministic. In such circumstances the distinction between past, present and future is blurred, since Laplace's calculator can, from the present, predict the future and retrodict the past with perfect accuracy. Such a world is one of frozen being; there is rearrangement but no true becoming. It is not the only possible world, nor the one that we in fact appear to inhabit.

(3) Thermodynamics introduced a macroscopic arrow of time (increasing entropy), but how the irreversible emerges from the microscopically reversible is still a matter of unresolved scientific debate. The identification of this arrow with the human "psychological arrow," pointing from past to future, is a matter of opportunist conjecture. Very few have been willing to tread the path beaten out by the chemist Ilya Prigogine[6] and the philosopher John Lucas,[7] in calling for a revision of fundamental physics to incorporate an intrinsic directed temporality at the microscopic level.

(4) Nevertheless it is characteristic of much twentieth-century physics that, at the least, it contains a prolegomena to the description of a world of true becoming. The intrinsic uncertainty of outcome ascribed by most scientists to quantum phenomena (so that only one of the outcomes potentially present "before" is realized "after" a measurement), and the intrinsic unpredictability enjoyed by the exquisitely sensitive systems of chaotic dynamics, may be held to combine to encourage such a view and to discourage a block universe account of a single unitary spacetime.

(5) It is true that scientific theories have provided no obvious accommodation for the basic human experiences of the privileged position of the present moment. Even the inalterability of the past and the openness of the future find no universally agreed understanding. The philosopher will reply, "so much the worse for scientific theories!" Their claim to completeness of description is

---

[6] I. Prigogine, *From Being to Becoming: Time and Complexity in the Physical Sciences* (New York: W. H. Freeman, 1980).

[7] See the article by Lucas in this volume.

thereby put in question. To deny the fundamental phenomenon of temporality on such a basis would be to allow a physical tail to wag a metaphysical dog, to capitulate to gee-whizzery in defiance of facticity, to embrace scientism.

(6) The metaphysician may appeal to the physics referred to in (4) to provide hints of how the task of describing a world of becoming, consonant with the insights offered by modern science, might be attempted. For a realist for whom "epistemology models ontology," the unpredictabilities encountered will be interpreted as signs, not of mere ignorance, but of the presence of a genuine openness to the operation of causal principles (such as human or divine intentional agency) not describable in reductionist physical terms.[8] Some speculative boldness is necessary, and though the details of such metaphysical schemes are necessarily tentative, they do deliver us from merely clinging to the vestiges of outdated thought by default, and they enable us to begin to take seriously the implications of a truly temporal account of the universe. If agents bring about the future by present action, that future is not "up there" waiting for us to arrive. A world of unfolding process must be known temporally and is not subject to the eternal knowledge appropriate to a static world of deterministic pattern.

(7) The philosopher will note that, while relativity abolishes the existence of a universal time, it by no means abolishes universally reconcilable experiences of temporality. For the latter, it suffices that observers can agree on the relative ordering of causally related events, and just such an ordering is indeed a relativistically invariant concept. The rates at which the observers' clocks tick off those intervals between events is a secondary consideration. It is important to recognize that the different planes of simultaneity that observers assign themselves are retrospective constructions. No localized observer has knowledge of any distant event till after it has happened.

(8) The same could not be said of any omnipresent observer, such as we may suppose God to be in his temporal relationship to God's creation. It has been argued that the combination of divine omnipresence with relativity leads to the necessity of supposing that God has a Boethian view of all that happens at once (*simul*), and so to the theological endorsement of a block universe as the true account of reality. Against that claim, the following argument may be presented:

> Omnipresence must be defined with respect to a particular spatial domain. It follows that an omnipresent observer will have to be assigned a special sequence of spatial frames of reference in which that omnipresence is successively experienced and which sequentially provides a corresponding definition of that observer's time sequence. Let us suppose that is so for God. (There might be held to be a natural choice, e.g., frames at rest overall with respect to the cosmic background radiation, although it is not clear at what physical scale one should seek such a frame). The Observer will experience each spacetime event *as and when it happens*, as his spacelike surface of omnipresence sweeps through spacetime. That is sufficient to

---

[8] See J. C. Polkinghorne, *Reason and Reality: The Relationship Between Science and Theology* (Philadelphia: Trinity Press International, 1991), ch. 3.

guarantee the Observer's knowledge is omniscient in an appropriately divine way. He will also experience all events in their correct causal sequence. That is sufficient to guarantee that the Observer's knowledge is appropriately temporal.

This crude model is sufficient to establish that it is not necessary for a divine Observer to know all that happens at once. It is proposed in order to establish this logical existence theorem, not as a serious picture of divine epistemology. It is very important to recognize that we are discussing a model of divine *knowledge*. One must distinguish that from a model of divine *action*. The way the Observer used his simultaneous knowledge as the trigger for providential agency must be expected to respect those self-limitations he has imposed as expressions of his will. If these laws of nature are relativistic, divine action will be contained consistently within that relativistic grain as the consequence of God's self-consistency. There is, of course, no end to the irrationality possible in a world whose Creator was prepared to act with the arbitrariness of a magician. Conundrums such as that arising from supposing God to answer the "first" prayer received on a subject, with the recognition that the perceived order of two spatially separated prayers could depend upon the frame of reference used, are resolved by recognizing that God would not act in that way. The theological undergirding of that last statement is the recognition that the Creator respects the nature of his creatures. This is so for divine omnipotence in relation to a relativistic physical universe quite as much as it is in relation to the free choice of moral beings. Neither will be overruled.

(9) Although the model given in (8) is crude, the theological motivation for finding some expression for the possibility of a temporal pole to God's experience is very strong. God knows things as they really are, and in the eyes of a Boethian God of single eternal knowledge, time is ultimately an illusion. The implications of that for a religion like Christianity would be destructive. Salvation history would have to give way to a theology of timeless Gnostic truth. Divine omniscience does not demand knowledge all "at once"; God knows all that can be known, but in a world of true becoming the future is not yet there to be known.

These considerations together persuade the opponents of a block universe that there is no need, either scientific or theological, to embrace such a view and every reason from basic experience to reject it. Temporality is real and that must be so, not only for us, but also for God, who knows all things as they really are.

A supporter of the block universe might respond to these points as follows:

(1) No one is claiming that relativistic physics is necessarily deterministic; in particular, the significance of quantum ideas is not in dispute. However, it is debatable whether this has much bearing on the status of the block-universe picture of reality.

(2) The block-universe idea certainly comes from the spacetime pictures of classical relativity, but this does not necessarily imply a deterministic view of reality. For example, quantum field theorists accept the intrinsic indeterminacy of quantum mechanics, but that does not stop them from

employing a fixed Minkowskian spacetime as an essential ingredient in their theories. Supporters of the block universe are unashamed Platonists and perceive the essential ingredient of relativistic physics to be the eternal reality of the spacetime manifold. This is not to deny the subjective experience of "becoming," nor that uncertainty is a central feature of modern physics. But this does not affect the central view on the true reality of the entire spacetime as a single mathematical entity.[9]

(3) There is certainly a problem with the origin of irreversibility. But the comments about Prigogine and Lucas raise another important issue. Do opponents of the block universe seek merely to *reinterpret* the existing theories of physics, or do they make the much stronger claim that their metaphysical views can be sustained only by *changing* the theories? There seems to be an uneasy oscillation between these two positions.

(4) There is no incompatibility between the block universe picture of spacetime and modern ideas of chaotic motion. On the other hand, there are problems with spacetime pictures in the context of quantum theory. The issue is not just a technical one (relativistic quantum theories certainly exist) but is part of the general debate about the interpretation of quantum theory, especially the notion of the "reduction of the state vector" and the associated a priori role ascribed to the act of measurement.[10] One sees this particularly clearly in theories of the quantum origination of the universe where any notion of an external time parameter is completely absent. One popular way of resolving these difficulties is to invoke the many-worlds interpretation of quantum theory. In this ultimate celebration of Platonic realism no special status is associated to the act of measurement, and there is a sense in which all possible outcomes of all events are "really there." This is quite consistent with a block-universe picture of the equal reality of all spacetime points, albeit with essential modifications introduced by the theory of quantum gravity. Not everyone agrees with the many-worlds view, and the issue is one that needs further investigation. However, it could be argued that this is a valid response by the theoretical physicists to the challenge in (6) not "to cling to the vestiges of outdated thought by default."

(5) The remarks about human experience are appropriate. Modern physics is concerned with the world of matter, not that of the human psyche, and someone who supports the idea of a block universe does not thereby necessarily subscribe to an ontological reduction of all human experience to the laws of microphysics. On the other hand, if the current laws of physics really are inadequate, should this not reveal itself by a failure in some well-defined extension of the current domain of applicability of these laws (cf. response (3) above)? Can philosophical objections to a block universe be developed into a prediction of this type? For example, does the work of Prigogine and Lucas fall

---

[9] This issue is considerably more complex in quantum cosmology because of the internal nature of time in theories of quantum gravity. However, the debate in this paper is mainly within the context of the classical theory of relativity which is far better established than is the corresponding quantized theory.

[10] See the appendix to the article by Isham in this volume.

into this category? This challenge is not intended to be facetious—the question is genuine. It would be a major achievement if conceptual worries about the nature of time could be transformed into a real change in the theories of physics.[11]

(6) This is an appropriate point at which to restate the heart of the disagreement between the two sides in the debate. This is contained in the use of the definite article in the phrase "If agents bring about *the* future...." A similar objection applies to the statement in (5) that "Even the unalterability of *the* future and the openness of *the* past..." (italics added). Which future is being referred to? If it is the future of the entire universe this seems to be equivalent to the idea of a "passing now" which is incompatible with both special and general relativity.

In these theories, the history of a point particle is represented by a world-line in the spacetime diagram, and to each point on this world-line there corresponds a particular division of the rest of spacetime into past, future and elsewhere (as in *Figure 1*). If desired, this changing division along the world-line can be regarded as a mathematical representation of a "passing now" for the entity whose world line it is, thereby appeasing the needs of the human psyche. The main objection is to the idea that the "nows" of different events can be related in some way, as seems to be implied by the reference to *the* future. A purely solipsistic interpretation of passing time would be acceptable to the supporters of a block universe, but their opponents seem to want more than that. If a special slicing of spacetime really exists, how is it determined, and where is the evidence for it? Surely the simplest solution is just to regard the moving-barrier image as an unfortunate by-product of consciousness. True reality is timeless. And if that was good enough for Plato, it should be good enough for us.

(7) The retrospective and artificial nature of the construction of planes of simultaneity is one of the main reasons for advocating the block universe picture! And, predictably, supporters of this picture have difficulties in understanding the meaning of the statement "... of any distant effect until *after* it has *happened*" (italics added). Events do not "happen": they are eternally "there."

(8) There is no objection to positing a special sequence of spatial reference frames for God if this is really felt to be necessary. It is hard to imagine how the choice of such a sequence could be related to the material content of the world, but what God does is up to God. Physicists have no direct insight into God's mode of experiencing physical reality, but that does not concern them *qua* physicists. It is true that, in practice, supporters of the block universe tend to be more sympathetic with the *simul totem* view of God's experience, but the issue is basically a theological one and is best left to the theologians.

Similarly, there is no objection to saying that God is aware of the experience of a passing now as felt by any living creature (stones and electrons presumably do not share in the psychological misconceptions of conscious entities). But if God, with God's single time dimension, is also deemed to act back on the world, then this action could correlate God's single time with events

---

[11] Cf. the article by Polkinghorne in this volume which discusses whether the laws of physics are the laws of nature.

in our material spacetime continuum, and that would be equivalent to introducing a preferred slicing of spacetime. You say that God acts on the world in such a way as to preserve the requirements of relativistic physics, and this may be correct. But is it possible to construct a model of such an interaction? It is not easy to add any external influence on the physical world and maintain the full fabric of relativity. As before, this question is not meant to be facetious. Can the debate be taken beyond the level of conceptual issues to the point where a genuine theoretical model is developed?

(9) Supporters of a block universe who are also religious do tend to translate their Platonic perception of mathematics into a mystical, Gnostic view of transcendent reality. This is very seductive, but the statements about the implications for a religion like Christianity are surely correct.

2 *Status of the Debate*

The debate continues; but why? Why do two groups of intelligent people find it so hard to agree on this particular question? And what is the disagreement really about?

The central problem is that, notwithstanding the final plea, the debate is essentially metaphysical in character and so appeals to issues beyond those that can be settled by the scientific answers of physics alone. All participants accept the great twentieth century discoveries of relativity and quantum theory, but they incorporate them into their worldviews by means of different general interpretations. Many philosophical issues which have been disputed for centuries lie just below the surface of the debate.

What is the scope of the laws of physics, and what is their relationship to the phenomenon of consciousness. What is the relation, if any, of the experience of the "passing now" to the openness of future events? How should one think of free will and determinism? How are we to interpret quantum theory and the collapse of the wavepacket? Are space and time "things" in their own right or are they merely a way of describing relationships between events? (Not withstanding the word *relativity*, both the special and the general theories of relativity are usually presented in a way that ascribes to spacetime the status of a thing-in-itself). Does the Platonic view of mathematical entities espoused by many practicing mathematicians extend to endowing the differentiable manifold of spacetime with a reality that implies an atemporal actuality for all spacetime events in the physical world?

The answers to these questions bear upon one's attitude to the idea of a block universe. Metaphysical schemes (including, of course, religious traditions) are total conceptions of reality, and adherents of different points of view do find it difficult to envisage the possibility of an alternative perspective. Dialogue can founder on the confrontation of assertion by counter-assertion. Nevertheless, the debate will continue, though with difficulty. Or, as the supporters of the block universe might put it, it has already done so.

THE INTELLIGIBILITY OF NATURE

P. C. W. Davies

1 *Doing Science*

Einstein once wrote that the most incomprehensible thing about the universe is that it is comprehensible. When the system of thought we now call science was developed by Galileo, Newton, and their contemporaries in Renaissance Europe, one could scarcely have guessed how fantastically successful this way of looking at the world would turn out to be. Today, we take the scientific method so much for granted that its success is tacitly assumed to be guaranteed. Scientists rarely question *why* science works. Still less do they ask why we human beings are capable of implementing the scientific enterprise. I want to argue that our ability to understand nature through the application of the scientific method is so surprising that it demands an explanation.

Physics is usually regarded as the archetype of a successful science. Physicists are most satisfied with their investigations when the phenomena being studied are adequately described by a precise set of mathematical laws. A central theme of this volume is an esquire into the nature and origin of these laws. There are several key questions to address. Are the laws merely a product of human creativity—an attempt by scientists to impose their own definition of order onto an external world so as to made sense of it? Or do the laws really exist "out there" somehow, perhaps as a product of Divine creativity? And if they do have some form of independent existence, are they truly prescriptive—"commanding" matter to conform to their mathematical dicta—or simply descriptive of an order that "just happens" to exist in nature as a brute fact? If the laws are prescriptive, do they enjoy an existence independently of the actual universe, that is, do they logically precede the physical world itself? Can the existence of transcendent, independent laws explain how the universe came into existence in the first place "from nothing" as the result of a lawlike physical process, or will the ultimate origin of the universe forever lie beyond the scope of lawlike scientific description?

I should like to start with some historical observations. As a matter of fact, science is the product of Western European culture. Opinions differ as to whether this was just a historical accident—that Europeans just happened to stumble on the importance of the scientific method first—or whether it was a more or less direct consequence of the European world view. Medieval and Renaissance Europe became strongly influenced on the one hand by Greek philosophy, with its emphasis on rational inquiry and the intelligibility of the world, and on the other hand by the Judeo-Christian-Islamic tradition of a created universe ordered by a Deity according to a meaningful plan manifested through prescriptive laws. Perhaps it is then no surprise that the early Christian scientists sought intelligible order in nature, regarding it as an indication of God's rational plan for the universe.

It is certainly the case that, at the time, European civilization was by no means dominant. Technology, mathematics and learning were more advanced in certain Eastern countries. Yet no systematic body of empirical science was developed in the East. This has been attributed by some[1] to the absence of the notion of a lawgiving deity, and hence of an infallible mathematical order in natural phenomena. However, a detailed historical study does not reveal any simple relationship between Christian theological doctrine and the emergence of scientific concepts.[2] Indeed, the history of early science presents a distinctly confused picture, with considerable muddle and misconception concerning many basic scientific concepts. Often there was no clear separation, as there is today, between sound science and mystical, astrological, and alchemical ideas.

While Western theology undoubtedly played some role in facilitating the emergence of science, other influences—sociological, technological, economic—cannot be ignored. The idealized picture of science emerging as a product of theology, and then finding application to technology and so being pronounced "successful," is clearly too simplistic. By the time science emerged as a recognizable discipline, European technology had already developed along certain specialized lines related to colonization and industrialization. It is fashionable to suppose that science drives technology rather than the other way about, but there is clearly a two-way influence. The development of surveying instruments undoubtedly influenced the Greeks to construct a world view based on perfect geometrical forms. In our own era, the rise of the computer has directed increasing attention to the study of complexity and nonlinearity in nature through the use of computer modeling, an emphasis that had led some scientists to replace the mechanistic paradigm of Newton with that of "nature as computational process."[3]

In Newton's Europe, the development of precision clockwork and astronomical navigation techniques for the very practical purpose of crossing the Atlantic had already sown the seed of the concept of time as a precise mathematical parameter for characterizing mechanistic activity. Newton systematized the concept into "absolute, true and mathematical time" which "flows equably," and drew upon the notion of a continuous flux of time in formulating his theory of fluxions (the calculus) from which the modern science of mechanics was born. How much Newton was influenced by technological imperatives and how much by metaphysical presuppositions may be impossible

---

[1] Joseph Needham, *The Grand Titration: Science and Society in East and West* (London: Allen & Unwin, 1969); and John D. Barrow, *Theories of Everything: The Quest for Ultimate Explanations* (Oxford: Oxford University Press, 1991), 295.

[2] See, for example, Gary B. Deason, "Protestant Theology and the Rise of Modern Science: A Criticism and Review of the Strong Thesis" in *CTNS Bulletin* 6.4 (1986): 1.

[3] Tommaso Toffoli, "Physics and Computation," *International Journal of Theoretical Physics* 21 (1982): 165.

for us to disentangle,[4] but it may be that the trigger for science owes as much to colonizing zeal as to biblical doctrine.

The identification of science with European culture has led to accusations of chauvinism. It is sometimes claimed that the so-called success of science is nothing more than the dominance of Western European culture over others. For example, certain scientific theories were banned in communist countries, and some radical feminists still claim that science as we know it is fundamentally male oriented in its conceptual structure. The appearance of "feminist" science, "Islamic" science or "Christian" science, and the rejection of scientific values altogether by some critics,[5] may be in part a response to this perceived chauvinism.

It is inevitable that the formulation of science will acquire slants and perspectives that reflect the prejudices of its practitioners. And most scientists have been, until recently, middle class males of European origin. This may provide some grounds for claims of chauvinism.[6] However, although science is a product of European patriarchal society (and there may still be some cultural relativism in the way it is constructed and taught), I contend that the success of science runs far deeper than the special interests of middle class males of European origin. Science *really has* given us a (partial) understanding of nature, and has led to genuinely new discoveries about the world, discoveries representing *true facts* that are just as true for other cultural groups as for the scientific community. The Z particle doesn't exist only in the imagination of scientists. It is simply there, for everybody, and it is almost certain that the Z particle would never have been discovered by a system of thought other than science. The same is true of neutron stars, superconductivity, and the structure of DNA. Whatever the cultural, social, or political reasons, the early scientists of Western Europe came across a way of looking at the world which is remarkably fruitful, and cannot be shrugged aside as a mere cultural construct. No attempt to build a theory of existence that ignores the power of the scientific world view can carry much conviction.

2 The Cosmic Code

To develop my argument that science is a surprisingly and mysteriously effective method of understanding the world, let me start by asking what, at rock bottom, "doing science" actually amounts to. It really means, I believe, recognizing that there is an order in nature, and categorizing that order in a readily communicable way. In practice it entails exploring physical processes using experiment and observation, and fitting the data to a mathematical scheme that is organized around certain fundamental laws.

---

[4] E. A. Burtt, *The Metaphysical Foundations of Modern Physical Science* (London: Harcourt Brace, 1925).

[5] Bryan Appleyard, *Understanding the Present* (London: Picador, 1992); and Mary Midgley, *Science as Salvation* (London: Routledge, 1992).

[6] Carolyn Merchant, *The Death of Nature* (New York: Harper & Row, 1989).

There are several things about this state of affairs that strike me as being significant. The most basic point is that the universe is ordered in the first place. Clearly it need not have been so. We can easily imagine a universe that is chaotic with varying degrees of severity. It is known, for example, that many systems subject to simple regular laws nevertheless display essentially random behavior. In the real world there are many examples of this form of chaos, but many important physical processes are nevertheless nonchaotic.[7] There are also laws of physics with an irreducibly random (strictly, indeterministic) component. However, the uncertainty associated with these laws is largely restricted to the atomic realm, and is unimportant for most physical processes on an "everyday" scale. Finally, there is the extreme possibility of no proper laws at all, just random, arbitrary events.

At the other extreme, the universe could have been so highly ordered as to be essentially featureless. One can imagine a world consisting of nothing but empty space, or space occupied by an inactive crystal lattice, or a uniform gas. The actual universe is poised, interestingly, between the twin extremes of boring over-regimented uniformity and random chaos. Physical phenomena are constrained in an orderly way, but not so much as to prevent a rich and elaborate variety of physical systems to emerge and develop. The laws of physics preclude cosmic anarchy, but are not so restrictive as to rule out some openness in the way systems evolve. In short, nature employs an exquisite mix of order and chaos that seems almost contrived in its ability to achieve complex but organized activity. Indeed, some scientists have suggested that this mix may even conform to a sort of principle of maximal diversity.[8]

Not only are the laws such as to permit the emergence of novelty and variety, they also give rise to a key quality in nature that is known as "self-organizing complexity." According to our present understand, the universe began in a more or less featureless state following the big bang, and the rich diversity of physical forms and systems that adorn the universe today has emerged since the beginning in a long and complicated sequence of self-organizing processes. Significantly, this advance of organized complexity has proceeded as far as life and consciousness—truly astonishing phenomena. Again, it is easy to imagine a world which is ordered, but not in a way that encourages the spontaneous appearance of such amazing organized complexity.

Clearly, then, the universe is ordered in a very special sort of way. Were this not the case it is doubtful if conscious beings would exist to contemplate the fact. But there is something yet more remarkable about human consciousness.

---

[7] Ian Stewart, *Does God Play Dice?* (Oxford: Basil Blackwell, 1989); and Joseph Ford, "What is chaos, that we should be mindful of it?" in *The New Physics*, ed. P. C. W. Davies (Cambridge: Cambridge University Press, 1989), ch. 12. For a recent discussion see *Chaos and Complexity: Scientific Perspectives on Divine Action*, ed. Robert John Russell, Nancey Murphy, and Arthur Peacocke (Vatican City State: Vatican Observatory, 1995; and Berkeley: Center for Theology and the Natural Sciences, 1995).

[8] Julian Barbour and Lee Smolin, "Extremal Variety as the Foundation of a Cosmological Theory," *Classical and Quantum Gravity* (not yet published).

For we are not merely *aware* of the world about us. We are able, at least to a limited extent, to *understand* it. Human beings are able to discern the basic laws on which the universe runs, the very same laws that have facilitated the emergence of our consciousness in the first place. The laws are therefore doubly special. They encourage physical systems to self-organize to the point where mind emerges from matter, and they are of a form which is apprehendable by the very minds which these laws have enabled nature to produce.

The remarkable and significant nature of this "mind-law" linkage is reinforced by the fact that the laws of nature are by no means obvious to us at a glance. When we see the apple fall we see a falling apple: we do not see Newton's laws. Indeed, it took many centuries of investigation of falling bodies before the underlying laws governing this particular process became clear. In daily life, the laws of gravitation and motion are badly complicated by incidental features such as air resistance that serve to mask the operation of the basic laws. But even without these complications, which we might think of as noise in the data, accessing the laws requires the scientist to do a number of unusual and rather special things, such as adopt a quantitative approach, select the appropriate variables (e.g., to recognize time as a key parameter), and know the relevant mathematics. Thus the laws of nature are in some sense hidden from us by noise, complexity, and arcane procedural protocols. Heinz Pagels summarizes this camouflaged accessibility by referring to "the cosmic code."[9] The laws of nature are, so to speak, written in code, and what the scientist effectively does is to decode nature and read off "the message" consisting of the underlying laws.

I hasten to add that words like 'code' and 'message' are intended to be used here in a deliberately provocative and informal manner. Pagels himself is at pains to point out that it is not necessary to conclude that the laws of physics have been designed by a demiurge to contain a message for us. Nevertheless, the cosmic code idea helps to clarify an important distinction between two types of knowledge of the world that are often conflated. Our primary form of knowledge comes directly through sense date: we *see* the apple fall. In daily life we make do almost entirely with this phenomenological knowledge. When the orchard keeper claims she "knows all about falling apples" she means she has such phenomenological knowledge, which may include spotting certain systematic regularities in the behavior of the falling apples. By contrast, the theoretical physicist may have a quite different knowledge of falling apples, having to do with Newton's laws of gravitation and motion, of inertia and curved spacetime, of relating the motion of the apple to that of the planets, of differential equations and boundary conditions that possess unique solutions, and so on. This latter form of knowledge, which we might call *theoretical knowledge*, does not come through our senses. We do not "see" the differential equations at work. Theoretical knowledge is quite different in character from phenomenological knowledge: it is indirect, abstract, hard to visualize, and requires advanced education and ancillary knowledge of other topics such as mathematics. In Pagel's terms, phenomenological knowledge is analogous to perceiving the signal; theoretical knowledge is analogous to decoding it. We are born with the

---

[9] Heinz Pagels, *The Cosmic Code* (New York: Bantam, 1983).

ability to acquire phenomenological knowledge, but we have to learn theoretical knowledge, just as we have to learn how to break codes.

It has been claimed[10] that all forms of knowledge about the world are theory-laden. It is certainly true that even what I have called phenomenological knowledge is contextualized against a background of interpretation. Thus the observation of the fall of an apple makes sense to us only because we have a prior model of the world that contains concepts such as 'apple', 'ground', 'movement', and so on. However, the physicist uses the word 'theory' in a much more specialized sense than this. A proper physical theory amounts to much more than a mere conceptual framework. It typically demonstrates the existence of specific mathematical interrelationships involving a deeper level of (often abstract) concepts.

Let me elaborate on this point. Professionally, physicists make a sharp distinction between what they call phenomenological modeling and a proper physical theory. It is a subtle distinction that is not readily apparent even to physics students. When Medeleev arranged the chemical elements into the famous periodic table he was able to display an underlying order in the proliferation of the elements. The table had certain predictive power, because elements in similar positions in the table have similar properties. It is also undoubtedly "theory-laden" in its structure. Medeleev's table must be seen, however, as a purely phenomenological exercise. It falls far short of a proper theory of the elements in the physicist's sense. The latter had to await an understanding of nuclear forces, the Pauli exclusion principle, chemical bonding, and so on. This latter, deeper, understanding is based on precise mathematical laws that relate the organization of the elements to many other branches of knowledge (e.g., astrophysics).

To take another example, in the 1960s large accelerator machines began producing many new types of subatomic particles. These particles were classified in much the same way as animals in a zoo. Each particle could be attributed a range of quantitative properties (mass, half-life, spin, etc.). Furthermore, certain family groupings could be discerned, and mathematical descriptions of some sophistication were developed, again, with some predictive power. At a glance, textbooks containing these early descriptions appear little different from other theoretical physics textbooks. Yet physicists were very well aware that these early descriptions were purely phenomenological models. A proper theory of particles had to await new branches of physics, such as quantum chromodynamics and the Glashow-Weinberg-Salam electroweak field theory.

These latter developments are recognized as proper "theories" in the sense I am trying to draw out, and are quite different in character, depth, abstractness, mathematical formulation, and scope from the phenomenological models they replaced. To be sure, there are branches of physics where the distinction between phenomenological and fundamental theories may be somewhat blurred, but these are in more "applied" areas (e.g., superconductivity) involving complicated systems.

---

[10] I. Barbour, *Issues in Science and Religion* (London: SCM Press, 1966), ch. 6.

Enlightening though the code-breaking analogy might be, it lacks an additional key quality of the scientific enterprise which I should like to emphasize. This has to do with the felicitous interconnectedness of physical laws. Here a better analogy is with a crossword puzzle: doing science is rather like solving nature's crossword. When we conduct experiments we are interrogating nature, and the answers we receive are in cryptic form (in code if you like). To make sense of them you have to be rather clever and "solve" them: the solution demands some intelligence. When you have solved a "clue" you get to fill in a missing "word" (e.g., a law or principle of physics). As time goes by, so more and more of the crossword gets completed by this method. Then you notice a marvelous thing. The various "words" are not just a haphazard conjunction of different entities. They interweave each other, as in a conventional crossword, to form a coherent and mutually-interdependent unity. So the order in nature has this additional beautiful tapestry-like quality that evokes adjectives such as harmonious and ingenious.

3 *The Mind of God*

The mystery that now confronts us is this: How did human beings acquire their extraordinary ability to crack the cosmic code, to solve nature's cryptic crossword, to do science so effectively? I have mentioned that science emerged from a predominantly Christian culture. According to the Christian tradition God is a rational being who made the universe as a free act of special creation, and has ordered it in a way that reflects his/her own rationality. Human beings are said to be "made in God's image," and might therefore be considered (on one interpretation of "image") to share, albeit in grossly diminished form, some aspect of God's own rationality. If one subscribes to this point of view it is then no surprise that we can do science, because in so doing we are exercising a form of rationality that finds a common basis in the Architect of the very natural world that we are exploring.

Early scientists such as Newton believed this. They thought that in doing science they were uncovering part of God's rational plan for the cosmos. The laws of nature were regarded as "thoughts in the mind of God," so that by using our God-given rationality in the form of the scientific method, we are able to glimpse the mind of God. Thus they inherited a view of the world—one which actually stretches back at least to Plato—that places mind at the basis of physical reality. Given the (unexplained) existence of rational mind, the existence of a rationally ordered universe containing rational conscious beings is then no surprise.

Today, few scientists or philosophers would be prepared to assign a fundamental role to mind in this direct way: minds are usually considered to be very complicated and derivative things. It is likely that mind emerged from matter as a result of a long and complicated sequence of self-organizing processes, and more specifically as a result of biological evolution through natural selection. I reject the hypothesis that mind should be treated as a sort of extra ingredient that has been mysteriously "glued on" to brains at some particular stage of evolution.

If we accept that minds emerge, two new questions arise. Firstly, is this emergence simply the result of a biological accident, or is it a ore or less inevitable product of the outworking of the laws of nature? Secondly, given that minds have emerged, is our scientific and mathematical ability really such a surprise, or might it be attributable to ordinary selective pressures in our evolutionary history?

My own belief is that the emergence of mind is in some sense inevitable. That is, if we could "run the cosmic movie again" with the same laws and initial conditions, then life and consciousness would arise again, somewhere and at some epoch. The general trend from simple featureless origin to emerging richness and complexity is, I maintain, assured. However, because of the intrinsic indeterminism and openness of physical processes, the details would be very different in the re-run. Thus any conscious beings would not be human. Indeed, I would not even expect Earth to exist in the re-run, let alone a particular species to inhabit it. Now I have argued that the emergence of complexity, life and consciousness severely constrains the laws of physics, so that minds are inevitable only because the laws of physics (and to some extent the initial conditions) are so special. In this respect, our existence in the universe as conscious beings is a fact of deep cosmic significance. On the other hand the existence of the specific species *Homo sapiens* is not especially significant, and undoubtedly has a large fortuitous component to it. We are certainly not playing a central role in the great scheme.

I have arrived at the belief that the emergence of life and consciousness is inevitable from a study of the processes of self-organizing complexity in the non-biological realm. To regard the origin of life and/or consciousness as either a miracle or a stupendously improbable accident seems to me highly unscientific— a sort of ultimate "just-so" story. It is more plausible to suppose that these transitions from part of the general self-organizing trend in matter and energy, so familiar from chemistry and physics.[11] If such a trend exists, then we can expect life to have emerged elsewhere in the universe too. A test of this hypothesis would be the detection of extraterrestrial intelligent beings.

4 *Biological Selection*

I now turn to my second question of the previous section. Given that minds have emerged, is it really such a surprise that they have evolved to the point where they can do science? One well-known counter-argument goes something like this: Evolution has selected for those individuals who can make sense of their environment and possess some predictive ability. It obviously has good survival value to be able to dodge falling objects, jump streams, figure out when a

---

[11] See, for example, Ilya Prigogine and Isabella Stengers, *Order out of Chaos* (New York: Bantam Books, 1984); and Paul Davies, *The Cosmic Blueprint* (New York: Simon & Schuster, 1988).

waterhole will be dry, and so on. What we call science is, according to this counter-argument, just a refinement and systematization of the biologically useful ability to spot patterns in the world.[12]

The problem with this counter-argument is that it ignores the important distinction between phenomenological and theoretical knowledge that I have been at pains to explain. When we duck to avoid the falling apple we do so not by computing its trajectory according to Newton's laws and working out a suitable evasion strategy, but by acting almost instinctively. That is, our brains are preprogrammed to take evasive action automatically. What we cannot attribute to instinct is due to experience. But in neither case do we employ theoretical knowledge; we use only phenomenological knowledge. We are able to make judgments about where the apple is likely to be at each moment by a comparison with other similar phenomena based on previous experience, not by relating the motion of the apple to a body of widely-encompassing theoretical principles. In other words, we are perfectly able to get by dodging apples (or catching them if that is more appropriate) by using phenomenological knowledge, without the slightest need for a theoretical understanding. Indeed, most people do not know the theory anyway, and many animals are a lot better at such activities than we are.

It might be objected that in the absence of a decent understanding of computational processes in the brain we cannot be sure that our brains are not actually using Newton's laws implicitly, and that in doing science we simply explicate something that is "already in our heads." To this I would say that the onus is on the objector to demonstrate that the brain is programmed to work in this way. After all, manufactured computers can be trained to perform similar tasks very well by giving them a few trial runs. Such computers are programmed entirely at the phenomenological level. Perhaps the most serious criticism of the foregoing counter-argument is that science is at its most successful in domains far removed from "life in the jungle." In physics, the most spectacular successes are in the fields of subatomic particles and astrophysics—the very small and the very large—neither of which seems remotely relevant to activities like jumping streams or catching apples. In biology, a knowledge of which could conceivably have survival value, it is in the field of molecular biology that scientific methods have been most conspicuously successful. Our intellectual abilities are likely to have been selected for long ago, before human beings knew about atoms, galaxies or nucleic acids.

Another counter-argument that is sometimes deployed is that because the human brain operates according to the laws of physics it is natural for it to express those laws. But this is about as silly as saying that George Washington was patriotic because all of his brain cells resided in America. It repeats the mistake that Gilbert Ryle, in his famous discussion of the mind-body problem,[13] calls "level confusion." When we talk about brains, we are referring to the level of description of physical particles (or cells) and their behavior. Talk about

---

[12] Ronald E. Mickens, ed., *Mathematics and Science* (Singapore: World Scientific Press, 1990).

[13] Gilbert Ryle, *The Concept of Mind* (London: Hutchinson, 1949).

minds refers to a higher level of description involving thoughts, emotions, sensations, and so on. Whatever theory you may opt for concerning the relationship between these levels, they undeniably represent two distinct categories. The situation is analogous to hardware and software descriptions in computing. Both descriptions refer to the same sequence of events, but at different conceptual levels. The concept of a law of physics is the abstract product of higher-level mental processes, akin to software. Individual brain cells (i.e., the hardware) may indeed obliviously comply with the laws of physics, but to take that to mean that collectively they have a tendency to produce physics-like thought patterns is absurd.

Another counter-argument, favored by some biologists, is that our ability to do science is an accidental byproduct of some other mental faculty that *does* have direct selective advantage for surviving in the proverbial jungle. However, if this were so it would imply that the whole edifice of scientific knowledge and understanding is the fortuitous consequence of some trivial quirk, the intellectual equivalent of wisdom teeth. I find this reasoning unconvincing, partly because I am always suspicious of an argument that appeals to special accidents. One might explain any human quality that way. More seriously, true biological accidents are random, and therefore obey the statistical principle: small accidents are much more probable than big accidents. One would expect only modest scientific ability to crop up spontaneously. It is hard to see how we can be quite *so* successful at science if it arises this way. It is worth noting that if the argument is nevertheless correct, then we are wasting our time searching for radio signals from extraterrestrial intelligent beings, because the chances are infinitesimal that this same evolutionary quirk has occurred elsewhere.

5 *Teleology Without Teleology*

I wish to argue that consciousness is a fundamental property of the universe, and not merely an incidental accident or byproduct of certain fortuitous evolutionary steps. Biologists have reacted strongly against talk of any "evolutionary progress," and are even skeptical of claims for a directionality in the advance of complexity. However, it is hard to deny that life on Earth is more complex today than at its inception, both in the level of complexity of the most complex organisms and in the diversity and richness of the ecology generally. Furthermore, the progressive advance of complexity is manifest in the wider context of physics and cosmology. As I have already remarked, the universe as a whole began in an essentially featureless state, consisting of a uniform gas of elementary particles, or possibly even just expanding empty space, and the rich variety of physical forms and systems we see in the universe today has emerged since the big bang as a result of a long and complicated sequence of self-organizing and self-complexifying physical processes. During a self-organizing process, a physical system jumps abruptly and spontaneously into a new, more complex and/or highly organized state. This organized state would occur with only negligible probability on the basis of random linear selection. If we were to mistakenly view such processes in terms of simple linear statistical processes we would be puzzled by what appeared to be the unreasonable ability for matter and energy to achieve complex organizational states.

Scientists are starting to discern definite mathematical principles that have the effect of encouraging matter and energy to develop along certain pathways of evolution towards states of greater organizational complexity.[14] These principles are associated with the nonlinear, open, far-from-equilibrium character of the systems concerned. It would be odd if such principles, which are so conspicuous in all branches of physical science, were for some reason inapplicable to biology. It seems likely that biological systems will also display strong self-organizing behavior. Such behavior may appear at first sight to represent an unusually efficient tendency for evolutionary progress, if biological evolution is regarded in the conventional Neo-Darwinian manner as driven by simple random linear gene shuffling. Computer simulations of gene networks[15] have confirmed the amazing ability of such networks to exhibit spontaneous order. A lucid survey of self-organization has been given by Peacocke.[16]

Consciousness should be viewed as an emergent product of a sequence of self-organizing processes that form part of a general advance of complexity occurring throughout the universe. It is therefore a product of the outworking of the laws of nature, laws that bestow upon matter and energy their self-organizing capabilities. This hypothesis leads to the prediction, which I have already mentioned, that life and consciousness should be widespread in the universe, and not restricted to Earth.

I distinguish here between "laws of nature" and "laws of physics," consistent with my emergentist position. According to this point of view, some physical systems evolve to the point where new physical qualities emerge that are consistent with, but cannot be reduced to, the physics of the underlying components. Such systems then have more than one level of description. For example, a physical system may be described as "living," even though none of its individual atoms can be so described. A brain may be described as conscious, even though none of its neurons can, and so on. To the extent that there exist lawlike regularities in the overall organization and behavior of the system at the higher level of description, I do not believe these lawlike tendencies (call them organizing principles if you like) can be reduced solely to the laws of physics. But this is by no means to claim that such higher-level laws violate or contradict the laws of physics that apply to the underlying components. To give two simple examples, Mendel's laws of genetics, and Feigenbaum's laws of chaos are both, I contend, legitimate lawlike regularities of the world entirely consistent with the known laws of physics, but neither can be described as consisting of "nothing but" the usual laws of physics in disguise. (This position is sometimes described

---

[14] Davies, *The Cosmic Blueprint*.

[15] Stuart Kauffman, "Antichaos and Adaptation," *Scientific American* (August 1991): 78.

[16] A. R. Peacocke, *An Introduction to the Physical Chemistry of Biological Organization* (Oxford: Oxford University Press, 1983).

as ontological holism.)[17] My "laws of nature" encompass both the laws of physics and these more general, higher-level, lawlike principles.[18]

The claim that the universe evolves in a progressive way towards states of greater organizational complexity is often resisted by scientists because it has a strong teleological flavor. It can suggest that the universe has been designed with an end goal in mind. However, the concept of design can have widely different interpretations. In its simplest and most naive form, the belief in design is the belief that a God within time thought about creating a universe, envisioned its various qualities in advance, selected a design that would achieve an especially desirable end result, and then set about creating the real thing. This sort of picture is, understandably, anathema to most scientists (and theologians too). Not least among its scientific shortcomings is the implied assumption that the workings of the universe are deterministic, so that God's design can unfold "according to plan," ultimately achieving the preordained end result.

In fact, as I have pointed out, the evolution of the universe is not deterministic: it is a subtle blend of chance and necessity. On the one hand there are laws, the existence of which constrain the universe to develop along certain pathways. On the other hand there is intrinsic indeterminism (e.g., in quantum mechanical phenomena), which introduces a genuine randomness and spontaneity into nature. This indeterminism ensures that the future is to some degree "open." Thus, the laws predispose matter and energy to evolve, for example, towards greater organizational complexity, including life and consciousness. But there is no fine-grained compulsion, no "fixing in advance" of the details of a particular system, let alone a particular species such as *Homo sapiens*.

Should this be called teleology? Certainly the universe develops as if it is being directed towards a specific end state, but in fact its future development is to a large measure open, so that the impression of detailed directedness can be illusory. On the other hand, we are not faced with cosmic anarchy. There is a general tendency towards stability, organizational complexity, increasing variety, and so on. These trends are a consequence of the lawfulness of nature. We are thus led to a more subtle notion of "design," in which a judicious selection of laws bestows upon the universe the potential to create richness and complexity *spontaneously*, but the inherent openness precludes detailed "fixing in advance."

The use of the word 'design' could be questioned here. One problem involves the implication of temporality. It is important to realize that time is itself part of the lawful fabric of the universe, and the laws to which I refer have a timeless, eternal character. Therefore it is misleading to think of the laws as devised in advance, in the temporal sense. They should be regarded as logically, rather than temporally, prior. Possibly a better word to use is "purpose." I contend that the laws of nature are such as to facilitate the evolution of the universe in a purposelike fashion. Again, it is often objected that purpose is a

---

[17] John D. Barrow and Frank J. Tipler, *The Anthropic Cosmological Principle* (Oxford: Oxford University Press, 1986).

[18] I have given a full discussion of these issues in my book, *The Cosmic Blueprint*.

concept that can be applied to human beings, and possibly other organisms or minds, but not to a universe. In response I would say that it is not necessary to imbue the universe with humanlike purpose as such for it still to exhibit purposelike behavior.

Let me expand on this point. It used to be claimed that the universe is a gigantic machine. Now machines are manmade things; what was really meant by talk of the cosmic machine is that the universe has certain machinelike qualities. This is true. It is, however, a mistake to suppose it is nothing but a machine. Today it is fashionable among process thinkers to argue for an ecological model of the universe, in which a comparison is drawn between the universe as a whole and a living organism or an ecosystem. The universe is not a living organism, but the organismic metaphor does capture some genuine qualities of cosmic activity, just as mechanism captured certain other qualities. But organisms have purposes, and to that extent we may legitimately talk about "cosmic purpose."

This more subtle concept of "design" or "purpose," which I have dubbed "teleology without teleology," can be better understood by analogy with a game. There are many games of chance, such as those involving the throw of dice. Such games vary greatly in the quality and richness of play, from those, (e.g., roulette) that are pure chance, to those (e.g., baccarat) in which the player has the opportunity to perform an elaborate variety of play. The rules of the game, which are analogous to the laws of physics, are fixed logically prior to the play itself, and the nature of the rules determines the richness of the game. But the details are not fixed by the rules of the game, and depend to some extent on chance. In a limited fashion one may say that the rules of the game have been designed with a certain overall intended result, but not the specific result of any particular game.

This volume is about God's action in the world, and I want to suggest that a universe which behaves as if directed towards a goal need not require a Deity who manipulates the specific "play " of this particular cosmic game. But it is suggestive of a Deity who has selected from the infinite variety of possible rules of the game (i.e., laws) a particular set which facilitates, and possibly even maximizes in some sense, the propensity for the game to develop rich and interesting behavior. (It is, of course, understood that by "select," I do not mean "choose in advance" in the temporal sense: the law-selection is timeless.) If I am permitted for the moment to describe this selection as a "plan," then the plan is at the level of general scheme. The fundamental role of chance in nature seems to me to constitute powerful evidence against the notion of a detailed plan.

6 *Evidence For Selection*

The above argument for selection at the level of laws becomes compelling only if it can be shown that the actual laws of the universe are not just "any old laws" but form a remarkably special set. A number of attempts have been made to argue for the specialness of the laws. The best known goes under the label "the anthropic principle."[19] Here one seeks to establish that even slight changes to the numerical constants appearing in the laws of physics would render the universe unfit for life and consciousness, and may even severely compromise its ability to self-organize and self-complexify. Here I wish to propose a rather different argument based on our remarkable ability to understand nature through science and mathematics.

I have argued that it is our ability to do science that provides the most striking evidence for mind being somehow linked into the basic processes of nature in an intimate way. It is easy to imagine a world in which intelligent life emerges happily oblivious of a cosmic code, and is quite incapable of unraveling the laws of nature. As I have explained, such life forms would enjoy good survivability even by operating entirely at the phenomenological level, like clever robots, for there seems to be no obvious need for intelligent life to *understand* the principles on which the universe runs. There must be many ways in which the universe might be put together to possess *intelligence* without *intelligibility*. It is precisely this contingent intelligibility, as revealed through science, that is so arresting.[20]

It might seem at first sight that the existence of scientific prowess is evidence for God's guiding hand in evolution (in the specific sense of continuing manipulation), but I regard it as another example of teleology without teleology. Just as the laws of nature encourage (but do not compel) physical systems to evolve towards life and consciousness, so they encourage the emergence of intelligent organisms with the ability to understand nature at the theoretical level.

Nothing more forcefully illustrates the foregoing link between mind and the underlying laws of the universe than the place of mathematics in our description of the world. "The book of nature," wrote Galileo, "is written in mathematical language," a sentiment echoed by astronomer James Jeans with his famous pronouncement that "God is a pure mathematician." Wigner wrote about the "unreasonable effectiveness" of mathematics in the natural sciences,[21] and there has been much discussion of this topic.[22]

The central oddity here is that mathematics is a product of higher mental activity, and yet it finds its most successful application to the most basic and elementary processes of nature, such as to the fundamental particles and fields which make up the primitive building blocks of the universe. This consonance between the highest and lowest levels is all the stranger for the fact that the emergence of the highest level (mind) from the lowest level (elementary particles and fields) depends delicately on so many special properties of the latter.

---

[19] Barrow and Tipler, *The Anthropic Cosmological Principle*.

[20] Thomas Torrance, *Divine and Contingent Order* (Oxford: Oxford University Press, 1981).

[21] Eugene Wigner, "The Unreasonable Effectiveness of Mathematics in the Natural Sciences," *Communications in Pure and Applied Mathematics* 13 (1960): 1.

[22] Mickens, *Mathematics and Science*; Barrow, *The World Within the World* (Oxford: Oxford University Press, 1988); and *idem*, *Pi in the Sky* (Oxford: Oxford University Press, 1992).

A further example of the "special relationship" that exists between these highest and lowest levels concerns the property of computability. Turing demonstrated[23] that mathematical operations can be divided into computable and noncomputable. The former can be explicitly implemented by a physical device (e.g., a brain, a computer) following a systematic procedure. There are, however, many mathematical operations that can never be carried out on such a device, however long it runs. It has been stressed by Deutsch[24] that the property of computability depends on physics, because the operation of a Turing machine (universal computer) depends on the structure of the physical world (e.g., the existence of discrete states). What we regard as computable—and indeed, why human beings can mentally perform arithmetic operations such as addition and subtraction—depends on the details of this structure. One can imagine a different world (e.g., one in which there was no classical limit to quantum mechanics) in which operations are computable that are noncomputable in our world, and vice versa. In other words, the nature of mathematics (specifically, the definition of computability) is not independent of the physical structure of the world, as is often assumed.

It is then all the more striking that the laws of physics in our world give rise to certain computable mathematical operations which in turn describe those same laws of physics. Not only are these laws captured by computable mathematics, but these mathematical equations are also astonishingly *simple* in form. It would be interesting to know whether this self-consistent loop is unique. (This loop has some elements in common with Wheeler's "self-excited circuit" and "closed loop of existence.")[25] Might there be other worlds with different laws and hence different computable mathematics in which that different mathematics nevertheless described those laws in simple form? And if there are these other worlds, would they permit the existence of conscious mathematicians? Or is ours the only cognizable world which admits a simple mathematical description? These tough questions may be amenable to investigation. The answers would cast important light on just how special the mathematical structure of nature may be.

I have claimed that our ability to decode nature through science and mathematics establishes a link between mind and cosmos, and suggests that consciousness is a fundamental and integral part of the outworking of the laws of nature. One can, perhaps, accept this link without seeking a theological underpinning, possibly by pursuing a Wheeler-type self-consistency argument.[26] A second position is that there exists an infinity of universes, each slightly different from the rest, and that the specialness of our world has been selected by

---

[23] A. P. Hodges, *Alan Turing: The Enigma* (New York: Simon & Schuster, 1983).

[24] David Deutsch, "Quantum Theory, the Church-Turing Principle and the Universal Quantum Computer," *Proceedings of the Royal Society of London* A400 (1985): 97.

[25] John Archibald Wheeler, "Beyond the Black Hole," in *Some Strangeness in the Proportion*, ed. Harry Woolf (Reading, MA: Addison-Wesley, 1980).

[26] Ibid.

our own existence. I have given reasons elsewhere[27] why I find this argument unconvincing. Finally one can argue, as I have done here, that the above mentioned specialness constitutes evidence for some notion of design or purpose in the universe operating through the laws of nature.

I should like to finish by addressing the question of whether, in my scheme, science can explain the emergence of life, consciousness, and intelligent beings who can come to know the laws that have produced them. I have argued that, given the laws of nature, evolutionary processes can do the rest, without the need to invoke a God who intervenes either sporadically or continually to guide evolutionary progress. As I understand the discipline of science, its job is to explain the world on the basis of laws. The question of the nature of the laws themselves lies outside the scope of the scientific enterprise as it is customarily defined. This does not mean that it is worthless to inquire into the nature of the laws. However, that inquiry, while it might be pursued in a scientific spirit, properly belongs to the subject of metaphysics and not science. So a scientist might claim, quite correctly, that the remarkableness of the above mentioned emergence occurs entirely in accordance with the laws of nature. But is it thereby explained? In the narrow scientific sense it is explained, but this limited notion of explanation is unlikely to satisfy many people. We want to know why the laws of nature are what they are, and in particular why they are so ingenious and felicitous that they enable matter and energy to self-organize in the unexpectedly remarkable way I have described, a way suggestive of design or purpose (in some suitably modified sense). To me, it points to a deeper level of explanation than just accepting the laws as a brute fact. Whether this deeper level can legitimately be called God is for others to decide.

Acknowledgments: Chris Isham, Nancey Murphy, and Robert Russell have provided detailed critical comments on an early draft of this paper. Their help has been of great importance in producing this revised version.

---

[27] Davies, *The Mind of God* (New York: Simon & Schuster, 1992).

QUANTUM COSMOLOGY, THE ROLE OF THE OBSERVER, QUANTUM LOGIC

A. A. Grib

1 *Introduction*

Quantum cosmology makes possible a scientific investigation of the problem of a quantum creation of the universe from nothing. In this problem two sciences—general relativity and quantum theory—come together. But I suspect that a full answer to the profound question of the origin of our universe needs a third participant—theology.

Classical cosmology leads us to the Big Bang theory of the origin of the universe. One of the difficulties with this theory is the existence of singularities. There are different views on this issue. Some scientists think that a cosmological singularity has the same meaning as does a singularity in other branches of physics: for example, in electrodynamics it signifies only the limit of the theory and the need to employ a deeper quantum theory. Others claim that, because general relativity is a theory of spacetime, 'singularity' has a real physical meaning and shows that time, like temperature, has an absolute zero and that there is an 'edge of spacetime'.

In any case one must speak seriously today about the *beginning of classical spacetime*, which is surely of some importance to theology. The existence of a singularity could be interpreted as saying that the universe is not self-sufficient and cannot explain its origin. A.A. Friedmann was the first scientist who in 1922 in Petrograd (now, again, St. Petersburg) wrote about the Big Bang and the beginning of the universe,[1] several years before Hubble's discovery of an expanding universe.

Friedmann was an Orthodox Christian and was sure of the correctness of his solution to Einstein's equations, in spite of the total absence of any experimental evidence to support it, and the negative attitude of Einstein himself to this solution at the time. For many years, the atheistic communist authorities in the (then) USSR prohibited their scientists from working in relativistic cosmology and, from their point of view, the authorities were right to do so. The reason is the following. In Orthodox Christian theology the so-called 'symbolic' interpretation of the Bible is popular. This means that the creation of the universe by God from nothing is a symbol—nobody can give a full account of what it means in human terms because it concerns the relation of the Transcendent and the Immanent. But in the 19th century, when scientists believed the universe was infinite in time, it became possible to say that this symbol was meaningless; and this was very important for the atheistic creed that the "creation of the universe

---

[1] A. A. Friedmann, *Zeit. Phys.* 10 (1922): 377.

by God is impossible." Now, in the post-atheistic era in Russia, we can say that the Big Bang theory of the beginning of the universe means that a creation of the universe by God from nothing is *not* impossible—thereby negating the atheistic formula.

As for quantum theory, there are the well-known words of Sir A. Eddington that "religion became possible after the discovery of quantum mechanics." What are the features of quantum mechanics that can lead to something connected with religion? To find the answer one merely has to read papers in philosophical journals in the USSR from the end of the thirties. In communist countries, only the atomic bomb saved quantum mechanics from the fate of relativistic cosmology. The key words that could play a fatal role in the life of a quantum physicist were *indeterminism, complementarity, the role of the observer,* and even *the collapse of the wave packet.* Why? A post-atheistic reading gives the following answer: 'indeterminism', 'complementarity', and the 'role of the observer' mean that consciousness may be important in the universe. The wave packet collapse is the point where 'indeterminism', 'complementarity', and 'the observer' come together.

In the problem of the quantum creation of the universe we have a confluence of the concepts of 'the beginning of the universe', 'indeterminism', 'complementarity', and 'observer dependence'. As for the singularity problem, in certain simple quantum models of the origin of the universe the singularity is absent[2] but in other cases a singularity is present, even in the quantum version of the theory.

The aim of this paper is to discuss quantum aspects of the problem of the creation of the universe. First we shall try to see where people writing papers on quantum cosmology make implicit assumptions about the existence of an observer who makes measurements of the quantum universe. We shall try to do this explicitly. Then we shall give a review of a little known version of the Copenhagen interpretation of quantum mechanics where an important role is played by consciousness. In spite of the fact that this version was developed by people such as J. Von Neumann,[3] E. Wigner,[4] F. London,[5] J. Wheeler,[6] and B. d'Espagnat[7] it is not very popular with physicists. There are two reasons for this. The first is philosophical (materialistic) prejudice; this cannot be considered very important. The second is that it has no consequences for most of physics. But this is not so if we go to biophysics and quantum gravity. Wigner tried to find some

---

[2] See the article by C. J. Isham in this volume.

[3] J. Von Neumann, *Mathematical Foundations of Quantum Mechanics* (Princeton: Princeton University Press, 1955).

[4] E. P. Wigner, *The Scientist Speculates* (London: Heinemann, 1961).

[5] F. London and E. Bauer, *La Theorie de l'Observation en Mecanique Quantique* (Paris: Hermann, 1939).

[6] J. Wheeler, "Beyond the End of Time," in *Physical Cosmology and Philosophy*, ed. J. Leslie (New York: Macmillan, 1990), 207.

[7] B. d'Espagnat, *Conceptual Foundations of Quantum Mechanics* (New York: Benjamin, 1976).

consequences from this interpretation for living systems in biology.[8] Following Wheeler, I think that this interpretation could be important for understanding quantum gravity and quantum cosmology.

Here I shall speak about the so-called 'quantum logical' interpretation (which also is not very well known) and the role there of consciousness. It was Von Neumann again who, together with Birkhoff, noted that the mathematical apparatus of quantum mechanics can be understood as a realization of a new quantum logic.[9] Wheeler, speaking about the quantization of gravity as a quantization of spacetime, asked: "What remains in physics when spacetime is absent?" And the answer is—logic! But this logic or 'pregeometry' need not be ordinary Aristotelian logic. It could be something very different from the familiar Boolean structure of our usual logic. D. Finkelstein—someone who has done much work in quantum logic—likes to say that one must speak about quantum geometry and even quantum set theory.[10]

But quantum logic cannot be the logic of the human mind. That is why a human observer must project the structure of the quantum universe onto his or her mind. A non-Boolean quantum logic is not isomorphic to a Boolean logic, and this projection results in a wave-packet collapse. From this point of view, the universe that *we* see and in which *we* live is, in Wheeler's words, a "participatory universe": we participate in constructing the universe.

The interpretation of quantum theory and quantum cosmology which is discussed here is sometimes called 'subjective' or 'solipsistic': a very bad word that is often used to condemn the whole affair! Nevertheless, it is more realistic than some other versions of the Copenhagen interpretation. This can be seen by comparing it with Lüdwig's discussion[11] of the idea that only classical objects are real and that 'atoms' and 'quantum particles' are merely words for describing connections between data for preparing and using classical apparata. The important thing is that the realism of the quantum logical interpretation is not 'naive realism' in the usual philosophical sense. The universe we see is the result of the projection of Ultimate Reality on human consciousness, and our consciousness is responsible for some features of this universe. I think this is one of Niels Bohr's "epistemological lessons of quantum mechanics"[12] which is applicable to quantum cosmology.

At the end of the paper I discuss some theological questions. The main result is not an identification of God with the Ultimate Observer in quantum mechanics, nor the resurrection of the dead with the collapse of the wave function of the universe at the Big Crunch. But, in more modest words: it is not impossible that God created the universe, and it is not impossible that due to

---

[8] Wigner, *The Scientist Speculates*.

[9] G. Birkhoff and Von Neumann, *Ann. Math.* 37 (1936): 923.

[10] D. Finkelstein, *Trans. New York Acad. Sci.* 11 (1963): 621.

[11] G. Lüdwig, *An Axiomatic Basis for Quantum Mechanics* I-II (Berlin: Springer-Verlag, 1985).

[12] N. Bohr, *Atomic Theory and the Description of Nature* (Cambridge: Cambridge University Press, 1934).

original sin humans spoil the whole universe; the resurrection of the dead is not impossible either!

2 *The Observer in Quantum Mechanics and Quantum Cosmology*

The main feature of quantum theory is complementarity. This is connected with the Heisenberg uncertainty relation which, for example, means that the coordinates and momentum of a quantum particle cannot be simultaneously determined exactly. Instead, all we can know are the statistical dispersions of these quantities governed by the Heisenberg inequality.[13] Analogous uncertainty relations are valid for the projections S_x, S_y, S_z of the spin along the three orthogonal coordinate axes. In the mathematical apparatus of quantum mechanics, properties for which the Heisenberg uncertainty relations are valid are represented by operators $\hat{A}, \hat{B}$ which are 'non-commuting', that is, such that $\hat{A}\hat{B} - \hat{B}\hat{A} \neq 0$. Bohr called such properties 'complementary'. Many analyses of the problem of trying to measure such properties by classical apparata led Bohr, Heisenberg, and other 'fathers' of quantum theory to the conclusion that the construction of these apparata must be such that it is impossible to measure simultaneously the coordinate and the De Broglie wave length of a quantum object. On the other hand, in a famous paper,[14] Einstein, Podolsky, and Rosen claimed that, in spite of the Heisenberg uncertainty relations, there do exist 'elements of reality' corresponding to complementary observables. In other words, coordinates, momenta, spin projections, and so on. exist independently of whether we observe them or not, and the uncertainty relations result from the interaction of our huge classical instruments with the small quantum particles. From this viewpoint it is possible to say that observers prevent themselves from knowing exact values of complementary observables by perturbing them with their measurements, yet objectively these exact values exist.

In his reply to this claim, Bohr insisted on the indivisible wholeness of the measuring apparatus with a quantum object so that it is impossible to speak about an 'independent reality' of complementary properties. In this claim he had support from the mathematical formalism of quantum mechanics. In this formalism the state of the particle is described by a wave function $\Psi(x, y, z, t)$ which is a vector in a Hilbert space. If this function is a simultaneous eigenstate of a collection of commuting operators $\hat{A}_1, \ldots, \hat{A}_m$, one can say that the associated properties $A_1, \ldots, A_n$ *exist* and could be measured in this state. But for any pair of non-commuting operators $\hat{A}$ and $\hat{B}$ it is impossible to find a wave function that is an eigenstate of both operators. Einstein, Podolsky and Rosen knew this fact and therefore insisted that the description given by the wave function is not complete. On the other hand, Bohr maintained that it *is* complete.

At that time, one of the other 'fathers', J. Von Neumann,[15] analyzed the mathematical formalism of quantum mechanics and was led to formulate his

---

[13] Von Neumann, *Mathematical Foundations*.
[14] A. Einstein, B. Podolsky, and N. Rosen, *Phys. Rev.* 47 (1935): 777.
[15] Ibid.

'projection postulate', or the postulate of wave packet reduction (or collapse) in a measurement. Suppose that a description by the wave function is complete, and at some moment t_o let the state $\Psi(x, y, z)$ be an eigenstate of some observable $\hat{A}$. Then, with the passage of time, the state changes according to the Schrödinger equation and becomes a new function $\Psi'(x, y, z, t)$. But if an observer then measures some property B, the result of this measurement is given by the following rule: find eigenfunctions u_n of $\hat{B}$ such that $\hat{B} u_n = b_n u_n$; then $|C_n|^2 = |(\Psi', u_n)|^2$ is the probability of finding a particular result b_n (which, for example, could be the position of a pointer). Moreover, the state of the particle *after* the measurement is u_n, so that if it is possible to make further experiments on this system one must use the new wave function u_n[16]. Thus the state *jumps* in the measurement process from Ψ' to u_n or, in other words, the wave function Ψ' collapses to one of its component functions u_n. Is there any cause for this jump? The answer is—no: it is totally spontaneous! Here we meet the *indeterminism* of quantum mechanics. There is no reason to select any particular u_n from the set $\{u_1, \ldots, u_m, \ldots\}$. Nevertheless it is important to note that without an observer measuring B there is no jump. So probabilities in quantum mechanics occur only because of measurements; without them there is no indeterminism.

Can this jump be described by a dynamical equation; for example, the Schrödinger equation? The answer given by Von Neumann was—no! This means that there is no physical interaction which is responsible for the jump. In his analysis Von Neumann came to the conclusion that in any experiment there is a fundamental difference or distinction between observer and observed, subject and object. For example, here is a particle, and there is an apparatus and a scientist looking at it. At first the distinction is between particle and apparatus: we use quantum mechanics for the particle but not for the apparatus. But the apparatus consists of quantum particles, and therefore we are entitled to use quantum theory to describe the physical interaction between particle and apparatus. Now the distinction moves to that between the scientist and the apparatus. The collapse of the wave function occurs always on the boundary where the distinction is made, and always leads to the same observable result. But Von Neumann proposed to move the distinction even further, to the eye of the observer. In the end the final observer is just the *abstract ego* of the observer—the one who is the subject of observation. From this point on, any observation or information includes the objective contents of the person whose observation or information this is. So it is this abstract ego, or ultimate observer, which is responsible for the collapse of the wave function. This is a strong form of the subjective interpretation of quantum mechanics.

F. London and E. Bauer[17] developed Von Neumann's idea in another direction. They analyzed the composite system constituted by the particle, the apparatus, and the observer with his or her consciousness. They proposed to

---

[16] In some cases (for example, a Landau measurement of momenta) the state changes more drastically after the measurement, but the significance of this is mainly technical and has no fundamental meaning.

[17] Einstein, et al., *Phys. Rev.* 47 (1935): 777.

describe the whole system by some wave function $\Psi(\vec{x}_p, \vec{x}_a, \vec{x}_m)$. Then they wrote the superposition used during the measurement process as:

$$\Psi(\vec{x}_p, \vec{x}_a, \vec{x}_m) = \sum_n C_n u_n(\vec{x}_p) v_n(\vec{x}_a) z_n(\vec{x}_m).$$

The special feature of the physical interaction during a measurement is that one has the same n for u, v, z. Then from the mathematical formalism it follows that, regarded as a subsystem of the whole system, the observer is described by a so-called *density matrix* ρ_m. Density matrices can be interpreted sometimes as 'mixtures' of states, meaning that with this or that (classical) probability one has a definite state described by a pure wave function. But if the whole system is described by $\Psi(\vec{x}_p, \vec{x}_a, \vec{x}_m)$ it is impossible to have this interpretation because then the whole will also, with this or that probability, be some mixture, which is not the case. This is where the idea of *consciousness* plays an essential role. According to London and Bauer the main feature of consciousness is introspection: in giving an account to myself of the state in which I am, I know that what I see now is white rather than black, and I know that I know. The hypothesis is that introspection leads to the metamorphosis of the density matrix ρ_m into a mixture, and a mixture into one of its elements u_k, v_k, z_k. Consciousness cannot be in the state described by a density matrix.[18] Consciousness means clearness and definiteness and, according to London and Bauer, this corresponds to a pure state.

This interpretation is different in some respects from that of Von Neumann. His claim concerning the 'abstract ego' could mean that it is impossible to describe consciousness using a wave function. This corresponds to the traditional idea of German philosophy[19] that "there is no object without a subject," and that the "subject is only one" because "the notion of number is not valid for the subject." It is only 'me' who is the subject of cognition; for 'me', all other persons are objects. So it is only 'me' who makes the collapse of the wave function. Nevertheless this 'me' or 'ego' cannot be identified with the 'me' regarded as Andrej because Andrej is an object. So anybody can use this philosophy and will always be right. There is no way to find an objective process of wave packet reduction. This process is not physical but logical, or epistemological.

But the London and Bauer interpretation leaves open the possibility of discriminating between systems with consciousness as something objective, and systems without it. Then in systems of the type 'particle + apparatus + person' there will be a collapse, but in 'particle + apparatus + table' there will not be. That is why Wigner, following London and Bauer, proposed the idea that

---

[18] Although perhaps ρ_m could correspond to subconsciousness. For a discussion of W. Pauli's thoughts on the problem of quantum theory and the unconscious see K.V. Laurikainen, *Beyond the Atom: The Philosophical Thoughts of Wolfgang Pauli* (Berlin: Springer-Verlag, 1988).

[19] See for example A. Schöpenhauer, *Die Welt als Wille und Vorstellung* I-II, (Altenburg: Brockhaus, 1818).

breaking the Schrödinger equation is possible for all living systems. He speculated on the existence of non-linear extensions of this equation in biophysical systems. From this point of view, wave packet reduction and indeterminism are connected with the property of life and freedom of will as a spontaneous activity of living systems. This distinguishes life from the usual physical world. Thus, the quantum properties of particles in the universe look so only from the point of view of living systems.

The usual objection to such interpretations is the belief that what apparata show they show independently of any observer. Since surely there was a time when there were no observers, what about the decay of a nucleus at that time? For example, R. Penrose[20] says:

> Those corners of the universe where consciousness resides may be rather few and far between. On this view, only in those corners would the complex linear superpositions be resolved into actual alternatives. It may be that to *us*, such other corners would look the same as the rest of the universe, since whatever we ourselves actually 'look' at (or otherwise observe) would, by our acts of conscious observation, be resolved into alternatives *whether or not* it had done so before. Be that as it may, this gross lopsidedness would provide a very disturbing picture of the 'actuality' of the world, and I, for one, would accept it only with great reluctance.

To end our investigation let us stress again the difference between the views of Von Neumann and those of London and Bauer. We are used to thinking of consciousness as some property of the brain, analogous—for example—to ferromagnetism. The oriented spins of electrons or ions produce ferromagnetism in a many-body system. A magnetic field is an objective physical field that can interact with atoms in the ferromagnetic itself. If consciousness is something like this, one can hope to find some *physical interaction* of it with the brain, and this is the road of Wigner looking to biophysics. But the other way of thinking is that consciousness is not something that belongs to this objective world at all—it is not a material field and it does not interact physically with anything. Only the *content* of consciousness is objective. Consciousness itself is totally external to the physical universe—a 'no-thing' that has no place and no time.[21]

When we speak about apparata as existing before we look at them, and ask about 'possible information' from them, we mean that if we looked at them, necessarily we should see either this or that. The same applies to processes that occurred before the origin of humankind. One is interested in 'information' about these processes because without information what will one speak about? But information always implies a subject of knowledge. To continue the story let us note that for many years physicists paid little attention to interpretive problems because so many thought Einstein was right. The blow to this position came in 1964 from J. Bell[22] when he proved his famous theorem showing that if

---

[20] Penrose, *The Emperor's New Mind*.

[21] M. Bitbol in *Symposium on the Foundation of Modern Physics, Joensun, Finland* (Singapore: World Scientific, 1990).

[22] J. Bell, *Physics* 1 (1964): 195.

properties of a quantum object exist in a way that is compatible with relativity theory then certain inequalities—Bell's inequalities—apply, and these can be checked experimentally (see Appendix A of this paper). In its usual form, quantum mechanics predicts a *breaking* of these inequalities. And the experiments (the main one is by Aspect[23]) show that they are indeed broken. In this sense Aspect's experiment, like that of Michelson, is one with philosophical implications: either (i) properties described by non-commuting operators do not exist unless they are observed, or (ii) if they do exist independently of observation, this existence is a quantum logical existence and does not correspond to the existence of objective events in spacetime. In either case, existence as we see it is the result of an observation.

Before explaining the quantum logical interpretation and the role of the observer in it, let us first say something about the general situation that arises from the experimental evidence for the breaking of the Bell inequalities. There are a number of interpretations (non-local hidden variables, a non-linear extension of the Schrödinger equation, a rejection of the Von Neumann postulate in the ensemble interpretation, etc.) of quantum theory which try to save objectivity as they understand it. But they all lead either to a breaking of some very important principle such as relatively or causality, or they are in conflict with the fact that quantum mechanics also describes individual particles. That is why only the Copenhagen interpretation is currently capable of accounting for the experimental evidence. But the Copenhagen interpretation can have two different meanings.

From the breaking of the Bell inequalities we learn that properties described by non-commuting operators do not exist as 'elements of reality' in the classical sense. However, when we speak about coordinates, momenta, spin projections of the electron and so on, we do tend to think that somehow a 'carrier' of these properties—the electron—exists objectively. But, in quantum field theory, the number of electrons is itself represented by an operator $\hat{N}$, and this does not commute with the operator representing the local density of electric charge. Hence even the existence of a *single* carrier arises only because we choose to *measure* the particle number. And if one thinks that quantum properties do not exist without measuring them, then it is possible to hold the philosophical position that only classical objects exist, and that quantum objects are merely a way of describing the behavior of these classical entities in certain extremal situations. From this viewpoint, a wave function is just a 'note-book' for describing our procedures for preparing and manipulating measuring apparata. There is no collapse of the wave function associated with a particle because there is no particle—we just rewrite the relevant information in our note-book.[24] An objection to this 'emptiness' of the quantum world view is that classical macroscopic bodies consist of quantum objects and not the opposite,

---

[23] A. Aspect, J. Dalihard, and G. Roger, *Phys. Rev. Lett.* 49 (1982): 1804.

[24] For this style of reasoning see Lüdwig, *An Axiomatic Basis for Quantum Mechanics*; and Wheeler's example of a game with an unknown word that is invented by the players during the course of the game. (*Physical Cosmology and Philosophy*)

and that (contrary to what is said by Lüdwig and by Penrose[25] when speaking about the Planck mass) there is no strict distinction between quantum and classical objects.

So the other version of the Copenhagen interpretation must be 'realist' in the sense that quantum objects and their properties exist, but this existence is not that of naive realism. This we call the 'quantum logical interpretation'. This is connected with the Everett-Wheeler-DeWitt interpretation, which is sometimes called the many-worlds interpretation although, in my view, a better phrase is the 'many points of view interpretation'.[26] All this is close to the Von Neumann-London-Bauer position where one tries to give some objective sense to wave functions, the properties of quantum objects and the role of an observer. But before making an analysis of this interpretation let us discuss first some fairly reliable facts from quantum cosmology.

According to modern cosmology, our universe is an expanding Friedmann universe. Galaxies, stars and so on, appeared from the primordial fireball at some period after the recombination era *circa* 300,000 years after the Big Bang. So in the very early universe there were only elementary particles, and the only classical field was the gravitational field. The natural language for describing this situation is quantum theory. Therefore our classical macroscopic world must have arisen from a quantum world. A key feature of this quantum world is that, for times significantly greater than the Planck time $t_{pl} = 10^{-43}$ sec, we have quantum particles in an external classical gravitational field, but for $t \approx t_{pl}$ gravity itself must be quantized. One can ask what are the complementary observables for the era $t > t_{pl}$? From calculations of processes of particle creation and vacuum polarization in the early universe[27] one can say that in the modern universe we measure mainly particles and the particle number $\hat{N}$, so that our present universe is particle dominated. On the other hand, in the early universe the classical field associated with the measuring apparatus giving us information was such that it corresponded to measuring a local observable—energy-momentum density—that does not commute with $\hat{N}$. It looks as if, during the evolution of the universe, the key observable has changed to a complementary one. The early universe was quantum field dominated and could perhaps correspond to some pre-Friedmannian[28] inflationary universe.

But, if it has any meaning at all, for a time before the Planck time one must quantize everything, including gravity. This means that we must turn to something like the wave function of the universe Ψ_u. This notion must be such that, as with Von Neumann and his changing line of demarcation between apparatus and object, we can speak in a quantum language about the classical gravitational field at later times. Thus, a classical spacetime must appear because

---

[25] Lüdwig, *An Axiomatic Basis for Quantum Mechanics*; Penrose, *The Emperor's New Mind*.

[26] Bitbol, in *Symposium on the Foundation of Modern Physics*.

[27] A. A. Grib, S. G. Mamayev, and V. M. Mostepanenko, *Quantum Effects in Intensive External Fields* (in Russian) (Moscow: Atomizat, 1980).

[28] Grib and Mostepanenko, in *A. A. Friedmann* (centenary volume) (Singapore: World Scientific, 1988).

of a quasi-classical form of the wave function of the universe. Even now, in our present era, the universe is described by this wave function. But what does complementarity mean in quantum cosmology? Who is the observer? What is meant by the collapse of the wave function of the universe? Can one find any connection between this collapse and the Big Bang or Big Crunch?

If we read carefully papers on the quantum creation of the universe we can see that there is often an implicit assumption about the existence of an observer who makes measurements. If we think our universe was created from nothing due to a quantum process similar to quantum decay[29] we have the same situation as in usual quantum theory: quantum decay, and the probability for it, appears only because somebody makes a measurement. For example, in α-decay one starts with a wave function that solves the Schrödinger equation with the appropriate interaction, and then projects it (does a wave packet reduction) onto an eigenstate of momentum (or of the free energy). The non-commuting observables here are the total energy and the momentum of the particle. If one decides to measure some observable that commutes with the total energy there will be no decay. In the quantum cosmological model mentioned above, one takes some solution of the Wheeler-DeWitt equation and then projects it onto the quasi-classical solution. The non-commuting observables are the Wheeler-DeWitt energy and the projector onto the quasi-classical function. The same situation arises in the Hartle-Hawking model. One takes some real wave function that solves the Wheeler-DeWitt equation and then projects it onto a (complex) quasi-classical wave function $\rho e^{is(a,\phi)}$ because only this leads to the existence of time and a classical evolution of the universe. From this trivial analysis one can say that, contrary to the absence of an observer making measurements, he or she is still present in quantum cosmological models. From this one can come to some very important epistemological (and, perhaps, theological) conclusions.

(1) It is very bad philology (but unfortunately used in several popular books) to speak about a 'self-originating universe'. The universe is really 'created' from nothing (the vacuum state) by some external observer who made a choice to measure this, and not that, commuting members of a set of complementary observables.

(2) One must be very careful about using the phrase 'contingent universe'. Although it is true that contingency in quantum theory is not due to the lack of any information by the observer, nevertheless it is not an *objective contingency* because it arises only from the act of making measurements. In the absence of any measurement we have a deterministic evolution according to the Schrödinger equation. So it is not correct to say that the universe could arise 'by chance' from some nothingness without any observer (or creator). In this sense, quantum contingency is subjective contingency.

---

[29] A. Vilenkin, *Phys. Rev.* D30 (1984): 507.

3 *The Quantum Logical Interpretation*

The main question in quantum cosmology is: Who is the observer? Can the observer be chosen freely? One idea is that this choice is determined by the anthropic principle. An external observer who is a human observer 'now' can only make such measurements as are consistent with his or her life and conscious state.[30]

But before any act of measurement there must have been some 'preparation' of the wave function of the universe. As in the example of quantum fields in a classical curved spacetime, the existence of a singularity in classical spacetime can be interpreted as signaling the need to find some complementary observable, a measurement of which will change the wave function. Near the singularity it is impossible to measure quasi-classical observables because of the ill-defined nature of the classical gravitational field. The Hartle-Hawking wave function is a speculative example of the possible form of such a 'prepared' wave function. It is defined in such a way that it has no boundary conditions in the Wheeler-DeWitt equation which, perhaps, is an aesthetic argument in its favor. Nevertheless, the main question is how we human observers acquire information about it if there are no classical ways of getting this information. So the uniqueness of this function is open to debate.

However, if we can say nothing about the pre-classical era, perhaps a good candidate for representing this 'nothing' is a vacuum state. This is done in certain 'third quantization' schemes[31] where one speaks about the creation of quasi-classical universes from a Fock vacuum corresponding to 'nothingness'. The opposite situation arises in the Big Crunch. Here, nothing becomes 'everything' because, in a closed Friedmann spacetime, all information can propagate to the future singularity, and there is no evidence that entropy (a memory of the universe) can somehow disappear at the final singularity. That is why there are reasons for thinking that the future collapse of the wave function of the universe will not be into nothingness.

The whole block-universe of events corresponding to the quasi-classical wave function will be reorganized in this future collapse. Surely, many speculations are possible here, including those of a theological nature. One of them is the idea of the resurrection of the dead (see the last part of this paper).

In a closed Friedmann spacetime there is nothing physical 'outside' the universe, and in the early universe there were no macroscopic bodies. Some people regard the Everett-Wheeler-Dewitt approach to quantum mechanics as being particularly relevant for quantum cosmology because they think there is no problem of an observer in this interpretation. However, the disadvantages of the interpretation are well known:

---

[30] See the speculation on biological and psychological conditioning of fundamental concepts by me in *Quantum Mechanics and Relativity (in memory of V. A. Fock)*, (Leningrad, 1980).

[31] W. Fischler, I. Klebanov, J. Polchinski, and L. Susskind, *Nucl. Phys.* B327 (1989): 157.

(1) In order to explain the measurement process one needs a special process of 'splitting' the universe according to some preferred basis in the Hilbert space. But, in a Hilbert space, all bases are equally valid and there is nothing in the general mathematical formalism of quantum mechanics that selects any particular one.

(2) If one understands 'splitting' literally one must speak about the transformation of one electron into many electrons when one world splits into may worlds. This would contradict conservation laws, for instance, the conservation of charge. To save the situation, even for a one-particle system, one must speak about many worlds which were identical before splitting and become different after it. But this means that the wave function describes an *ensemble* of particles (or worlds) so that we have an ensemble interpretation of quantum mechanics. The criticism of this is well known: if the wave function is a description of an ensemble what describes the elements of the ensemble? In particular, in what sense are the elements identical?

(3) It is not clear in what sense these different universes 'exist' because, as we see from the EPR experiments and the breaking of the Bell inequalities, they are not worlds of events but only of potentialities.

(4) Observers are still present in the many-worlds interpretation because it is the observer's ability to see all the copies that leads to a definite value of the observable arising in some 'chosen' world.

What we call the 'quantum logical theory' is free from all these criticisms. According to this interpretation, the properties of a quantum object exist even when they are not observed, but this existence reflects a 'quantum logical reality' in which conjunctions and disjunctions do not satisfy the distributivity law. Thus:

$$a \wedge (b \vee c) \neq (a \wedge b) \vee (a \wedge c)$$

where $\wedge$ means 'and', and $\vee$ means 'or'.

It is important to emphasize that non-distributive logic is not just a hypothetical construction. For example, suppose that in some town a bank is robbed an it is known that there are two antagonistic groups of criminals, let us call them group α and group β. Then non-distributivity means that the statement "'The bank is robbed' *and* 'this is done by α or β'" is not the same as the "'the bank is robbed' *either* 'by α' *or* 'by β'."

Now consider the case of an electron—a point-like particle H with definite momentum p_x which goes through a screen with two holes z_1 and z_2. The interference picture corresponds to a situation in which the electron goes through the holes 'z_1 or z_2'. According to quantum logic, this is not the same as going through *either* z_1 or z_2. To clarify this point, a quantum object can be viewed as the quantum logical lattice of its properties. Due to the non-distributivity, 'or' is less distinct from 'and' than it is in classical logic, which is why, from this point of view, Everett-Wheeler-DeWitt universes exist as 'these *or* those universes', which is not the same as 'either these universes *or* those universes'. In quantum logic, an electron does not split into many electrons; rather it has, for example, 'this *or* that spin projection S_z' *and* 'a definite projection of S_x', but it is still only one particle.

As to the problem of the observer, we take the subjective variant of the Copenhagen interpretation due to Von Neumann, London and Bauer, and Wigner. Here the projection leading to a definite value of an observable does not arise from any physical interaction in the Schrödinger equation but from the special property of consciousness. In particular, we recall the claim by London and Bauer that the property of consciousness which makes it different from unconsciousness is 'introspection': I know that what I see now is white and is different from black, and I know that I know.

The idea of introspection can be developed in quantum logic by saying that the logic of human consciousness is distributive and Boolean while the logic of the universe is non-Boolean. In order to grasp the non-Boolean world the human mind uses apparata that provide information about a Boolean substructure (a sublattice). A very important role is played here by time. A non-Boolean lattice is not isomorphic to a Boolean lattice. This is why, to get information about non-Boolean physical reality, the human mind must 'move' in time, so that at one moment it sees one Boolean substructure (a set of commuting operators) and at the next moment it sees another Boolean substructure corresponding to the operators which did not commute with the previous ones. One could even say that time is 'invented' by the Boolean mind in order to grasp non-Boolean reality. This process of 'Booleanization'[32] of the non-Boolean world by time leads also to the possibility of going from non-commuting observables A, B to the commuting observables that represent the frequencies of their occurrence in an infinite number of samples of the system where each sample is the many-worlds interpretation. A Boolean mind uses different apparata to study the quantum world. What is an apparatus? It is not necessary for it to be macroscopic. The only important thing is that it can give information about some set of commuting observables corresponding to a Boolean sublattice of the non-Boolean lattice chosen by the human mind.

But is the Boolean mind human? The well-known paradox of Wigner's friend shows that, because different human observers see the same world, one is drawn to the idea of a *World Consciousness* or an *Ultimate Subject*. Surely this Ultimate Observer is not in any sense a part of the universe. The role of the brain is simply to enable human observers to participate in the World Consciousness.

To illustrate the paradox of Wigner's friend consider the Hasse diagram of the quantum logical lattice for a spin 1 system (see Appendix B). Let a Boolean-minded observer try to project onto his mind some non-Boolean structure. The structure of the lattice is such that:

$$1 = 1 \wedge (4 \vee 5 \vee 6) = 1 \wedge I.$$

---

[32] Grib and R. R. Zapatrin, *Int. J. Theor. Phys.* 29 (1990): 113.

Then, if this observer sees that 1 is true he/she will say that $1\wedge(4\vee 5\vee 6)$ is true. This is because 'and' is defined in such a way that if $a\wedge b$ is true then a is true and so is b. But if $4\vee 5\vee 6$ is true, this Boolean-minded observer will say that either 4 or 5 or 6 is true because he/she cannot imagine that in a non-distributive logic it is possible to have a, false, b, false but $a\vee b$, true. Because of this Boolean-mindedness, the observer cannot grasp $4\vee 5\vee 6$ at the same time as 1, and therefore claims that at some other moment there is some probability, for example 1/3, that he will see a true result from the collection (4, 5, 6). Thus, although there is no probability in non-Boolean logic, it appears as a result of the Boolean-mind projection, as does the related notion of indeterminism. But what happens if two Boolean-minded observers follow the same procedure with the lattice? Why is it not possible that if one says '4 is true', the other can say '5 is true'? There is nothing from which one can conclude that because $4\vee 5\vee 6$ is true it must be 4 that is true but not 5. This is the analogue in our case of the Wigner friend paradox. So the absence of this paradox in real life (at least in the world of physics) can be considered a 'proof' of the existence of a Boolean Mind in which we all participate.[33,34]

The role of the observer in quantum cosmology arises from two different kinds of measurement.

(1) The observer creates a quasi-classical wave function for gravity and for some fundamental matter fields (in simple models this is often taken to be some scalar field that can generate an inflationary expansion). Because of the quasi-classical nature of this function the possibility arises of defining a (quasi-classical) time for the macroscopic universe and a classical evolution in this time.

(2) For the excitations which correspond to the usual particles an observer obtains the standard Schrödinger equation and can measure different non-commuting observables in the normal way. But the measurement of these excitations does not change the quasi-classical nature of the wave function.

To measure non-commuting observables an observer must 'move' in time, which is where irreversibility appears. This suggests an idea of why we all move in time: from the perspective of quantum physics our body consists of fundamental particles which are non-Boolean in nature. However, the unity of our constitution as mind and body is such that only by moving in time can there be mental experiences of physical embodiment.

---

[33] E. Squires, in *Symposium on the Foundations of Modern Physics, Joensun, Finland,* 390.

[34] The other possibility of resolving Wigner's paradox is to invoke the idea of communication: communication is possible only between observers who see the same world, but this idea is somewhat less developed.

If the universe has a quantum non-Boolean structure it cannot be a universe of space and time. Events appear only because of measurements. For example, the Minkowski spacetime of conventional special relativity has the structure of a Boolean set. Without an observer there are no events. So, like secondary properties such as 'greenness', the Boolean event structure depends on an observer.

Sometimes people ask: if events are due to an observer, what was the universe doing before the appearance of humankind? Comparing it with 'greenness' one can ask: were plants green before the appearance of the human race? We know that greenness depends on the structure of the human eye, so if someone could have looked at plants at that time they would have appeared to be green. For events, the answer is that what we call the 'past' is itself a projection of a non-Boolean structure onto our Boolean minds. In particular, because of one's idea of time and causality, a present-day observer extrapolates what is seen now and posits the existence of earlier 'events' that led to this current state of affair. But this extrapolation is possible only up to the singular point. The appearance of singularities in classical cosmology signals the need for a wave function collapse to acquire knowledge about the ultimate universe which is non-Boolean and which cannot be described using only commuting observables. The quasi-classical wave function arises from the collapse of something else.

At what time did this collapse occur? Because the wave function of the universe defines the history of the universe it is difficult to say that it occurred at some particular moment. Rather it occurs 'now' for any human being, and time is defined in such a way that there is enough time for the universe to produce human bodies from elementary particles—another reason for the anthropic principle.[35]

4 *Some Theological Speculations*

Let us consider now the ontological status of the past, present and future, and the issues of eschatology and apocalypse.

An interesting feature of classical relativity is that our classical universe consists not of stones or particles but of events in four dimensions. Past events exist in the same manner as present events. There is no annihilating force which transforms past events into nothing. From this point of view, human life is just a four-dimensional subset of spacetime that does not disappear at a person's death. Again, it is a peculiar feature of our body that it is too large and too empty to reflect any rays of our past, and hence we do not see our past. That is why we do not feel ourselves to be four-dimensional creatures with a 'jet of past events' but see only a three-dimensional projection moving into the future. So, because of the act of 'Booleanization', the past is crystallized as some definite set of events.

---

[35] This idea of collapse leading to the appearance of time is close to the idea of the Russian theologian S. Bulgakov on original sin as a product of 'wrong time'. See S. Bulgakov, *Coupina Neoplatonimaya* (in Russian) (Paris: YMCA Press, 1927). The quotation is in Appendix C.

But are we really unaware of the existence of past events in a four-dimensional block-universe? A deep, and as yet unsolved, question is: What is memory? What does it mean to 'remember' something? H. Bergson in his *Matter and Memory* had the idea that in the act of memory we 'see' in our consciousness past events as they 'are' and not merely some signs of them written into our brains. A brain works like an eye, making possible a focusing on these events. According to Bergson, in our normal active life it is not biologically useful to pay much attention to past events. Relativity gives the idea of the 'finite eternity' of any human life. Specifically, such a life is eternal in the block-universe because its events are not annihilated. Surely, the main question is what becomes of consciousness at the moment of death when apparent movement into the future ceases? Does it become a four-dimensional contemplation of the whole life, like an eternal memory identified with life itself?

For the macroscopic universe, the future is defined as a classical future of events, but it is undefined for a quantum system before it is observed. Nevertheless, even the classical future of a closed Friedmann spacetime is defined only up to the future singularity. Near this point[36] all the past events of the universe can be seen again through gravitational waves and the convergence of all light cones near the singularity. It looks like a flash of eternal memory of the universe.

But quantum cosmology says even more apocalyptic things about the Big Crunch! If, as explained earlier, the future singularity necessitates a reduction of the wave function of the universe, several exciting possibilities arise. As we have said before, the wave function relates to the history of the universe, so its change near the final singularity will influence the whole past history. The wave function of the universe, which now has the quasi-classical form $\Psi_q = \rho e^{iS(a,\phi)}$, collapses into some new wave function after the Big Crunch: $\Psi_q \to \Psi_b(a, \phi)$. This new function $\Psi_b(a, \phi)$ is defined on the same set of events (a, ϕ)—where a is the scale factor of the universe and ϕ is the matter—but the probabilities for (a, ϕ) will be different than those given by $\Psi_q(a, \phi)$. However, the most important difference is that these (a, ϕ) will not be ordered in time because time can be defined only in the quasi-classical case. So the same combinations of atoms, the same space volumes corresponding to human bodies, cars, buildings, and so on, will come again. All events in spacetime will be reorganized or 'resurrected' but not in such a way that they will be in time. So this provides one way of talking about the resurrection of the dead using ideas drawn from physics.

It is not that the dead arise from their graves, as it were, but rather that sets of events of any life can arise again, together with the consciousness of those who were alive. Here one can agree with the thought expressed by Tipler: if this is possible, and if there is a God or an 'evolving God' (in Tipler's terms), humankind must have some message (a revelation) in order to understand that

---

[36] See for example, F. J. Tipler, "The Omega Point Theory: A Model of an Evolving God," in *Physics, Philosophy and Theology: A Common Quest for Understanding*, ed. Robert John Russell, William R. Stoeger, and George V. Coyne (Vatican City State: Vatican Observatory, 1988), 313.

this future eternal life depends on the temporal life; only thus can one be 'saved' from sin and the judgment of eternity. But this is one of the central ideas of the Gospels: the idea of salvation in its apocalyptic sense! The four-dimensional subsets of spacetime corresponding to different lives of human beings much resemble the so-called 'books of life' of the Scriptures. The idea of the collapse of the wave function of the universe after a Big Crunch corresponds to these lives coming into a new existence where different weights will be given to different events. Some of the events could be annihilated (i.e., have zero weight), which is very close to the idea of the Last Judgment. How can we know in *this* life what will, and what will not, be important for the eternal life after the Big Crunch? The only sure answer is revelation. Our speculations certainly cannot be considered as any proof of theological conjectures. But, as we said in the introduction, their meaning could be that eternal life and resurrection of the dead are not improbable, and it is not impossible that these theological notions can be given a striking new sense!

Appendix A: Bell's Inequalities and the Non-Existence of the Properties of Quantum Objects as Events Without an Observer

Consider some particle with properties A, B, C, each of which can take two values ± 1 (for example, the spin-projection S_x, S_y, S_z of a spin-half particle). Then suppose that some particular set of values of these properties 'exists' even if nobody observes them, that is, an observer registers them as events ($S_x = +1$, $S_y = -1$, $S_z = 1$, etc.) that can occur without the act of observation. Then it easy to see that if one has some number $N(A^+B^-C^+)$ of particles the following equalities are valid:

$$N(A^+B^-) = N(A^+B^-C^+) + N(A^+B^-C^-)$$
$$N(B^-C^+) = N(A^+B^-C^+) + N(A^-B^-C^+)$$
$$N(A^+C^-) = N(A^+B^+C^-) + N(A^+B^-C^-)$$

from which one deduces $N(A^+B^-) \leq N(B^-C^+) + N(A^+C^-)$ which is an example of a Bell inequality.

The experimental evidence is that there are quantum situations in which inequalities of this type are not satisfied, which implies either that such properties do not exist in the absence of observation, or that distributivity is not true:

$$A^+B^- \wedge (C^+ \vee C^-) \neq A^+B^-C^+ \vee A^+B^-C^-$$

Moreover, there is no 'objective chance' or 'self-origination' of these properties as events in spacetime without an observer. To see this let us analyze the Einstein-Podolsky-Rosen experiment with two particles with spin 1/2 prepared in a singlet two-particle state, so that if the value $S_z = 1/2$ for the particle 1 is measured, it can be said with certainty that the other particle has $S_z = -1/2$. But does this 'certainty' correspond to an 'event' for particle 2 that occurs without any observer?

Consider the diagram of Minkowski spacetime:

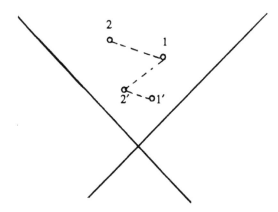

Event 1 is inside the light cone and represents an observation of $S_z = 1$ for the particle 1. Then if $S_z = -1$ for particle 2 is an event let it be the point 2. But then for some observer simultaneous with 1 the point is 2', not 2. So the property $S_z = -1$ should 'pre-exist' before the point 2. But there is another observer in whose frame the point simultaneous to 2' is 1', which is 'pre-existent' before 1. But, if the observer's decision to measure S_z is not important, why do we insist on S_z and not S_x or S_y? They also pre-exist as events with a distributive structure, and then we come to the Bell inequalities, which are violated. So only when the other observer at 2 'looks' at the particle 2 do we have an 'event'. Without an observer there is no actual event in reality but only the *objectively existing potentialities* of the quantum logical reality.

Appendix B: Quantum Logic (An Example for a Spin-1 Particle)

Let 1, 2, 3, 4, 5, 6 be the logical atoms (YES-NO questions) corresponding to the questions respectively: is $S_x = +1, 0, -1, S_z = +1, 0, -1$? (See diagram below) These properties are all exclusive in the sense that $1 \wedge 2 = 1 \wedge 3 = 1 \wedge 4 = \ldots = 3 \wedge 6 = \ldots = 5 \wedge 6 = \emptyset$, where $\wedge$ means 'and', $\emptyset$ means 'always false', and I is 'always true'. Two lines going down the diagram intersect in an 'and', and two lines going up intersect in 'or'. So $1 \vee 2 = 9$, $4 \vee 5 = 12$, etc. Imagine that the lines are wires with an electrical current going from $\emptyset$ to I. Then if 1, 2, ... 12 are lamps we see that if 1 is bright then $8 = 1 \vee 3$ is bright, and 9 is bright. I is bright in any case. If 'bright' is interpreted as 'true', and 'dark' as false, we see the following properties of quantum logic:

Boolean Logic	**Quantum Logic**
1) Distributive	1) Nondistributive
$a \wedge (b \vee c) = (a \wedge b) \vee (a \wedge c)$	$2 \wedge (3 \vee 4) = 2 \wedge I = 2 \neq (2 \wedge 3) \vee (2 \wedge 4) = \emptyset \vee \emptyset = \emptyset$
	because lines from 3, 4 intersect 'up' only in **I**.

2) There is a classical probability measure $p(a \vee b) = p(a)+p(b)$, $a \wedge b = \varnothing$	2) No classical probability measure but there is an amplitude of probability—a wave function $p(3 \vee 4) = p(\mathbf{I}) = 1 \neq p(3) + p(4) = 1/6 + 1/6$ if all outcomes are equivalent.
3) If a—false, b—false, then $a \vee b$ is false.	3) If a—false, b—false, $a \vee b$ can be true. For example 3—false, 4—false, but $3 \vee 4$ is true.
4) Negation is such that if $a \wedge b = \varnothing$ then if a—true, then b is nontrue.	4) Negation defined through the orthogonality is such that in spite of $a \wedge b = \varnothing$, a—true, b 'can be' true, 3—true, 4 can be true.
5) There is no need for time in Boolean distributive logic. "All men are mortal. Gaius is man. He is mortal." No need for time.	5) There is need for time in order for a Boolean mind to contemplate non-Boolean structure. There are Boolean sublattices (two pairs of our lattice—left and right) corresponding to atoms 1, 2, 3—left, 4, 5, 6—right. One can use Boolean logic for each of these parts. At one moment an observer measures one part (left), at the other—right part. So he/she measures S_x at t_1 and S_z at t_2. Thus he/she must 'move in time' in order to grasp the whole non-Boolean structure.
6) There is no chance or indeterminism.	6) Indeterminism arises from the difference between the structure of a non-Boolean lattice and a Boolean interpretation of it. Let an observer say 1 is 'true'. Then, because of incompatibility he/she will say 2 is 'false', and the same for 3, 4, 5, 6. But $4 \vee 5 \vee 6 = I$ is 'true'. According to Boolean logic this can be so only if one of $4 \vee 5 \vee 6$ is 'true'. But it is impossible for a Boolean mind that at the same moment of time either $4 \acute{U} 5 \acute{U} 6$ is 'false' but some of them is 'true'. The resolution of the paradox is that at some other moment one of $4 \vee 5 \vee 6$ *becomes* true. At this moment 1, 2, 3 become false as incompatible with the true property $4 \vee 5 \vee 6$. For Boolean mind it is impossible to predict what of $4 \vee 5 \vee 6$ will be 'true', and this is the reason for indeterminism.

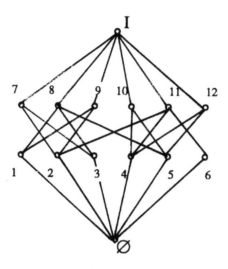

Appendix C: Father Sergius Bulgakov on the Original Sin in 'Coupina Neopalimaya' (in Russian) (Paris: YMCA Press, 1927).

In original sin humanity ceased conversing with God. As a result, the Great Fall became a cosmic catastrophe; instead of a cosmos of peace the universe become the World of evil and suffering.

As sinless, Adam has no individuality in our sense, since individuality is a limitation which contradicts the general, especially humanity in general. In this sense Adam was not an individual but a hypostasis; for sinless humanity is 'humanity in general', free from any limitation of individuality. Individuality is the result of the Great Fall.

Then we come to the paradoxical consequence that the individuality of Adam has no specific meaning, so that any of his descendants is in the same condition. The individuality of Adam, as one from many having its own fate and place, arises only because of the Great Fall.

If babies are sinful, then where and when is the sin committed? It is the sin of human freedom against its nature. It is beyond the limits of individual historical fate, on its 'edge'.

The negation of temporal predestination does not deny the non-temporal existence of all existents in eternity, in God, as the creative foundation of all that takes place in human life on the 'edge' of time: not in eternity, but yet not in time.

In this timeless moment, on the edge of time, in the creation of the soul every person participates *personally* in the Original Sin of Adam (humanity in general).

Acknowledgments: The author is indebted to M. Heller, C. J. Isham, and P. C. W. Davies for stimulating discussions.

DIVINE ACTION, HUMAN FREEDOM, AND THE LAWS OF NATURE

William P. Alston

1 *Preliminaries*

The primary goal of this paper is to discuss some of the basic issues concerning the relation of divine action to laws of nature and to human freedom. A secondary goal, in line with the emphasis of the this volume, is to consider what bearing contemporary cosmology has on these issues.

Let me be explicit about the presuppositions of the discussion. First, I am taking *seriously* and *realistically* the idea of God as a personal agent, an agent Who performs actions in the light of knowledge and in order to realize the divine purposes.

I am taking it *seriously* in that I do not construe talk of God's doing this or that as "symbolic," "pictorial," or "poetic" ways of making points about human life or the natural world or history, or about anything else that leaves God and God's activity out of account. To put it crudely, if what we say about divine action is true, then God really does what we are saying God does! To put it less crudely, in attributing to God one or another action we are, on my construal, making truth claims, the truth conditions of which involve God actually engaging in the appropriate sort of activity. If I say "God forgave my sins" or "God guided the Israelites to Canaan," then what I say is true if and only if God is the agent of the actions thus specified. There is no more basic way of specifying what I am asserting.

Let me hasten to add that in thinking of divine action talk as *literally* attributing actions to God I do not thereby commit myself to the thesis that (some or all) action terms can be predicated univocally (with exactly the same meaning) of God and human beings. That is as it may be. I have argued elsewhere that although there are very significant differences between concepts of human and divine action, and corresponding differences in the meanings of the terms that express those concepts, there is a significant overlap in these concepts, giving us a "univocal core" to talk of divine and human action.[1] But whether or not this is so, it is my assumption here that we do have the conceptual resources to make attributions of actions to God that have truth conditions in the way just mentioned. Again, in saying that we can, and sometimes do, literally attribute actions to God, I am not denying the obvious fact that much talk of God, including talk of divine action, involves figurative uses of language. "He

---

[1] See my "Can We Speak Literally of God?" and "Divine and Human Action," both in *Divine Nature and Human Language: Essays in Philosophical Theology* (Ithaca: Cornell University Press, 1989).

chastened them with God's terrible, swift sword." I am only assuming that there are also literal ways of ascribing to God the actions in question.

Let me also be more explicit as to the general force and scope of my assumptions. I am not just reporting on how I construe divine action talk, but neither do I claim that everyone means or understands it in the way I specify. Rather, I make the following double-barreled claim: (1) Throughout the Christian tradition, talk of divine action has mostly been understood in this way. (2) This is the understanding of it I will be assuming for purposes of this discussion.

I also said that I was taking divine action talk *realistically*. That may already be implied by the above discussion of truth conditions, but in any event I will make it explicit. In opposition to the waves of anti-realism sweeping over the contemporary philosophical and theological establishment, I take it that theology, and religious discourse in general, as well as many other areas of discourse including science, history, and commonsense talk about the world, aims (in its statemental aspect) at giving an accurate portrayal of an independently existing reality, one that is what it is regardless of how we think or talk about it, and regardless of what "conceptual schemes" we use to grasp it. There are facts of the matter about God's purposes, plans, activities, and so on; and what we say about God is true or false depending on their relations to those facts. This would still seem to me to be the case even if I thought, as I do not, that we are wholly incapable of determining what the truth is about any theological matters.[2]

So much for my first major assumption. The second is that I follow the Christian tradition in taking seriously the idea that God's activity is not confined to the creation and conservation of the universe, but that God is active at particular times and places within the universe. God is not just present to creation in a blanket way, keeping everything from lapsing into non-existence, but God also enters it in a more particularized role, bringing about particular states of affairs at particular times and places. That is, I am assuming with the tradition that there are occasions on which God brings it about that some particular outcome is different from what it would have been had only natural, created factors been operative. I will follow a widespread usage in speaking of such actions as *divine interventions*, even though I find the term "intervention" not wholly felicitous. (God is intimately present to creation anyway and doesn't need to *intervene*.)

Talk of divine intervention in the natural order immediately brings to mind super-spectacular miracles like the parting of the waters of the Sea of Reeds, or walking on water, or the resurrection. And I certainly want to include items of this sort in the picture, though I don't want to encumber this paper with the assumption of the historical accuracy of any particular biblical miracles. But of equal importance are the many occasions on which human beings take themselves to be in communication with God, receiving messages from God and speaking to God in turn, being aware of God's activity toward them—

---

[2] I will not be able to go into epistemological issues in this paper. See my *Perceiving God* (Ithaca: Cornell University Press, 1991).

upbraiding, forgiving, comforting, commissioning, or whatever. As these events are almost universally regarded by the participants, they too involve God's doing something at a particular time and place to bring something about that would not have been brought about apart from this particular action of God's. If God did not bring it about that I was aware of a certain message as coming from God at that moment, then I was mistaken in supposing that God had "spoken" to me on that occasion.

Finally, if anyone feels the need for some explanation of how we can "literally and realistically" think of God as acting, try this. God is a personal agent, possessed of intellect and will. These are, of course, incomprehensibly more perfect than human intellect or will, but nevertheless the terms "intellect" and "will" do apply across those differences in degree of perfection. For God to bring it about that the waters are parted or that I receive a certain message is for God to will that the waters be parted or that I receive that message, whereupon, since anything God wills necessarily comes about, the willed state of affairs eventuates.

So much for the background against which this paper is written. The specific focus in the paper is on certain external problems concerning the assumptions I have been detailing. That is, I will be concerned with how these assumptions fit, or fail to fit, with other things we think we know or have sufficient reason to believe. The focus will be on foreign relations rather than domestic problems, a familiar maneuver for hard pressed administrations. I have, in fact, examined internal problems concerning how to construe divine action in a number of previous works.[3] Here I will concentrate on problems that arise with respect to the relation of traditional convictions about divine action to convictions concerning (1) the reign of natural law and (2) human freedom.

2 *Divine Action and Natural Law*

I begin with questions about divine action and natural law. Here I can be relatively brief. The traditional view of divine action adumbrated above is obviously in direct contradiction with the doctrine that the universe exhibits a closed causal determinism, that every happening is uniquely determined to be just what it is by natural causes within the universe. (I will refer to this doctrine simply as "determinism.") As made explicit in the previous section, the traditional view is that God acts to bring about, at particular times and places, states of affairs that would not have been realized (in that form) if only natural causal factors had been operative. This directly contradicts the thesis that *every* state of affairs is determined to be just what it is by (exclusively) natural causes. We can't have it both ways. Efforts have been made by philosophers and theologians to save something of the notion of particular acts of God while

---

[3] In addition to the works mentioned in n. 1, see my "God's Action in the World," in *Divine Nature and Human Language*; *idem*, "How To Think About Divine Action," in *Divine Action*, ed. B. Hebblethwaite and E. Henderson (Edinburgh: T & T Clark, 1990); and *idem*, "Divine Action: Shadow or Substance?" forthcoming.

affirming determinism.[4] For example, it is held by both Tillich and Macquarrie that even though every event in the created world is determined by natural causes and God does not "break through" at any points to specially influence the outcome, still some of these events are different from others in that we are able to see especially clearly in those cases a relation to divine purposes.[5] But it is clear that these interpretations differ widely from the traditional reading, and the stark contradiction of the latter with determinism remains unmodified.

Why should determinism have taken such a strong hold on contemporary belief? There can be no doubt but that it is due to the extraordinary success of modern physical science in discovering purely natural causal determinants of phenomena. Reflection on these achievements gives rise to a heady intoxication with the idea of the universe as a single, closed, deterministic system, in which the total state of the universe at one moment is a determinate function of its state at any other moment, a vision classically expressed by Laplace around 1800. But now, of course, the parent has turned on its child. From within the scientific holy of holies itself (physics) developments have occurred which seem to provide a definitive refutation of determinism. I refer of course to quantum mechanics, according to which the physical factors operative at the sub-atomic level determine at most a high probability of a certain outcome; the precise outcome is most definitely not uniquely determined by those factors. In putting it in this way I am assuming a realistic interpretation of quantum theory, parallel to my previously announced realistic understanding of theology. There are those who would understand the deviation of quantum theory from determinism as a reflection of the inadequacy of our grasp of the physical facts. The facts themselves are uniquely causally determined; it is just that our theories are so far incapable of coming to terms with this. This sounds to me like whistling in the dark, or, to change the metaphor, clinging to a discredited theory by any means available. To be sure, the future of physical theory cannot be predicted. (If it could, it would already be present.) It is certainly conceivable that future developments will lead to a deterministic physical theory of the sub-atomic. But for the present, the weight of evidence is against determinism, and hence, so far as I can see, that relieves the traditionalist about divine action of that cause for concern.

Let me explain how I am thinking of quantum theory's changing the ball game with respect to divine intervention. Let's say that God wants to alter my consciousness in a certain way (so as to communicate a message to me) or to alter the course of a disease. If we assume determinism, then either God couldn't do it, or God would have to, at least temporarily, suspend natural causal determinism in order to do it. This alternative depends on whether we think of

---

[4] See, e.g., Paul Tillich, *Systematic Theology*, vol. 1 (London: Nisbet & Co., 1953), especially 124-130; John Macquarrie, *Principles of Christian Theology*, 2nd ed. (New York: Scribner's, 1977), especially 250-253; Gordon Kaufman, "On the Meaning of 'Act of God,'" in *God the Problem* (Cambridge, MA: Harvard University Press, 1972); and Maurice Wiles, *God's Action in the World* (London: SCM Press, 1986).

[5] For a criticism of similar moves by Wiles, see my "Divine Action: Shadow or Substance?"

determinism as true independently of God or as an order established by God, and hence subject to cancellation at God's pleasure. But on quantum theory there is no such problem. The relevant physical laws only provide for a large probability of one outcome rather than another in a given situation. And the highly improbable can sometimes happen without violating probability laws. Hence God can, consistent with quantum theory, do something to bring about a physically improbable outcome in one or more instances without any violation of physical law. And even if these interventions are all on the sub-atomic level, they can, if properly chosen (and presumably God would be in a position to do so), snowball so as to make a difference to macroscopic states of affairs.[6] Hence, divine activity that makes specific differences at particular times and places is quite consistent with a quantum theory of physical causality. Perhaps God designed the universe to operate in accordance with probabilistic laws so as to give room for God to enter the process as an agent.[7]

However, I do not think it is necessary to appeal to quantum theory to defend the compatibility of divine intervention with physical science. I believe that divine action at particular times and places is quite consistent with physical laws of a deterministic form. Whether this is so depends on the right way of thinking of such laws. If we suppose that divine intervention in a physical process would involve a violation of physical law, it is because we are thinking of physical laws (of a deterministic form) as specifying *unqualifiedly* sufficient conditions for an outcome. Thus, a law of hydrostatics might specify as an unqualifiedly sufficient condition for a body sinking in still water (of sufficient depth) that the body be of a density greater than the water. A man standing upright in the middle of a deep lake without sinking would be a violation of that law. But we are never justified in accepting laws like this. The most we are ever

---

[6] This is in line with the recent emphasis of "chaos theory" on the way in which in certain kinds of systems a small initial deviation from a certain pattern can mushroom over time to produce large scale deviations. See, e.g., James P. Crutchfield, J. D. Farmer, N. H. Packard, and R. S. Shaw, "Chaos," *Scientific American* (Dec. 1986; reprinted in *Chaos and Complexity: Scientific Perspectives on Divine Action*, ed. Robert John Russell, Nancey Murphy, and Arthur Peacocke [Vatican City State: Vatican Observatory, 1995; and Berkeley: Center for Theology and the Natural Sciences, 1995]).

[7] Not surprisingly, this view of the bearing of quantum theory (construed as involving objective, irreducibly probabilistic laws) on the possibility of divine intervention does not command universal assent from those who have discussed the issue. For a variety of perspectives see William Pollard, *Chance and Providence* (London: Faber, 1958). There is a response to this by Ian Barbour, *Issues in Science and Religion* (New York: Harper & Row, 1966). See also Arthur Peacocke, *Creation and the World of Science* (Oxford: Clarendon Press, 1979); John Polkinghorne, "The Quantum World" in *Physics, Philosophy and Theology: A Common Quest for Understanding*, ed. Robert John Russell, William R. Stoeger, and George V. Coyne (Vatican City State: Vatican Observatory, 1988); and Robert J. Russell, "Quantum Physics in Philosophical and Theological Perspective," in *Physics, Philosophy and Theology*.

justified in accepting is a law that specifies what will be the outcome of certain conditions *in the absence of any relevant factors other than those specified in the law.* The laws we have reason to accept lay down sufficient conditions only within a "closed system," that is, a system closed to influences other than those specified in the law. None of our laws take account of all possible influences. Even if a formulation took account of all influences with which we are acquainted, we can never be assured that no hitherto unknown influences are lurking on the horizon. A man standing upright on the surface of a lake will sink, *unless* he is being supported by a device dangling from a helicopter, or *unless* he is being drawn by a motor boat, or *unless* a sufficiently strong magnetic attraction is keeping him afloat, or. . . . Since the laws we have reason to accept make provision for interference by outside forces unanticipated by the law, it can hardly be claimed that such a law will be violated if a divine outside force intervenes; and hence it can hardly be claimed that such laws imply that God does not intervene, much less imply that this is impossible. No doubt, that is not the sort of outside force scientists normally envisage, but that is neither here nor there. If we were to make the rider read "in the absence of outside forces of the sort we are prepared to recognize as such," our confidence in all our law formulations would be greatly weakened; we have no basis for supposing that science has at this point identified all the factors that can influence natural phenomena. Thus, even if physical laws take a deterministic *form*, the above considerations seem to me to show that they by no means rule out the possibility of direct divine intervention in the affairs of the physical world.[8]

To put the matter in another way, if we construe physical laws of a deterministic form in the way I have just suggested (as holding only within a closed system), then the fact that many such laws have been scientifically established does nothing to support the doctrine of determinism. To bring that off, it would have to be possible to treat the universe as a whole as a closed system vis-à-vis our body of physical laws. That, in effect, is what is envisaged by the Laplacian formulation of determinism. Laplace thought of the universe as a whole as a closed system of Newtonian mechanics, closed to any other factors. And that dream has, of course, long since been shattered. Apart from the replacement of Newton by Einstein, we now know of a number of physical forces undreamed of by Newtonian mechanics, such as electro-magnetic forces, to take only the simplest example. Moreover, we are without an up-to-date substitute for the Laplacian dream as a single closed system for the universe. That is, after the abandonment of the idea that the universe is a closed mechanical system, we have not found any other way of thinking of the universe as a closed system, the structure of which is delineated by a single unified theory. The universe, as envisaged by contemporary physics, seems to me to be highly permeable by divine influence. And until that situation changes, there is not much to be said, in my opinion, for the thesis that particular divine actions would violate physical law.

---

[8] This line of argument was originally suggested to me by C. S. Lewis' *Miracles* (New York: Macmillan, 1947), ch. VIII, though this presentation is my own.

Thus, whether some physical laws are irreducibly probabilistic, or whether they are, in principle, deterministic in form, nothing in physical science would be contradicted by the supposition that God sometimes brings out particular outcomes that are different from what they would have been had only natural factors been involved.

3 *Divine Action and Human Free Choice*

Now for the consideration of problems concerning the relation of divine action to the free actions of creatures, in particular human beings. As a background to this discussion I will make explicit what conception of human freedom I am presupposing. It will be what is often called a "libertarian" conception. There are a number of versions of this, but they all agree in being "incompatibilist," that is, in taking human free action (choice, will, decision . . .) to be incompatible with determinism. To say that I had a free choice (a "real choice") as to whether to accept the invitation to present a paper at the conference, is to imply that nothing other than my choice itself uniquely *determines* me to choose one way rather than the other. Nothing apart from the choice itself strictly determines what it will be—not causal factors operating on my volition, not divine concurrence, providence, or foreknowledge. When I act freely I am the ultimate originator of that action, or of the choice or volition that issues in the action. I make that choice by virtue of my capacity to do so. Thus I can be said, in a sense, to cause the choice, but not by any sort of "event causality," in which one occurrence or state gives rise to a succeeding event or state. What is involved here is what is sometimes called "agent causality," the exercise by an agent of the powers it possesses *qua* agent. This exercise can be, and is, influenced by various factors. In the case at hand my response to the invitation can be influenced by my desires, fears, aspirations, and attitudes, by social pressures and group standards. But none of these *determine* what the outcome will be. Even if they are all pushing in the same direction it is possible for me to buck them and make a contrary choice. In the ensuing discussion I will be thinking of human beings, and possibly other creatures, as possessing free choice in this sense. To be sure, I am not so situated with respect to every thing I do. I lack this kind of free choice with respect to whether I stay in a room in which I am securely tied to a chair. But the assumption is that with respect to many alternatives a human being is free in this sense.

Many thinkers in the Christian tradition have supposed that libertarian freedom is incompatible with basic theological principles—divine omnipotence, sovereignty, or whatever. But I see no merit in this. Being omnipotent God can determine what my actions will be if God chooses to do so. But, just as surely, God can make the opposite choice and refrain from any such determination, leaving the ultimate decision up to me in certain cases. To deny that God *can* voluntarily limit Godself in this way would itself be to deny God's omnipotence. As for divine sovereignty, if that is a matter of God's determining all the details of creation, then it does, of course, rule out human free will. But the question is as to whether we should think of God as sovereign in that sense, rather than simply as having the power to order things as God sees fit. It is sometimes suggested that it would be inconsistent with God's dignity or status to allow creatures to have the final word as to how a given matter should be resolved. But what if God should have sufficient reasons to allow this? I think that a case can be made for supposing that God does. If granting such powers to human beings is necessary for the realization of the divine aim at making it possible for human beings to develop into persons fitted for eternal loving communion with each other and with God, then that would be a sufficient reason to grant us such freedom.

Again, highly influential models of divine grace have represented God as manipulating human beings in such a way as to be incompatible with their exercising free choice.[9] But there are other theologically attractive models of divine grace and sanctification that represent God as influencing the human agent without determining its choices.[10] I cannot pursue these issues further here. I merely wanted to indicate that there are reasons for regarding the belief in this kind of creaturely free will as theologically acceptable.

Now I want to turn to the ways in which human freedom complicates the picture of divine action at particular points in the world. But first I must explain the way in which the discussion will straddle the fence on a basic theological issue concerning creation. That issue stems from the distinction between considering the universe to have an infinite, beginningless existence, and considering it to have begun to exist some finite length of time ago. In the modern period prior to this century it was the accepted view that the universe is temporally infinite, in a backwards direction at least. But with the development of the "Big Bang" cosmology scientific orthodoxy has swung around to the view that the universe began some finite number of years ago. Call this the distinction between an infinitist and a finitist view. Here is the reason this alternative has an important bearing on the doctrine of creation. If the universe has a beginning in time we can think of creation in the most obvious way as a matter of God's bringing the universe into existence at a certain moment, which can be dated relative to the present moment. If, on the other hand, the universe has always existed, divine creation cannot be so construed. It would have to be thought of as an eternal dependence of the world on God for its existence. No matter how far back we go in time there is both the universe and God. The only reason they are not equally fundamental metaphysically is that the universe depends, at each moment, for its existence on God and not vice versa. There is an asymmetrical relation of dependence between the universe and God. On this infinitist view there is no "moment of creation"; or, if you like, every moment is a moment of creation. On the finitist view there is a sharp distinction between creation and preservation. Creation takes place at the beginning. From then on God is preserving the universe in existence but not creating it. On the infinitist view this distinction is conflated. At every moment the universe depends on God for its

---

[9] To be sure, even if divine grace sometimes overrides human free choice, that by no means implies that human beings do not exercise free choice anywhere.

[10] For some discussion of such models see my "The Indwelling of the Holy Spirit," in *Divine Nature and Human Language*.

existence in the same way it does at every other moment. This is, at once, creation and preservation. You could say that creation becomes preservation.[11]

I need not choose between these interpretations here, for the points I wish to make can equally well be made in terms of either alternative. I will explain this is as I go along.

Against this background I return to the ways in which human freedom complicates the picture of divine intervention. The most basic point is this. Apart from creaturely freedom what appears to us as divine responses to particular developments in the history of the world could all be confined to the initial creative arrangements. Suppose there is no creaturely freedom. In that case everything that happens in the world can be rigidly determined by the divine plan. Either God separately determines every detail of the universe, or God wills that what happens at a certain juncture will be uniquely determined, in accordance with laws of nature that God has ordained, by what has happened at other junctures. In either case everything in creation unrolls in accordance with God's eternal purposes, including what appears to us as *ad hoc*, on the spot reactions to particular situations. Thus God could have arranged things in such a way that Moses would feel enjoined by God to issue a certain warning to the Egyptians, whereupon Moses does issue this warning, whereupon the Egyptians ignore it, whereupon they suffer a plague. The initial arrangements could be such that Benson prays to God to restore Curtis to health and shortly after Curtis is restored to health. And similarly for innumerable other cases. It was all contrived from the beginning that the scroll of history should unroll in such a way that events succeed each other in just the way they would if God were intervening in all these situations by way of special acts. Thus, if there is no creaturely freedom, the deterministically minded can recognize an unlimited number of cases in which God is really responsible for the fact that things have come out in one way rather than another in the world, including those kinds of cases that are commonly so taken, while clinging to the position of natural causal determinism.[12]

Now for the bearing of the finitism-infinitism alternative on this issue. I have just been using the term, "God's initial creative arrangements," a term that seems to presuppose that there is a first moment of creation. For if there isn't, how could any divine arrangement be "initial"? If the universe has always been in existence, there are no "initial arrangements." No matter how far back we go, there are earlier stages of the universe. Nevertheless, and this is the basic point, we can still recognize an order of priority in the divine activity. Some creative decisions and actions of God are "initial" in the sense that they presuppose no other divine decisions or actions. Consider, for example, the distinction between

---

[11] I have a presented this alternative in the simplest way, as a choice between a temporally finite and a temporally infinite universe. A more adequate discussion would take into consideration more *outré* views, such as Stephen Hawking's position that the universe is temporally finite but without an initial moment.

[12] To be sure, I have already pointed out that determinism is a lost cause. Here I am only pointing out a motive that a determinist would have for embracing this position.

God's will that there be free creatures and God's will that the Israelites be led out of slavery in Egypt. The former is, or can be, in an important sense, part of God's "initial creative arrangements," while the latter is not, even if the universe is temporally infinite. For the former decision does not necessarily presuppose any other divine decisions, while the latter presupposes that God has set things up so as to permit a situation in which the Israelites are in slavery in Egypt. Thus, on the finitist alternative, "initial" can mean *temporally first*. On the infinitist alternative it will mean something like *presupposing nothing else of the same sort*. But in either case we can make a distinction between God's most basic creative activity and all the rest.

Thus if there is no creaturely freedom everything that happens in creation, including apparent divine interventions, can be completely determined by God's most basic creative decisions. But suppose that human beings, and perhaps other creatures, are endowed with libertarian free choice. In that case is it possible that all divine action (other than conservation) should be squeezed into the original act of creation? So far as I can see, this is possible only if we can credit God with "middle knowledge."[13] This is knowledge that is intermediate between divine knowledge of necessary truths and divine knowledge of actual contingent states of affairs. It is knowledge as to what free choice a certain free creature would make in a situation of a certain sort, if the creature were to make a choice between certain alternatives in such a situation. If God eternally has complete knowledge of this sort, and hence has it independently of any creative activity, God could utilize this knowledge in making original creative decisions. God could adjust free creatures to situations in such a way as to guarantee the outcomes that are in accordance with divine purposes. Any unwanted free choice by a creature could be prevented either by not actualizing that particular individual or by seeing to it that the individual is not in a situation in which it would make that choice. Thus all the apparent on-the-spot reactions of God to particular situations could have been written into the script from the outset just as much as on the alternative of natural causal determinism. As for the plagues in Egypt, God could have chosen possible individuals that would react in just the ways Moses and the Egyptian leaders did react. And in the case of the prayer for recovery from illness, God could have chosen to actualize a person who would pray for Curtis' recovery in just the situation in which Curtis was going to recover. On this alternative there is no more need for particular *ad hoc* divine actions at particular times and places than there is on the deterministic picture.

Let me nail down the point that simple foreknowledge of actual events would not suffice to do this job. To make it possible for God to arrange all the details of the world to God's liking in God's original creative choices, while endowing creatures with libertarian free will, God has to have knowledge of all possible outcomes—of free choices as well as everything else—*knowledge that God can make use of in making creative decisions*. This knowledge has to be *prior* to the act of creation, prior in the sense that it does not presuppose that God

---

[13] This concept was developed by the sixteenth century theologian, Luis de Molina. See his *On Divine Foreknowledge (part IV of the Concordia)*, trans. with intro. and notes by Alfred J. Feddoso (Ithaca: Cornell University Press, 1989).

has made those creative decisions in one way rather than another. But foreknowledge of actual events, including free creaturely actions, is *posterior* to God's act of creation. God could not have knowledge of what actual creatures do in actual circumstances unless there are actual creatures in those circumstances. (This is a particular application of the general principle that I can't know that p unless it is the case that p.) But there are no creatures and no circumstances apart from God's creative decisions. Foreknowledge of the actual comes onto the scene too late to be of any guidance to God in deciding how to shape creation.

To be sure, God could make some creative decisions—to bring into existence a limited number of individuals, including free creatures, and to institute a natural order that had certain features. God could then use foreknowledge of what these individuals will actually do in that order for a period of, say, 5000 years in making decisions as to how to order things after that. But on this scenario, God would still have to make use of middle knowledge in arriving at those initial creative decisions. Since this requires middle knowledge anyway, God might as well make full use of this capacity and do the whole job for the entire creation in the initial creative arrangements.

So, given creaturely free will, the possibility of God's restricting God's activity in the world (over and above conservation) to the original creative act depends on God having middle knowledge. And this has been a highly controversial topic from the time it was originally suggested by Molina.[14] Those who contest the attribution of middle knowledge to God do not, of course, deny that God knows facts of the form "If Cedric were to decide freely between A and B in situation S, he would choose A," if there are such facts. Divine omniscience implies that God knows whatever facts there are to be known. The controversy is precisely as to whether there are or can be any such facts, "counterfactuals of freedom" as they are sometimes called. More specifically, the question is as to whether the supposition that there are such objective facts is compatible with the supposition that the agents in question are genuinely free in the libertarian sense. Let's restrict ourselves for the moment to actual human beings, for example, Winston Churchill, and consider whether there can be a fact of the matter as to what choice he would have made in a situation that never arose. Suppose that Churchill was made Prime Minister in 1932, and consider a situation in 1933 in which he has a free choice as to whether to appoint Sir Samuel Hoare as Foreign Minister. How can there be a fact of the matter as to what he would have done? If he had actually been in such a situation there obviously would have been a fact of the matter as to what choice he did make. And we can even suppose that at an earlier time there was a fact of the matter as to what choice he will make when the moment of choice arrives. But when it is a situation in which Churchill never finds himself during his entire existence, how can it be a fact either that he would make the one choice or that he would make the other. Where we are speaking of causally determined events, there undoubtedly are facts of these sorts. Even if the pencil I am now holding is never, in its whole career, released in still air in my

---

[14] See R. M. Adams, "Middle Knowledge and the Problem of Evil," *American Philosophical Quarterly* 14 (April, 1977): 109-117; and William Hasker, *God, Time, and Knowledge* (Ithaca: Cornell University Press, 1989), ch. 2.

study, it remains true that if it were to be so released it would fall to the floor. There is no problem about the objectivity of facts as to what outcome would be causally determined by a given situation, just because deterministic causal laws, in conjunction with initial conditions, imply a unique outcome.[15] But just because causal determination is lacking for free decisions, it seems that there cannot be any objective fact of the matter as to what a free agent would decide in situations that agent never encounters. In so far as the agent has habits, tendencies, inclinations, desires, or aversions that are relevant to the alternatives in question, it may be that she would have been more likely to make the one choice. If Sandra is specially keen on Mozart, then if she were faced with a choice between going to the Salzburg festival or to the Bayreuth festival, we can say that she would be more likely to choose the former. But if it is a truly free choice that is in question, we would never be in a position to say categorically that she *would* choose the Salzburg festival. In short, if the choice in question is genuinely free, then there is nothing in the nature of things to make a counterfactual of freedom either true or false. Since nothing outside the choice itself determines what the choice would be, and since, by hypothesis, the choice itself is never made, there is no foothold in reality for any answer to the question, "What would she choose if she were to find herself in that situation?" Please note that the point is not just that we are often (if not always) in the position of not *knowing* the correct answer to such questions. That is obviously the case. The point, rather, is that there is no correct answer. There is no fact of the matter.

Thus far I have been discussing situations that, by hypothesis, never arise. But now let's consider God, (temporally or logically) prior to creation, considering what various possible agents would freely choose in various possible situations. Here we cannot assume that the agents in question never find themselves in those situations. Whether they do or not depends on God's creative decisions. If we (God) are thus behind a "veil of ignorance" so far as the actualization or non-actualization of the situations are concerned, will that make a difference to whether counterfactuals of freedom have an objective truth value? Obviously not. The veil of ignorance may affect one's epistemic situation with respect to the question of what free choice Cedric would make in a certain situation, but it can hardly affect whether or not there is a fact of the matter about what Cedric would choose. So long as Cedric is not actually in that situation, there can be no objective fact as to what choice he *would* make, even if there is a real possibility that he will be confronted with the situation. It still remains true that since nothing determines his choice, other than the choice itself, there can be no fact of the matter as to how he would choose.

Nor will the case for objective truth values for counterfactuals of freedom be strengthened if we drop the restriction to actual free creatures, and let the counterfactuals range over merely possible creatures as well. Quite the contrary. There is even less of a basis, if possible, for a fact of the matter as to

---

[15] It is often taken as a distinguishing characteristic of lawlike statements, as contrasted with "accidentally true generalizations" like "all the books now on my desk have red covers," that the former entail counterfactual conditionals as to how things would have behaved in situations that have never developed.

what a never existent human being (my third son) would have chosen had he been faced with a choice between going to Harvard or going to Yale than there is for a fact as to what an actual human being (myself) would have chosen if faced with these alternatives.

Thus there are solid reasons for denying that there are *facts* as to what choice an actual or possible free creature would make in a given situation. And if there are no such facts, even an omniscient being could not know them; even an omniscient being can know only what there is to be known. But if God lacks middle knowledge, and if God has included creatures with free will in creation, then, as we have seen, there is no way in which God could make provision in the initial creaturely arrangements for what appear to us to be on-the-spot divine reactions to particular developments in the history of the universe. If things are to turn out so that it is just as if God reacted to particular free creaturely choices and/or situations affected by those choices, then this appearance will have to be a veridical one. It will have to be the case that God does indeed "intervene" in the world at those times and places and bring about results different from what would have eventuated had only natural causal factors been operative. God will have to make specific reactions to specific local circumstances, just as it has always seemed to people in the Christian tradition. The escape sought by the deterministically minded of cramming all divine action into creation and conservation is blocked. Since we have seen in the previous section that deterministic constraints lack any force they might once have had, this is hardly cause for lamentation. Moreover there are other reasons for taking this result as an occasion for rejoicing. Since the conviction of God's active involvement in the course of events (other than just by keeping everything in being) is so intimately woven into the warp and woof of the Christian life, ripping it out would do perhaps irreparable damage to that life. Nowhere in this more true than in the communion and dialogue with God that is at the heart of Christian spirituality. If we are not genuinely interacting with the divine when we pray, listen to God's voice, ask and receive strength and guidance, then the Christian tradition has been a snare and a deceit, and we are of all people the most miserable.

But we must be careful to claim no more than what the argument warrants. We have by no means shown that none of the apparent divine on-the-spot reactions could have been arranged in the initial act of creation. What we have shown in this section is that, given creaturely free will and the absence of middle knowledge, *divine reactions to free creaturely actions and whatever has been influenced by them* could not have been pre-programmed in the original creative decisions. But that may leave vast stretches of the creation untouched. So far as we can tell, most of the spatio-temporal extent of the universe up to this point has been unaffected by free human activity. That leaves open the possibility that some or all of it has been affected by the free actions of other creatures, but we have no solid reasons to suppose this to be the case. Thus we should take seriously the possibility that with respect to much that has happened in cosmic history, God could have confined God's own activity to the initial creative arrangements. To be sure, quantum indeterminacy is relevant here also. If the basic physical laws God instituted do not render future states of the universe rigidly determined by previous states, that in itself might prevent God from seeing to it that everything comes out in accordance with God's plans, just by God's choice of an initial state plus natural laws. But if quantum indeterminacy on the sub-atomic level is compatible with determinism, for all practical purposes, on the macro level, and if God's more important purposes extend only to the latter, then God could insure the realization of the divine purposes (apart from creaturely free will) just by the initial physical arrangements, even if these involve probabilistic laws of quantum theory.[16]

However, this possibility may not be of any real religious interest. The putative divine interventions at particular points in the course of events that are taken special notice of in the Christian tradition, and other theistic religions, virtually all have to do with divine reactions to free creaturely actions or to situations affected by them. This is true of all forms of divine-human communication and interaction, including messages and commissions delivered by God to particular human beings, divine responses to prayer, divine appearances to human experience, divine punishments for sin, divine arrangements for influencing the course of human history (the Exodus, the Incarnation and all that is involved in that, the Resurrection, and so on). Even "purely physical" miracles like the parting of the waters of the Sea of Reeds or the burning of the bush that is not consumed are construed as responses to a particular juncture in human affairs. It is the conviction of divine intervention in human affairs that is crucial to the Christian tradition, not God's dealings with nebulae, galaxies, and black holes. Thus, if it can be shown that the furtherance of divine purposes requires on-the-spot reactions to free human action and what is affected by them, that will be of enormous significance for the way we think about such divine action as is religiously significant.

4 *The Relevance of Physical Cosmology*

I have been going into these issues concerning the relation between divine action, on the one hand, and the laws of nature and human freedom on the other, partly because of their intrinsic interest, but also partly, in line with the emphasis of volume, in order to consider whether cosmology has any significant bearing on the issues. Does it make any real difference to the issues of this paper how we answer various cosmological questions?

I will begin by making a couple of related points about the relevance of cosmology to philosophical theology, points that I would think are obvious but still deserve mention here. Contemporary scientific cosmology has made a major difference to arguments for the existence of God. The dramatic shift in the weight of opinion from a temporally infinite to a temporally finite universe has

---

[16] I have not forgotten that in section 2 I speculated that God might influence macro outcomes by suitably influencing sub-atomic outcomes, thus assuming that natural causal determinism does not reign, even for all practical purposes, on the macro level. Obviously, I can't have it both ways. But note that I am only considering the view just formulated as a possibility. It is not a part of my thesis in this paper.

markedly altered the thinking about the cosmological argument. If the physical universe came into existence out of nothing some finite period of time ago,[17] there is a very simple argument for an external cause of the existence of the universe that we lack if the universe did not begin to exist at a certain time. We can simply argue that any beginning of existence must be caused by some agency already in existence.[18] I am not suggesting that there is no viable cosmological argument if the universe is temporally infinite. Many such versions have been developed, and I consider some of them to have a certain amount of force.[19] Nevertheless, developments in contemporary cosmology have rescued the idea that God can provide an explanation for the beginning of the universe and made it a live option once more.

Again, new forms of the teleological argument have been spawned by considering the "fine tuning" of the relative magnitude of various physical constants that is required if the universe is to develop in such a way as to make life possible. This point has been developed at length in the recent literature, and I refrain from adding to it here.[20]

But what bearing does contemporary cosmology have on the issues discussed in this paper concerning the way to think about divine action and its relation to laws of nature and to human freedom? There are strong reasons for supposing that cosmology has no important bearing at all on these issues. At least at first sight it seems irrelevant whether we take the universe to have come into existence *ex nihilo* at the Big Bang, or whether we suppose, for example, that there is an infinite series of expansions and contractions; whether we give a more classical or a more quantum account of the origin of our universe; whether we opt for the standard Big Bang model or the "inflationary universe"; and so on. The choice between these cosmological models makes little if any difference on what is the best way to construe the nature of divine action: whether we should think of God as carrying out specific *ad hoc* actions at particular times and places, how specific on-the-spot divine actions fit into the lawfulness of the universe, or how the account of divine action has to be shaped to allow for free creaturely action. How could the physical details of the way in which the present state of the universe has developed from an inferred initial state constrain the way in which we think of God as active, especially how we think of God as active in the very recent stages of the universe, where human beings are on the scene?

But on closer inspection it appears that some differences in physical cosmologies carry with them differences in the ontology of time, and that those

---

[17] I am not suggesting, of course, that all cosmologists accept this. I am only speaking of a certain "weight of opinion."

[18] This principle would not be universally accepted, though I must confess that I see no reasonable alternative. Be that as it may, my point here is the modest one made explicit in the last sentence of this paragraph.

[19] See William Rowe, *The Cosmological Argument* (Princeton: Princeton University Press, 1975), for a development of this point.

[20] See John Leslie, *Universes* (New York: Routledge, 1989), and the references contained therein.

differences do make a difference to some of the issues of this paper. I will devote the rest of the paper to a consideration of their possible bearings on human freedom and divine action, in that order.

5 *Cosmology and the Ontology of Time*

The strongest candidate for a difference in the ontology of time that is relevant to human freedom is the difference that is associated, in philosophy at least, with the name of J. M. E. McTaggart. McTaggart made a distinction between the A and the B series of temporal positions.[21] Briefly put, the A series involves the distinction between past, present, and future, while the B series involves the relations of earlier than, later than, and simultaneous with. These series need not be taken as incompatible. In our ordinary ways of thinking about time we take it for granted that they are both realized. And no one who assumes temporality of any sort doubts that it involves the B series, though the special theory of relativity holds that there is no unique order of such relationships. The issue concerns the A series. Let's spell out what it would take for the A series to be objectively real in a fundamental way.

If the distinction of past, present, and future marks some irreducible, objective feature of reality, this has two important consequences. First, different moments successively enjoy the status of being the *present* moment, the one at which things are actually going on; and this is an irreducible feature of reality. It is an objective fact about the universe that my writing of this sentence is *happening now*, is in the *present*, while my eating breakfast this morning has already happened, is in the past, and my eating dinner this evening has not happened yet, is in the future. If one were to set out to give an exhaustive description of reality, the temporal side of that would not be exhausted by an account of the temporal relations of events to each other (the B series)—whether event F is earlier than, later than, or simultaneous with, event G. One would also have to include in the description what is happening *then*; what is present, what is past, and what is future. (The details of this description will, of course, vary depending on when the description is being given.) Second, the ontological status of a given event changes in the course of the world process. An event *actually exists* only at the moment at which it is happening, only when it is *present*. At other moments it is either past—over and done with, or future—yet to come. Thus what actually exists in the most basic sense is constantly changing. Even on this view there may be a sense in which all events are real; one can recognize a kind of reality possessed by the past and the future. They are, after all, significantly different from sheer nothingness. But the crucial point is that their ontological status changes in the course of time and does not remain the same eternally.

Let's call the view that the A as well as the B series is a fundamental feature of reality the "A-B view," and the view that only the B series is ultimately

---

[21] See *The Nature of Existence*, vol. II (Cambridge: Cambridge University Press, 1927), bk. V, ch. 33.

real the "B view." The B view will deny the two cardinal commitments of the A-B view laid out in the last paragraph. It denies that it is an objective feature of the universe that certain things are going on now; and it denies that the ontological status of events changes with the passage of time. It is this latter denial that gives rise to the popular designation of "block universe" for the universe on the B construal.[22] On this view the universe is "frozen"; everything is there eternally. Nothing really changes. All that is objectively real is a matrix of temporally related events,[23] each of which unchangeably has whatever reality it possesses. Since the A-B view regards process as metaphysically fundamental, it is sometimes called a "process view."[24]

The issue can also be put linguistically in terms of the status of temporal "token-reflexives." Token-reflexives are expressions, a particular use of which can be fully interpreted only in the light of information about that particular use. They include demonstratives like "this" and "that" (including compounds like "this book" or even "the book"), pronouns, tenses of verbs, and other temporal expressions like "now" and "present." We can't understand a particular utterance of "I am hungry" without knowing who is speaking and when. We can't understand a particular utterance of "The glass is chipped" without knowing which glass the speaker is indicating on that occasion. We can't understand a particular utterance of "She's over there" without knowing the place being designated by the speaker. Knowledge of the general meaning of words in the language is not sufficient in these cases, whereas when token-reflexives are not involved, such knowledge is often sufficient.[25] In these terms, it is clear from the above exposition that the A-B theory implies that temporal token-reflexives are indispensable for a complete description of the world. Since one of the irreducible facts about the world is what is going on *now* (as contrasted with what *was* going on in the *past* and what *will be* going on in the *future*), such facts cannot be stated without temporal token-reflexives. Since such facts vary in content from one time to another, we have to be speaking *from a given temporal*

---

[22] See the article by Isham/Polkinghorne in this volume.

[23] This makes it sound as if the B-view presupposes a relational conception of time rather than an absolute conception. But it need not be so. The relations featured by the B-view hold between events on both the relational and the absolute conceptions of time. The difference between these conceptions is not over whether the relations hold but over whether they are ultimate (the relational conception) or whether they are derivative from the position of each event in the absolute time.

[24] I am not entirely happy about calling these views different views of *time*. It is not as if they necessarily differ on how to think of the temporal dimension. The difference, rather, concerns how to think of change or process. Should we think of it as something ultimate (A-B view), or should we think of it as simply a matter of the relations between events at different moments (B-view)? But the views are traditionally called views of time, and I will go along with that.

[25] I have to say "often" because there are other sources of variation in the content of what is said. For example, where a word has more than one meaning in the language we have to use contextual indications to determine in which sense it is being used on this occasion.

location in order to state such a fact. But on the B-view temporal token-reflexives are dispensable, as far as stating basic metaphysical truths are concerned. Since temporal reality is exhaustively describable by making explicit temporal relations between events, there is no need to say anything the content of which is partly constituted by the temporal position from which it is said. The fact that writing this sentence is later than eating my breakfast on January 25, 1992 AD is a fact that can equally be stated from any temporal perspective whatever.

If the A-B view does not take facts about what is going on *now* and what has already happened as irreducibly real, how does it construe them? Surely it is a fact that the 1992 presidential campaign is going on now! The short answer is that it proposes to construe them as subject-relative. To say that that campaign is going on now is just to say that it is simultaneous with that utterance. To say that the 1988 presidential election has already happened (is in the past) is to say that it is earlier than that utterance. These relations of simultaneous with, earlier than, and later than, make up all there is by way of objective temporal reality. The A-series facts are what get expressed when someone notes some B-series relationship from a certain temporal position. From the standpoint of moment m, some events appear to be present, some past, and some future. But the only reality involved in that is the network of temporal relations between events. All else is the way things appear to an observer, speaker, or thinker.

How are these metaphysical differences related to concepts of time in physical theories and to ways in which time is treated in physical theories? The first thing to note is that nowhere in physics is anything made of the A-series. Physics takes an impersonal view and speaks of temporal relations between events. Its theories are unconcerned with what is happening *now*, as contrasted with what has already happened.[26] Does that mean that physics supports the B-view? It does in that physics claims to be able to carry out its jobs successfully without any attention to the A-series. But to suppose that this is a conclusive argument for the B-view is to suppose that only what needs to be taken account of in physical theories is objectively real; and that leaves out a great deal of what many of us would regard as enjoying that status—persons and aesthetic objects, not to mention furniture, mammals, and flowers.

How about more specific features of the treatment of time in physical theories? Special relativity denies that there is a unique temporal framework. But this implies non-uniqueness for the B-series as well as the A-series, and thus would not seem to favor one over the other. However, I am inclined to think that a conviction of the objective reality of the B-series can survive relativization better than the like conviction for the A-series. According to special relativity facts about what is happening now will have to be facts about what is happening now in some framework. That may well seem further from the pre-relativity A-B view than the relativistic version of the B-view is from its pre-relativistic cousin.

---

[26] So far as I can see, this point has nothing to do with whether any physical theories provide a basis for an "arrow of time." Even if some do, that only gives a nomological basis for dealing with relations of earlier than and later than, and that can be expressed in the B-view.

But there is still no doubt that there are intelligible relativized versions of both views.

As for general relativity, it is sometimes thought that the treatment of space and time together in a four-dimensional manifold is incompatible with the A-B view. But, so far as I can see, this all depends on whether the theory has some objective basis for differentiating the temporal dimension from the three spatial ones. If it does, then it is not clear that the A-B view, in a relativized form, is ruled out. If, however, the theory is so construed that it is arbitrary how three dimensional spatial slices are linked up in what is called a temporal dimension, that is a different matter.[27] Now there is no unique answer to the question of what physical reality corresponds to our intuitive sense of a temporal dimension. Thus the question of whether there are objective facts as to where we are now in the world process along that dimension cannot even be raised. This really does rule out the A-B view. But we must realize that it also rules out the B-view as usually understood. For there would also be no objectively specifiable order of relations of earlier than, later than, etc.

6 *The Ontology of Time and Free Choice*

Thus, features of some physical theories would seem to have some implications for the ontology of time, though the matter is far from clear. Now I turn to the possible bearing of the distinction between the B-view and the A-B view on human freedom. It is intuitively plausible to think that we enjoy libertarian freedom only on the A-B view. If all parts of time are equally real from any temporal perspective, then all of 1992 is already real from the perspective of any earlier time. But that means that it was already settled 5, 50, 50,000, or 50,000,000 years ago what I would do in 1992, including any putatively free actions. And that means that I have no real choice as to whether to perform them when the time comes. Anything else is impossible for me, and so I do not freely do what I do.

But this line of argument suffers from a confusion of eternal, timeless existence with existence *at every time*. The B-view is distinguished from the A-B view by virtue of denying that certain events exist now in a basic sense in which they do not exist at any other moment. The entire matrix of moments and what goes on at those moments has its basic mode of existence in such a way that it is unaffected by the "passage of time." More precisely, the B-view denies that there is any such thing as a *passage* of time, and hence denies that there is anything to affect the ontological status of events in the way claimed by the A-B theory. But this fundamental mode of existence in which all times and all events are eternally on a par is an eternal or timeless mode of existence. There is no implication that all events exist *at all times* or, indeed, *at any time*. The only sense in which an event exists at a moment is the sense in which that is the moment at which it happens. The Battle of Waterloo existed during a certain stretch of time in 1815, not at any other. It is certainly not going on now, or at any time in the third

---

[27] See the article by Isham on quantum cosmology in this volume.

century! The entire complex of temporal moments and what they contain in the way of events enjoys an immutable eternal mode of existence. None of them ever gain or lose this basic mode of existence. But each event does *happen* or *occur* at a given moment, and not at any other. That is what it is to be temporal. Thus the "preexistence" of my putatively free actions is only a bogus worry. There is no intelligible sense in which the B-view implies that what I (supposedly) freely do now was already real 5,000 years ago. What I do now happens when I do it, and that's that.

This way of averting the threat to libertarian free will from the nature of time is interestingly similar to a way of averting the threat to free will from divine foreknowledge. If God's own mode of being is atemporal (see below), then it is not the case that God knows what God knows *at one or another time*. Rather God knows what God knows eternally. Hence, it is not true that God knows what I do at this moment *before* I do it, nor does God know this at any other time. God knows it because God sees me doing it when I do it, God's eternal existence being all at once simultaneous with every temporal moment. Again, it is only if we confuse an atemporal fact with a fact that obtains at every time that we fall into the trap of supposing there to be a conflict with free will.

Not everyone will accept my reading of what is and is not implied about the time of existence of events by the A-B view. And in any event some will feel that libertarian free will requires an open future in a sense in which that is available only on the A-B view. In supposing that it is strictly "up to me" whether I accept the job offer, am I not supposing that my acceptance or rejection is still open? And isn't that to suppose that the world process has not yet arrived at the point of settling it? And doesn't that require the A-B view? So far as I can see, the answer is "No." It's being up to me whether I accept the job requires only that my decision not be *determined* by anything that obtains prior to it. And that is certainly allowed (though not implied) by the B-view if the above reading of that view is correct. On that reading the B-view does not imply that my decision already exists at times before it is made. And it clearly does not imply that my decision is causally determined; it has no implications at all for the causal structure of the world. Hence it leaves open the possibility that my decision is uniquely determined by nothing outside itself. That possibility remains even if the A-B story about the ultimacy of process is false; and that possibility seems to me to be enough for libertarian free will.

Thus, my message on the relation of this issue in the ontology of time to the free will issue is a deflationary one. Even if cosmological or other physical theories have import for that ontological issue, it remains to be shown that they have a bearing on the free will issue. There are other possibilities to be explored, but this is as far as I can push the discussion here and now.

7 *Cosmology and the Temporality of God*

Finally, I will say something about the possible bearing of cosmology on the question of whether God's existence is better thought of as temporal or atemporal. This issue is obviously important for how we think of divine action. If

God is a temporal agent, living through a succession of temporal moments, then we can much more straightforwardly think of God as forming plans, doing things to implement them, reacting to situations as they arise, and so on. It is generally agreed on all hands that our basic model for thinking about divine action is human action,[28] and human action prominently features temporal succession. Whenever we react to a perceived situation, a temporal sequence is involved. First we note the situation, then we think about how to respond to it, then we set a chain of events in motion. If God is atemporal, we will have to think of divine action in a radically different fashion. I do not believe that this poses insuperable obstacles. Even if God does not live through successive moments, God's eternal acts of will can have temporally ordered consequences in the world, and so God can be said to act at the times occupied by those consequences. And even if God's realization that the Israelites did so-and-so and God's will to respond in a certain way are not temporally successive, they can be ordered asymmetrically in other ways (e.g., by the latter presupposing the former but not vice versa), so that it can be significant to speak of God's *responding* to the situation in Israel and to speak of God doing what God did *in the light of* that situation.[29]

Be that as it may, does physics in general and/or cosmology in particular have any implications for this aspect of the nature of God? I believe that it does. For one thing, the whole development in relativity physics of treating time as an integral aspect of a single spacetime manifold provides support for the view that a being cannot be temporal without also being spatial. Time cannot be split off from spacetime and realized apart from space. Hence, if God does not exist spatially, neither can God exist temporally. Nor does this point require that time not be objectively distinguishable from the other dimensions in spacetime. Even if it is, physical theories seem to imply that it has no reality out of that four-dimensional complex, in which case we get the above conclusion about the divine being.

Moreover, as has been noted by several papers in this volume, including those of Drees, Isham, and Russell, more than one quantum cosmological theory envisages a kind of emergence of time in the "early" stages of the universe from a more undifferentiated four-dimensional manifold. In that case not only is time essentially tied up with physical, spatial existence, it is tied up with the latter only under certain, not invariably realized conditions. That makes time even more special to certain creaturely conditions, and even less suitable as a mode of existence for the Creator. As the above named writers have noted, this lends powerful support to the idea that time was created along with the universe, indeed, along with a certain form of the universe. And that tells against its being

---

[28] This does not imply anything about the extent to which divine action exhibits the features of human action. Whenever we use X as a model for thinking about Y, there will be some differences between X and Y; and the degree of difference will differ widely in different cases.

[29] For further discussion of this point see my "Divine and Human Action" and "Divine-Human Dialogue and the Nature of God," both in *Divine Nature and Human Language*. See also Eleonore Stump and Norman Kretzmann, "Eternity," *Journ. Philos.* 78 (August, 1981).

a metaphysically necessary form of every kind of existence, including the divine existence.

It may be felt that the view of God as enjoying a timeless mode of being is incompatible with the thesis defended here that God does really act in the world at particular times and places. If God is not in time how can God do something at a particular time. The answer is that it is the state of affairs brought about (directly) by God's activity that is located at a particular time and place. This leaves the divine activity itself (from God's side) free to have whatever status (temporal or otherwise) that is dictated by the divine nature.[30]

8 *Summary*

The main concern of this paper has been with the ways in which considerations of laws of nature and human freedom impinge on the question of how to think of divine action in the world, particularly those alleged divine actions at particular times and places that are thought of as divine "interventions" in the natural order. My contention has been that the reality of both physical laws and human freedom is compatible with the reality of divine interventions in the world. Indeed, in the case of human freedom I have argued that unless we recognize divine middle knowledge God could not, in general, bring off what looks to us like *ad hoc* interventions without actually doing it that way.

As for the bearing of physical cosmology and other physical theories on these issues, the message of the paper is mostly deflationary. With respect to the main concern of this volume—physical cosmology—I cannot see that a decision on the main outstanding issues in that field will materially affect how we should think of divine intervention in the world and how that is related to laws of nature and human freedom. I have been able to discern some relation of issues in physics to the ontology of time, and the question of whether God is temporal, but here too, so far as I can see, there are no important implications for the main topics of this paper. I think it not surprising that this should be the case. Important implications of widely separated fields for one another are the exception rather than the rule in the intellectual realm, despite the current vogue of interdisciplinary studies.

---

[30] For more on this see Stump and Kretzmann, "Eternity."

CONTEMPORARY PHYSICS AND THE ONTOLOGICAL STATUS OF THE LAWS OF NATURE

William R. Stoeger, S. J.

1 *Introduction*

William Alston writes, ". . . I believe that divine action at particular times and places is quite consistent with physical laws, even of a deterministic form. Whether this is so or not depends on the right way of thinking of such laws. To suppose that divine intervention in a physical process would involve a violation of physical law depends on thinking of physical laws (of a deterministic form) as specifying *unqualifiedly* sufficient conditions for an outcome."[1] What then *is* the right way of thinking of "the laws of nature"? This is an absolutely crucial question for a number of the key issues we are confronting in this series of conferences.

The concept of the laws of nature which we take as operative will seriously affect many of the conclusions we reach about reality as we seem to know it from the sciences, and will especially affect key philosophical issues deriving from that concept. Certainly, as we approach the horizons and limits of scientific knowledge, the consequences flowing from the transcendence, the necessity and independent reality or fundamental character with which we endow the detailed descriptions and laws given by physics, chemistry and biology will strongly determine any conclusions about the reality which lies beyond those limits and horizons; the weight and power we give the Anthropic Principle, and the way we approach the difficult issue of purpose and natural tendency in the physical and biological worlds; the degree to which an ontological reductionism can be justified; the way in which we conceive the proximate and ultimate origins of the laws themselves; our critical or uncritical acceptance of the reality of other universes with alternative sets of laws as a valid explanation of ours; the application of retroduction (abduction) and the principle of sufficient reason in situations where the consequences are observable, but the possible explanatory antecedents never will be—the resolution of all these important questions depends heavily on the ontological status we attribute to the laws of nature. The way we conceive these laws of nature will also, as Alston points out, have a very important, if not determinative, bearing on how we understand and model God's action in the universe.

In this paper I shall first employ a critical description of, and phenomenological reflection on, the physical sciences themselves to determine what concept of the laws of nature is really justified, and to purify our notions of some common illusions. In doing this we shall better see the degree of depth we

---

[1] See the article by W. Alston in this volume.

are able to achieve in using them to describe and explain material reality. I shall not be able to do this thoroughly—that would be a project involving years and volumes. But I do want to point to what seem to me critical bits of evidence, and the directions in which they seem to take us, in clarifying the weight we are able to place on the laws of nature.

Second, I shall pose five key philosophical questions concerning the laws of nature and suggest answers to them on the basis of our phenomenological reflections in the previous section.

Third, I shall focus on several key philosophical problems raised by the laws of nature: realism, explanation, inference, and identification. Our findings will reinforce our principal arguments and conclusions in the preceding sections.

In this discussion I shall argue, as others have done before me, that, although the laws of nature do reveal and describe fundamental patterns of behavior and regularities in the real world, we cannot consider them the source of those regularities, much less attribute to them the physical necessity these regularities seem to manifest. Nor can we ascribe to them an existence independent of the reality whose behavior they describe. Instead I claim that they are imperfect abstract descriptions of physical phenomena, not prescriptions dictating or enforcing behavior. Thus, a Platonic interpretation of these laws is unjustified. Nor can we postulate the pre-existence, or ideal existence, of other alternative sets of laws which may be possible—or might actually obtain—in other universes. That there are such pre-existent laws is a philosophical stance independent of any evidence. Other universes may in fact exist, but what we observe here does not bear one way or another as evidence for such existence. Nor does the character of the laws emerging from the natural sciences—the particular cases in point here pertain to physics and cosmology—find possible explanation in such ensembles of other universes, or in other sets of laws, of which our set of laws is just one instance.

Throughout this paper I shall often refer to the underlying regularities and relationships of physical reality, as they are in themselves, which the laws of nature describe. These underlying realities and processes do not constitute a Platonic world, but rather the real world as it is in itself, independent of our knowledge of it. If our long experience of gradually improving our theories and laws of nature is any indication, this network of regularities and interrelationships is incredibly richer and more intricate than any models of it that we can construct.

Although we cannot really uncover these realities "as they are in themselves," we know that they exist by arguments from the principle of sufficient reason. What explains and governs them? I am not dealing directly with this question in this paper—though, again, by the principle of sufficient reason, there must be an explanation. The answer may be a Platonic one—or a theological one with Platonic overtones (in terms of the relationship of physical reality to God). From the scientific point of view, however, we cannot answer this question—and we obtain only vague hints of an answer from philosophy. What I wish to deny is that our theories and laws of nature provide any such explanation of why these underlying regularities and interrelationships are as they are, rather than some other way. They do not even adequately reflect the way the world is in itself; we are always finding that, though they reveal the structure of physical reality as we encounter it in great detail, they are incomplete even in their description—and certainly incapable of explaining it in any real sense of that word.

Fourth, I shall indicate briefly how our concept of the laws of nature affects: our view of the limitations and strengths of the sciences; the account we can give of the origin of the laws of nature (in the two senses this phrase now takes: their epistemological and their ontological origin); the answer we give the question of reductionism; the prospects for constructing a "theory of everything"; and the prospects for speaking sensibly about God's action in the world. A crucial aspect of our discussion will be the impact of this concept of the laws of nature on the conclusions we arrive at when we apply the Anthropic Principle. As Alston says, much depends on thinking correctly about the laws of nature, what they contain and what they really imply—not only how we characterize or describe God's action in the world, but also, and perhaps more importantly, how we conceive the scope and the horizons of the scientific endeavor itself and its relationship to other disciplines.

2 *The Laws of Nature in the Natural Sciences*

Do we really have in our laws of nature a complete description of the fundamental structures, processes, and regularities which constitute physical reality? Do they give us the underlying necessities which *govern* it? (These are two separate questions!) Or are they, rather, approximate descriptions which reflect the underlying regularities and constraints in the physical, chemical, and biological world, but not the independently existing enforcers of that behavior? There is a long historical debate between these two alternatives. In attempting to contribute to this debate, and to provide adequate justification for my answers, I shall discuss a number of complementary aspects of scientific theory and the philosophy of scientific knowledge. The conclusions we shall reach from these different perspectives will converge on an answer to the questions concerning the ontological status of "the laws of nature."

First of all, in some contexts we speak of the theories that make up the natural sciences. In doing so, we emphasize their provisionality, and their incomplete, imperfect, and approximate character. To be sure, there are many different levels of justification for the theories we hold and use—in some cases a theory is so thoroughly confirmed and functions so basically as a building block for the structure of a discipline, that there is little practical doubt about its validity. Nevertheless, this does not rule out the possibility that it may be replaced or subsumed by a different or more complete theory—one which explains more or is applicable in a wider range of situations and conditions. This is what happened to Newtonian mechanics and gravitation; they were subsumed as limited cases of special and general relativity, respectively. In other contexts, strangely enough, we speak of the more fundamental—and even sometimes the less than fundamental—descriptions of theories as "laws," or "laws of nature," as if they were complete and perfect explanations of the regularities and interrelationships we find in physical reality. This seems to endow them with an immutability which perfectly reflects the immutability of those regularities and interrelationships themselves, giving them a fixity and an absolute character which their status as "theories" does not seem to support. Going beyond this we often even speak of these laws as if they were *prescriptions*, sources of physical necessity, not just imperfect representations of regularities.

We have the almost continual experience in the sciences of replacing or subsuming laws and well-confirmed theories by more comprehensive ones, which more adequately describe the relevant aspects of phenomena under a wider range of conditions. This experience leads us to conclude that these laws and theories are actually "models," or approximate descriptions, albeit very accurate and detailed ones. The movement from the old theory to the new theory is very often such that the first one is correct in some limit—for instance, Newton's theory of gravitation is still correct, or gives correct answers and predictions, for relatively weak gravitational fields (such as that of the Earth) and slow velocities (much less than the speed of light). At the same time, however, the new theory will usually be articulated in terms of radically different concepts, entities, and features—giving us a completely revised description of the underlying reality. And, within certain limits, operations or equivalencies relating the fundamental concepts of the more expansive theory with those of the restricted theory will be established. Despite these relationships the difference between the two theories represents a definite and often radical paradigm shift. This new theory, in its turn, may be replaced by another one—for example, in the case of general relativity by a quantum theory of gravity.[2]

The support for both the essential shift of paradigm and the limiting correspondences between the new theory/laws and the old theory/laws are, of course, found principally in a whole complex of controlled experiments and observations which are carried out to test the theories, and repeated again and again to higher and higher precision under different experimental conditions. The new theories and laws really do describe what occurs in a larger range of circumstances and conditions than do the older ones. They also predict new phenomena, which may eventually be detected and measured. If they are not, this increases the pressure for further theoretical modifications.

But does this mean that the older theory does not describe reality as well as the newer theory *within* the narrower range of its applicability? It will often be simpler in formulation and more usable in those cases. These simpler original laws still hold to a very good approximation in a broad range of well-defined circumstances. Thus, for those cases, we have in effect at least two theories and two sets of laws which work equally well—and we have an apparent uniqueness problem. Which one better represents physical reality under these conditions? It depends on *our* criteria for "better" and on what constitutes a "good representation," doesn't it? Which one better represents physical reality as it is *in itself*? The improved theory? Not necessarily. For to warrant such a claim we must either assume or demonstrate that we have some privileged knowledge

---

[2] See the articles by G. F. R. Ellis/W. R. Stoeger; and C. J. Isham in this volume.

of physical reality independent of the phenomena which we observe and by which we can decide which theory better describes reality. But our observations and measurements of those phenomena are actually, to a greater or less degree, "theory laden," through the design of the experiments and observations and the interpretations of the results. Much more could be said here on this issue. These few comments, I think, expose some of the illusions to which our lawlike descriptions and explanations of natural phenomena are subject.

There is another somewhat similar set of experiences in the physical sciences (and perhaps in biology as well). The pervasive phenomenon of phase transitions and spontaneous symmetry breaking in both microscopic and macroscopic physics—and the related phenomenon of bifurcations in dynamical systems—challenges us to link the detailed specification of behavior under one set of conditions (say, when water is a liquid) to a completely different description of behavior under another set of conditions (when the water has frozen to ice). The characteristics of a physical or chemical system in a broad range of conditions are often specified by an "order parameter," such as temperature. As the temperature rises above or slips below certain thresholds, the system undergoes sudden radical changes—losing some of its structure, or gaining new structure which manifests very different physical or chemical behavior. This new behavior requires different physical or chemical descriptions and follows new laws. Even the theoretical entities which characterize the situations on each side of such a phase transition may be different. In addition, important parameters which were constant on one side of the transition are no longer constant on the other side. A mode of interaction which dominated on one side no longer even exists on the other side, and what was very regular, predictable behavior before may now be completely chaotic and unpredictable. The laws governing the dynamics—and the theories from which they flow—are significantly, even radically, different in the two different temperature (or other order-parameter) regimes.

However, theories of symmetry breaking and phase transitions often provide the link between the two different regimes, describing and even explaining at a certain level how the higher temperature entities and their characteristics and dynamics are related to and flow into lower temperature entities, characteristics, and dynamics. Baryon conservation is a law at low temperatures, but it may be severely violated at very high temperatures, and these more complete theories must specify exactly how the two regimes fit together and how more general processes which violate baryon conservation at high energies yield only to baryon-conserving ones at lower energies. Similar cases of important transitions have proliferated in contemporary physics and chemistry: gas to liquid, liquid to solid, laminar flow to turbulent flow, unmagnetized regimes to magnetized, normally conducting or nonconducting to superconducting, and so on.

Though it is clear that these unified theories and laws certainly do more and more adequately describe and, even at a certain level, account for what we observe—so much so, in fact, that we can use such knowledge to construct new devices, harness the dynamic characteristics of nature to improve our abilities to communicate, navigate, travel, control disease, generate and store energy, and so on—they do not seem to give us more than detailed and reliable *models* of certain aspects of nature and of its regularities. They do not give us a complete and thorough understanding of the natural processes—nor an immediate and unambiguous intuition of nature's secrets as they really are in themselves, if indeed that concept has any sense. We are always peering into nature and nature's secrets from the outside, using instruments and ploys which put us into contact with them only indirectly. Nor do the increasingly detailed and comprehensive theories and laws answer the ultimate "Why's"—why they are that way, rather than some other way. Why does mass-energy curve space? Why do electro-magnetic fields obey Maxwell's equations? What determines the values of the fundamental constants—why do they possess these particular values, and not some other ones? Intermediate answers can sometimes be given by the sciences to questions such as these, but only in terms of another entity or structure or relationship of which we must ask the same question. We seem to end up with an infinite regress of "Why's."

The theories and "laws" of the sciences—no matter how well-confirmed—are always idealized in some way. They involve constructions with which we confront the phenomena in a very sophisticated and controlled way, but always in a way which abstracts from certain concrete and sometimes very important features of the actual world. Before they become adequate or acceptable to the scientific community at large and part of the solid foundation for a new and more ambitious program of research, they typically undergo many stages of testing, modification, and fine-tuning. We build toy model after toy model. Eventually we venture to construct models sophisticated enough to do justice to the complex of phenomena we are interested in describing and explaining—only to find that more careful and intricate alterations are necessary to account for all the relationships and regularities we discover, as well as the precise values of all the parameters—for what is observed and what is not observed.

The conceptual or theoretical entities which we use to model physical reality inevitably have internal structure—correlative with the multi-layered features of its phenomena. Quarks and anti-quarks make up protons, neutrons, and mesons; electrons, protons, and neutrons make up atoms, and so on. And these hierarchies of theoretical entities and their structuring are thoroughly supported by experiment. Again, the models are constructed that way. But often when a model or paradigm changes significantly the fundamental theoretical entities—or hierarchies of entities—are replaced by other more adequate ones. For example, phlogiston was an operative theoretical entity for centuries, until the chemical and thermodynamical revolution (which also included the discovery of oxygen) led by Joseph Black and Lavoisie in the late seventeenth century discredited it. So were the ether, and, earlier, the different concepts of absolute space and absolute time. Certain fundamental particles or structures are often postulated by high energy physics only to be abandoned or replaced by others upon further experimental and theoretical work. Will the Higgs particle eventually turn out to be one of these? How about superstrings?

Furthermore, the definition or characterization of key entities, parameters, or concepts in a given theory is not always unique. A given choice may not even be gauge invariant! (That is, the definition of or mathematical expression for a key parameter such as density may, under certain circumstances, change radically in transforming from one coordinate system, or spacetime slicing, to another—that parameter then is not gauge-invariant.) There are slightly different, inequivalent concepts of entropy, of mass, of angular momentum, and even of energy. Within some theories certain of these inequivalent definitions are shown to be equivalent, *for that theory*! This illustrates the dependence of these definitions on the context established by the theory, or by the meta-theoretical presuppositions upon which the given theory is built.

Then there are often free parameters or constants in our theories or laws which can either be chosen arbitrarily or are determinable only by experiment. In other words, the theory or law does not specify them—it merely specifies that one is needed. Other aspects of these laws or theories are also often left undetermined—for example, the masses of some particles, or the orientation of some fields in a phase transition involving spontaneous symmetry breaking. And these indeterminates often remain undetermined by the transcendent theories which unify different interactions and regimes. The particular values assumed by these constants or parameters are left unexplained by the theories themselves. And often there is no clue as to their origin. This is another indication of the incomplete, approximate, and descriptive character of even the best theories and laws. They represent but they do not represent completely or fully. They reveal but they also hide aspects of physical reality. With Paul Davies we can be amazed at how much the "algorithmic compressibility" of physical phenomena allows us investigate, model, and even explain key aspects of reality.[3] If we are given a data set of, say, 10,000 entries representing times, positions, and velocities of a pendulum which is swinging, we can give all the information contained in the data set in much shorter (more "compressed" form) by using the equation ("algorithm") for a pendulum given by Newtonian mechanics along with initial conditions—the position and velocity of the pendulum as some initial, or other, time. In other words, we can "algorithmically compress" the data representing this particular physical phenomenon. Many physical phenomena, however, are not algorithmically compressible. Describing these always takes as much information as is in the full data set. But we are at the same time sobered by the limitations we face in perfecting and completing our models and representations.

Essential to the formulation and the understanding of these models, theories, and laws are various theoretical entities and concepts—mass, energy, temperature, entropy, electron, photon, nucleon, and so on. Some of them have definite characteristics. Particles have mass, spin, charge, magnetic moment, and less familiar attributes such as "charm," "color," and "flavor." Fields have either a scalar, vector, or tensor character, spatial and temporal behavior, invariances,

---

[3] P. C. W. Davies, "Why is the World So Comprehensible?" in *Complexity, Entropy and the Physics of Information*, ed. Wojciech H. Zurek (Redwood City, CA: Addison-Wesley Publishing Co., 1990), 61-70; see also *idem, The Cosmic Blueprint* (New York: Simon & Schuster/Heinemann, 1987), 224.

and sources and particles related to them. Spaces have topological characteristics such as compactness and connectedness, and geometric properties such as curvature, torsion, symmetries, dimension, and so on.

What is the relationship between these abstract theoretical entities and the complex of phenomena they describe and, to some degree, explain? It is really that of a model to the very rich and full patterns of kinematic, dynamic and structural behavior we observe. The model represents in an idealized and imperfect way the structures we find manifested in the phenomena and the detailed qualitative and quantitative relationships which appear to exist among them. However, it leaves a great deal out too—and oftentimes it leaves out sophistications and precisions we desperately wish to include, but do not know how. It is an illusion to believe that these incredibly rich representations of the phenomena are unconstructed isomorphisms we merely *discover* in the real world. Instead they are *constructed*—painstakingly so—and there is no evidence that they are isomorphic with structures in the real world as it is in itself. In fact, as we shall see below, this is not the sort of relationship for which there could ever be evidence. One hopes that the best of these models at least mimic the key structures and relationships we are interested in dealing with in a given case. But they are the result of imaginative and conceptualized abstraction and idealization, guided by continual experiment and careful observation and measurement.

There is, of course, the deeper question: What is the relationship between these theoretical entities and the entities of reality itself—those which in some sense cause the phenomena? Or do these theoretical entities really exist in the manner in which the theories and laws describe them? It is unclear whether or not we can even ask this question meaningfully, much less answer it. This is simply due to the severe limitations of our abilities to know and to understand—which are much greater than most brands of critical realism (let alone naive realism) recognize. However we try to answer these two related questions, our detailed experience as scientists shows us that we cannot facilely identify these theoretical entities and features with independently real entities and features. They *do* have a more or less well-supported foundation in reality as we observe it and experiment with it, but precisely as models and representations. Even in the strongest cases they remain idealizations for structures and relationships we never *directly* observe, and certainly never even indirectly observe as they are *in themselves*. Using them, however, we are able to investigate and manipulate the physical and chemical realities which are "out there"—and so perfect the models and our ability to apply our understanding and knowledge of physical reality through them.

These theoretical entities also possess remarkable internal consistency and correspondence with all that we observe, and a certain durability as descriptive and explanatory instruments. They are endowed with characteristics which have observable correlates. Of course, this is by no means surprising. Often, especially these days, a great deal of work by many specialists over long periods of time has gone into constructing them precisely this way! But to maintain that they correspond directly to the entities, structures and relationships of physical reality as it is *in itself* is a very long jump from that—and one which I believe is impossible to justify.

But if interrelated phenomena are modeled, and to some extent explained, by abstract idealized entities with carefully specified relationships among them—particles of various sorts, fields, and the interaction among them—can we not say that these entities really exist and that the relationships really obtain? After all, if all the observations and experiments we perform support the qualitative and quantitative details of a well-confirmed set of theories and we begin to build on those *as if* those theoretical objects and relationship really exist, can we not say that they do really exist? Something that behaves like an electron under all circumstances must be an electron! This is really the philosophical problem of "natural kinds," or "real and abstract predicates."[4] There are many subtleties involved in this issue, and it is not clear how one resolves it in a concrete case without depending on fundamentally unjustifiable assumptions. The main problem is how to relate the abstract entities of language and models to ordinary objects—how can we determine whether or not a predicate or a category we use as fundamental in our laws and theories corresponds to a real distinction, a real category, or a natural classification? (This is the definition of "a natural kind.") To be satisfied that we can give an answer to this question means that we at some stage are able to know that we are using "correct language";[5] or that we, as a matter of fact, are using the correct model. But we never really know that! We know that we are using an adequate model to a good approximation for a given set of conditions and requirements. But change the conditions and requirements and our confidence in its adequacy is often shaken.

Imagine the Newtonian gravitational physicist who is convinced that Newton has provided the correct theory of gravity—after all it works and gives very precise predictions for what occurs in all situations in our solar system. Then our gravitational expert wants to show that light is immune to gravitational influences, as it should be according to Newton's gravitational theory. But upon precisely measuring the behavior of light rays from a star grazing the limb of the sun he or she discovers that they bend! Back to the drawing board!

We can only know that we are using correct language if we have some reason to believe that we can in our knowledge distinguish between natural and unnatural classes of objects and that this special insight is "one of the guiding factors in science."[6] But that does not seem to be the case, and there is no justification for believing that it is. In fact, science tends to weed out a large number of what we would call "natural kinds" in its search for explanatory models, in favor of abstract entities we have no way of knowing are natural, and could not intuitively suspect are natural except possibly after long, careful study.

Alternatively, we could distinguish between natural or unnatural kinds if there were some reason to believe that successful theories—through modification made in light of experimental and observation data and theoretical

---

[4] See, for instance, Bas C. Van Fraassen, *Laws and Symmetry* (Oxford: Clarendon Press, 1989), 51-55; and Mary B. Hesse, "Lawlessness in Natural and Social Science," (unpublished).

[5] Van Fraassen, *Laws and Symmetry*, 51.

[6] Ibid., 52.

criteria—tended to be formulated in "correct language" reflecting the natural classes of entities at every level of structure and process. But it is not possible to justify this claim either—as we just saw in our example above. The success of a theory is relative—one which is very successful in describing and accounting for a certain range of phenomena in terms of certain abstract entities, characteristics, and structures may turn out to be very unsuccessful in accounting for phenomena in another range. The new theory, as we have already mentioned, may introduce completely new abstract entities, dominant characteristics, and structures. Both cannot be formulated in terms of natural classes.[7]

As a matter of fact, there may not be a supportable distinction between natural and unnatural kinds or classes. To justify such a division seems to rely once again on an assumption that at some level we have privileged access to, and knowledge of, physical reality as it is *in itself*. It assumes an epistemological realism of a rather strong type—that we have isomorphic knowledge of physical reality for at least some levels of its multi-layered structure. As we have already seen, this seems to be unjustified on many different grounds—certainly from an analysis of our perceptions and knowledge of ordinary phenomenal reality it cannot be justified. Reality is just not that accessible, nor is our knowledge of it. Objects are by and large constituted by interrelationships of various sorts, and our knowledge of them is constructed according to certain schemas or representations.[8] Finally, language is not that simple. The idea of natural kinds seems inconsistent with the "family resemblance" or "cluster" notion of word and concept denotation in the later philosophy of Ludwig Wittgenstein, which I believe is very close to the truth about the character of language.

Let us return to our reflections on the construction of successful scientific models of certain key aspects of physical reality as we observe it. We become aware that one of the reasons why large networks of phenomena are algorithmically compressible is that they are determined by a hierarchy of structures in which relatively simple relationships dominate the processes at every level. Our ability to idealize and construct successful models and laws is due to the fact that there are levels of reality at which the dominant behavior of systems is simple and uncomplicated. Certain phenomena can be objectified as definite entities with a definite location, or range of locations, within a definite interval of time at a definite level of size or energy. These entities have certain stable characteristics, which are measurable and relevant to what we observe. And they interact at this level sometimes in a particularly simple and straightforward way—as the planets do around the sun, and sometimes in a much more complex way but still deterministically and according to general patterns, as in a chaotic system with, say, a strange attractor. In these cases, as Davies has pointed out,[9] the relatively unchangeable laws describing behavior can be isolated from the always changing "states of the system," which in turn can be

---

[7] Ibid., 53-54.

[8] Michael A. Arbib and Mary B. Hesse, *The Construction of Reality* (Cambridge: Cambridge University Press, 1986), 320.

[9] Davies, "What are the Laws of Nature," in *The Reality Club #2*, ed. John Brockman (New York: Lynx Publications, 1988).

isolated from boundary conditions and initial conditions, which further constrain and determine the dynamics of the system.

There are many other situations, such as the dynamics of biological systems, which are too complicated for such descriptions—even of a statistical sort. We cannot cleanly isolate the laws from the states of the system, nor from the boundary conditions. In other cases we may not be able to localize dominant influences or interactions in the system as we normally do. Strong nonlinearities may afflict the dynamics. The system is much less algorithmically compressible, and the normal idealizations do not work. The infant science of the physics of complex systems is beginning to tackle such situations.[10]

There is another peculiarity which sometimes arises in our theories and which also gives evidence of their purely representational and idealized character. Oftentimes we employ entities or constructions in our models which we come to suspect as having no direct correspondence with the phenomenal realities we observe, but which are essential for modeling and explaining them, their characteristics, frequency, evolution and, most importantly, their relationships and interactions with one another. Examples are: ensembles of systems in statistical mechanics, and in quantum cosmology; the wave-function in quantum mechanics; fundamental particle spin; mathematical structures such as symmetry groups and superstrings; the spacetime manifold itself in cosmology, which can expand and contract, and into which we can dump energy, as during inflation; the fundamental constants. These entities "exist," at most, only in the sense that they indicate relationships between theoretical entities which model the complex of physical phenomena and the idealized events which model the interactions and transitions. To be sure, these theoretical objects are justified within the model and by its success in describing the phenomena. But there is a tendency to reify these entities and constructions in a way which isolates them from their defining relationships—for example, the fundamental constants, the expanding spacetime manifold. They are abstractions of real relationships, but when they lose contact with these, we can end up giving them an independent status which can lead us to conclusions or speculations at the limits of science which are unfounded and even inconsistent with key aspects of the models themselves.

So far we have discussed the laws of nature and the theories which enshrine them as imperfect descriptions and explanations of the regularities we encounter in physical reality. Can they in any way also be considered *prescriptions*—sources of physical necessity, or at least imperfect descriptive models of such sources? There is really no justification for considering them as such. This is essentially the problem of "identification," which we shall briefly treat in section 4. There is no fact about the world to which we can point that endows the law with prescriptive force.

There is a sense with some well-established fundamental theories, or components of theories, in which there is prescriptive force for other laws or theories, but *not* for regularities and interrelationships of physical reality itself. For instance, if I construct a theory of gravitation which does not yield

---

[10] See n. 3 above.

conservation laws of mass-energy and total angular momentum, I will have good reason for abandoning it immediately. These conservation laws are considered fundamental in any adequate theory because they are observed to be fulfilled in all cases—even in the most extreme and primordial ones. And they are not only universally fulfilled, but are also essential for the power and the applicability of the theory. So, laws like the conservation laws are so fundamental that they must be realized in any acceptable theory. In a sense, they prescribe essential components of the theory. However, they do not prescribe how physical reality itself behaves—they rather describe some of its fundamental and unchanging regularities in its continual transformations.

There seems to be a deeper reality, so to speak, which prescribes and enforces this conservation of mass-energy and of angular momentum, and explains it as well. But we cannot observe it. I do not mean this in a Platonic sense, but in the sense that the "laws" are only imperfectly explanatory, precisely in being unable to reveal and model the sources of necessity and order or to enforce them. Their deficiency in this regard discloses the difference between our laws of nature and the underlying regularities and relationships of reality itself which are to a large degree hidden. A physicist will immediately jump up and say, "The deeper reality that enforces these conservation laws are the symmetries under which the Lagrangian describing the system is invariant—for every invariance there is a conservation law." And this is very true. But this still does not explain why real systems should be that way. It merely transfers or translates the question into new language. We are then faced with explaining how symmetry groups and their invariances apply to physical reality, allowing certain regularities and suppressing others. What instantiates them?

Can we support the idea that models and laws somehow have an independent existence outside of our minds and outside their approximate manifestation in the physical phenomena we observe? From what we have seen so far, the answer is "no." At least there is no reason, either scientific or philosophical, to think that this is the case. We use these models and laws to probe the phenomena and to understand and describe the physical realities manifested through them. So we are here far from the Platonic position. We can say, then, as we have stressed already, that the theories and models and the laws they encompass have a definite basis in reality as we observe and experiment with it. But to go further and maintain that they constitute the underlying pattern or plan of physical reality as it is in itself is thoroughly unjustified. The most we can say is that there are regularities and interrelationships in reality as it is in itself—a fundamental order—which are imperfectly reflected in our models and laws. These successfully model this fundamental order to a certain level of depth. But they are imperfect, and have some key deficiencies. We can never know that underlying order in itself, nor the physical realities in themselves (from the inside, so to speak), nor in any detail the realities and relationships which determine them—the sufficient reason(s) determining what exists and what does not exist; why certain metaphysical possibilities are not realized and others are. The real sources of necessity, possibility, and explanation are only dimly perceived through our models and laws, if at all. We are able to include in our models—the best and most successful ones at least—certain intermediate sources of necessity, possibility, and explanation. But we are often reminded that these laws and sources of necessity and possibility pertain first and foremost to the models, and not to the physical realities themselves. They pertain to the physical realities only in so far as the model in some way corresponds with them as they are measured and observed. And when we push deeper to ultimate grounds of explanation, necessity, and possibility we always find ourselves in an infinite regress.

These conclusions may sound somewhat Platonic. But there is no question here of an ideal order which supports or determines our physical world and its regularities. We are instead pointing to the fact that there are important aspects of the world for whose existence and operation we have philosophical and scientific evidence (by the principle of sufficient reason, for instance), but which we are unable to model scientifically, and for which we cannot properly account.

And so the laws we discover are essentially descriptive—not prescriptive—and only weakly explanatory and revelatory of the grounds of possibility and of necessity, and of the relationships which flow from them. This becomes more obvious when we deal with complex or indeterministic systems, and when we reflect on our experience with quantum mechanics and quantum field theory. In the biological sciences in particular—whether the functional disciplines or the evolutionary disciplines—laws are rarely spoken of seriously, and then only in an analogous sense. There universality, explanatory power, simplicity, and predictability become much more problematic and uncertain.[11] Casting our glance back to the physical sciences from this perspective, we realize that the seeds of the unraveling of our certainty and our understanding of the laws of nature are already growing in the theories and models we construct there. Some of our grander illusions are banished, and though we realize how far we have come, we also painfully realize how much we have yet to understand and how serious are the limitations under which we labor. The illusion that we are somehow discerning reality as it truly is *in itself* is a pervasive and dangerous one; and it dies very hard. Two related illusions also die hard: that the theories we construct, and the fundamental laws which flow from them, are the patterns for reality as it really is in itself, and that these laws in some sense control, govern, and ground that reality—not merely describe it. A careful analysis of the character of our scientific knowledge will alert us to the presence of these illusions and liberate our thinking from their domination. Intermediate and proximate explanations abound and are modeled reasonably accurately. But the more profound grounds of explanation, necessity, and possibility remain forever veiled.

---

[11] E. Mayr, "Is Biology an Autonomous Science?" in *Toward a New Philosophy of Biology* (Cambridge: Harvard University Press, 1988), 8-23.

3 A More Philosophical Analysis of the Laws of Nature

We have surveyed and reflected upon some of the key phenomenological aspects of physical theories relevant to our understanding of what sort of objects the laws of nature are; we now pause to take stock of where we find ourselves. What are the key questions concerning the ontological status and weight carried by the laws of nature? Here we formulate the principal ones and attempt to give answers based on our reflections above. We shall be covering more or less the same ground as above, and some repetition is unavoidable. But we do so from a somewhat different perspective, in an attempt to articulate both the questions and the answers more clearly.

The first and simplest of the important issues is: To what extent do the most successful and well-confirmed physical theories and the laws they embody describe what really occurs in reality? In itself this question is already extremely difficult—not least because any answer to it depends heavily on certain philosophical commitments, one might almost say prejudices. Certainly, I believe that a very strong case can be made for saying that such theories—which because of their success have in practice been elevated to the position of laws— closely model in great qualitative and quantitative detail the fundamental patterns of order, causal influence, and constraint we observe at different levels of the physical and chemical world. They have, as a matter of fact, been carefully and painstakingly molded, modified, and refined to do so in confrontation with continual observation, experiment, and application involving greater and greater precision. There is, in this context, a certain critical care and zeal by the scientific community to get right all conceivably relevant aspects of the phenomena involved as well as their interrelationships—allowing for what is observed, and for the frequency at which it is observed, and forbidding what is never observed, always coherently explaining and founding these allowances and prohibitions or constraints in the structure of the theory itself or in the structure of a more basic, underlying theory, which in turn is subjected to careful experiment and refinement. Thus, these theories *do* have a very strong basis in observed reality.

But these theories and laws only describe reality as *we observe it*—with all the power of critical intersubjectivity brought to bear on the phenomena being studied but also with all the deficiencies implied by our limited perspectives and our distant exterior relationships with the reality we observe. Thus, they describe reality as it is for us and in relation to us. This becomes even more obvious when we enter realms in which we are concerned with the entire observable universe as a single entity or in which a quantum description of physical reality is essential. This means that these theories and laws do not describe physical reality *as it is in itself*. But, *if* what we observe physical, chemical, and biological reality to be closely reflects certain important features possessed by that reality in itself, then our theories and laws, to the extent that they adequately model and interrelate these observed phenomena, reflect and model key aspects of that reality itself. Notice, however, that there is an important "if" here, which seems impossible to justify.

Notice, too, that the phenomena, and therefore the theories and laws, are not able to reveal all features of observed reality nor even its most fundamental features. Some of them are hidden from, or inaccessible to, our probing. This is simply because, in making our observations and measurements, we focus on those stable and characteristic features we are capable of: (a) observing at energies or other ranges of detection available to us; (b) discerning as being universalizable and relevant for the questions and the context we are concerned with; (c) isolating, simplifying, and modeling in a straightforward way by concepts such as mass, velocity, energy, and so on. which possess an isomorphic, or at least a homomorphic, relationship with the observed features; (d) seeing as relevant to our interests in performing the experiments and constructing the theories; and (e) anticipating heuristically in the theory or at least in the design and interpretation of the experiment or observation. But there may be, and certainly are, many aspects of observed reality which are neglected or simply missed, and which are equally important for characterizing it. These features might be recognized and utilized by some other set of observers with different interests, theories, or observational abilities than ours.

Certain parts of the reality in which we are immersed are such that we can successfully model them in all their detail—they are algorithmically compressible. But that fortunate and mysterious trait of reality should not delude us into thinking that we have a direct access to reality as it is in itself—if indeed this term has meaning. More careful reflection and analysis of our experience indicates we do not have such access. What we know has been bought with great effort and with careful, painstaking research and struggle. But we are unable to delve into "the heart of things."

Other areas of reality are not so malleable to description by lawlike or theoretical characterizations, involving predictability, simplicity of description, explanatory power, or even universalizability. There is a level at which reality seems to be well modeled by rigid, specific descriptions—an ordered substrate where everything can be modeled and fit together neatly. But beyond a certain point this breaks down—even with purely physical systems. And so we have the physics of complex systems, where order and chaos nourish one another with a strange reciprocity, and where behavioral descriptions mix intimately with the state of the systems, even where the entities themselves are uncertain and shadowy. Fluid flows exhibiting strong turbulence and the dynamics of complex information processing systems are good examples. Here it is difficult to model or isolate constitutive entities so as to algorithmically compress the reality we observe. And of course, as we move into the science of complex chemical molecules, into biology, neurophysiology, psychology, economics, politics, and sociology, these problems increase and prevent us from describing phenomena in anything like the lawlike and rigidly predictable way to which we are accustomed in physics and mathematics. Finally, we should recall the difficult problems bearing on these questions coming from quantum theory, and what they have taught us about our knowledge of an independent physical reality.

The second issue is: Given that for the most successful and well-confirmed theories there is a strong reflection of what we observe in reality at certain levels of abstraction, do these descriptions of it really prescribe how reality behaves or can behave? They describe how it does behave under carefully controlled circumstances, and they attribute such behavior to certain causes, processes, and well-modeled influences. But do they *prescribe* such behavior? This is a subtle but important point, which we began to address in section 2. Do they force or constrain the objects in question to behave in such and such a way, and not otherwise? I think we can show that they do not.

They seem to do so, or give the illusion of doing so, simply because they are very accurate and carefully refined descriptions of a hierarchy of closely intertwined and isolatable phenomena, which are characterized by levels of rigid behavior that is easily and successfully generalized in lawlike descriptions. However, the laws cannot be said to be the source of the behavior nor of the constraints on behavior—they model or describe it. In fact, the laws are not always obeyed without fail—but obeyed often enough so that the instances which deviate from them follow no understandable pattern or can be suspected of being due to outside influences, improper isolation of the experiment or observations, etc.

One of the reasons the laws of nature came to be seen as prescriptive is that they were originally thought of as divinely instituted, and hence necessarily prescriptive. They were seen as the instruments God employed in governing the physical world, as were God's laws governing human conduct.[12] This metaphor cannot be taken literally with regard to the laws of nature we formulate in trying to describe reality. It may still be applicable in some sense to the patterns of regularities and necessities inherent in reality as it is in itself. But, as we have seen, these are not directly accessible to our probing.

In this regard, it should be emphasized that the lawlike descriptions derive from a study of the objects in question themselves—they are not arrived at *a priori*. In a way, saying that something is a "law of nature" is simply a way of indicating that it is so fundamental to the description of the detailed workings of physical, chemical, or biological systems that it never is observed not to hold when those systems are properly isolated and simplified and certain conditions are fulfilled. But there seems to be little support for the position that the law is the cause of the regularity observed or that it forces physical, chemical, or biological entities to behave in the way they do. It is rather a very tight, detailed, and specific *description* of that regularity and of its fundamental character. What enforces this behavior or regularity is certainly not the law itself. It may be something that is hidden from our eyes, and explains or grounds the law. Sometimes, this is eventually revealed in an intermediate, proximate way in the next level of physical process and structure—what was originally a fundamental law at one level comes to be seen as the consequence of structures, relationships, and regularities at a deeper level. Determinative connections with the infrastructure are revealed. Examples include the explanation for the laws of chemical reactions and bonding in the details of atomic structure (electron energy levels and the application of quantum mechanics to them), and the explanation of conservation laws in symmetries and invariances under symmetry groups.

But these deeper explanatory connections are never capable of giving the ultimate source of the regularities—one might be tempted to say "necessities"—imperfectly described and explained by the laws of nature. They

---

[12] Nancey Murphy, private communication.

provide intermediate, highly detailed descriptions which closely link phenomena which originally seemed unconnected, but they do not completely explain why the reality is that way rather than some other way. Rather, they explain that, *since it is this way*, it has to have these relationships with what appear to be more fundamental realities. And this "has to have" is also an intermediate type of necessity—not an ultimate one—with an infinite regress lurking just around the corner. It flows from other more fundamental regularities and patterns which are always observed as being connected with the derived ones, and we tailor our descriptions and models accordingly. The models then give the impression of imparting necessity. But the apparent necessity does not come from the models; it is hidden in the observed regularities and in the entities which adhere to them. Its ultimate source is not accessible to our probing.

Turning this around, however, does not always seem to work. We cannot say that since these deeper relationships hold, the derived relationships must be such and such. First, the implication from the more fundamental to the derived relies on the model itself. The model's lawlike character as such, and not just which laws it entails, is exactly what one is seeking to explain—and this lawlike character as such cannot be explained merely by stating the specific laws it entails. Second, it often happens that the more fundamental entities or relationships in the model are not sufficient conditions for the emergence of all of the derived realities in it, together with their detailed characteristics and interactions (for example, in spontaneous symmetry breaking). Third, though they may be in some sense necessary conditions, to say this is equivalent to what was pointed out above, that this is in virtue of the model itself, assuming that it holds. But we are interested in explaining *why it* holds, and why not some *other* model. Fourth, *even if* these problems were resolved satisfactorily, there would be no ultimate explanation of its lawlike or regularity-endowing character within the complex of models developed to describe the multi-layered reality we observe.

The third issue is: To what extent can the laws be endowed in a justified way with an existence independent of the objects they govern? This is tricky in some ways because, although they derive from observing the objects, the objects themselves do not necessarily have a clearly independent existence as such—they depend on the theory, or on levels of theory, such as quarks in fundamental particle theory. If there were not these objects, there would not be these laws; but without the theories, there would not be the objects we have constructed within the theory which model reality by these laws and relationships.

We can therefore assert that there is, often, a symbiotic existential constitutive relationship between the theories and the objects. At any rate, from some of the tentative conclusions we reached in section 2, it seems highly unlikely that we can ever justify postulating an independent existence for the laws, or even make sense of such a suggestion. They belong, first and foremost, to the models and therefore rely for their meaning on the details and characteristics of those models and the theoretical entities which constitute them. Of course, as we have already seen, the laws are intended as descriptions of the regularities and interrelationships which we discern in physical reality. But we cannot confuse our laws with the possible realities which determine these features.

And what about that deeper question: do the underlying regularities and relationships within nature, which our laws attempt to describe, have a temporally or logically prior existence to the realities to which they pertain? In light of our analysis, it is clear that this question does not even make sense. It rests on the confusion to which we alluded above, between our laws and the entities and relationships which provide a sufficient reason for the regularities and relationships in nature as it is in itself. Furthermore, it seems to rest also on the unwarranted philosophical assumption that the regularities or order in nature must have an existence prior to its realization—just as a blueprint is prior to a building. But I have claimed that the regularities and relationships are given in the reality, not before it. This is true even in the emergence of novel entities or structures, as for example when the temperature of the universe decreases through certain critical thresholds. In such cases, the character of the emerging realities depends importantly on the laws governing more fundamental entities and structures. This is because they depend not so much on those laws but rather upon the more fundamental constituents which already exist and which have properties that allow them to combine in certain ways and at certain rates, while forbidding their interaction in other ways. Again, it is similar to the question of our laws being derived from the phenomena—not vice-versa. This is the case at least epistemologically, and ontologically, as far as we can see, it must also be nearer the truth.

A fourth issue is: Does it make any sense in light of all this to postulate other sets of laws that really exist and describe alternative patterns of behavior of alternative sets of entities, or even ensembles of other universes? Because we can without contradiction imagine that things were very different—that things did not have to be the way they are—this possibility seems attractive. There is no metaphysical necessity for any of the physical, chemical, or biological laws we have discovered—or as a matter of fact for the mathematical laws we have. Things could have been different! But is that sufficient justification for postulating other universes with other sets of laws? Or even the much less extreme case of other values for the fundamental constants? This issue, of course, is connected with the questions of initial conditions in cosmology and "fine tuning," which attempt to give some explanation why the universe is as it is and not otherwise. But it seems that such attempts are irretrievably flawed, simply because other conceivable laws really cannot be considered to exist independently of the regularities and relationships they describe. And there is no evidence for such parallel universes of entities obeying such alternative relationships and regularities—or for any parallel universes at all, though we cannot rule them out. In fact, there is as yet no idea of what would constitute evidence for the existence of parallel universes, or how we would recognize some feature of ours as evidence for them. We can perform a retroduction to some nonunique, unobservable ensemble of universes which would be a possibly sufficient—but by no means a necessary—condition for the existence of our universe today. But it really would not be an ultimately sufficient condition or explanation.

We could legitimately ask why that ensemble should exist, with that probability measure, satisfying those statistical laws, and having constituent universes with just those sets of physical laws, so that one of them would evolve into ours but we would not solve the problem by this approach. We would merely introduce another intermediate explanation, or an intermediate set of sufficient reasons or conditions of possibility, which, it should be noted, are *not* unique, and which in turn would require a further unknown sufficient condition or reason. Again, too, as we saw above, the laws and entities are intimately tied up together—you cannot have one without the other, strictly speaking. You can certainly construct imaginary universes with imaginary entities which obey very different or only slightly different laws, without apparent inconsistency or contradiction. But that by no means indicates that these universes have to be instantiated or realized. And, even if they were, this would not, as we have just seen, significantly improve the explanation for why our universe exists.

Connected with this question is the problem of possibility. What are the grounds of possibility? The only unequivocal ground is existence—"ab esse ad posse valet illatio." (The inference is from being to possibility.) But are there other grounds which are not so demanding? How else can we establish possibility? There is another way: Even though something does not exist, in principle we can show that the conditions of its possibility are fulfilled—the necessary conditions for its existence. But the difficulty here is that we can never show that *all* its conditions of possibility are fulfilled, because this implies a complete knowledge of that "something," and this is beyond our limited abilities. With our well-tested models and with the laws deriving from them, we describe the regularities of physical reality. Certain patterns of behavior are described and accounted for in great detail and with remarkable precision. Certain extremely precise values are assigned to the different fundamental constants from experiments. We then say that in *our world* certain things are possible, and other things are impossible, according to whether or not they are consistent with these laws, theories, and constants. But we never know for sure whether or not we have indicated *all* the conditions of possibility. Even after we find the object of our quest, and thus know that all the conditions of its possibility have been fulfilled, we cannot be sure that we have specified them all. In fact, we almost never are able to do so.

But is it legitimate to transcend this level of analysis and say, "Other universes are possible in which the objects and the laws they obey in their behavior are very different"? Certain behaviors are not possible in our universe. They might be possible in another universe, and those other universes might be possible. But what is the ultimate ground of that possibility? This is an interesting and important question. The usual rationale is simply that the existence of such universes is not contradictory, nor illogical—their existence does not involve any apparent logical or metaphysical inconsistencies. We can therefore imagine their existence—and on a detailed analysis of models and theories of physics and chemistry, we can see in a preliminary way that many different modifications of these theories would not damage their internal consistency, for example, changing the values of fundamental constants. These changes would render the physics and chemistry of these alternative universes very different from ours.

Hence the ground of possibility usually given is "our model of all internally consistent models"—but note, this is not an argument which is directly related to the data or the phenomena of our observable universe. Such calculations are unobjectionable, as long as we realize that the possibilities we infer from our set of all internally consistent models may reflect actual possibilities very poorly. But then the usual point here is that we certainly *cannot* infer existence from possibility. We also cannot infer real *physical* possibility itself from the internally consistent modifications allowed in a model of what does exist. The grounds of real possibility, like the grounds of real necessity, are impossible to capture adequately in models because these grounds are intimately tied either to existence or to the fulfillment of conditions. And models can never dictate their own instantiation and can never give all the conditions of possibility.

A fifth issue is: To what extent does it make sense to seek a scientifically based explanation for the laws of nature in a single overarching fundamental law of the universe—say a superstring theory based on the symmetry group $E8 \times E8$? We may be able to construct such a theory and demonstrate that it describes all the fundamental interactions of physics, thus providing the "ultimate" intermediate explanation at the level of the natural sciences in the sense that all other laws derive from it. But it still will be an intermediate explanation—it will not explain itself. And it will depend on current notions of what is fundamental, which may change. Other "ultimate" intermediate explanations can be imagined. We always run into these problems of the uniqueness and the finality of the explanation. It seems we can always imagine other explanations which would suffice and be internally coherent, but no one of them seems to be the final, necessary, unique and self-evident primordial law. We cannot even imagine what would count as such. Things could have been different without contradiction. Furthermore, if the lawlike descriptions we construct in the natural sciences are essentially the result of generalization from many instances, how can we sensibly develop a "law of universes" which specifies the range of allowed sets of laws for different universes and the constituent entities they would each possess, along with the processes governing their evolution? We only know of one instance, and can examine only one instance. Of course, we can theoretically construct subsets of such universes to give the illusion that we are generalizing. But then we are not generalizing from known instances, but only from imagined ones. And specifying the relationships (for example, the probability measures, etc.) now becomes arbitrary—within the confines of consistency. What we said above in discussing the grounds of possibility is also relevant here. Again, too, a plausible explanation might be constructed, but it would not be unique, and would never be subjected to either falsification or empirical confirmation.

4 *Key Philosophical Problems with the Laws of Nature*

In the course of our phenomenological reflections on characteristics and functions of the laws of nature in the sciences, we have already dealt to some extent with the principal philosophical issues. Here, I shall merely highlight them briefly and refer to some of the other philosophical work in this area.

There is, first of all, the general epistemological problem of the relationship between our knowledge of physical reality and physical reality itself—the problem of realism. We cannot even begin to treat adequately the issues involved in realism here. What I believe we end up with, as a result of our reflections, is an "empirical realism," according to which our models, laws and explanations have a firm basis in reality as we and every set of physical observers experience it. It is weakly objective, in the sense that it is intersubjective. It is not knowledge of physical reality as it is itself. And it is impossible to conceive of it as being knowledge isomorphic with reality. This is due to the vagueness and imprecision of language and models, the cultural and social conditioning to which they are subject, and most importantly to the fact that we do not have any privileged access to physical reality as it is in itself which would allow us to determine whether or not our knowledge of it is, or could be, isomorphic. The issue of "natural kinds" which we discussed above is relevant here. Our brand of realism is close to that of, say, Bernard D'Espagnat as he propounds and justifies it in great detail in his masterful *Reality and the Physicist*.[13] He bases his position almost entirely on an analysis of the limits of our knowledge of the microscopic world through quantum mechanics and quantum field theory. I believe the character of knowledge as evident in the science of the macroscopic realms pushes us in the same direction. Mary Hesse arrives at a similar position[14]—and perhaps at an even weaker objectivity than I think is possible. However, I largely agree with the importance of the factors she introduces into the conversation, though not always with the way she marshals them in supporting her conclusions.

Second, there is the enormous problem of explanation. I have discussed this above at some length. Essentially, it is that the laws of nature really do not explain the regularities in physical reality—even though many people think that they do. They *describe* them in some detail. To say that the reason two masses attract one another is the operation of the law of gravity is not a satisfactory explanation. In some ways it is the avoidance or temporary postponement of an explanation. The person who really wants an explanation of why two masses attract one another will retort, "And why does the law of gravity operate?" Now, as we have seen, it is true that the theories and laws we have are intricate and sophisticated enough to provide proximate explanations and connections with other more fundamental structures and laws. So, for instance, in answer to the inquirer's more persistent second question, I could say that "the law of gravity operates because material objects—however we describe them—determine a spacetime manifold, the amount of mass-energy within a certain volume of space curves the spatial slices of the manifold in that region, and objects, including massless ones, subject to no forces, will follow geodesics in that spacetime. All this is given by Einstein's field equations." And that would, in a proximate sense, be an explanation. But, in another sense, it would not. My inquisitor could justifiably complain that I was just presenting even more difficult things that required explanation. "Why does mass-energy curve space?" "Why does the

---

[13] Cambridge: Cambridge University Press, 1989, p. 280.
[14] Hesse, *Lawlessness*.

totality of material objects determine a spacetime manifold?" "Why do objects in the absence of forces follow geodesics?" And so we can imagine an infinite regress of laws and relationships that require explanation. This is an essential characteristic of human knowledge in general, and of scientific knowledge in particular. We deal always with contingent realities, and the inverted tower of explanations for them eventually disappears in the mists of the epistemological abyss. Could there be a turtle at the very bottom? For a more orderly and sophisticated treatment of this issue, see Bas C. Van Fraassen, *Laws and Symmetry*.[15]

There are several other important issues connected with laws—for instance, the definition of laws in terms of counterfactuals or in terms of "partial definitions."[16] But there are two important and intimately connected issues which Van Fraassen highlights, and which I believe are crucial to any account of the status of the laws of nature. They are the problem of inference and the problem of identification.[17] The problem of inference is simply that, in any acceptable account of real laws, "that it is a law that A, should *imply* A." This is essentially the problem of necessity with which we have been toying in one way or another throughout this paper. One obvious way to translate this straightforward sentence would be: "It is a law that A = it is necessary that A"—and then to appeal to the fact that necessity implies actuality. But, Van Fraassen asks, is the concept "necessary" univocal?[18] "And what is the ground of the intended necessity, what is it that makes the proposition a necessary one?" It is worth noting here that the problem is not one of *logical* necessity, but rather one of the relationship *between* logical necessity and physical necessity. We cannot resolve this issue here, but I think we can see that it must be resolved in one way or another before anyone who envisions the laws of nature in prescriptive terms can claim to have an adequate account of the laws of nature.

This takes us to "the problem of identification," that is, the problem of *identifying*, or pointing out, "the relevant sort of fact about the world that gives "law" its sense."[19] If we insist that necessity is a given, we essentially refuse to answer the problem of identification and thus cannot provide an adequate basis for the law we promote. Van Fraassen concludes that it is extremely difficult, if not impossible, to resolve both the question of inference and the question of identification, without introducing a number of other insoluble problems. Much of the first half of Van Fraassen's book is taken up with exposing the intricacies of this dilemma. On the one hand, if we solve the problem of inference by defining exactly what we mean by "necessary," indicating what makes the proposition a necessary one, and linking logical necessity to physical necessity in an adequate way, we have an impossible task in identifying something about the world which grounds this link and corresponds with what we said makes the proposition necessary in this sense. If, on the other hand, we identify some facts

---

[15] Van Fraassen, *Laws and Symmetry*, 94-128.
[16] See, for instance D'Espagnat, *Reality and the Physicist*.
[17] Van Fraassen, *Laws and Symmetry*, 38-128.
[18] Ibid., 39.
[19] Ibid.

about the world that "gives" the law its sense, we find that it becomes impossible to define univocally what necessity means, without making other unjustifiable assumptions.

One way around this dilemma is to abandon the idea of "laws" as enforcing, or even describing, physical necessity, and to treat them as descriptions of the regularities and relationships we find in nature—more or less as we have done. This is not exactly what Van Fraassen does—it is impossible to do justice to his treatment here. But some of his reasons for eliminating "laws" in the more rigorous sense from his concept of what science provides us are the same as mine.

5 *The Impact of the Status of the Laws of Nature*

What are the consequences of conceiving the laws of nature in this more modest way—as key elements of descriptive, not prescriptive, models which only imperfectly represent and explain the underlying regularities of physical reality? The consequences are not particularly noticeable within the sciences themselves, but become very obvious as we begin to philosophize and speculate beyond the secure pale of scientific rigor and understanding.

First of all, the idealized, abstract, and simplified character of the laws and theories must be highlighted, along with the limitations of the natural sciences and of human knowledge generally. We are reminded that the models are one thing and the physical realities they represent another—and that the links between them are very tenuous and imperfect. Confusing the two leads to the illusion that the necessities and possibilities of our constructions are all there is to the necessities and possibilities of reality.

Second, from what we saw in our extended discussions in section 3, the prospect of constructing a general law of the universe which is unique and ultimate—a theory of everything—vanishes as an illusion. Our knowledge of physical and material reality will never be that complete, or of a character consistent with the demands of a unique and ultimate theory of everything. In particular it is weak and uncertain—inherently so, I believe—in its identification and representation of the grounds of necessity and possibility in physical reality. We grasp it only "from without"—and only from a particular perspective with the help of theoretical and experimental instruments with which we operate upon it, or upon its manifestations. The theories and laws we construct to model the universe usually describe relative necessities—they are conceived as holding as long as all the conditions and constraints are fulfilled. If they do not hold under these conditions then new laws must be developed which do. But the laws and theories themselves are not necessary—and it seems clear that any proposed universal law of the universe, or theory of everything, would be incapable of being necessary. Laws are contingent. They do not contain within themselves the reason why they, and not some other laws, obtain. They do not explain themselves. And this contingency of the laws of nature extends to the underlying regularities and relationships in physical reality itself which they purport to describe and to some extent explain.

Third, any move towards an ontological reductionism—or a related determinism, which has already been put to flight—is scotched. Such a view of physical reality is thoroughly inconsistent with the limited character of our theories and of the laws of nature which are embedded in them, and contradicted by our detailed understanding of both macroscopic and microscopic phenomena—quantum mechanical and quantum field behavior, chaotic and non-equilibrium phenomena, the physics of complex systems, and so on.

Fourth, the question of "the origin of the laws of nature" now becomes several questions: the origin of "the laws" as we formulate them; the intermediate, proximate origins of the underlying, somewhat hidden regularities and apparent necessities which they model and describe; and finally the ultimate origins of those same underlying regularities and interrelationships in reality as it is in itself. The answer to the first is epistemological. The laws derive from our constructions, guided by careful and repeated experiment and theoretical modeling and analysis. The answer to the second, presupposing that our laws to some approximation successfully represent key aspects of the regularities and relationships in nature, is that they originate in the relationships of fundamental entities—in a multi-layered universe. We know some important aspects of those relationships from our successful models, but we are ignorant of those aspects to which our models are blind or insensitive. The answer to the third question, if one can be given at all, is strictly ontological. Whether one can construct such an answer in a valid way depends on a number of other crucial issues—such as whether or not the principle of sufficient reason, or something like it, can be employed at this level.

Fifth, the philosophical weight that can be born by the anthropic principle becomes significantly diminished. We can still "fine tune" initial conditions to some extent. But fine-tuning laws and constants, and having ensembles of universes with different laws and constants, no longer has much significance. The laws are descriptive and are derived from physical phenomena in the universe. The possibilities, and the necessities, inferred by generalizing them without inconsistency to other sorts of laws have only very tenuous contact with anything we can identify as "real possibilities" and "real necessities." In other words we have no way of either knowing, estimating, or guessing how closely these possibilities we generated by generalizing our laws and theories correspond with the what is really possible. Furthermore, whether any one of those modifications is in fact realizable depends on hidden features of reality to which the models or theories are not particularly sensitive. These are problems with the anthropic principle in addition to the ones pointed out by others.[20]

Finally, this conception of the laws of nature provides a somewhat different context within which to discuss God's action in the universe than the more strongly realist one with which we are familiar. God can be conceived as acting through the laws, but the ones through which God is acting principally are not "our laws," but rather the underlying relationships and regularities in nature

---

[20] E. McMullin, "How Should Cosmology Relate to Theology?" in *The Sciences and Theology in the Twentieth Century*, ed. A. R. Peacocke (Oxford: Oriel Press, 1981), 40-47.

itself, of which "our laws" are but imperfect and idealized models. And these underlying interrelationships and regularities possess aspects which we are unable to represent adequately—for example, the grounds of possibility, of necessity, and of existence itself. Much more can be said about this. It is important to point out, too, that God's relationship to these underlying realities will be essentially different from our relationship with them. We experience them only "from the outside," as it were, that is, only in some of their outward manifestations and only in some of their important relationships. God experiences and knows them intimately from "within" in all their relationships and connections—including those which determine their possibilities and necessities and their grounding in God. We need to model physical reality, its relationships and its laws, representing them through language and theoretical concepts. God has no need to do that.

Some of the other things we have discussed in arriving at our conception of the laws of nature in the course of this paper also will impact how we speak of God's action in the world—in particular, the abstract and idealized character of the laws. In a definite sense they deal only with some of those simple aspects of reality which admit of generalization. And those aspects by no means determine everything about reality. There are many other important factors which involve the particular, the special, and the personal in ways which are not subsumable under general laws.[21] As we know from revelation, much of God's action as we experience it is focused into the channels of the particular and the personal. But, of course, God is always acting through the deterministic and indeterministic interrelationships and regularities of physical reality which our models and laws imperfectly describe. God is the ultimate source of them and continuously acts in and through them. The laws of nature are impoverished but very useful and reassuring reflections of the rich and interconnected regularities which drive our evolving and life-giving universe.

My primary point has been to emphasize that an improved and more "down to earth" appreciation for the true character of the laws of nature, such as I have suggested, provides a more fruitful scientific and philosophical context within which to discuss God's action in the world, as well as a number of other closely related philosophical and theological issues.

Acknowledgments: My appreciation and thanks go to Ernan McMullin, Robert Russell, and especially Nancey Murphy, for their helpful criticisms of earlier versions of this paper, and for stimulating me to clarify several key issues.

---

[21] See Albert Borgmann's concept of "the deictic" in *Technology and the Character of Contemporary Life* (Chicago: University of Chicago Press, 1984), 302.

IV

THEOLOGICAL IMPLICATIONS 1:
TIME AND QUANTUM COSMOLOGY

THE TEMPORALITY OF GOD

J. R. Lucas

1 *Introduction*

Traditional orthodoxy denies the temporality of God, following the teaching of Boethius, himself much influenced by the Neo-Platonism that St. Augustine never entirely threw off. In this century, however, the evident contradiction between the theologians' doctrine and the witness of the Bible has made Christian thinkers increasingly uncomfortable. Karl Barth, in a striking phrase said that "The theological concept of eternity must be set free from the Babylonian captivity of an abstract opposite to the concept of time,"[1] and the process theologians have sought to give an account of God in which what he does is far more important than what he is. Without adopting all the tenets of process theology, I want in this paper to argue for the fundamental temporality of God on theological grounds, and then deal with the difficulties that have been raised by the special theory of relativity and quantum cosmology.

2 *God as Person*

To deny that God is temporal is to deny that he is personal in any sense in which we understand personality. To be a person is to be capable of being conscious, and to be conscious is to be aware of the passage of time. Some, influenced by Aristotle or latter-day verificationists, have denied this, and have claimed that since we can only measure time by physical change, in the absence of physical change there is no time, and that the experience we have of the passage of time in the absence of external change is due to the neurophysiological processes that are necessarily going on in our brains.

The objection fails. It may be true that for human beings consciousness is only possible if there are appropriate brain processes, but it is not true of logical necessity: many people understand what it is to be conscious without knowing anything of neurophysiology, and some thinkers have been dualists without being involved in evident inconsistency. The concepts of consciousness and brain process, therefore, are not analytically linked. Nor are the concepts of time and change. In Prior's tense logic different formulae are needed to express the thought that before the Big Bang nothing happened and the thought that

---

[1] Karl Barth, *Church Dogmatics* II/I, ed. G. W. Bromiley and T. F. Torrance (Edinburgh: T & T Clark, 1957), 611.

before the Big Bang there was no time at all.[2] Shoemaker has described a conceivable situation in which we should have convincing reasons for holding that during some interval of time no events have taken place.[3] Even when nothing happens, we have some subjective sense of the passage of time, and that is enough to show that the concepts not only are distinct, but might be applied differently in some conceivable situation. We should not, therefore, seek to identify time with change, but recognize that it is always meaningful to raise the possibility that a change has occurred faster or slower than on other occasions, or that time has elapsed without there having been any change. Although it is open to us to maintain that awareness is itself a change, it is not logically required to be a physical change. And our intuitive sense that to be conscious is to be aware of the passage of time remains, so that if anyone denies that God is temporal, he denies also that he is conscious as we understand the term.

Even if we could form a concept of consciousness that was not temporal, we could not form one of agency. To act is to bring about change. Things are different, if we act, from what they would have been otherwise. God is reported in the Bible as having done many wonderful things, most notably in raising Jesus from the dead. Some of God's actions can be dated with a fair degree of precision: some time between 29 and 33 AD God did something remarkable.

Traditional theologians and modern deists try to account for all God's actions as the temporal consequences of a timeless decision. Apart from its inherent implausibility, such an account cannot accommodate God's responding to the actions, and especially the petitions, of free human agents. If man has free will, and can do things which cannot be foreknown with certainty by God, then God cannot have by some timeless decision anticipated what man may do or ask for. If we believe that when Jesus prayed *Abba*, the Father heard his prayer and answered it, we must date the hearing and the answering to the same time as Jesus praying and receiving the answer. Even if we can make sense of the God of the deists, we cannot, if we accept man's free will, think that such a deity spake by the prophets, was the Father of Jesus, could communicate with us, or take action in the light of what we had freely decided to do.

Some modern theologians agree that it is incoherent to say that God is timeless, but hesitate to say that he is temporal, preferring to say that he is trans-temporal, beyond time. We can make sense of that locution. Certainly God is not limited by time as we are. No future time is too remote for him to care or think about: no time is too long ago for him to remember and judge. If we distinguish the present *interval* from the present *instant*, we can say that for God the present *interval* encompasses the whole of time, and that the passage of time for him is not a matter of regret for the loss of the transitory goods of this world, nor the present age a period of tedious waiting for the redemption of time. Life for us is

---

[2] See more fully, J. R. Lucas, *A Treatise on Time and Space* (London: Methuen, 1973), §2, 10, 11.

[3] S. Shoemaker, "Time without Change," *Journal of Philosophy* 66 (1966): 363-381. See further, W. H. Newton-Smith, *The Structure of Time* (London: Routledge, 1980), ch. 2.

perpetually poised between the "not yet" and the "already gone." Not so for God; no time is remote and inaccessible to his understanding, the whole *aeon*, the whole age, is his today. In this sense certainly we can say that God is beyond time, and his eternity transcends our temporality. But more is meant. There is some sense of ontological dependence. Time depends on God, not God on time. If we simply say that God is temporal, it is easy to read this as saying that God exists in time, and is subject to time. Such language jars. The theist wants to stress the ultimacy of God, and traditionally has expressed this by saying that God created time. But that locution is unfortunate: it uses a temporal predicate to account for the existence of temporality, and naturally invites the question of what God was doing before he created time. I prefer to express the ontological dependence of time on God by saying that time stems from God rather than was created by God. But so long as there is no suggestion of God's being timeless, no exception need be taken to saying that God is beyond time.

3 *The Relativity of Time*[4]

God's temporality raises problems for the physicist, since God is absolute and modern physics deeply relativistic. It is generally thought that it is a problem peculiar to the twentieth century, since Newtonian physics posits an absolute time, independent of space and motion, but this, though true, skirts the underlying problem, which arises from the tension between the physicists' search for greater and greater universality, and the obstinate particularity of a personal God. The physicists frame their laws in terms of underlying symmetries, and are essentially concerned with the view from nowhere: persons have a personal standpoint, and each views the world from nowhere. If God is a person, he too breaks symmetry, and there are not merely many human frames of reference, but a divine one, which is, at least in theological terms, preferred.

Contrary to much recent thinking, there is nothing wrong with preferred frames. What both Newtonian mechanics and the special theory established were certain symmetry principles, according to which there were no physically significant differences between different systems so far as those theories were concerned. There was no method within Newtonian mechanics of picking out one system as being absolutely at rest from any other moving uniformly with respect to it. But this did not stop Newton from wondering whether the center of mass of the solar system might not be at rest absolutely, nor later physicists trying to determine the velocity of the earth with respect to the ether. If the Michelson-Morley experiment had yielded a positive result, physicists would have had no difficulty in accepting the conclusion that although the laws of Newtonian mechanics were covariant under the Galilean transformations, the Maxwell equations were not, so that the conjunction of Newtonian mechanics with electromagnetism did determine an absolute frame of reference.

---

[4] The argument of this section is developed in more detail in J. R. Lucas and P. E. Hodgson, *Spacetime and Electromagnetism* (Oxford: Oxford University Press, 1990), 2.9, 3.9, 67, 117-121.

The same applies to the special theory. By itself it cannot pick out any one rather than any other inertial frame of reference as being absolutely at rest, but if some further information were forthcoming, it might be possible. The analogue to the putative positive result of the Michelson-Morley experiment is some mode of instantaneous transmission of messages. Suppose the experimenters at Duke University succeeded in establishing telepathic communication with rational beings in the vicinity of Alpha Centauri, and were able to send messages and receive answers back promptly. By comparing the dating system based on telepathy with that based on sending and receiving wireless signals we could determine whether or not Earth was at rest with respect to the conjunction of electromagnetic theory and telepathy. Once again we should have both Absolute Simultaneity and Absolute Space. There is nothing wrong with these concepts. Contrary to the current understanding of Einstein's principle of relativity, we can make sense of there being a preferred frame of reference. Einstein's principle of relativity is concerned only with electromagnetic phenomena, and tells us that, *so far as those phenomena are concerned*, all inertial frames of reference are equivalent.

Although the electromagnetic phenomena studied by the special theory do not pick out any preferred frame of reference, the standard model for the general theory—the Robertson-Walker model—does. The background radiation picks out a preferred frame for the universe as a whole. Locally, of course, the time we need to ascribe to distant events in order to give a coherent account of physical phenomena, especially gravitation, may be very different; and the Robertson-Walker model is only an approximation, fitting our universe in the large, but clearly not when small objects (like the sun) are being considered. Nevertheless, the actual practice of those working in the general theory belies the claim that the modern scientific world-view cannot accommodate the absolute time that the temporality of God implies.

Quantum mechanics argues similarly. The time-dependent Schrödinger equation takes time as a fundamental category, not at all open to question. Moreover, the collapse of the wave-function into one of its *eigen*-functions seems to indicate the present, marking off the open, probabilistic future, expressed by a superposition of wave functions, from the definite black-or-white past. We wonder when Schrödinger's cat died, and reckon that it must have been at some definite moment of time. On this rather realist construal, there is a tide of wave-collapse spreading through the whole universe, which represents the cosmic present. Even if we start from a more phenomenalist standpoint, and believe that it is the observer who forces the dead-or-alive cat superposition into either a dead- or an alive- wave function, a theist will still have a cosmic present determined by the observation of the cosmic observer. God is always about in the quad, and keeps even the sparrows under observation, so he must certainly know when a cat dies.

It follows that there is no reason for a theist to feel queasy, in the light of the special theory, about God's time, which he needs to do if he is to make sense of the biblical record and believe in a God who can hear our prayers and answer them. In that case he must know what we are doing when we are doing it. He must have up-to-date temporal knowledge of what is going on in the world.

He must know when a flare on Alpha Centauri occurs, and whether it occurs before or after a prayer offered up in a service at St. Peter's, Rome. God's omniscience may properly be restricted so as to exclude future contingents, which are not yet there, so to speak, to be known; but since a sentient being on Alpha Centauri could observe the flare, and know when it occurred in his own experience, and the priest in St. Peter's know when he offered up his prayer, it would be impossible for any divine being not to know both and still be omniscient. A God who responds to prayer must have temporal knowledge, and an omniscient being with temporal knowledge must know what things are occurring throughout the universe at any particular time—God's present time—and therefore that they are simultaneous, really simultaneous, absolutely simultaneous. And this, though not playing any part within the special theory, is no more inconsistent with it than the rest-frame of the ether would have been in mechanics had the experimental evidence supported it.

4 *Quantum Theory*

Although quantum mechanics at one level and on one interpretation argues in favor of absolute time, it also raises the question whether time is, as we ordinarily assume, infinitely divisible, indeed whether the concept of time has any meaning when we get down to the fundamental level of analysis. Our first thoughts are to ascribe to God perfect power to discriminate different instants of time. We are unable to notice time differences of less than a twentieth of a second, but we can improve on the naked eye with the aid of cameras, and God, we naturally suppose, is the perfect photo-finisher. But this is to take too epistemological a view of Heisenberg's uncertainty principle, which is not just a limitation on what we can discover about non-commuting observables, but locates the problem in the question rather than the absence of an answer. The lesson of quantum mechanics seems to be that our concept of time cannot be idealized indefinitely far, and that it makes no sense to ask of two events less than a chronon apart which actually came first. If so, we should not idealize God's time as being perfectly exact, with God being able to discriminate temporal differences to within whatever level of exactitude we care to think of, but should think of divine perfection in some different way. But it would not be a very serious difference. So far as macro-time is concerned, everything would be the same, and it is macro-time that is important for our concept of God as a conscious being and a rational agent. God needs to know things at least as well as we do, and to be able to do things, bringing about the effects that he intended by means of antecedent causes: he does not need to have direct awareness of infinitesimals in order to be the ground of our being and the focus of our worship.

5 *Explaining the Big Bang*

Quantum cosmology adds to this difficulty the further ones of whether we can talk intelligibly about the moment of creation, and the possibility of the fundamental theory of everything being one in which time features only as an imaginary parameter, or does not feature at all. Some of these difficulties arise from the methodology adopted by quantum cosmologists, and the ideals of explanation they pursue. The Big Bang is an embarrassment, because it is a singularity, and the laws of physics break down at that point, and because it suggests that physicists are introducing God, as a sort of *Deus ex machina*, to provide an explanation when physics itself cannot. Theologians, too, feel shy about invoking this type of explanation, and murmur that they don't want a God of the gaps. But they should not be shy. God, if he is God at all, is God of the gaps as well as of everything else, and there are bound to be gaps. Nor is it necessarily a sign of defeat for scientists to switch from one type of explanation to another when the first type runs out. Nevertheless, we can sympathize with the urge quantum cosmologists feel, which impels them to seek for some physical explanation of the Big Bang; and we can recognize in the way their ideal of explanation develops a shift somewhat similar to the switch in the Schoolmen's search for the first cause, who in their search for the ultimate explanation of all things were impelled to distinguish temporal from logical priority, and identify God not with the temporally most antecedent condition, but with the logically most profound one.

In our ordinary thinking, explanations in terms of antecedent conditions and covering laws—deductive nomological explanations as they are currently called[5]—which can be viewed as the successor of Aristotle's "Efficient Causes," occupy pride of place. But no deductive-nomological explanation of the Big Bang is available, since *ex hypothesi* there were no conditions antecedent to the Big Bang in terms of which, together with some covering law, an explanation could be constructed. Worse, if we extrapolate backwards to the Big Bang, we retrodict that at that instant spacetime would have had zero volume and infinite curvature, which is as much to say that the laws of physics would not have operated then. Not only is the Big Bang inexplicable according to that canon of explanation, but itself contravenes the laws on the strength of which we are supposed to believe in it.

These difficulties disappear if we seek some other type of explanation. The pre-eminence accorded to deductive-nomological explanation is, as

---

[5] See C. G. Hempel, "The Function of General Laws in History," *Journal of Philosophy* 39 (1942; reprinted in *Readings in Philosophical Analysis*, ed. H. Feigl and W. S. Sellars [New York: Appleton Century Crofts, 1947], 459-471): 35-48. See also C. G. Hempel and P. Oppenheim, "Studies in the Logic of Explanation," *Philosophy of Science* 15 (1948): 135-176. For a modern account see Peter Lipton, *Inference to the Best Explanation* (London: Routledge, 1940), 29-31.

Collingwood points out, a fairly recent development, and although it is often taken to be the paradigm of explanation, it has also been trenchantly criticized.[6]

Deductive-nomological explanation provides answers to only one type of question, and often the question actually being asked is of a different type: sometimes it is not a *Why Necessarily?* question, but a *How Possibly?* question;[7] or, more important for our present argument, not a *Why Precisely?* question, but a *Why Generally?* one, where we are concerned with general features of the world around us, without seeking to show that they could have been predicted in precise detail by a sufficiently informed Laplacian calculator. Two steps are taken in the passage from the deductive-nomological to this more general type of explanation: there is first a shift in interest. If I am a biologist, I am not interested in the precise accounting for the exact position and momentum of every atom, even if that were feasible. Such a wealth of information would only be noise, drowning the signal I was anxious to discern, namely the activities and functioning of organisms, and their interactions with one another and their ecological environment. The Laplacian account not only gives us too much detail, but is modally flat: it gives us a full account of what would ensue from one particular initial condition, but fails to bring out the crucial fact that if these conditions were altered in all sorts of ways, the features of interest to the biologist would be exactly the same. Just as a full account of causality needs to express the counter-factual would-be content of a causal law, so an adequate account of organisms must accommodate their permanence and persistence and high degree of invariance under change of circumstance. We need to look from the top down rather that from the bottom up in order to discern the features that are of interest to us.

In the same spirit, and with powerful metaphysical support from the indistinguishability of sub-atomic entities, quantum cosmologists seek a top-down explanation of the Big Bang in terms of some extremal principle, and apply quantum, rather than classical, principles. Two such explanations have been attempted in recent years: one by Hartle and Hawking, the other by Vilenkin.[8] In each case quantum indeterminacy removes the embarrassment of a zero-volume universe with the associated infinities and singularities, and a top-down explanation takes the focus off the initial conditions as the crucial component of the explanation. The universe is not explained as a consequence of what it was like at the time of the Big Bang or some time earlier, but as being the best, in a somewhat *recherché* sense of "best," instantiation of a number of rational *desiderata*.

---

[6] For example by R. G. Collingwood, *An Essay on Metaphysics* (Oxford: Oxford University Press, 1940), chs. XXIX, XXX, 285-297.

[7] W. H. Dray, "Explanatory Narrative," *Philosophical Quarterly* 4 (1954): 15-27; see also *idem*, *Laws and Explanation in History* (Oxford: Oxford University Press, 1957).

[8] J. B. Hartle and S. W. Hawking, *Phys. Rev.* D 28 (1983): 2960-2975; A. Vilenkin, "Quantum Cosmology and the Initial State of the Universe," *Phys. Rev.* D 37 (1988): 888-897.

"Best" for Hartle and Hawking is a state of minimum excitation for a spatially homogeneous and isotropic universe closed in all spacelike dimensions. The set of all possible universes of this type can be considered as a phase-space, a "minisuperspace" they call it, and the Euclidean-functional integral prescribes, analogously to the principle of Least Action, one reasonable candidate for the ground-state wave function. Whereas with a deductive-nomological explanation the initial condition is of the same type as the subsequent state of the universe we are trying to explain, the constraints which determine a single solution as the only one satisfying some extremal principle are of a quite different type. They therefore do not give rise to an infinite regress, and may be argued for on purely rational grounds. It seems reasonable to impose a number of symmetry conditions, such as homogeneity and isotropy, and to argue (though the argument is not compelling) for a closed topology so as to avoid there being any adventitious boundaries. Quite apart from obviating the embarrassment of a point-singularity, a top-down approach offers a prospect of giving us a Theory of Everything, which a deductive-nomological explanation never could. Hawking himself says:

> If spacetime is indeed finite but without boundary or edge, this could have important philosophical implications. It would mean that we could describe the universe by a mathematical model which was determined completely by the laws of science alone: they would not have to be supplemented by boundary conditions.[9]

The Vilenkin scheme is similar in general outline. It seeks to pick out *this* universe from a possible set of universes by imposing certain boundary conditions on solutions to the quantum gravity equation so as to exclude universes that begin at a spacetime singularity. Once again, in a manner of speaking, our universe is singled out by the boundary condition that there are no boundaries. But there are some universes in which the quantum gravity state function oscillates as it ought to if it is to be reasonably classical, and others in which it does not, and the latter can be described by means of giving the time parameter imaginary values; there is an internal boundary between the region in which the quantum gravity state function oscillates and time seems to be real, and the region in which the quantum gravity state function oscillates seems not to be real. A path from the latter to the former region represents a possible history of a universe which ends up developing properly, but with no spacetime singularity at the Big Bang, though with an "antecedent" state in which time is imaginary, and which does not have the Lorentzian signature of Minkowski spacetime, but the positive-definite signature and topology of Euclidean four-dimensional space.

Once again we need to be cautious in interpreting the scheme. It is essentially a non-deductive-nomological, top-down explanation, explaining a particular in terms of the general—why this particular universe in terms of its best satisfying some rational *desiderata*. It is not an explanation in terms of a genuinely antecedent state, as a deductive-nomological explanation would be.

---

[9] Hawking, "The Edge of Spacetime," in *The New Physics*, ed. P. C. W. Davies (Cambridge: Cambridge University Press, 1988), §4.5, 69.

The *explanans* is not temporally prior to the *explanandum*. Nor should every linearly ordered path in the minisuperspace be taken to represent a possible world-history of a possible world. Linear orderedness is one mark of time, but directedness is another: directedness is preserved in the Lorentzian signature of Minkowski space, but not in the positive-definite signature of Euclidean space, in which it would be perfectly possible to have a continuous rotation of any one direction, in one sense, into its opposite. What we have is not the breakdown of time at the Big Bang, but at that juncture, the breakdown of the identification of temporally oriented directions in some curved version of Minkowski spacetime, we have changed our type of explanation. In seeking for the first cause, we were led to earlier and earlier states of the universe, but could not explain, in deductive-nomological terms, the originating event itself. By asking a different question, however, we may be able to find a satisfactory explanation, not of why the first event took place in terms of an even earlier event, but of why the whole lot of events took place at all.

6 *The Reality of Time*

So far, so good. But difficulties remain. It may be that quantum mechanics and quantum cosmology lead us to explain the temporal phenomena of this fleeting world in terms of some underlying theory. That could be the case, and if it were so, it would be reasonable to think of this underlying reality as more real than the world of change and chance which we experience. The physical sciences have long sought omni-temporal laws holding at all times and in all places, and accorded them a higher degree of reality than the boundary conditions. We cannot rule out further developments in that direction. In a different vein I have spent many years trying to work out a modal derivation of time, seeing it as the passage from possibility through actuality to necessity. Quantum cosmology might succeed in telling us why time had to exist. In theology too we draw distinctions between an ultimate self-subsistent, self-explanatory ἀρχή (*arche*) and the persons of the Godhead, acting and reacting in history, and working through rational agents who are drawn towards the reasonable and the true. A deep theory of time would be like a sort of metaphysical trinitarianism—very different from the trinitarianism of Christian theology, I hasten to add—which allows the possibility of explaining the temporal in terms of something non-temporal. Such a possibility cannot be ruled out, either in natural science or in theology. But it is subject to a severe constraint. Time may be explained, but cannot be explained away. We may explain why it is that our universe is a temporal one, and why God is a temporal one, and why God is a personal God who acts in time and has temporal experience: but the test of the explanation is that it should explain just that, and we could not accept as satisfactory an explanation which made out that *au fond* time was unreal. It would not save the phenomena: also it would make nonsense of our Lord's passion, and the cry of dereliction from the cross.

Hartle and Hawking work with imaginary time, as does Vilenkin: they use the transformation $t \to -i\tau$, thus again transforming the $(3 + 1)$ Minkowski spacetime into four-dimensional Euclidean space. In itself that might be an unobjectionable notational device, but it is taken to be much more than that. Hawking comments:

> In the very early universe, when space was very compressed, the smearing effect of the uncertainty principle can change this basic distinction between space and time. It is possible for the square of the time separation to become positive under some circumstances. . . . The question then arises as to the geometry of the four-dimensional space which has to somehow smoothly join onto the more familiar spacetime once the quantum effects subside.[10]

His answer (somewhat different from Vilenkin's) is that there is no sharp boundary between real and imaginary time:

> What happened at the beginning of the expansion of the universe? Did spacetime have an edge at the Big Bang? The answer is that, if the boundary conditions of the universe are that it has no boundary, time ceases to be well-defined at the North Pole of the Earth. Asking what happens before the Big Bang is like asking for a point one mile north of the North Pole. The quantity we measure as time had a beginning but that does not mean spacetime has an edge, just as the surface of the Earth does not have an edge at the North Pole, . . .[11]

It is unfair to press analogies too hard, but the analogy with the North Pole does help to show what is going amiss. Although on the surface of the Earth the direction "north" is the same as "towards the North Pole," there is an underlying and deeper sense in which it is determined by the axis of the Earth's diurnal rotation. In that sense we can cope with someone asking for a point one mile north of the North Pole. We distinguish. In the superficial sense, one cannot get more towards the North Pole than the North Pole itself, and the question is nonsense: but the persistent questioner, who is trying to articulate a deeper sense of direction, can be answered in terms of the astronomical north; a point one mile north (that is, astronomically north) of the North Pole is a point at latitude 90°N, and altitude 5,280 ft. In something of the same way, some of the physical parameters we correlate with time during most of the universe's history—entropy or the spatial volume of the universe—cannot be extrapolated back before the Big Bang, so that we cannot ask quasi-temporal questions of them. But that does not mean that we cannot ask temporal questions at all. For time is not defined solely by reference to physical parameters, in respect of which it is a concomitant of change, but is linked with other non-physical concepts too, notably mental ones. The theist can perfectly intelligibly ask what God was doing before the Big Bang, and even the atheist can hypothetically imagine a disembodied intelligence at the time of the Big Bang and wonder what his experience would have been like.

---

[10] Ibid., 68.
[11] Ibid.

7 *The Temporal Experience of God at the Creation*

St. Augustine reports the witticism that God was occupied before the creation of the Universe in thinking up suitable punishments for those who pry into matters that do not concern them, and most theologians have shied away from addressing the question seriously. But if we can ascribe temporality to God before the creation, we shall have to face the question why God did not create everything a year sooner.[12] Richard Swinburne faces the question but suggests that time ceases to be a magnitude at all, so that although time existed before the Big Bang, there was not any stretch of time during which God was idle or might get bored. But in spite of there not being any processes by means of which time could actually be measured, it would still be a magnitude, of which we could ask "How long?" This follows from purely topological considerations. So long as we hold that for any two distinct temporal instants there is another between them, and that the past and the future are always separated by a present which is a boundary between them, temporal instants must not merely be dense, but have the order-type of a continuum. And we must allow that there always is a present, for it is its anchorage in the present that makes time real. So we cannot evade the question of what God was doing before the creation by claiming that then time was merely a dense ordering without any elapse of interval, so that questions of "How long?" and ascriptions of tedium would be entirely inapposite.

But though the question cannot be evaded, it ought not to worry us unduly. Leibniz' question seems damaging only because we impute a homogeneity to time as a necessary condition of our being able to measure it, not as an inherent feature of time itself. God is thought of like Buridan's ass, unable to create the universe at any one time, because there would be no sufficient reason for choosing that time rather than any other. But quite apart from its being somewhat asinine to starve rather than break a symmetry, there is not any essential symmetry in the time of personal experience. Although once we allow that there was time before the creation, the question "Why not sooner?" does arise, it loses its edge, and can be answered by saying that if the creation was to occur, it had to occur at some time or other, and that that time was as good as any other. Similarly, it may at first seem very damaging to those of us who are infected with the Protestant work ethic, and feel that one ought always to be doing something, and that manufacture is the only proper sort of doing. But there is something belittling about the workaholic's view of God. The ancient Jews saw nothing wrong with God resting on the seventh day, and it is one merit of the doctrine of the Trinity that it does not condemn God to solipsistic idleness until the creation gave him some toys to play with. For us time goes both too slowly and too quickly, and we often seek to kill time, and often repine at its slipping away from our grasp. But these are the limitations of finite and imperfect humanity: for God there is no sense of the future being slow in coming, or the past being lost in the mists of time. Although time is magnitude,

---

[12] Leibniz' Third Letter to Clarke, §6; reprinted in H. G. Alexander, ed., *The Leibniz-Clarke Correspondence* (Manchester: Manchester University Press, 1956), 26-27.

so that we must think of its stretching back everlastingly before the Big Bang, it is a stretchy stretch, in that though the present instant is always distinct from those that are past or are yet to come, God's *specious* present can extend indefinitely far back, and except for the free decisions of the free agents he has created, indefinitely into the future too.

GOD AS A PRINCIPLE OF COSMOLOGICAL EXPLANATION

Keith Ward

The purpose of this paper is to explore the way in which the formulation of a Christian doctrine of creation is helped by modern cosmology, and the sense in which the doctrine of creation adds an important explanatory element to the form of explanation given by quantum cosmologists. The paper falls into six sections. After a discussion of the classical Christian view of creation by a timeless God, I maintain that the concept of a temporal God provides a more adequate analysis of the notions of omnipotence and of creation. I then develop a model of creative temporal freedom which is meant to steer a middle way between determinism and sheer randomness, and suggest that there are two important ways of explaining the nature of the universe—the nomological and the axiological. The first of these, often taken as an ideal in physics, leaves one with an ultimately blind necessity; the second leaves one with an apparently fortuitous contingency. Taking both together, the most adequate explanatory concept seems to be that of a being which is necessarily existent, and which brings events into being by a process of creative emergence for the sake of the values they realize or make possible. I conclude by suggesting that this is in fact a concept of God which both modern cosmology and modern Christian doctrine help to shape, which shows the contribution a doctrine of creation by God can make to the explanatory theses of quantum cosmology.

1 *Creation and Timelessness in Classical Christian Thought*

Christians assert that this cosmos is created by one self-existent being of supreme perfection. To say that the universe is created is to say that it is brought about intentionally, that its existence is the expression of a consciously formed purpose. It is not to say, nor does it imply, that the universe has a beginning in time. Theologians have usually stressed that the universe is not brought about in time, in the sense that there exists an uncreated time within which the physical universe comes to exist at a particular point. Augustine makes the point with particular clarity; and although I will be critical of his view in this paper, his discussion forms a classic statement of the Christian position. When Augustine asked the question, "What made God create heaven and earth then, and not sooner?"[1] his reply was that "no time passed before the world, because no creature was made by whose course it might pass."[2] That is, as he understood it, time is a certain sort of relation between objects, and where no objects exist

---

[1] Augustine, *City of God*, bk. 11, ch. 4 (New York: E. P. Dutton, 1957), 314.
[2] Ibid., ch. 6, p. 317.

which are related by that relation, then the relation does not exist either. The temporal series extends back, as a relation, T, between objects, to an object which simply has no relation T to anything preceding it. Since "beforeness" is a relation between two objects, the first object is not related by "beforeness" to anything. There is nothing "before" the temporal universe; neither any objects nor any time. For Augustine, God brought about time and space as well as all the things that are in them. Just as God did not create space at a certain place, but non-spatially caused all places to exist, so God did not create time at a certain time, but non-temporally caused all times to exist. Whatever reservations one may have about a doctrine of non-temporal causation, it does not seem self-contradictory to say that A wholly depends for its existence upon B, though there is no temporal relation between A and B. At any rate, this is a view Augustine is committed to. As far as this doctrine goes, there might or might not have been a first moment of time; both possibilities are consistent with the dependence of all times, however many there are, upon God. Augustine did accept, like most classical theologians, on grounds of revelation, that this universe had a beginning in time; though he also suggested that there might have been an infinite number of universes before this one, in which case, of course, time would have no beginning.[3]

If this is the case, it follows from the doctrine of creation that God, the creator, is a non-spatio-temporal being. As the creator of the whole of spacetime, God is beyond spacetime. This is an important part of the traditional belief that God is transcendent. This statement, however, like most statements in theology, immediately needs to be qualified. God is not, for any standard theistic view, outside spacetime in the sense that God is excluded from every space and time. On the contrary, God is omnipresent.[4] That is to say, every object or event standing in spacetime relations only exists by the originating and sustaining power of God. Without such a Divine activity of bringing into being and sustaining in being (these being simply different aspects of the same act in God), nothing could exist even for the smallest moment. That relation of *holding in being* is the most intimate one that can be conceived; it is a relation of immediate and total origination, and it obtains between God and every created object, whenever it exists.

Thus it is wholly inadequate to think of God having created the universe at some remote point in time—say, at the Big Bang—so that now the universe goes on existing by its own power. This popular misconception, that "the creation" is the first moment of the spacetime universe, and that the universe continues by its own inherent power, wholly misconstrues every classical theistic tradition. It is irrelevant to a doctrine of creation *ex nihilo* whether the universe began or not; that the universe began was usually accepted because of a

---

[3] Ibid., 359.

[4] See Thomas Aquinas, *Summa Theologiae*, 1a, 8, 2: "As the soul exists wholly everywhere in the body, so God exists wholly in each and every thing." (New York: McGraw-Hill), vol. 2, 117.

particular reading of Genesis 1.⁵ The doctrine of creation *ex nihilo* simply maintains that there is nothing other than God from which the universe is made, and that the universe is other than God and wholly dependent upon God for its existence. Creation is, properly speaking, the relation which holds between every point of spacetime and the creator, such that each moment exists in total dependence upon the sustaining being and will of the creator. God is not nearer to the beginning of time than to any other point of time. On the contrary, God is present and active, in the most intimate and fundamental way, at every point of spacetime. God is not excluded from any space or time. Though God, as creator, infinitely transcends spacetime, precisely because God *is* creator, God is present and active at every point of spacetime.

The classical form of theism which is found in most of the Church Fathers and which is especially clearly formulated by the thirteenth-century theologian Thomas Aquinas, asserted that God is eternal, in the sense that God is timeless.⁶ That is, God contains no *internal* temporal relations—no "before" and "after," no past to be remembered or forgotten, and no future which is as yet undecided—and God has no *external* temporal relations—does not exist before, after, or at the same time as any other thing whatsoever.

The creation of the universe can then be represented as the total dependence of spacetime, in all its parts from beginning to end (if time has a first and last moment), upon a timeless being, in whom there is no change, coming into being, or passing away whatsoever. As Augustine put it, "all that ever God created was in (God's) unchanged fixed will eternally one and the same."⁷ One must not think of God wondering whether to create a universe, deciding it would be a good idea, choosing which one to create from a selection of possibilities, and then bringing it into being. Since God is in the strictest sense immutable, being wholly timeless, there is never a time when God has not yet decided to bring about a universe, or has not yet decided which one to bring about, or has not yet brought it about. God timelessly brings the universe into being. There is, in the divine being, an immutable intention to bring precisely this universe into being, an intention which is actualized by the very same act by which it is formed. As Aquinas puts it, "As the divine existing is essentially necessary, so also is the divine knowing and the divine willing."⁸ God's willing the universe is not something that God came to a new decision about; it is part of the immutable Divine being from all eternity.

2 *God, Temporality, and Omnipotence*

This classical view seems to match rather well the "block view" of spacetime which some physicists hold—the view that every point of spacetime is in some

---

⁵ Ibid., 1a, 46, 2: "That the world has not always existed cannot be demonstratively proved but is held by faith alone," vol. 8, 79.

⁶ Ibid., 1a. 10. 2; vol. 2, 139.

⁷ Augustine, *City of God*, bk. 12, ch. 17, 361.

⁸ Aquinas, *Summa Theologiae*, 1a, 19, 3; vol. 5, 17.

sense existent, and that the sense of passage through time which is characteristic of humans does not point to a fundamental and irreducible characteristic of unobserved physical reality.[9] For the classical view clearly entails that God knows every moment in time, from first to last, in one timeless act of knowing. God does not need to *fore*know anything, since God knows it directly, in a timeless manner. God stands in need of no inferences or predictions, but knows all spatio-temporal events, from first to last, in one unchanging intuition.

Despite this pleasing congruence, however, there are deep theological difficulties with the idea of a wholly timeless God, and it may be doubted whether the idea is really entailed by a doctrine of creation. What the doctrine of creation requires is that God should transcend spacetime, as its intentional cause. That does entail that God cannot be limited in spacetime. But it does not entail that God cannot be in time at all, in any sense; and it does not entail that the divine being contains no analogates of past, present, and future. In fact, both the doctrine of omnipresence, which makes God present to every time, and the Christian doctrine of incarnation, which identifies God, in some sense, with a human being who lives in time, seem to imply a form of divine temporality. If God is present at every time, the clearest thing to say is that, in creating spacetime, God creates new temporal relations in the divine being itself. That is, by the same act by which God creates events in spatio-temporal relations, so God thereby creates the relation of being co-present with many times. Thus one is able correctly to say, at each time, "God *now* exists"—which, strangely, is not true on the classical view, since for that view God, properly speaking, has no temporal properties.

The classical view also has to invent complex logical epicycles to explain how it is true that God lives, suffers, and dies in Jesus Christ, and yet the divine nature itself has no temporal relation to Jesus at all. A more straightforward reading might be that God is the subject of temporal properties, and is co-temporal with Jesus in an additional way to that in which God is co-temporal with everything; namely, by being active in the life of Jesus in a way in which God is not directly active at other times.[10]

There is an even greater advantage in ascribing to God a certain sort of temporality, which is not that of our spacetime. That is, only if some form of open future exists in God can it be possible for God to do new things, to be genuinely and radically creative and to come to new decisions. The classical idea of omnipotence involves the thought that God can do anything possible that is compatible with God's own existence and prior decisions. But it would be an odd definition of "omnipotence" which denies to God the possibility of ever doing anything new. This would seem to deny the applicability of the concept of "power" to God at all. If God is, as the classical tradition has asserted, purely actual without any potentiality at all,[11] then it follows that God can never do

---

[9] See the article by C. J. Isham/J. C. Polkinghorne in this volume.

[10] Such a revision of the classical immutabilist view is expounded in Thomas Morris, *The Logic of God Incarnate* (Ithaca: Cornell University Press, 1986).

[11] Aquinas, *Summa Theologiae*, 1a, 3, 1: "In the first existent thing everything must be actual; there can be no potentiality whatsoever," vol. 2, 21.

anything other than God does (there is no potentiality in God for doing anything else). This may well seem to be a restricted notion of power, and a richer concept of omnipotence may be the idea of a God who can bring into being an infinite number of new things throughout endless time. One may even argue that the assertion of omnipotence entails that any necessarily omnipotent being must necessarily possess contingent properties. For if it is possible to create some contingent worlds, then the creation of any such world must itself be contingent (i.e., some other contingent world could have been created). So any being which can create contingent worlds must logically be capable of contingent acts. So it must necessarily be contingent in some respects. Far from necessity and contingency contradicting one another in God, certain necessary properties entail certain contingent ones—as the necessarily possessed divine property of being able to create any world compossible with God entails the necessary possession of contingent properties by God. Thus only a God who is necessary in the possession of some properties (including existence, omnipotence, goodness, and omniscience) and contingent in the possession of others (the capacity to freely create some of many possible universes) can be maximally great.

Asserting a form of temporal passage for God in this way obviously requires the logical possibility of speaking of time without space. But if a non-spatial being like God can exist at all, then there seems no good reason why God should not change; and that in itself entails a sort of time without space. If one admits the real possibility of a being which has no spatial extension—and I do not find it plausible to assert as a necessary truth that the existence of anything should be extended in space—then it is a separate question whether such a being can be in two contradictory states, one causally related to the other. If that is possible, and again there is no good *a priori* argument why it should not be possible, then one has a close analogue of time, as that which allows contradictory states of the same being, standing in causal relation to one another, to exist. Thus there is no good *a priori* argument against the possibility of change in a non-spatial being, and therefore of temporality in such a being. Though the time with which physics deals is a relation between physical objects or events, there is another form of temporal relation of which we are aware, which is the relation between different states of some consciousness. One need not deny that all human states of consciousness are related in some way to neurological changes to see that relations between successive conscious states are not analytically or necessarily related to relations between successive physical states. Most philosophical materialists would accept that such physical/mental state identity as there may be is contingent. So it is logically possible, so far as one can see, that temporal relations can exit in conditions of non-physical change. Given this bare and apparent logical possibility, one can feel as confident as one can about anything in this area in suggesting that, if God is creatively free, then God is temporal, though God transcends the physical spacetime system of this universe. This seems to many contemporary theologians a more adequate account of biblical talk of God than does the classical view that God cannot ever do what God has not changelessly willed.[12]

The assertion of a temporal God is compatible with the assertion of omnipotence, and it is not necessary to follow the many process thinkers who deny that a temporal God is omnipotent.[13] Their case is that, if creatures are to share in the creative freedom of God, then God cannot determine all their actions. A truly open future permits the existence of creaturely freedom in a strong sense—that "if a person is free . . . no antecedent conditions and/or causal laws determine that he will perform the action, or that he won't."[14] Moreover, the agent can actualize any one of a number of possible future states, even while its character at a given time remains precisely the same. So there will be many states—the results of free creaturely acts—not determined by God. That, however, need not cause one to deny that God is omnipotent. All one needs to say is that God *could* determine all things, but decides to let creatures determine some things themselves for a good reason—perhaps so that a community of creative love can exist between creatures and God, as the Judeo-Christian revelation asserts. God retains total power, but is not compelled to exercise it at all times. It is relevant that God can ensure that the Divine purposes are realized at some stage; God is not doomed to be perpetually frustrated by creaturely choices. Yet it may be an important fact, which classical theologians were not fully able to take account of, that creatures are free in this creative sense and that God is free to respond to them in creative freedom too.

3 *A Model of Creative Freedom*

The best analogy for such creative freedom is something like the writing of a piece of music. This is an analogy which has been deployed effectively by A.R. Peacocke,[15] and the point of it is to suggest a form of causality which is neither wholly deterministic nor completely arbitrary. At any given point in composition, the note that is to come next is not *entailed* by any past or present note or any combination of them. A number of possible notes might exist but there is literally nothing which determines that one of them should. One the other hand, the note that will next exist is not arbitrarily assigned, as if a computerized random generator could do as well as Mozart. So what governs the choice of the immediately successive note? Simply the creative, directing activity of the composer, who generates musical patterns because of their beauty. It is important that there is an intentional activity at work; that is what prevents the process from

---

[12] See Keith Ward, *Rational Theology and the Creativity of God* (Oxford: Basil Blackwell, 1982), ch. 7.

[13] See A. N. Whitehead, *Process and Reality* (New York: The Free Press, 1978), 346: "He does not create the world, he saves it . . . leading it by his vision of truth, beauty, and goodness."

[14] A. Plantinga, *God, Freedom and Evil* (London: Allen & Unwin, 1975), 29.

[15] A. R. Peacocke, *Creation and the World of Science* (Oxford: Clarendon Press, 1979), 105.

being arbitrary. It is mind-directed or intentional. It is important that there is an intentional activity at work; that is what prevents the process from being arbitrary. It is mind-directed or intentional. It is equally important that the intention is to create a unique and original form of beauty. It is concerned with the generation of order and sensory and affective appeal. So one has a mind concerned with the production of beauty, working to continue an unfinished pattern. Central elements of this situation are (1) the pattern itself, (2) the active concentration of a mind with certain skills and capacities, (3) canons of beauty and taste which can be modified but which provide models of creative activity and (4) human pleasure in certain sorts of sensory pattern which form the basis of selected goals (i.e., realized states of value).

Deterministic models of causality work with a basically deductive scheme, for which the future can only unfold what is already contained in the past (one cannot have more in the conclusion of a deductive argument than there is in the premises). In cases of creativity one has what might be called a teleological model of causality. For such a model, the future builds on the past, by adding to it in a way which seeks to realize certain goals in the process of creation. At any moment of the temporal process, one has an incomplete pattern, a creative agency and a general goal (itself emerging from the recognition of desires of certain sorts) setting the direction of creativity. It is consciousness which recognizes certain states as desirable, which accordingly sets certain goals of activity and which realizes these goals through continuing an apprehended process. At each moment its successor moment is generated by a directive agency informed by its apprehension of the past, endowed with characteristic skills and dispositions and oriented by its conception of values.[16]

What this means is that one cannot consider time as a series of quite discrete events, each complete in itself, one succeeding another either by some form of quasi-entailment or quite arbitrarily. At least as far as conscious and free agents are concerned, and as far as they are conscious and acting rationally, each temporal moment is filled with an awareness of the past and oriented towards a possible future, as part of a dynamic process in which future events are emergent—genuinely new and yet continuous with, and to some extent developmental of, what has gone before. And this flow towards the future is essentially directed to the realization of values, which are what give the temporal process intelligibility.

Whitehead has, in my view, very helpfully emphasized the value of creative emergence and therefore of temporality, a value which gives a much more positive reason for the creation of a spacetime universe than the traditional account provided. He has given creaturely freedom and creativity a much more important place in the explanation of how things have come to be as they are; and he has suggested a form of causality which fits the facts of human experience quite well. It is a consequence of a Whiteheadian view that, where there are many developing sentient beings with limited knowledge, their aims will often conflict, and from such conflict suffering and destructive processes

---

[16] Whitehead has deployed these notions effectively in *Process and Reality*, though I am restricting their range to conscious agents, as he does not.

will flow. Natural and moral evil are inevitable in a world with general structure. However helpful that may be with the problem of evil, many theists feel that a process view nevertheless undesirably limits the power and sovereignty of God. Whitehead was basically an atomist, believing that the real causal impulses of the world derive from its atomic constituents, the "actual occasions." He did give God a place as the "poet of the world," or the lure towards greater harmony and beauty.[17] But he was not able to give a convincing account of how such conflict as there is could ever be overcome and the Divine purposes ever by wholly fulfilled.

On the account I am suggesting, human freedom and creativity remain basic constituents of the universe, but God has a much more active and positive role to play, as Christian revelation claims. God is the master creator, laying down the pattern, influencing creaturely choices towards a goal whose general nature is given by God but whose specific character is, in many respects, determined by creatures. God creates our capacities; he influences them towards good. But we express them in specific ways; we can and should do so by free reliance on God, who will ensure that their potential for good is realized, though in a uniquely creative way that only we can specifically determine. To use another analogy, that of the master chess player, used by Geach and Polkinghorne,[18] God enters into the game to ensure that it is concluded in the way that he desires. In this way, he ensures that all the evil caused by the misuse of creaturely freedom will be ordered to good, though the tragedy of its occurrence cannot be simply obliterated by anything God does.

4 *Nomological and Axiological Explanation*

In the light of this discussion, one might suggest that there are two complementary models of intelligibility by means of which one can understand a world containing free actions. They might be termed the nomological—a model familiar to physicists—and the axiological—often found in the social sciences. For the first model one understands the genesis of the new by appeal to general principles or laws of regular succession. That there are simple, elegant, and mathematically stateable laws which describe the structures of the universe is a surprising and remarkable fact,[19] and their formulation enables scientists to understand and predict many of the processes of nature. For the second model one understands the genesis of the new by appeal to values which are to be realized by beings of certain distinctive capacities and modes of awareness. This model does not provide general laws which facilitate prediction. It provides an

---

[17] Whitehead, *Process and Reality*, 346.

[18] P. T. Geach, *Providence and Evil* (New York: Cambridge University Press, 1977), 59; and John Polkinghorne, *Science and Providence* (London: SPCK Press, 1989), 98.

[19] "The physical universe is put together with an ingenuity so astonishing that I cannot accept it merely as a brute fact," Paul Davies, *The Mind of God* (New York: Simon & Schuster, 1992), 4.

idea of the values which are appropriate to beings of a certain nature and enables one to understand why such beings act as they do. The first model ends with the postulation of ultimate brute facts; the second with the postulation of ultimate values or desirable and worthwhile states.

The two models are not wholly disconnected, however. For the first model is strongly evaluative, in seeking the simplest, most elegant, and mathematically fruitful postulates. This is in fact an aesthetic criterion, answering the question about why ultimate brute facts are as they are, by suggesting that they are the most beautiful (elegant or simple) set of facts which give rise to the interestingly complex universe we apprehend. One can either say that it just happens that such a value—of ultimate mathematical beauty—is instantiated in our universe; or one can turn to the second model for intelligibility and say that such beauty exists because it realizes a state of great intrinsic value.

On the other hand, if one asks why the values that are realized in this universe should be just as they are—why these values are realized and not say others, which perhaps we cannot fully imagine—then one has to refer to the natural capacities and desires of beings within the universe. The values of sensitivity, creativity and freedom, wisdom and justice which form the objective good of this universe are comprehensible as fulfillments of the natures of the sort of sentient social animals which have evolved in a physical universe of a particular structure. The values which this universe realizes are those states which unfold the potential implicit in its ultimate natural constitution. Thus to see what values this universe expresses one needs to discern the natural tendencies and capacities of beings in the universe, which in turn refer back to the ultimate brute facts of the universe. This insight is enshrined in the Christian tradition that true human good is to be found in appropriate fulfillment of natural human inclinations; that goodness is rooted in the nature of things, and is not some sort of arbitrary decision or purely subjective expression of feeling.[20]

Ultimate brute facts and ultimate values are not disconnected. Using the first model of intelligibility alone leaves one with ultimate brute facts which are just there for no reason at all. Clearly may physicists are unhappy with this ultimate non-rationality, and have sought some sense of quasi-logical necessity which will dissolve away the last recess of arbitrariness.[21] But they are thereby resorting to the principle of sufficient reason, which would make all things necessary implications of some initial state which is itself necessarily existent. For some quantum cosmologists, one can speak, albeit controversially, of an initial state (of quantum fluctuations in a vacuum) which is necessary, in that it instantiates the only set of states compatible with quantum theory.[22] But of course one still has to ask in what sense such a theory could be necessarily existent, or give rise to a physical instantiation of itself. Traditional theism comes

---

[20] "Reason of its nature apprehends the things towards which man has a natural tendency . . . and therefore to be actively pursued," Aquinas, *Summa Theologiae*, 1a, 2ae, 94, 2; vol. 28, 81.

[21] "We would prefer a greater sense of logical inevitability in the theory," Steven Weinberg, *The First Three Minutes* (New York: Basic Books, 1977), 17.

[22] See the article by C. J. Isham in this volume.

surprisingly near to such a hypothesis, in holding that there is a being, God, which contains the basis of all possible universes, which exists necessarily, and which generates this universe as one contingent realization of the exhaustive set of possibilities in its own being. The theistic model suggests how a mathematical theory could be existent (as conceived in the divine mind) and how a universe could be generated from it (by an intentional act); but it might be seen as simply a more pictorial way of putting the quantum cosmologists' proposal that this universe originates from a necessarily existent being, or is itself necessarily existent in its deep structure. However, although this model of intelligibility does eliminate arbitrariness from the universe, which is very satisfying for a physicist, it also seems to reduce the phenomena of freedom and creativity, of value-realization, to relative insignificance. One main advantage of introducing the concept of God is that it enables the aspects of necessity and creativity to be held together in one coherent, non-arbitrary, and non-deterministic form of explanation.

5 *The Idea of Divine Necessity*

The possibility of God as a necessary being could, of course, be denied. One clear form of denial is given by Richard Gale. He argues that, if a necessary being exists, it excludes the possibility of a world without a God, or of a world with unjustifiable evil in it, or of a world which is evil overall. For a necessary being is a being which exists in every possible world, by definition. Yet these other conceivable states, says Gale, are intuitively logically possible; therefore God cannot be a necessary being—there are possible worlds in which God does not exist.[23] Indeed, Gale holds that "if he is so conceived [as having necessary existence], it follows that he does not and cannot exist."[24] This is certainly false and a careless slip on Gale's part. If God is conceived as existent in every possible world, what follows is merely that no world is possible without God. Now to this Gale simply retorts that he can imagine a world without God; therefore at least one world is possible without God, therefore God cannot be a necessary being. But this is only so if it follows from the fact that Gale can imagine X that X is possible, and I see no reason to accept that. One can imagine an infinite number of worlds without God; all one has to do is think of all possible worlds and subtract God from them, leaving them otherwise the same. But does it follow from this that such imagined worlds are existentially possible? This would require the axiom: "Whatever I can imagine is possible." Far from being true, this seems false in many cases. Mathematicians imagined squaring the circle for centuries; but it is not possible. I can imagine that there is an even number which is not the sum of two primes; but such a thing is probably not possible. The plain fact is that what one can imagine is not a reliable guide to logical possibility in complex cases. And if logical possibility means lack of

---

[23] Richard Gale, *On the Nature and Existence of God* (New York: Cambridge University Press, 1991), 224-237.
[24] Ibid., 202.

contradiction, it is almost certainly untrue that any state at all which is not self-contradictory might actually exist. We simply have no idea of the conditions of the possibility of real existence. The Humean ontological argument, that what one can conceive is actually possible, is no better than the Anselmian ontological argument, that since we can conceive a necessary being, it must exist. If a necessary being exists, then unjustifiable evil is not existentially possible in any world; and we would see this if and only if we could clearly conceive the conditions of real existence and possibility. So Gale should have said, if God is conceived as having necessary existence, it follows that may things we think we can imagine are in fact so sketchily, incompletely, or confusedly conceived that they cannot even possibly exist. Is that very surprising? I should have thought it was a fairly familiar philosophical experience, as anyone should agree who has met a range of absolute idealist, sense-datum theorists, central-state materialists, and phenomenologists. At least some of them are talking about things that are not even possible—Gale even thinks theists are, or might be, such people. I conclude that, for all we know, there is a possible being which, if it is possible, is actual. The argument to it mentioned here is not an argument from concepts alone, an ontological argument, but from the postulate of the absolute intelligibility of the cosmos, the completion of that search for understanding which is such a natural and valued trait of the human mind. That requires, if taken to its fullest conclusion, the grounding of all actual states in a reality which necessarily exists and thus uniquely and fully answers the question, "Why is there anything at all?"

There have been those, such as Spinoza, who have held that the universe emanates from the necessarily existent being, God (by the principle of sufficient reason), by some internal necessity. As I have mentioned, that seems to be a position espoused by Weinberg, though without explicit use of the idea of God. Such a move, however, is in conflict with the existence of real creativity and freedom either in God or in the universe. So the alternative, and traditional Christian, view is to give the reason for the ultimate brute physical facts of this universe being the way they are as the realization of those ultimate values which are implicit in the physical structure. This appeal to the second model of intelligibility does explain why the ultimate brute facts are as they are; but it leaves unanswered the question of how values can have causal power. The ideal intellectual resolution of the quest for universal intelligibility would therefore, it seems, be found in a being which exists by necessity as the ground of all possibility, and therefore of all necessary values, and with the power to bring any compossible states about. Since it brings states about for their value, it can be properly termed good. It is thus omnipotent, being able to actualize any state; omniscient, knowing what states can be actualized; and benevolently good, actualizing states for their value. Further, insofar as it can choose its own states, it will choose maximally good states for itself; so that it will itself be of supreme intrinsic value, a maximally valuable being.

Can the necessary being choose its own states? If it is necessarily existent, then it must possess its essential nature in every possible world; so it cannot choose its essential properties to be other than they are. Those essential properties will include omnipotence and omniscience and unlimited benevolence, and thus meet the standard definition of the minimal essential properties of a maximally valuable being. Now I have argued that an omnipotent being will necessarily possess some contingent (therefore non-essential) properties. These are possible states of God which may or may not be actualized. It is these states which any omnipotent being will maximize to become a maximally valuable being. There probably is not one maximal state of the greatest degree of the greatest number of the greatest kinds of value; but a number of incommensurable valuable states. In that case the values of a maximally valuable being will be maximal in the sense that they will comprise a set of compossible values which are always greater in degree and extent than any others in fact and which are endlessly realizable in new combinations. God may not be a being of one statically supreme state of intrinsic goodness, but a dynamic reality of the endless realization of intrinsic values, which are underivatively rooted in the divine creativity itself. As necessarily existent, and having both necessary and contingent properties, the general nature and existence of the contingent properties being necessarily rooted in the necessary properties, and their specific actualization being subject to the necessary property of being freely creative, God is not an arbitrary or random brute fact. As the foundation of value, God can provide a reason for actualizing a set of possibilities as expressive of some set of ultimate values. And as free and creative in its particular actions, God can avoid both the morally dubious determinism of sufficient reason and the rationally dubious randomness of brute factuality.

6 *The Idea of Creative Emergence*

Thus, the best possible intelligible explanation of the universe, its maximal explanation, is to be found, not in a set of physical brute facts and not in a form of "personal explanation" which is rooted in the merely given nature of some personal being, but in a necessarily existent source of all possible and actual beings which creatively brings to be a universe in which creatures can cooperate in realizing a distinctive set of values. This source, this self-existent, supremely perfect and freely creative being is God. In this sense, theism is the completion of that search for intelligibility which characterizes the scientific enterprise. The universe can be seen as intentionally brought into being to realize a distinctive set of values, and the laws of nature are the general principles which set the conditions for the unfolding of natural tendencies which bring those values about through a temporal process of emergence. The discovery of the remarkable amount of "fine-tuning" in the physical structure of the universe greatly enhances our view of the universe as one elegant and closely interconnected whole.[25] The universe is both rational and contingent; for necessity and freedom are both key characteristics of it, and require precisely that sense of creative but non-arbitrary emergence which is the mark of the divine creative act itself.

---

[25] Polkinghorne, *Science and Creation* (London: SPCK Press 1988), 22-23.

This model of causality is different from the classical philosophical model of sufficient causality—the entailment model, which fails completely to account for the fact that the universe, as we know it, contains qualities which did not exist in its primordial constitution. And it is different from the Humean regularity model, which sees laws of science simply as statements of observed regularities between events. It is a model of *creative emergence*, accounting for the existence of things by seeing them as creative realizations of envisaged intrinsic values (states valued for their own intrinsic qualities, not their consequences or effects).

The Christian doctrine of creation cannot be construed properly either by the sufficient causality model or by the regularity model. At least for orthodox belief, the universe does not flow from God as a deductive entailment of the divine being. And there is certainly no regular observable succession between gods and the production of universes. But it could be construed by a creative emergence model, according to which this universe is a novel imaginative expression of specific intrinsic values. The particular values chosen are a subset of the many values which exist eternally in God, and they are chosen precisely by the sort of creative spontaneity in God which it is itself a very great value to possess. This universe exists for the sake of the values it can realize and it is a realization of those values in ways which have not previously existed, which might have taken a different particular form, but which partly because of their novelty and individuality are of inherent worth.

If this is true, there is a teleological explanation for the universe; it exists as a dynamic patterned process, realizing distinctive values through emergent creativity. Rational creatures may reasonably be seen on this model as cooperating in this creative emergence of the natural order. They are sub-creators with specific roles in the general creative pattern which issues from God. We would understand this explanation fully if we could see what values were to be realized, the condition of their realization, the nature of the God who brings them about, and the way in which God is able to structure events to ensure their final realization. But since we understand none of these things very well, the explanation remains a very general one, postulating simply that this universe does exist in order to realize specific values, and probably appealing to revelation to say in more detail what those values are. It still has great explanatory force to say, if it is indeed true, that the universe has a goal of value, which is realized in the process of its temporal existence.

The sort of explanation which cosmologists typically seek is in terms of some fundamental set of initial conditions and laws of behavior which will give rise to a universe like this. It is clearly possible that such a set of conditions and laws could simply happen to exist, as one improbable set among an indefinitely large number of other equally improbable sets. Richard Swinburne defines an "ultimate explanation" as one in which some set of conditions and laws exists which entails all states of affairs in our actual universe, but which is not itself explainable at any time.[26] They are "ultimate brute facts." He hypothesizes that

---

[26] Richard Swinburne, *The Existence of God* (Oxford: Clarendon Press, 1979), 75.

there is a prior probability that the set of ultimate brute acts will be simple rather than complex. However, this seems counter-intuitive, since, in the array of all possible universes, there seems no reason why there should be more simple universes than complex ones. On the contrary, since complex universes may be built in various ways out of simple ones, if probability applies at all in such an area, it would seem more probable that a complex set of brute facts would exist, since there are more complex than simple possible universes.

Swinburne's "weak" argument for the probability of God—that the probability of a simple God is low, but not as low as the probability of a complex universe—thus fails to get off the ground. It is not more probable, *a priori*, that a God exists and would create a universe like this, than that this universe just exists. Indeed, bringing in God just seems to complicate things unnecessarily, since any particular state of affairs is hugely improbable anyway. It is true that the existence of some ultimate brute facts would explain the nature of this universe, making clear why things should be as they are, given certain initial conditions. But those brute facts remain inexplicable (in principle, since Swinburne dismisses the idea of a self-explanatory cause as incoherent). And it is possible, as some quantum creationists maintain, that an adequate ultimate explanation is available in purely mathematical terms, without bringing in God.

Yet what such explanations omit is any consideration of emergence in the universe, particularly of the emergence of value, consciousness, freedom, and creativity. Since things do exist, a maximal explanation must combine both the element of necessity which is found in sufficient reason explanations and the element of free creativity found in value-realization explanations. That is precisely the sort of explanation the doctrine of creation postulates. What modern cosmology does is to show the elegant structure and the parameters of natural necessity, the unity and necessity of nature. It sets the notion of divine action in its broadest and most all-embracing context. Christian revelation claims to show the purpose of this unitary and intelligible natural order, as the creation of a community of love which is also a participation in the divine love. It is a striking fact that many theories of modern cosmology and the central assertions of Christian revelation cohere and illuminate each other in this way, disclosing an intelligibility and purpose inherent in the structure of the universe which point to its foundation in a deeply interfused spiritual reality.

The laws of the physical universe should not be set in opposition to God, as inviolable general rules which God, as an external power, would have to violate in order to act. The laws give the structure within which the purposes of God are brought to realization; and at every point there is the possibility of a creatively free initiative or response both from creatures and from God. The existence of a general rational structure in nature in no way inhibits the possibility of discontinuous emergent events which disclose the underlying character of the divine presence and prefigure the consummation of value which is the goal of creation. It is better to construe miracles as such transformations of the physical to disclose its spiritual foundation and goal than to think of them as violations of inflexible and purposeless laws of nature.[27] In overturning old mechanistic and deterministic models of laws of nature, modern science helps move towards a more satisfactory formation of a conception of divine action as the creative realization of purposes potential in the physical structure itself, not as interferences from some alien power wholly outside the system. Not all potentials can be realized, so there remains much scope for creative choice. Nor are potentials realized in a wholly predetermined way, so there remains much scope for imaginative elaboration and specification. Yet the model of the universe as directed towards a future which is partly open to creative choice is very different from the mechanistic model of it as predetermined in every detail, and it allows for divine and human action in much more intelligible and specific ways.

Modern cosmology may lead one to see the physical cosmos as an emergent value-oriented totality which is the optimal solution of a highly ordered and elegant set of boundary conditions. To the extent that it does so it is consistent with and even suggestive of the belief that the physical universe is intentionally brought about by a self-existent being of supreme value. It emphasizes that the maximal form of explanation is that postulated by theism; and it helps to show the way in which the universe can be understood as the creative act of God. In these ways theism can be seen as an implication of the scientific attitude itself, and the pursuit of scientific understanding may be seen as converging upon the religious quest for self-transforming knowledge of God rather than as being opposed to it. As modern science sprang from the context of Christian belief, so now it seems to be leading back to its roots, the apprehension of the physical cosmos as the visible expression of the mind of God.

---

[27] See David Hume's vastly popular definition of miracle as "a violation of a law of nature" in "Of Miracles," in *Enquiry Concerning Human Understanding*, section 10 (first published 1748).

THE TRINITY IN AND BEYOND TIME

Ted Peters

1 *Introduction*

In order to pursue the extensive question of "God's action in the universe," it may be helpful to divide the overall theme into component subthemes. The particular subtheme of the present paper is the relation between eternal and temporal action. Specifically the problem to be addressed is this: *how can an eternal God act, and be acted upon, in a temporal universe?* What makes this problematic within Christian theology is the habit in the classical tradition of understanding eternity to be the polar opposite of time. Eternity is said to be timelessness. In time, things pass. In time, we experience the disjunction of past, present, and future. In time, one event succeeds another. Yet, eternity is said to be immune from passage, succession, and disjunction. As mutual opposites, we ask here, how can a timeless God interact with a temporal world? This question is vital because, without some interaction, Christian claims that God is creator and redeemer of the world become incoherent.

As part of the problem and perhaps also the door to the solution we find the doctrine of God as Trinity. The first thing to note is that the eternal immanent Trinity already includes relationality and dynamism. The begetting of the Son and the proceeding of the Holy Spirit combined with the givenness of the Father obviate any notions that God is a sleepy monad, an undifferentiated or simple unit. Rather, there is within the eternity of the divine life action and movement, separation and reunion. In classical thought, of course, this movement is described in non-temporal terms as the eternal *perichoresis* which is conceptually prior to the creation of the world. Although the two sendings of the Son and the Spirit appear to us to be temporal and sequential, our theological forbearers denied this, arguing that time and sequence belong to the created world only, not to God.

The second thing to note is the economic Trinity: the trinitarian fashion by which God relates to the world. God enters the world in the incarnation of Jesus Christ, taking unto the Godself the limitations of temporal and spatial finitude. God becomes one physical being among others, subject to the same laws of physics and biology that govern the rest of the creatures.[1] Then, as Spirit,

---

[1] The subjection of the divine to temporal finitude in the incarnation is my constructive proposal based upon an explication of the biblical symbols. I recognize that this presses beyond many classical formulations wherein the divine attributes are said to communicate themselves to the human Christ but not the reverse. I prefer an Irenaean interpretation in which God assumes what is saved, in this case assuming the finite and temporal form of the created creature.

God continues to influence the course of events within history. This Spirit, so Scripture promises us, has the power to transform this world and will do so at the end of the age, at the advent of the eschatological kingdom of God, where we will rise from the dead and be united with the already risen Jesus Christ in everlasting glory.

What is important for our theme of God's action in the universe is that in the economy of salvation we find God on both sides of the fence, both as eternal and as temporal. It is *God* who was in Christ "reconciling the world to himself" (2 Cor. 5:19). The Christian claim regarding the economic Trinity is that God has entered our world of time and space, and this worldly involvement of the divine life is one step in a sequence of steps that leads to the final redemption of all that exists. God works, and the divine working takes time and takes place in time. What is necessary to solve the problem of the relation between eternity and time is to define the two as complementary rather than mutually exclusive. Somehow, eternity, though retaining its transcendence, needs to incorporate rather than eschew a temporal dimension. The way to achieve this, I recommend, is to posit an integration of the economic and immanent Trinity so that the temporality of the world is taken up into the eternity of the divine life proper.

En route to this goal, it will be appropriate to investigate the nature of time in our universe. We need to ask what understandings of time we find in physical cosmology, and what implications these understandings might have for our project. With this in mind, the discussion that follows will open with an attempt to define time. Three overlapping definitions will be offered. The focus will then turn to the nature of the present moment, sandwiched as it is between past and future. The present moment within time will then be compared to the notion of the eternal present as it is understood by classical thinkers such as Boethius as well as by contemporary philosophical theologians who take special relativity theory into consideration. At this point we will become aware that classical formulations of the eternity-time relationship are inadequate because they deal only with abstract philosophical notions of the Godhead. They ignore relationality within the Trinity and God's entry into the temporal course of events. We will then turn to time's arrow in human subjectivity, giving special attention to its tie with thermodynamics, Big Bang cosmology, and Stephen Hawking's quantum gravity depiction of creation. This will be followed by a discussion of the wholeness of time, a construct made necessary by the idea of creation understood as a whole. Here we will reference the thought of theologian Wolfhart Pannenberg and buttress the argument with support garnered from the scientific notion of downward causation. We will then turn to a more direct examination of issues regarding our understanding of God as Trinity. Here we will explicate the insightful contribution of Robert Jenson, who stands on the frontier of the discussion of trinitarian theology as it has proceeded over the last half century.

The claims this chapter makes for itself are preliminary. Its objective lies one step beyond basic assemblage—that is, its task is to assemble and begin to synthesize what appear to be relevant resources in the natural sciences and contemporary theological discussion for dealing with the question of God's eternity and temporality. It is offered as a first step.

As such a first step, the chapter's conclusion is presented as a hypothesis for further consideration. What I will suggest is this: the dynamic inter-relationships of the three persons, which Christians assert regarding the internal life of God, namely, warrants our expectation that the created and redeemed order will be temporal in character and, furthermore, the temporality of the world will be eschatologically taken up into God's eternal and everlasting reality.

2 *Time: What Is It?*

To say that all of history will be taken up eschatologically into the divine eternity makes a dramatic claim regarding God's relation to the universe, including the notion that time has an effect on eternity. In what follows we will ask what this might mean. As we do so, we might note that the word "time" is equivocal, referring to more than one thing. It seems to refer to at least three things. First, the word "time" can refer simply to passage, to the quality of one-after-the-otherness that characterizes the successive events which constitute actual existence. Second, the term "time" frequently refers to the measure of this passage. Third, "time" may refer to world history, to the realm of creation which is defined in part by passage. To speak of "temporal affairs" or "this passing life" is to speak of the transitory or ephemeral world in which we live, contrasting it with the divine eternity. Time here is shorthand for our natural reality. Theologically, time refers to God's creation subject to decay, the very reality we hope will be eschatologically redeemed. To answer the question, "What time is it?" we normally look at our watch or clock. To measure the passage of time we refer to things that repeat themselves such as the earth's daily rotation, the earth's yearly orbit around the sun, the rotation of neutron stars, the swinging of a pendulum, quartz crystal oscillation, or atomic oscillation. Such repetition provides the parameters that make measure of passage possible. What is amazing is that when these various modes of measure are compared, they seem to repeat in consonance with one another, although in varying degrees of accuracy. An enduring issue is whether time is merely a human psychological phenomenon or belongs to extra-subjective natural processes. To think of time as measure may indicate that it is merely psychological and, hence, possibly illusory. Yet, our measuring devices are based upon natural rhythms, giving credence to the view that both passage and the measure of passage belong to nature proper. The natural world is in itself temporal. Thus, temporal passage is both subjective and objective, both conditioning the human perspective as well as belonging to the physical world in which we live.

3 *The Present Tension Between Past and Future*

Human time consciousness begins with our awareness of present perception demarcated by memory of the past and expectations for the future. This present is dynamic, continually lopping experiences off into past memories while actualizing what were previously future potentials. This is time's passage. Time here has an arrow, a one directional movement from what has been to what is now and toward what will yet be.[2]

There is a sense in which we can say the past and the future are not actual. The present is actual. The present provides the perspective for apprehending past and future realities. What makes the past past and the future future are their respective relations to the present. Christian understanding of the present combines the prophetic and sacramental understandings of time. The vision emerging from the Hebrew prophets places the present moment between promise and fulfillment. God's promise lies in the past, the fulfillment in the future. Here, God's presence is not actual. Christians employ a communal memory to look back toward a past reality, to the event of the crucifixion and resurrection of Jesus Christ. Although what we remember is what happened to Jesus, it has significance for us. That past event constitutes for us in the present a promise: as Christ rose from the dead two millennia ago, so also at some point in the future we too will rise into God's new creation. Our present moment is situated between the times, between God's past and God's future.

This prophetic tension between past and future is complemented by the sacramental sense of divine presence in the human present. We are not alone. The Holy Spirit comforts us with divine company, with mystical nearness. But such divine presence is characterized normally by an inobtrusiveness in ordinary affairs. It is an acausal presence that lets things be. There are times when the reality of the Holy Spirit seems more intense, when we are more aware. We can locate ourselves in time and space at those moments, even if the sacramental presence itself seems immune from such location.

---

[2] The human perspective involves more than mere passage that is constantly altering memory and expectation. It also involves the constitution of the human self. Memorable experiences are not just items stored in the memory; rather, they are influences that shape who we are in a determinate way. Edward Farley says they become "sedimented" and thereby determine personal being as its *being as* or *I-as*. Similarly, the grasp of expectations for the future include the possibilities of wrenching us loose from past sedimentations and determinations. Because of self-formation, temporality plays a different role for human consciousness than it does for say a mountain or some other object. Edward Farley, *Good and Evil* (Minneapolis: Fortress Press, 1990), 68-69.

That which permits self-formation and also history-formation is the openness of the future. J. R. Lucas says, "the future is not already there, waiting, like a reel of film in a cinema, to be shown: it is, in part, open to our endeavours, and capable of being fashioned by our efforts into achievements, which are our own and of which we may be proud." *The Future: An Essay on God, Temporality and Truth* (Oxford: Basil Blackwell, 1989), 8.

Whether in the prophetic form of past promise and future fulfillment or in the sacramental form of present presence, Christian consciousness apprehends God in temporal terms. The discreet succession of events, even divine events, never disappears from our awareness of God. Our human encounters with God are timely, as all human encounters must be. Our relationship with God is timely, as all human relationships must be.[3]

4 *The Eternity-Time Relationship: How Do We Formulate the Problem?*

But, a difficulty arises when we note that, according to the patristic theologians of antiquity, this timeliness allegedly belongs strictly to human experience and not to divine reality. The ancients sought to drive a wedge between human temporality and divine eternity. God is not subject to time, insisted Gregory of Nyssa, Augustine, and Boethius. Time and space belong only to the creation. "The world's Creator laid time and space as a background," wrote Gregory, "to receive what was to be; on this foundation He builds the universe."[4] Time and space provide the framework or container, so to speak, within which human activity is carried on. Yet God is outside the container and therefore not limited to temporal and spatial constraints.

What is vital in this view is that God as eternal is not subject to passage, decay, and death. "In the eternal nothing can pass away," wrote Augustine, "but the whole is present."[5] These considerations led to the formulation by Boethius which has become the insignia of the classical position: eternity is the complete possession all at once of illimitable life (*aeternitas igitur est interminabilis vitae tota simul et perfecta possessio*).[6] Boethius proceeds to emphasize that the eternal God possesses illimitable life in such a way that nothing future is absent from it and nothing past has flowed away. God is always present to the Godself; and all that was, is, and shall be within time is also present. Boethius contrasts the human present as a "now" that is "running along" with the divine present as a "now" that is "remaining, and not moving, and standing still."[7]

---

[3] God is temporal, argues Arthur Peacocke, while putting forth a doctrine of *creatio continua* in which God creates continually through necessity and chance. God limits divine omnipotence and omniscience for the good of the world as an act of divine love. This makes God subject to disappointment and suffering. All this adds up to change and time within the divine life. "For if God suffers with creation in some sense analogous to that of human suffering, God must be conceived as being changed through this interaction with the world." *Theology for a Scientific Age: Being and Becoming—Natural and Divine* (Oxford: Basil Blackwell, 1990), 128.

[4] Gregory of Nyssa, *Against Eunomius*, 1:26.

[5] Augustine, *Confessions*, 11:11.

[6] Boethius, *The Consolation of Philosophy*, V:6; cf. *The Trinity*, IV.

[7] Boethius also adds "*semper*" to "eternity" to produce "sempiternity" referring to "the perpetual running resulting from the flowing, tireless now." The dictionary defines "sempiternity" as everlastingness. Eleonore Stump and Norman Kretzmann complain that Boethius' etymology is misleading here, because he

Eleonore Stump and Norman Kretzmann analyze and critique the Boethian understanding of divine eternity. They see four components in his definition: (1) the eternal has life, which distinguishes it from abstract truths such as mathematical principles; (2) the eternal is illimitable—that is, it cannot be limited; (3) eternity has duration even if it is a beginningless, endless, and infinite duration; and (4) to speak of "complete possession all at once" signifies that eternity is atemporal—that is, without sequence. There is an apparent incoherence here, say Stump and Kretzmann; because ordinarily we cannot combine coherently the ideas of duration and atemporality.[8] The incoherence can be rendered merely apparent and genuine coherence achieved if, Stump and Kretzmann argue, their particular interpretation is accepted. Their interpretation is called "E-T Simultaneity." It presumes that eternal existence and temporal existence are of different orders. The entire life of any eternal entity (E) can be fully realized and fully present and, hence, coexistent with any temporal entity (T) in any given moment within its passing existence. Hence, my own birth occurring an unmentionable number of years ago plus my present consciousness and my future death are sequential and not simultaneous for me, yet they can be perceived as simultaneous from the perspective of eternal eyes. Every temporal moment appears simultaneous to the one eternal perspective, to God's perception or knowing. Key to the Stump and Kretzmann interpretation is the affirmation that it makes sense to speak of eternity as "fully realized duration."[9]

Paul Helm is not convinced by the Stump and Kretzmann argument. Helm finds it difficult to accept the notion that the eternal entity can be exempt from any sequence yet at the same time enjoy a sense of "present." Is it not the case that the present is determined by its relation to temporal indexicals such as "yesterday" and "tomorrow?" Does not the Stump and Kretzmann notion of eternal duration abandon the idea of timelessness in the strict sense? Why is it required that eternity have any duration at all? Helm points out that E-T Simultaneity does not assert the simultaneity of two events. Rather it conjoins one temporal event with an observation from an eternal viewpoint. Helm finds the interpretation of Stump and Kretzmann to be simply a "rewording" of the problem. "Nothing is illumined or explained by it."[10] Despite his criticism, Helm himself holds out for an atemporal divine eternity, saying, "God's eternal existence has no temporal relations whatever to any particular thing which he creates. This does not mean that there are no relations at all between the eternal

---

associates "sempiternity" with "eternity" in a context in which he has been distinguishing between sempiternity as unending passage with eternity as exempt from passage. "Eternity," *The Journal of Philosophy* LXXVIII:8 (August 1981; reprinted in *The Concept of God*, ed. Thomas V. Morris [Oxford: Oxford University Press, 1987]), 221, n. 3.

[8] Ibid., 224.
[9] Ibid., 237.
[10] Paul Helm, *Eternal God: A Study of God Without Time* (Oxford: Clarendon Press, 1988), 33.

God and his creation, only no temporal relations."[11] In sum, God relates to the world eternally, not temporally.

Helm formulates what he sees as the problem: How can something which is an event in time be wholly present to an eternal—that is, timeless—entity? As I study this debate, it appears to me that this formulation of the problem lacks the precision necessary to guide the inquiry in the proper direction. It restricts us to relating two abstract notions, eternity and time. What it leaves out is the actual claim of the Christian revelation, namely, that God is to be understood as a Trinity and, as such, has entered the history of time through incarnation and inspiration. If we were dealing with strictly an abstract monotheism, or perhaps even restricting our consideration to the first person of the Trinity, then the abstract formulation might be adequate. But when we add the claim that the eternal has entered the temporal in the past and promises redemption yet in the future, much more than a mere divine audience for what happens on the earthly stage is at stake. God is an actor. This would seem to require a formulation of the problem that would include: does God's involvement in the course of temporal events imply that the nature of eternity is affected by time? Or, to ask it another way: does God relate to the world temporally?

5 *The Special Theory of Relativity*

The conceptual significance of Albert Einstein's special theory of relativity is that we lose the sense of a unique universal present and we end up with multiple inertial frames of reference for the relation between past and future and with multiple rates at which time passes. How might this affect our understanding of God and contribute to the formulation of our problem? Although a few scholars have sought in preliminary fashion to draw out some of the implications of Einstein's work for thinking about God, there is as yet little consensus regarding what it all means. Let us look at a selection of philosophical and theological explorations.

Stump and Kretzmann make Einstein work for them. They ask us to imagine a train traveling at a high rate of speed, at six tenths the speed of light. One observer is stationed on the embankment next to the track. A second observer sits in the train. Two lightening bolts strike the train, one at each end. Suppose the observer beside the track sees the two lightening bolts as simultaneous. The passenger also sees both bolts but, because he is traveling toward the light ray emanating from the bolt that strikes the front of the train and away from the one striking the rear, he will see the front bolt before he sees the rear one. The result is a rule: events occurring at different places which happen simultaneously in one frame of reference will not be simultaneous in another frame of reference. This permits us to say both that the two lightening bolts strike simultaneously and that they strike in sequence. It is both true and false that these

---

[11] Ibid., 36.

two lightening flashes occur at the same time.[12] There is no privileged observer to determine whether in any final sense the two events are simultaneous.

Via analogy Stump and Kretzmann move to support their theory of E-T Simultaneity. They are dealing with two modes of existence, one eternal and one temporal, neither of which is reducible to the other. This gives them two observers and two reference frames. But one of the *relata*, the eternal observer, always exists in the same present. E-T Simultaneity combines an eternal present and a temporal present, not two temporal moments. The eternal present is capable of simultaneity with all moments in all frames of reference.

One wonders if there might be limits to the analogy, noting that inertial frames of reference in special relativity apply to *temporal* frames of reference; whereas the notion of eternal duration seems to imply an atemporal frame of reference. Should the analogy then apply? Regardless, the bottom line seems obvious: God is not limited to one temporal frame of reference. God's viewpoint is not just one temporal perspective among many.[13]

The Stump and Kretzmann proposal illustrates what is at issue. The particular challenge posed to theology by the special theory of relativity has to do with a privileged frame of reference, a universal divine viewpoint. Does God have an inertial fame of reference that includes all others so as to permit omniscience?

Theologians tend to answer affirmatively. Ian Barbour suggests that the limitations set by the speed of light on the speed of physical signals between distant points should not apply to God. This is because God is immanent at all points of the cosmos and in all events. God is neither at rest nor in motion relative to other systems. This means God must influence each event in terms of the pattern of events relevant to its particular causal past, which, of course, is uniquely defined for each frame of reference.[14]

In a similar way, Holmes Rolston identifies God with the whole of the cosmos in panentheistic fashion. He suggests thinking of God as a superspatial "Great Universal Mind" who stands astride the whole, imparting to it intelligibility. He does not identify God directly with ether or with a spacetime plasma, but rather places God one or more orders of being beneath them. This

---

[12] Stump and Kretzmann, "Eternity," 228-230.

[13] Lucas to the contrary does claim that God has a *unique* frame of reference. He therefore believes that the traditional understanding of God's simultaneous apprehension of the course of events is not challenged. "The divine canon of simultaneity implicit in the instantaneous acquisition of knowledge by an omniscient being is not incompatible with the Special Theory of Relativity, but does lead to there being a divinely preferred frame of reference . . ." *The Future*, 220. What is important to Lucas is to assert that God can know past and present as fixed, but God can know the future only as open. ". . . unless quantum mechanics is logically inconsistent, an almighty God could have chosen to make a world that exemplified the laws of quantum mechanics. And if He did, then He would be unable to have detailed knowledge of its future developments" Ibid., 227.

[14] Ian Barbour, *Religion in an Age of Science*, The Gifford Lectures, 1989-1991, vol. 1 (San Francisco: HarperSanFrancisco, 1990), 112.

permits him to say that spacetime is God's creation while also saying that God is the very substrate of the world. Particles, waves, matter in motion, stars, planets, and persons are all warps in spacetime. They also constitute wrinkles in God. As wrinkles in God, they are his creation. Whether designated Brahman or *sunyata* or Tao, God is "in, with, and under" the energy pit out of which all comes, the prime mover lurking beneath the scenes. As Rolston develops this idea of God, he moves God further and further away from the course of events and risks confining God to an isolated eternity. "God is not a spatiotemporal entity," he writes. "God is pure spirit. Having no velocity or mass, God has no time."[15] Not having space and time, God can gather past, present, and future into a single whole. Not being physical and not being subject to spacetime, God can communicate instantly between various quadrants of the universe at speeds exceeding that of light. Omnipresence means instant communication.

These proposals have not taken us very far beyond where Gregory, Augustine, and Boethius had left us. God's eternity still transcends the temporal movement of the creation. What is added by reflections on relativity is that we now have multiple temporal movements instead of just one. God's interactions with the created order now have to be tailor-made to specific frames of reference, but this marks only minimal change from the previous view. What remains is the theological problem: How can a timeless God experience or act upon a temporal world?[16]

Rolston offers a clue but does not follow up on it. "If God is anywhere known," he writes, "it will be as God 'comes through' in our spacetime, relative to our local existence, as God is, so to speak, locally incarnate."[17] The key is local incarnation—that is, the entering of God into the created order and taking up residence within a single inertial frame of reference. Through such an incarnation we would then find God on both sides of the ledger, both eternal and temporal, both universal and particular, both acting and acted upon. This, I take

---

[15] Holmes Rolston, *Science and Religion: A Critical Survey* (Philadelphia: Temple University Press, 1987), 65.

[16] John Polkinghorne tackles the challenge of relativity for a temporal God when he writes, "The God who simply surveys spacetime from an eternal viewpoint is the God of deism, whose unitary act is that frozen pattern of being." *Science and Providence: God's Interaction with the World* (Boston: Shambhala, 1989), 79. Therefore, there must be temporality within the divine life. But how? When confronting the twins paradox in relativity theory, Polkinghorne offers a fascinating theological interpretation. He asks: Which clock does God use, the one on earth or the one on the spaceship? He answers: Both. God is the omnipresent observer, experiencing the course of events as they happen within their respective inertial frames of reference. And no frames of reference are deleted. God experiences them all. "That totality of experience is presumably the most important thing to be able to say about God's relation to world history. He would not miss anything and his action would always be causally coherent . . ." Ibid., 82. Polkinghorne helps by offering an understanding of divine temporality that accounts for the effect the world has upon God, but he is less clear on just how God affects the world.

[17] Rolston, *Science and Religion*, 62.

it, belongs properly to the trinitarian claim. Could our conception of God better handle the question of temporality if we make use of the trinitarian claim?

6 *Entropy and Time's Arrow*

As already mentioned, the overall movement of physical time is in one direction, toward the future. If there were any doubts, the second law of thermodynamics seems to have eliminated them. Following on the heels of the first law regarding the conservation of energy, the second law sets a boundary condition so that heat may never flow spontaneously from a cold body to a hot one. The natural movement is always from hot to cold, never the reverse. Alternatively put, the movement is from order to disorder. Entropy is the measure of disorder. In closed systems which have no input of matter or energy from outside, entropy increases uniformly until a state of equilibrium has been reached. At equilibrium no more energy can be degraded. The upshot of the second law is that time is asymmetrical: entropy increases in the direction of the future, never the past. The temporal movement of the physical world is irreversible. Time has an arrow.[18] It is unidirectional.

The second law applies inexorably to closed systems. In open, non-linear, and far-from-equilibrium systems which interact with their wider environment and receive energy input, however, creative fluctuations can take place. High energy input at first increases randomness and chance, that is, it increases chaos. Chaos can lead to previously undetermined possibilities and a bifurcation point may be reached. At this point the system may disintegrate into further chaos. Or, it may leap to a new, more differentiated and higher level of order. The fluctuating chaos becomes the source out of which new order emerges. Ilya Prigogine calls this "order through fluctuation."[19] In sum, chaos in an open system can be creative.

---

[18] Noting how the image of the arrow of time is used in physical cosmology, Robert Jenson's use of the term may be unnecessarily confusing. He speaks of three arrows of time, one for each: past, present, and future. Ascribed to the Trinity, the three arrows refer to the Father as given, the Lord Jesus as the present possibility of God's reality for us, and to the Spirit as the outcome of Jesus' work. *The Triune Identity* (Philadelphia: Fortress Press, 1982), 24. The weakness here is that Jenson too quickly exports the human experience of a present detached from past and future into the realm of the divine, bypassing the arrow of time indicative of the creation. There is only *one* arrow of time in cosmology (i.e., from the past through the present toward the future). Moreover, Jenson ignores the problem of the light cone and the ambiguous meaning of the present in relativity theory. See R. J. Russell, "Is the Triune God the Basis for Physical Time?" *CTNS Bulletin* 11:1 (Winter 1991): 7-19.

[19] Ilya Prigogine and Isabelle Stengers, *Order Out of Chaos: Man's New Dialogue with Nature* (New York: Bantam Books, 1984), 178. James Gleick describes the second law of thermodynamics as a "piece of technical bad news from science that has established itself firmly in nonscientific culture." It is taking the blame for the disintegration of societies, economic decay, the breakdown of manners, and such. Such

We do not assume the universe as a whole is an open system receiving energy input from outside. If there were such an energy source, our notion of the universe would expand to include it, and then we would be right back where we started. It seems prudent, then, to think that the law of entropy might apply to the cosmos as a whole.[20] If so, then the entire universe would find itself in a monodirectional movement from hot to cold, from order to disorder. It seems to be headed for a future equilibrium in which all its energy will be forever dissipated. Although chaos will be creative of new order in local subsystems, cosmic time is irreversible and all things will eventually wind down to a state of maximum entropy.

When combined with Big Bang cosmology and applied to the universe as a whole, the second law suggests a finite history for the cosmos. We can press our thoughts backwards to a point perhaps 15 or even 20 billion years ago, to the beginning of time, to t=0, when everything began. It began with a singularity of maximum density and heat. After the bang, matter shot out in all directions. The universe is now expanding and cooling. Oh, yes, certain far-from-equilibrium systems such as galaxies are temporarily creating order out of chaos, but the overall movement is toward greater entropy.

This applies whether the universe is open or closed. If it is open, the universe will expand forever; after another 65 billion years or so the energy of the cosmos will be so dissipated that no new order will be able to arise. The end will be death by dissipation. If closed, the expansion will reach a limit. Then gravity will reverse the direction of movement, causing all matter to recollapse back into a singularity of unimaginable temperatures. This singularity could explode once again in an additional Big Bang or a long series of Big Bangs, giving us an oscillating universe. Even so, the law of entropy might hold, so that each successive expansion could have less to burn than the previous. Although the oscillating closed universe is more complex than the open one, its end too will be death by dissipation. Despite internal reversals, the overall arrow of physical time seems to move in one direction.

---

secondary or metaphorical incarnations of the second law are misguided. If we are looking for scientific models for understanding our culture, Gleick recommends that we pick on the laws of chaos. Here we find the source of creativity. *Chaos* (New York: Penguin, 1987), 307-308. Chaos and the second law belong together. We find local eddies of creative order and complexity within a larger flow toward dissipation.

[20] Willem B. Drees is not so sure the second law of thermodynamics applies to the universe as a whole. Entropy presupposes an environment, such as background radiation. In an expanding universe, the notion of an environment is problematic. See *Beyond the Big Bang: Quantum Cosmologies and God* (LaSalle, IL: Open Court, 1990), 32. The context of Drees' doubt is his wider argument that a cosmological argument for God's existence based on the Big Bang will not work. The context of my argument is not one of cosmological proof. Rather, it is assessing evidence for describing physical reality as temporally directional. It seems to me that the lack of an environmental precedence for entropy in an expanding universe would not affect the temporal direction.

The tie between thermodynamics and an arrow of time is complicated somewhat when we note that, within our expanding and dissipating universe, entropy has a dialectical relationship to depth. By "depth" we refer to growth in organized complexity. The fundamental laws of nature are comparatively simple, yet the ongoing exercise of these laws from the beginning through the fifteen-billion-year history of nature has produced greater and greater complexity, including life and consciousness. The future will bring not only degeneration by entropy but also the generation of further "depth."[21] Entropy will be the price paid for depth. This causes some physical cosmologists to ask: Which will win ultimately, entropy or depth? Time's arrow continues to fly in only one irreversible direction, to be sure; but thermodynamics may give insufficient reason to think pessimistically about the world's entropic future.

What is significant for our discussion here is the affirmation that the concept of temporal irreversibility is not merely the product of human subjectivity. It is more than just a projection of the human consciousness of past and future. We do not superimpose a mental idea of time on an otherwise timeless natural world. Rather, nature itself is speaking and telling us that it is temporal and monodirectional. "We are becoming more and more conscious of the fact that on all levels, from elementary particles to cosmology, randomness and irreversibility play an ever-increasing role," writes Prigogine. "*Science is rediscovering time.*"[22]

The irreversible temporal movement of natural processes elicits theological reflection in a number of ways. Most importantly, it helps justify a historical understanding of nature. This leads potentially to a common framework for scientific and theological discourse. For nearly two centuries now theologians have tackled the question of how God acts in the world as a historical question. It is virtually axiomatic to say that "God works in history." God's actions are events. Contingent historical events in which divine action is identified become the basis for Christian claims regarding ultimate reality. Thus, Wolfhart Pannenberg can assert that "history is the most comprehensive horizon of Christian theology."[23] Pannenberg is inspired in part by physicist C. F. von Weizsäcker, who believes nature has a history. "Man is indeed a historic being," says von Weizsäcker, "but this is possible because man comes out of nature and because nature is historic herself."[24]

With this in mind, some attention needs to be given to just how God acts in the history of natural events. We might pose the question: Does God necessarily act in every natural event? Or, are natural events normally *sui generis* so that we can say that God acts in some events but not others? Arthur Peacocke opts for the former, contending that God's activity is a constituent component of

---

[21] "Depth" is a term introduced by Paul Davies at the 1991 VO/CTNS conference in discussing the complexity of natural systems.

[22] Prigogine and Stengers, *Order Out of Chaos*, xxviii, author's italics.

[23] Wolfhart Pannenberg, *Basic Questions in Theology*, trans. George H. Kehm, 2 volumes (Philadelphia: Fortress Press, 1970-71), I: 15.

[24] C. F. Von Weizsäcker, *The History of Nature* (Chicago: University of Chicago Press, 1949), 7.

all natural activity within the universe. He holds that "we must conceive of God as creating within the whole process from beginning to end, through and through, or he cannot be involved at all."[25] God is continually creative, and God prompts creativity within nature. The dialectic of chance and law are the means whereby creativity is elicited. Building upon Prigogine's notion of creative fluctuation whereby order emerges from chaos and applying it to biological systems, Peacocke asserts that the interplay of chance and law make possible the emergence of living structures and propels them through evolution. He emphasizes that the history of nature is open; it is a fabric of turning points, open at every step to new choices and new direction. Randomness and chance are responsible for this openness, and God is responsible for randomness and chance. "It is as if chance is the search radar of God," he writes, "sweeping through all the possible targets available to its probing."[26]

With regard to time, Peacocke acknowledges that time in relativistic physics is an integral and basic aspect of nature. Matter, energy, and spacetime constitute together the created order. "Hence, on any theistic view," he says, "time itself has to be regarded as owing its existence to God, something Augustine long ago perceived."[27] This "owing its existence to God" is the central core of our understanding of nature as creation. "Thus," he continues, "the fundamental otherness of God must include the divine transcendence of time."[28] Note how Peacocke places together transcendence of time with God's otherness. Yet, this otherness and transcendence is not all there is to God. God is also immanent in the natural processes. Peacocke strongly advocates a doctrine of *creatio continua*, continuous creation. God is continually producing new forms of emergent matter within the flow of nature's history. God creates continuously, and this creative activity takes place within time.

Carrying the discussion a step further, we might now ask: can we conceive of the eternal God enveloping time? Is it proper to ask: Does time have edges, a beginning and an end, beyond which we might peer into eternity? A thermodynamic cosmology seems to indicate a beginning and an end. It originates with a very dense, very hot singularity, and at the moment it begins to expand time begins. Dare we ask: Did God act at the beginning? Did God exist prior to the beginning? Was it God who created the original singularity? Did God create that singularity out of nothing, *creatio ex nihilo*? Was it God who lit the fuse on the original dynamite that exploded with the Big Bang? Did the first moment of cosmic history drop from divine eternity to follow its independent path of successive events?

Then, what about the future end? Will God be there to sweep up the dissipated debris after its all over? Will God then take the history of the cosmos

---

[25] Peacocke, "Theology and Science Today," in *Cosmos as Creation: Theology and Science in Consonance*, ed. Ted Peters (Nashville: Abingdon Press, 1989), 34.

[26] Peacocke, *Creation and the World of Science* (Oxford: Clarendon Press, 1979), 95.

[27] Peacocke, "Theology and Science Today," 33.

[28] Ibid., 34.

up into the eternal, or will its dispersed remnants forever lie there in memoryless equilibrium? Does God have sufficient patience to wait until the end, or might God intervene earlier to stop the inevitable? In either case, does the eternal reality have what it takes to redeem what becomes lost in the epic of temporal passage? Most generally, should we look beyond the edges of time for God's eternity?

7 *Stephen Hawking and the Edge of Time*

"No," would be the answer to these questions if posed to Stephen Hawking. Hawking is fully aware that if time has edges then God-talk might begin to make scientific sense. He fully recognizes that the concept of the absolute beginning implied in the standard Big Bang model might lead to speculation regarding the existence of God. The notion of a beginning means that there is a front edge to the cosmos, and the acknowledgment of an edge requires us to ask what lies beyond the edge. ". . . If there is an edge," he once said in an interview, ". . . you would really have to invoke God."[29] Yet Hawking wants to avoid this. So he goes searching for a theory that will eliminate the edge. He does this by challenging the hypothesis that there once was a singularity prior to the cosmic expansion.

The Hawking argument begins with the assertion that general relativity is not enough to solve the problem of the nature of the originating singularity, because relativity applies only to the macro-universe. Since the alleged original singularity was in fact arbitrarily small and infinitely dense, we need to resort to another theory which applies more appropriately, namely, quantum mechanics. The central thesis of Hawking's work is this: The theories of relativity and quantum mechanics combine into a single uniting *quantum theory of gravity*. Hawking's declared objective here is to enable us to describe the universe by a single mathematical model which would be determined by the laws of physics alone. Incorporated into the unification is Heisenberg's uncertainty principle which implies, among other things, that the subsequent course which the developing universe would follow was not fixed by its original boundary conditions. In fact, there is no boundary condition for either time or space. There is only a curved spacetime, which is finite; but it does not take us back to a point of absolute zero, before which there was no time.

Hawking illustrates the point by distinguishing between real time and imaginary time. Real time is the time we experience as sequence. Real time applies to the domain of spacetime beginning with the inflation. Prior to inflation, however, only the quantum world exists where the distinction between space and time breaks down and they converge. Phenomenological or physical time—that is, real time—disappears from the mathematical formulas. In its place the equations for the quantum universe treat time on a par with space through the use of the imaginary variable, "ict," where $i=\sqrt{-1}$, c is the speed of light, and t is

---

[29] Stephen Hawking, "If There's an Edge to the Universe, There Must Be a God," in *Dialogues with Scientists and Sages: The Search for Unity*, ed. Renée Weber (New York: Routledge, 1986), 209.

time. In so doing, time cannot be distinguished from a direction in space nor is there a difference between forward and backward in time. Hawking introduces imaginary time as necessary to implement his proposed unification of gravity with quantum mechanics and to rid spacetime of its initial singularity.

> In real time, the universe has a beginning and an end at singularities that form a boundary to space-time and at which the laws of science break down. But in imaginary time, there are no singularities or boundaries. So maybe what we call imaginary time is really more basic, and what we call real is just an idea that we invent to help us describe what we think the universe is like.[30]

Hawking himself, then, draws out some theological implications of his proposal. On Hawking's reading, they are anti-theological implications. The universe needs no transcendent creator to bring it into existence at t=0, nor does it need God to tune the laws of nature to carry out a divinely appointed evolutionary purpose.

> ... the quantum theory of gravity has opened up a new possibility, in which there would be no boundary to space-time and so there would be no need to specify the behavior at the boundary. There would be no singularity at which the laws of science broke down and no edge of space-time at which one would have to appeal to God or some new law to set the boundary conditions for space-time. One could say: "The boundary condition of the universe is that it has no boundary." The universe would be completely self-contained and not affected by anything outside itself. It would neither be created nor destroyed. It would just BE.[31]

One major effect of this quantum gravitational vision is to untie the relation between time's arrow and the essential nature of the universe. Real time is reduced to an epiphenomenon of imaginary time. Although Hawking is making a reductionist move, he by no means reduces time's arrow to a mere psychological phenomenon produced solely by human consciousness. Temporal passage remains physical and supra-subjective, but only in *some* regions of physical activity within the universe, while the cosmos at the quantum level is judged to be atemporal.

If we wished to press the point, Hawking's atemporal model of the universe begins to look somewhat like the timeless eternity with which we began this discussion. It transcends while enveloping regions of finite temporal passage. And it does so without God.

Why does Hawking follow this road? He is motivated by the hope of removing an inexplicable singularity from the mathematics of physical cosmology. Yet, one cannot help but wonder about another possible factor, namely, religious skepticism. Hawking admits that his denial of an original singularity is based only on a proposal for a future theory. He grants that no strong case can be made. Might this be a smokescreen blurring an otherwise anti-

---

[30] Stephen Hawking, *A Brief History of Time: From the Big Bang to Black Holes* (New York: Bantam Books, 1988), 139; cf. 143.

[31] Ibid., 136.

religious agenda? Might we view Hawking as belonging to that subculture of natural scientists who, on the one hand, drive as big a wedge as possible between rational science and allegedly irrational religion; while, on the other hand, invoking scientific discoveries to buttress their belief that belief in God is out of date? It is relevant to note that Carl Sagan, writing in the introduction to Hawking's book, *A Brief History of Time*, somewhat smugly advertises Hawking's argument for "the absence of God" on the grounds that there is "nothing for a Creator to do."

We might ask in theological terms: Just what kind of God are Hawking and Sagan rejecting? What they are rejecting, it seems, is the "God-of-the-gaps." What they object to is the God affirmed by the kind of physico-theology which once sought to find a divine explanation wherever scientists failed to give us a natural explanation. In the case of the Big Bang in particular, many of us are tempted to think of God as the one who set the original boundary conditions, who brought time and space out of a prior nothingness. The point of the Hawking proposal is to shut this door to divinity. To be more precise, the God which is being rejected here is the God of deism. According to deism, God brought the world into existence at the beginning and then departed to timeless eternity, leaving the universe to run according to its built-in natural laws. God has only one job to do for the deists, namely, create at the beginning. Thus, if the Hawking proposal holds, God is not needed.

This challenge should give us pause to recall that Christians along with their monotheistic comrades, Jews and Muslims, are not deists. They are theists. The theistic belief is that, in addition to what God did at the beginning, God is still active in world events today. The temporal trinitarian understanding of God we are trying to develop here is theistic in that it affirms an ongoing personal concern for the world on the part of an active divine life.

C. J. Isham has taken the Hawking proposal into account and offers a potentially fruitful line of argument: the Christian doctrine of divine creation still makes good sense, even as *creatio ex nihilo*. Instead of viewing it simply as crossing time's front edge from pre-temporal existence into temporal existence at t=0, God's divine work of creation should be seen as ongoing, as continual. The reasonableness of divine creativity does not depend on positing time's edge. The initial event of creation need not be thought of as crossing such an edge. Rather, the initial event by which the macro-universe comes into existence should be seen as continually arising from within a more comprehensive quantum realm within which spacetime is a single block reality. Isham infers theologically that all "times" are co-present to God and the ongoing indeterminacy of quantum processes represents the continuing activity of God's bringing something out of nothing. What is at stake for the theist is to understand God as a contemporary factor in world events, not merely a past factor. This affirms that God's creative work is not restricted to a one-time event in the ancient past. Rather, it continues in the present and we can expect more things yet in the future. Today's research in quantum physics challenges us to think through what this means.[32] What is

---

[32] Isham, "Creation of the Universe as a Quantum Process," in *Physics, Philosophy and Theology: A Common Quest for Understanding*, ed. Robert John

significant about pondering time's finitude is that it elicits questions about infinity, questions about eternity. Even Hawking and Sagan in their skepticism cannot avoid entering into theological discussion. Finite time cannot readily explain itself. It takes eternity to understand it.

And, despite Hawking, if we could speculate about two edges to time, one at the beginning and one at the end, it would be a small jump to conceive of a transcendent eternity that embraces temporality at both ends. This ascribes to time a sense of totality or completeness. Like a story, once told it becomes a discreet whole. Might the idea of the whole of time be worth pondering? It is to the possibility of thinking about the whole of time and its relation to divine eternity that we now turn.

8 *Eternity as the Whole of Time: Plotinus and Pannenberg*

In our own era, Wolfhart Pannenberg is seeking to retrieve the insights of Plotinus for the purpose of understanding the relation between eternity and time. What he finds in Plotinus is the notion of eternity as the whole of time.

Plotinus followed Plato by defining time as the moving image of eternity. But he could not reason as Plato did. Plato claimed that the circular motion of the heavenly bodies provides the basis for temporal consciousness and calculation. Plotinus, in contrast, could not make time dependent upon motion, because he thought motion was dependent upon time. Time is prior, he thought. So, rather than appeal to the sun and the stars in motion, Plotinus appealed to the human soul (and to the world soul) for mediation between eternity and time. What the soul can apprehend is that the eternal is the whole of life, namely, "life that is fixed within Sameness, because the whole is always present in it—not now this, then another, but all simultaneously" in the sense of "completion without parts."[33]

The human soul for Plotinus, of course, has fallen from eternity into time and now experiences temporal reality in terms of parts and intervals which are alien from eternal reality. Instead of a single undivided moment, the human soul undergoes the passage of moment after moment. What was a whole is now fragmented into parts. Yet, what attracts Pannenberg to Plotinus is the notion that there is a future return to wholeness. He interprets Plotinus to be saying that the eternal whole is present in the sense that it hovers over the parts as a future whole or totality. The whole becomes the future goal of all striving within the realm of the temporally finite. The path to this goal is time. "In short," writes Pannenberg, "when the theory of time is oriented toward the eternal totality, the consequence is a primacy of the future for the understanding of time."[34]

Russell, William R. Stoeger, and George V. Coyne (Vatican City State: Vatican Observatory, 1988), 375-408. Cf., Peacocke, *Theology for a Scientific Age*, 133f.

[33] Plotinus, *Enneads*, III:vii:11,41. Cf., Wolfhart Pannenberg, *Metaphysics and the Idea of God* (Grand Rapids: William B. Eerdmans, 1990), 76-77; 97.

[34] Pannenberg, *Metaphysics and the Idea of God*, 77.

Although definitely sympathetic to the notion of eschatological wholeness, I am less certain than is Pannenberg that Plotinus can be garnered for support. Plotinus' idea of eternity is immune to temporal succession and also to the part-whole dialectic. He states that eternal life is "instantaneously entire, complete, at no point broken into period or part."[35] Eternity as "Ever-Being" can have "no this and that; it cannot be treated in terms of intervals, unfoldings, progression, extension; there is no grasping any first or last in it."[36] If, eternity cannot grasp any first or last, we must ask Pannenberg, how can it envelop and enfold the temporal history of the world without annihilating it?

To make matters more difficult, it appears that Plotinus repudiates the dynamic of the future that Pannenberg seems to find so fruitful. On the one hand, it seems at first that Plotinus is amenable to Pannenberg's interpretation. Plotinus does admit that for engendered temporal beings to have life they must have a future. On the basis of this observation, Plotinus speculates that there might be an engendered All or totality, and as such "a thing of process and change." Such a totality would keep hastening towards its future, dreading to rest, seeking to draw all things into itself. Such a whole understood as totality would seek perpetuity by way of futurity. It appears that this is what Pannenberg has latched onto when pressing Plotinus into the service of eschatological wholeness.

Yet, on the other hand, this generated All is not what Plotinus identifies with eternity. Rather, eternity has to do with what he calls "the primals." The primals exist in an unbroken state of blessedness with no aspiration for things to come in the future. They are whole now. "They, therefore, seek nothing, since there is nothing future to them, nothing external to them in which any futurity could find lodgement."[37] The primal existents, which are incapable of increment or change, make up the resident plurality within the eternal. And the eternal is the divine.

> Eternity, thus, is of the order of the supremely great; it proves on investigation to be identical with God: it may fitly be described as God made manifest, as God declaring what He is, as existence without jolt or change; and therefore as also the firmly living. . . . Thus, a close enough definition of Eternity would be that it is a life limitless in the full sense of being all the life there is and a life which, knowing nothing of past or future to shatter its completeness, possess itself in tact for ever.[38]

This is not a doctrine of eternity understood as everlastingness, as reality with an unlimited future. It is rather eternity understood as timelessness, as an unqualified now with everything in full manifestation.

It seems to me that it would be better for Pannenberg to say Plotinus is at least partially mistaken and proceed to develop a concept of eschatological wholeness in which unlimited futurity plays a decisive role. Eternity, I should think, must include everlastingness if it is to be understood as living and creative.

---

[35] Plotinus, *Enneads*, III:vii:3.
[36] Ibid., 6.
[37] Ibid., 4.
[38] Ibid., 5.

The eternity in which we place our hope consists in the integration of parts and whole and the consummation of temporal history, not in their virtual annihilation into an everpresent now.

This, I believe, is what Pannenberg is rightly pursuing. Though wrongly attributing it to Plotinus, Pannenberg's eternal whole consists in the integrated unity of finite being. It is not a simple eternal unity that stands in opposition to a plurality of temporal events. It is rather a consummate unity.[39] Pannenberg wants to see the "participation in the interrelatedness of everything divided by time at the moment of eternity."[40] God's eschatological action unites all time and space into the one thing that is the creation, and the creation is taken up into the everlasting reality that is God.

9 *Holism and Anti-Reductionism*

Thus, there seems to be a theological warrant for a (w)holistic principle. It may be relevant to note that methodological holism plays a significant role in current physical and biological theory. It takes the form of positing *downward* causation as an explanation for the behavior of complex organized systems and the emergence of living organisms. The principle that the whole is greater than the sum of its parts makes it anti-reductionistic. The behavior of a living organism, though dependent upon its physical and chemical components and their respective laws, cannot be adequately explained on the basis of those physical and chemical or even biological laws. A qualitatively new factor has entered the picture, a higher principle of ordering that derives from the organism as a whole. Although in complete harmony with the underlying physical and chemical and biological processes, the holistic principle frees the ordered system or organism for independent behavior that cannot be reduced to those underlying processes.

There is more than anti-reductionism going on here. Holism recognizes that the higher levels of systems organization and even the level of individuality in living selves act downward on the lower physical processes, thus pressing these physical processes into the service of the higher level ends. Water molecules remain H_2O even when whirled about by a vortex. Metabolism remains what it is, even though a person's decision either to play tennis or to go to sleep will have an impact on its rate. "The component description does not contradict the holistic description," writes Paul Davies; "the two points of view are complementary, each valid at their own level."[41]

---

[39] After showing appreciation for Augustine's concept of divine duration (*Dauer*) as a time-bridging present, Pannenberg ties eternity to the future of God. God is eternal, he says, because there is no future beyond or outside of God. The eternal God is the absolute future, and as the community of Father and Son and Spirit he is the free origin of both the Godself and the creation. *Systematische Theologie*, 3 volumes (Göttingen: Vandenhoeck & Ruprecht, 1988), I: 443.

[40] Pannenberg, *Metaphysics and the Idea of God*, 90.

[41] Davies, *God and the New Physics* (New York: Simon & Schuster, 1983), 62. Davies adds that with the doctrine of holism we no longer need the idea of a life-

Again we raise the question of the cosmos as a whole. Can we extrapolate the holistic principle to cover the cosmos as a whole? If we do, we cannot simply posit holism in the form of downward causation, because the universe is not causally unified, as we saw in our review of special relativity. However, we might be able to posit a temporal holism, according to which all parts of spacetime are related to one another in that they are all related to God's eternity.[42] To do so would in no way violate the contingency of natural events or what we experience as human freedom, yet it would provide a way to conceive of God's creating and redeeming relationship to the world. We could conceivably begin by asserting that God creates the world, meaning the whole of the world's temporal history (or histories). Then we could assert that God saves the whole of creation, adding that our individual salvation is a component part of this larger event.

If we are to make this conceptual move, we should make three qualifications. First, we need to avoid the temptation to vitalize the world. Because the most complex system we know of empirically is that of the living organism, especially the human organism, we will be tempted to extrapolate and describe the cosmos as an even greater living organism. We might then want to ascribe to God the distinct principle of life. The problem with such a vitalization is that it adds an unnecessary component, a life-force or soul or something which goes beyond what the notion of holism warrants. In addition, if we were to divinize the whole, we would risk falling into pantheism. The problem with pantheism is that God's transcendence would be threatened. God is more than the animating principle of the world. God is more than the world.[43] God must be more than the world if redemption is possible.

The second qualification has to do with time. The whole of which we speak theologically is the whole of history in its eschatological consummation. This whole is more than simply a higher principle of systems organization that explains what is happening at any given moment. Eschatological wholeness has to do with fulfillment, with the consummation of the entire history of creation that preceded it. Wholeness requires an eternal reality that sweeps up all that has

---

force to account for the seemingly miraculous jump from inanimate to living matter. Ibid., 64. Arthur Peacocke would agree. ". . . it is possible to be anti-reductionist without being a vitalist." *God and the New Biology* (San Francisco: Harper & Row, 1986), xv.

[42] There is another logical possibility, namely, to postulate a denial that the universe is an inclusive whole. We could try to claim that currently existing organisms such as human beings constitute the highest form of systematic wholes we know in the actual world. This would make the universe an agglomeration of such wholes without it itself being a whole. It may be that the theistic hypothesis that God transcends the creation is decisive in pressing the question of the wholeness of reality. In this case, the creation would be a whole primarily due to its opposition to God, not because it is a system of a higher order. Even if this theistic hypothesis is the origin of the question regarding the whole of the cosmos, it is a question worth pursuing scientifically to see if it might yield confirmation or disconfirmation.

[43] Cf., Pannenberg, *Metaphysics and the Idea of God*, 142-143.

been into a divine presence that affirms its original spatio-temporal actuality while redeeming it for the good of the grand whole. It requires a completeness of time.

The third qualification is the addition of *prolepsis*. This is the trinitarian dimension. The scientific model of holism makes sense regarding the dialectic of whole and parts. The whole is dependent upon the parts, yet the parts function according to the bidding of the whole. In top-down causation, causation is both upward and downward. So far, so good. Yet, whereas in the scientific model the parts are present within the whole and exhaust the physical being of the whole, in trinitarian theology we want to add that the whole is also present among the parts. In the incarnation of Jesus Christ, the whole is present as one part among others. The eternal has become temporal. The infinite has become finite. And the work of the Holy Spirit anticipatorily binds the part to the whole, the present to the future, the expectation to its fulfillment.

10 *Where Have We Been?*

We began with the problem of how an eternal God can act in, and be affected by, time. The problem arises when eternity is defined as a state of timelessness that contrasts sharply with the world of time in which we humans and the rest of nature find ourselves. What is necessary is that, while reserving for God some temporal transcendence, we modify our concept of eternity so that it does not stand in total exclusion of what is temporal. Eternity needs to transcend time, but it need not exclude or annihilate time. If God's creating and redeeming action is to be efficacious in this world, then there must be some manner in which God's eternal being can be affected by the temporal course of worldly events. If eternity is to be salvific, it needs to be eternity "for us."

For something to be "for us," it needs to have a temporal dimension.[44] This acknowledgment has led to asking about the nature of temporality within and beyond human experience. Thermodynamics and Big Bang cosmology seem to support the notion that nature as a whole is characterized by an arrow of time. This is challenged in part by Stephen Hawking's quantum gravitational pre-theory of creation, in which the arrow of time is limited to regions of the universe such as ours, and does not apply globally to the universe itself. According to this

---

[44] I believe that positing a temporal dimension within the divine life does not necessarily eliminate all God's immutability. Thus, I am cautious about Maurice Wiles who, after defending the idea of *creatio ex nihilo* by an omnipotent God who grants finite power to the created order, goes on to affirm divine immutability. We ought not to think of the classic commitment to "God's unchangingness and impassability" as "just a foolish mistake to be denied outright," he says. "There is an important truth in that old conviction. It is the truth that God is never, as we are, affected by forces or events which come entirely from outside the sphere of his own influence." *God's Action in the World*, The Bampton Lectures for 1986 (London: SCM Press, 1986), 24-25. Wiles is certainly correct in keeping the old conviction, but he risks divine isolation when protecting God from becoming affected by the changing world.

view, nature itself takes on the qualities of a timeless eternity enveloping time, but it does so without the benevolent activity of a creating and redeeming God. What this leaves open in terms of physical cosmology is whether or not time has edges. What this leaves open theologically is whether we can speak of God's eternity as transcending edges which may or may not exist.

This has led to the question of the whole of time, of temporal holism. This question issues from the theological agenda, because the unity of temporal history is implied by the notions of a single God in creative and redemptive relation to a single world. Eschatology points to a future, divinely initiated event that will not simply put an end to temporal history; it will unify it and fulfill it. The theological vision seems to warrant a principle of cosmic holism. The scientific idea of downward causation appears to have sufficient consonance with such a holistic principle that we might try—by analogy—to press it into theological service. It will have limited application, however, because, although the idea of temporal holism serves to bind creation's history into a unity, it does not say enough about God's action in history. To proceed further, we need to turn more directly to God's personal and tri-personal character.

11 *The Personal and Temporal Trinity: Lucas and Jenson*

J. R. Lucas holds that "reality is through and through temporal. Equally, God is temporal, though not merely that . . . we cannot say that He is timeless, or that for Him there is no difference between future and past."[45] Why? On what grounds does Lucas say this? Because God is personal. If God is "personal then temporal, and if temporal then in some sense in time, not outside it."[46] The personal character of God and the nature of temporality go hand in hand.

> Time is not a thing that God might or might not create, but a category, a necessary concomitant of the existence of a personal being, though not of a mathematical entity. This is not to say that time is an independent category, existing independently of God. It exists because of God: not because of some act of will on His part, but because of His nature: if the ultimate reality is personal, then it follows that time must exist. God did not make time, but time stems from God.[47]

The way in which we experience God as personal is threefold. God is personal in each of the three persons of the Trinity: Father, Son, and Spirit. We need to proceed by attending to the way Trinity and temporality belong together.

To pursue this further, we will look at the contribution to trinitarian thought offered by Robert Jenson. That temporality is decisive for Jenson's understanding of God is indicated by describing the divine in terms of event.

---

[45] Lucas, *The Future*, 209.
[46] Ibid., 213.
[47] Ibid.

"There is one event, God, with three identities," he writes.[48] Jenson continues to travel the trail blazed earlier this century by Karl Barth, according to whom the threeness in God must be understood not as three instances of one deity but as three *events* of one deity: God is God, and then God is God again and again, each time in a different way.[49] With this as his point of departure, Jenson attempts to go further than Barth in emancipating the Christian Trinity from captivity to Hellenistic thinking, especially that of an atemporal and impassable eternity. God certainly is eternal, believes Jenson, but he argues that God's eternity embraces time and takes temporal events up into the divine life.

The problem with the idea of eternity understood as timelessness is that it depicts God as unrelated to the world. This notion comes from ancient Hellenistic culture and Greek philosophy. Jenson, like Barth and so many other theologians in our century, believes Scripture-based thinking is antagonistic to Hellenistic metaphysics. The chief reason is that the Greeks thought of temporal movement as a loss of perfection. Time destroys. Cronos devours his children. So Greek religion became a quest for the rock of ages, for an eternity which was immune to temporal change and deterioration. The defining character of the gods was immortality, immunity to destruction. We are in time, but the gods are not. This makes our mortal situation desperate. Therefore, the religion of late antiquity became a frenzied search for mediators, for beings of a third ontological kind who could connect time and timelessness. Not so in the Old Testament. Here Yahweh was experienced as eternal, but his eternity was understood as a faithfulness through time. The prophets spoke of God making and fulfilling promises. Yahweh constantly challenged the past and everything guaranteed by it from a future that is his own freedom. Time destroys, but Yahweh is faithful. In Yahweh we find freedom and a future. Unlike the other gods of ancient Asian polytheism or the one inclusive being of mystical monism, Yahweh does not transcend time by being immune to it. There is a continuity over time which strains forward toward the future when Yahweh's identity will become fully revealed. This continuity is not a constancy of being, an ontological immutability. Rather, it is a continuity which Jenson describes as "personal." It is established through Yahweh's words and commitments, by faithfulness of his later acts to the promises made in his earlier acts. So, Jenson concludes, whereas the Hebrew Yahweh was faithful *through* time, the Greek gods' eternity was due to their abstraction *from* time. Yahweh's eternity is intrinsically a matter of relation to Yahweh's creatures, whereas the Greek gods' eternity is the negation of such relation.

This contrast received attention already in the ancient discussion. The fourth century Cappadocian theologians (Basil, Gregory Nazianzan, Gregory of Nyssa) in particular resisted the Hellenization of the Bible's God. Whereas divinity for the Greeks at the time of Nicea consisted of a motionless center with

---

[48] Jenson, *Triune Identity*, 114; see also *idem*, "The Triune God" in *Christian Dogmatics*, ed. Carl Braaten and Robert Jenson, 2 volumes (Philadelphia: Fortress Press, 1984), I: 140.

[49] Jenson, *Triune Identity*, 136ff; 179f. This is not intended to be a modalistic sequence.

time circling round about it; the God of Gregory of Nyssa, in contrast, is eternal in that he envelops time, is ahead and so before it. The Hellenic God stands still, so that we may ground things in it. Gregory's God keeps things moving.[50]

Inspired by the work of Gregory and the other Cappadocian fathers, whom he favors somewhat over Athanasius, Jenson develops a future-based concept of time which he dubs "temporal infinity." The essential mode of divine temporality is unbounded futurity, he says. This means time is neither linear nor cyclical and, therefore, we ought not understand reality in terms of enduring or cyclical substances. It is the oncoming of the future that creates time for us, because it forces the present reality to go beyond itself. It is the reality of the future that temporalizes us. It is the structure of the still outstanding eschatological future which determines the character of time and becoming in our present experience. What we experience as endurance is given not by persisting substances but rather by the reinterpretation which the future gives to all past occurrences. What is significant here is that what we experience as the temporal movement of events is taken up into the divine relationality proper. This is the ongoing transformative work of the Spirit.

When responding to the work of Stephen Hawking, Jenson argues that the dynamic relations within the trinitarian life of God constitute the condition for the possibility (*Bedingung der Möglichkeit*) of temporality within the created world. The constancy of God—the divine *Dauer*—unites the world of time. It even unites real with imaginary time.

> ... my argument runs: if there is the triune God, then his "*Dauer*" is the condition precisely of the oneness of real time and imaginary time; and that is to say, God's *Dauer* is the condition of the subsistence of what we can call, and only can call, time.[51]

Jenson answers "yes" to the question, "does God have time?" In doing so, he goes on to argue that one quality of the creator's time, namely, God's faithfulness or *Dauer*, is necessary for the creation to be temporal.

But the tie that binds Trinity and temporality is found in Christology: in Jesus Christ we find God on both sides of the line that divides time and eternity. Jenson depends in part upon the pioneering work of Jürgen Moltmann when interpreting the dynamics of shared suffering. In Jesus we find God crucified. God is crucified because God has become historical. He has entered the finitude of a single spacetime frame of reference. Jesus' suffering is shared with the world around him and, in turn, shared with the Father who sent him. This

---

[50] Ibid., 57, 165.

[51] Jenson, "Does God Have Time?" *CTNS Bulletin* 11:1 (1991). Robert Russell objects at this point, stressing that in the Hawking model, all of time has a "real" and an "imaginary" component (in the mathematical sense), while in the Big Bang theory there is never an imaginary component. Hence, the choice is between standard Big Bang (real time) and Hawking (real and imaginary time) models. Thus, Russell argues that we must choose between the "standard cosmology, with its initial singularity and globally defined real time ... and Hawking's quantum cosmology, with its lack of an initial singularity." Russell, "Is the Triune God the Basis for Physical Time?"

suffering at first signals a split in God, the experience of dying within the divine source of life. Amidst the suffering of the cross, Jesus experiences among other things the agony of being forsaken by the Father. The Father in turn experiences the anguish of being separated from the Son. Yet in the dramatic event of Jesus' surrender for the sake of sinful humanity, Jesus and the Father experience a new bond of unity, a unity provided by the Spirit. The experience of suffering, separation, and reunion occurs within the divine life and, at the same time, between the divine life and the world. Moltmann writes, "God suffers with us—God suffers from us—God suffers for us: it is this experience of God that reveals the triune God."[52] By saying that the historical event of Jesus Christ is constitutive of the divine life proper, Moltmann is here extending the line which runs from Karl Barth through Eberhard Jüngel and Wolfhart Pannenberg to Robert Jenson.

On this count, Jenson and Moltmann both endorse Rahner's Rule. Formulated by Karl Rahner, the rule states that the economic Trinity is the immanent Trinity and the immanent Trinity is the economic Trinity.[53] But Jenson suspects a dilemma, a dilemma between two rules. The first is Rahner's: the economic Trinity is the immanent Trinity, and vice versa. Jenson subscribes to Rahner's Rule because it accounts for trinitarian identity and definition determined by the course of events in salvation history. But there is a second rule: the legitimate theological reason for distinguishing the immanent and the economic trinities is the freedom of God. It must be the case that God in himself could have been the same God he is, and so triune, had there never been a creation or any saving history of God within the creation. So here is the dilemma: are these two compatible? Yes, says Jenson, they are compatible if we think of the identity of the economic and immanent Trinity as eschatological, that is, if the immanent Trinity is the eschatological reality of the economic.[54]

---

[52] Jürgen Moltmann, *The Trinity and the Kingdom* (San Francisco: Harper & Row, 1981), 4.

[53] Karl Rahner, *The Trinity* (New York: Herder & Herder, 1970), 21-22. Rahner writes that "God relates to us in a threefold manner, and this threefold, free, and gratuitous relation to us *is* not merely a copy or an analogy of the inner Trinity, but this Trinity itself, albeit as freely and gratuitously communicated." Ibid., 34-35. Moltmann follows by saying, "I found myself bound to surrender the traditional distinction between the immanent and the economic Trinity, according to which the cross comes to stand only in the economy of salvation, but not within the immanent Trinity. That is why I have affirmed and taken up Rahner's thesis that the economic Trinity is the immanent Trinity and vice versa." *Trinity and the Kingdom*, 160.

[54] Jenson, *Triune Identity*, 140; cf., Braaten and Jenson, *Christian Dogmatics*, I: 154. See also Moltmann, *The Trinity and the Kingdom*, 161. This leads to a doxological approach that is warranted in part by the difficulty of conceiving of eschatological temporality. John Polkinghorne speculates, "It may be that the time of that new world bears no relation to the time of this world and that we pass to experience of it directly at death. Or it may be that that time will only begin to run when this world's process will have been concluded by some great act of God, so that

Instead of interpreting Christ's deity as a separate entity that always was, Jenson interprets it as a final outcome. And as the final outcome it thereby becomes eternal. "Truly, the Trinity is simply the Father and the man Jesus and their Spirit as the Spirit of the believing community. This economic Trinity is eschatologically God himself, an immanent Trinity. And that assertion is no problem, for God *is* himself only eschatologically, since he is Spirit."[55] To take seriously the biblical acknowledgment that "God is Spirit" is to recognize that the Spirit is equally principle and source with the Father. The Spirit's witness to the Son and the Son's saving work are equally God-constituting. What is more, through the identification of the Son and the Spirit with the economy of world history, all of cosmic history is likewise taken up eschatologically into the divine eternity.

12 *Where Should We Go?*

Jenson to this point has made only some of the connections we need to make regarding time as understood in physical cosmology. Yet, his explication carries the conversation to the point where this would be the next logical step.

The direction future investigations into this problem should take, I suggest, is to expand this very discussion of the Trinity. In particular, such discussion should focus on the temporal components of God's incarnation in Christ and the loving work of the Holy Spirit in history to bring peace and unity to the world. If we wish to affirm that the work of the economic Trinity is in fact the work of God, then we will have to say that God has entered someone's inertial frame of reference within which past and future can be experienced in relation to an observer's present. The work of the trinitarian God in history is by no means a mystical cancellation of time. It is rather a divine participation within the limits and possibilities of temporal passage. The incarnate Jesus is who he is by virtue of his past, by virtue of his having become a spacetime event. Jesus is who he is also by virtue of his future, the expected new creation which will complete the work begun in his Easter resurrection. Without that past and future, Jesus is not the Christ, the Messiah. Without the Christ, the history of creation does not hold together as the redemptive object of God's love.

Also needing further thought is the idea of the eschatological consummation, wrought by the Holy Spirit, as the final temporal event that renders all previous history, both natural and human, a single whole. It is not enough to posit an end to creation, whether by the thermodynamic heat death when the Big Bang peters out or by a divine intervention that puts a stop to the process. More than a mere end is warranted. The salvation we hope for includes a strong sense of fulfillment, a confirmation and preservation and enhancement and eternal enjoyment of what has been meaningful. The past cannot simply be

---

the new world comes into being by the transformation of the old. In that case the dead await their resurrection. Who can say?" *Science and Providence*, 90.

[55] Jenson, *Triune Identity*, 141; Braaten and Jenson, *Christian Dogmatics*, I:155.

canceled and replaced by a subsequent present moment. Nor can the future be eliminated. The co-reality of past and future must give form and content to the ecstasy of the eternal "now."

With regard to theological method, two things ought to be said. First, this eschatological hope for redemption is drawn from the interpretation of biblical symbols. It comes strictly from theological resources, not scientific evidence. The prospects of heat death due to the law of entropy might even be interpreted as a possible disconfirmation of theological affirmation. We will have to admit that at the present stage of inquiry there exists very little consonance between theology and physics regarding the ultimate future of the cosmos. The conversation between the disciplines should be pressed further, however, to see if it may yield any mutually beneficial explanatory fruitfulness. The second methodological concern has also to do with our continuing agenda: we need to investigate the viability of beginning with our understanding of God as Trinity and moving toward positing how the world must be. If we were to proceed this way, one of the points of departure would be the two sendings within the immanent Trinity: the Father generates the Son, and the Holy Spirit proceeds from the Father (and the Son, according to the Latin tradition). Dynamism exists within the divine life proper. There is a one-after-the-otherness here, even though classical trinitarian thought speaks of it as an eternal *perichoresis* so as to render the sendings timeless. Be that as it may, when we assert that this trinitarian God is the creator of the world, we should expect that world to be similarly dynamic in character. Time's arrow should not come as a surprise. Furthermore, where it is challenged, Christian theology should keep the hypothesis of time's arrow alive, exploring the evidence for confirmation.

Without indulging in emmanationism, I believe it is safe to say that creation is an act continuous with that of generating the Son, and consummation is an act of the Holy Spirit that unifies the creation with the divine life that is continuous with the bond of love the Holy Spirit maintains between Father and Son. Through creation and incarnation the divine life becomes economically trinitarian. Through spiration and consummation the economic Trinity becomes the immanent Trinity. And this trinitarian movement brings the history of the creation right into the divine life proper.

FINITE CREATION WITHOUT A BEGINNING: THE DOCTRINE OF CREATION IN RELATION TO BIG BANG AND QUANTUM COSMOLOGIES

Robert John Russell

1 *Introduction: Should Cosmology be Important to Theologians?*

With the rise of contemporary scientific cosmology this century, and especially with the triumph of the Friedmann-Lemaitre-Robertson-Walker (FLRW) cosmology over the steady state theory in the mid-1960s, there has been growing interest in the possible theological significance of the singularity at "t=0" conventionally referred to as the "beginning of the universe" or "Big Bang" some 12-15 billion years ago.[1] During the past two decades, the original Big Bang model has been modified by inflationary scenarios to correct for a variety of technical difficulties, such as the flatness, matter/antimatter ratio, and horizon problems. Still, even with inflation, if one assumes general relativity to be the correct theory of gravity, and if one makes certain reasonable assumptions about the mass-energy of the universe, then the initial singularity at t=0 is unavoidable (i.e., it is an "essential singularity"), as shown by the theorems of Stephen Hawking and Roger Penrose.[2]

This situation raises a number of difficult questions, many of which require more than a strictly scientific resolution and involve interdisciplinary research: Does the initial singularity in inflationary Big Bang cosmology point to a fundamental feature of nature or is it merely an artifact of this particular cosmological model? If t=0 is of lasting scientific status, should it, in principle, bear any significance for Christian doctrine, or is it theologically irrelevant regardless of its scientific standing? If it is theologically significant, should t=0 be seen as only one aspect of God's action throughout nature and history as creator and redeemer, or does t=0 represent the sum of God's action as in deism or uniformitarianism? One of the purposes of this paper is to explore these questions in critical interaction with other scholars who have written on the relationship between Big Bang cosmology and Christian theology.

Meanwhile, attempts are being made to obtain a quantum mechanical treatment of gravity which will replace general relativity and from which to develop quantum cosmology as a successor to Big Bang cosmology. An important, though highly speculative, approach is the Hartle/Hawking proposal,

---

[1] Here *t* stands for cosmological time. For a technical introduction see the article by Bill Stoeger/George Ellis in this volume.

[2] See S. W. Hawking and R. Penrose, "The singularities of gravitational collapse and cosmology," *Proc. R. Soc. London A* 314 (1969): 529-548. This publication includes a full list of references on the singularity theorems.

with its surprising elimination of the singularity, t=0, from the scientific model. According to the Hartle/Hawking proposal the past of the universe is finite (as it is in the Big Bang model) but, unlike the past of the Big Bang, it is *unbounded* (there is no initial singularity, t=0). Although research in quantum gravity is still in a *highly tentative* stage, and the Hartle/Hawking proposal is only one of several avenues being explored, it already raises some very interesting interdisciplinary questions. For example, what effect will the move to quantum cosmology have on the ways theologians have appropriated Big Bang cosmology, particularly regarding the theological significance of t=0? How will the change in the meaning and status of temporality in the move from Big Bang to quantum cosmology affect the theological understanding of God as the creator of temporality? It is to questions like these that this paper is also addressed.

Accordingly, my paper is divided into two sections. In the first (section 2) I focus on inflationary Big Bang cosmology and the problem of t=0. The theological reaction to t=0 has been mixed. Some theologians, such as Ted Peters, have welcomed it as evidence of divine creation; others such as Ian Barbour and Arthur Peacocke have dismissed it as irrelevant to the creation tradition. As I see it the argument on both sides has been shaped largely by the seminal work of Langdon Gilkey in his *Maker of Heaven and Earth*.[3] Although I will be quite critical of Gilkey's resolution of the problem, we owe him a great deal for his lucid insistence on its importance as such. If Gilkey is correct, the epistemological problems surrounding t=0 are well worth *our* pursuing because they are *inherent* to the theological problem of divine action, or as he later calls it the "travail of biblical language." Thus they are not a minor issue best forgotten; instead they raise the fundamental problem of the rationality of theology in an age of scientific epistemology.

Nevertheless, the resolution which Gilkey offers, and which I find inadequate, begins with his interpretation of the traditional distinction between what can be called "ontological origination" and "historical/empirical origination." Gilkey, citing Thomas Aquinas, seems to view these as *strictly dichotomous alternatives*. One then either rejects the latter as theologically irrelevant (Gilkey's position) or elevates the latter into the essential meaning of the doctrine of creation *ex nihilo* (the position Gilkey rejects). In the first case science, in so far as t=0 is concerned, plays no role in theology; in the second case it plays a normative role.

I will criticize both extremes by seeking to undermine Gilkey's assumption that the alternatives should form a *strict* dichotomy. Instead I believe that historical/empirical origination provides an important corroborative meaning

---

[3] Langdon Gilkey, *Maker of Heaven and Earth: The Christian Doctrine of Creation in the Light of Modern Knowledge* (Garden City: Doubleday, 1959; reprint, Lanham: University Press of America, 1985). In recent lectures Gilkey insists that his views have changed dramatically since writing *Maker of Heaven and Earth*, and that he now stands at a considerable distance from the Neo-Orthodoxy reflected there. However, given the influence *Maker of Heaven and Earth* has had on a generation of scholars I believe it is reasonable and valuable to assess it on its own merits and in its historical context, independently of its author's current position.

for ontological origination, although it is neither its essential nor even its central meaning, a view, incidentally, which I take to be more in keeping with that of Aquinas. I then argue that an important way of relating historical/empirical origination to ontological origination is through the concept of finitude. This abstract concept, initially closely connected to ontological origination, can take on an important historical/empirical meaning in the context of cosmology, where the past temporal finitude of the universe is represented by the event, $t=0$. Hence I will argue that $t=0$ is relevant to the doctrine of creation *ex nihilo* if one interprets arguments about historical origination as offering confirming, but neither conclusive nor essential, evidence for the central thesis of ontological origination. In this way science plays a more vigorous role in the doctrine of creation than many scholars today allow *without* providing its essential meaning, and certainly not its foundations. In particular, taking a cue from the writings of Ian Barbour, Nancey Murphy and Philip Clayton, I will frame my approach in terms of a Lakatosian research program in theology. Creation *ex nihilo* as ontological origination will form the core hypothesis of this program, with $t=0$ entering as confirming evidence through the use of a series of auxiliary hypotheses involving the concept of finitude deployed in increasingly empirical contexts of meaning.

In the second part (section 3) I discuss the Hartle/Hawking proposal for quantum cosmology. How should their claim that the universe, though having a finite age, has no beginning event $t=0$, that is, that the universe is finite but unbounded in the past, affect the theological arguments in part one? To answer this, I will first critically discuss the positions developed by C.J. Isham, Paul Davies and Wim Drees regarding the theological significance of the Hartle/Hawking proposal. Next I will present Hawking's own theological views and offer a counterargument to them. My constructive position will then be that the Hartle/Hawking proposal can have much to teach us theologically, even if its scientific status is transitory. Given their work, we should distinguish between the theological claim that creation is temporally finite in the past and the further claim that this finite past is bounded by the event $t=0$. This leads to the important recognition that the first claim (about finitude *per se*) is sufficient for developing the empirical meaning of *creatio ex nihilo*. Hence we can set aside arguments *specifically* over $t=0$ and yet *retain* the historical/empirical sense of the past temporal finitude of creation, as developed in section 3 below on the Big Bang model. I attempt to capture this insight with the phrase, *finite creation without a beginning*. Moreover, this theological insight can be maintained whether or not the Hartle/Hawking proposal stands scientifically and thus it suggests that we can in fact work with "speculative proposals" at the frontiers of science instead of *necessarily* restricting ourselves to well-established results, as most scholars cautiously advise. I view this generalization of the meaning of finitude as an additional auxiliary hypothesis to our research program, and following Lakatos again, look for novel predictions it might entail.

To do so I analyze the temporal status of the universe in terms of the relation between quantum gravity and general relativity. The variety of ways time functions here (external, internal, phenomenological) and their implications for the temporality of the universe lead to important new directions for understanding God's action as creator and the doctrine of creation. From one perspective the universe has both a temporal and a timeless domain connected by a transitional domain. From this perspective we should inquire into God's relation as creator to each domain. Here the generalization of the concept of finitude to include an unbounded finitude might allow us to claim the occurrence of the transition domain as a Lakatosian "novel fact" of our research program. Alternatively, if quantum gravity is to be the fundamental theory replacing general relativity, God's relation to the universe as a whole will need to be reinterpreted in terms of the complex role and status of temporality and finitude in quantum gravity. In any case, God's activity as creator is not limited to a 'first moment' (whether or not one exists) but to the entire domain of nature, returning us to the general problem of divine action in light of science. I close by pointing, then, to the need now to rethink the current models of divine agency and of the relation between time and eternity in terms of a more complex understanding of temporality from a trinitarian perspective informed by quantum physics and quantum cosmology.

2 Theological Reflections on Big Bang Cosmology

2.1 Gilkey and his Successors on Theology and Cosmology.

2.1.1 Langdon Gilkey: *The Critical Distinction Between Ontological and Historical/Empirical Origination* We start with someone who was strongly influenced by Neo-Orthodoxy and who, in turn, directly influenced many scholars in theology and science. Here I refer to Langdon Gilkey who, in 1959, wrote the ground-breaking text on the doctrine of creation, *Maker of Heaven and Earth*. Gilkey begins his detailed discussion of creation and time by reminding us that the idea of an "'originating' activity of God" has always been a part of the Christian creation tradition. It has, however, taken two distinct forms: "ontological origination" and "historical/empirical origination."[4]

1) Ontological origination (Origin as ontological dependence): ". . . God originates the *existence* of each creature out of nothing, whatever its position in the time scale."

2) Historical/empirical origination (Origin as historical narrative): "(I)t has meant 'originating' in the sense of founding and establishing at the beginning, starting the whole sequence of things at a first moment. For most Christians, creation lies not only at the base of the existence of things; it also lies *back there* in time, at the beginning of the time series. Creation was an act done 'once and for all' at the beginning of the universe: it was the first event in the history of the world." (Gilkey's italics)[5]

---

[4] These are not Gilkey's exact phrases, but I believe they capture his intent. Gilkey does refer to Anthony Flew's use of the term "absolute ontological dependence" in his discussion of this distinction.

[5] Gilkey, *Maker of Heaven and Earth*, 310.

Gilkey attributes this distinction to Thomas Aquinas.[6] According to Gilkey, in the *Summa Theologica*, Aquinas affirmed that the created world is ontologically dependent, drawing on philosophical arguments in several ways. Aquinas also argued for a historical interpretation of creation based on revelation, that is, that creation has a finite age, although he asserted that philosophy on its own could never prove this revealed truth. Hence Thomas could assert both a historical and a philosophical understanding of creation, but in Gilkey's opinion, he kept their biblical and philosophical warrants separate.

Gilkey supports the ontological interpretation of origination, but he finds *two* obstacles which tend to undermine the historical interpretation. The first has to do with science's challenge to the traditional, static understanding of cosmology, the second to the legitimacy of either scientific or theological talk about a moment of absolute beginning.

a) According to Gilkey, creation in the historical narrative sense, with its presupposition of a static cosmology, has been severely challenged by modern, evolutionary science. Instead of a creation all at once, we are called to reinterpret the meaning of God's creative act as a continuing process of bringing things into being in time. This in turn means that creation and preservation are ". . . different aspects of the simultaneous activity of God, who continually gives to all that arises existence and form, molding the new as well as preserving the old."[7]

b) The second problem undermining the historical narrative sense of creation involves the status of the "beginning." Gilkey acknowledges the important role such a traditional idea had in breaking the prevailing Hellenistic belief in a cyclic view of time and in establishing time as purposive and developing. Nevertheless, Gilkey is highly critical of it. His primary argument is *a priori*: knowledge about a first moment of time cannot be a valid part of theology since theology does not contain *any* "facts" about the natural order. Revelation does not "inform us of its character or its constitution."[8] Since knowledge about a cosmic beginning would be just such a "fact" about the universe, it "cannot be a part of religious truth."

Gilkey is also dubious about science making a claim regarding a first cosmic moment, and again for *a priori* reasons. "Since science presumably would have to assume that there was 'something there' out of which that moment arose, albeit an existence in a very different form, science could not include the concept of an absolute beginning of the process itself."[9]

Thus Gilkey recommends that we embrace the ontological/historical-empirical distinction as a genuine dichotomy and abandon historical-empirical language as myth, choosing instead to theologize strictly in terms of ontological dependence. "The idea of a beginning to time has a great theological and cultural

---

[6] Ibid., 313, where he cites Aquinas' *Summa Theologica*, I, 46, a. 2.

[7] Ibid., 312. One clearly sees the influence of Freidrich Schleiermacher here in Gilkey's willingness to view creation and preservation as equally containing the entire meaning of the tradition.

[8] Ibid., 314.

[9] Ibid., 313.

value; but . . . we have been forced to deny that there can be for theology any factual content to this idea."[10]

Gilkey's views have had wide-ranging effects on theologians wrestling with creation in the context of the sciences, as will become evident in the writings of Arthur Peacocke, Ted Peters, and Ian Barbour.

2.1.2 *Arthur Peacocke: t=0 is Unimportant Theologically* Although science in general, including contemporary physics, cosmology, and evolutionary biology, play a crucial role in Arthur Peacocke's theology, the problem of t=0 finds marginal significance at most. This seems to stem from the fact that, without specifically referring to Gilkey's writings in this regard, Peacocke makes essentially the same distinction Gilkey makes between ontological and historical origination.

One sees this clearly in his 1978 Bampton Lectures, *Creation and the World of Science*.[11] Here Peacocke explores at length the historical and contemporary discussion surrounding the doctrine of creation as both *ex nihilo* and *creatio continua*, and the important relation of science to the doctrine. Use of the ontological/historical distinction pervades Peacocke's discussion of *ex nihilo*:

> The principal stress in the Judeo-Christian doctrine of creation . . . is on the dependence and contingency of all entities, and events, other than God himself: it is about a perennial relationship between God and the world and not about the beginning of the Earth, or the whole universe at a point in time. . . . Time, in modern relativistic physics . . . has to be regarded as owing its existence to God. . . . It is this "owing its existence to God" which is the essential core of the idea of creation.[12]

The results of science are simply not germane to the core of that doctrine:

> Scientific cosmology . . . cannot, in principle, be doing anything which can contradict such a concept of creation. From our radio-telescopes . . . we may, or may not, be able to infer that there was a point (the "hot big bang") in space-time when the universe, as we can observe it, began, and, perhaps, what happened on the other side of that critical point. But, whatever we eventually do infer, the central characteristic core of the doctrine of creation would *not* be affected, since that concerns the relationship of all the created order, including time itself, to their Creator. . . .[13]

Arguments very similar to these are found in his Mendenhall Lectures, *Intimations of Reality*[14] and in his article, "Theology and Science Today," in *Cosmos as Creation*.[15]

---

[10] Ibid., 315.

[11] A. R. Peacocke, *Creation and the World of Science*, The Bampton Lectures, 1978 (Oxford: Clarendon Press, 1979).

[12] Ibid., 78.

[13] Ibid., 79.

[14] Peacocke, *Intimations of Reality: Critical Realism in Science and Religion* (Notre Dame: University of Notre Dame Press, 1984). See in particular pp. 62-63.

2.1.3 *Ted Peters: t=0 is Crucial Theologically* Ted Peters seems to accept Gilkey's distinction but opts, almost exclusively, for a historical/empirical interpretation of origins. In this sense, Peters is representative of many theologians who take Big Bang cosmology to have direct and unambiguous implications for the doctrine of creation.

In several recent articles, Peters has focused on the importance of the *ex nihilo* tradition in Christian theology in terms of its being in "consonance" with the problem of t=0 in Big Bang cosmology.

> It simply makes sense these days to speak of t=0, to conceive of a point at which the entire cosmos makes its appearance along with the spacetime continuum within which it is observed and understood. If we identify the concept of creation out of nothing with the point of temporal beginning or perhaps even the source of the singularity, we have sufficient consonance with which to proceed further in the discussion.[16]

Peters gives both a theological reason and a scientific rationale for such a move:

> To reduce *creatio ex nihilo* to a vague commitment about the dependence of the world upon God—though accurate—does not help much. It simply moves the matter to a higher level of abstraction. We still need to ask: just what does it mean for the world to owe its existence to God? One sensible answer is this: had God not acted to bring the spacetime world into existence, there would be only nothing. Furthermore, it makes sense to talk about the temporal point of origin. The assertion that the cosmos is utterly dependent upon God is familiar to theologians, but such an assertion lies outside the domain of scientific discourse. The ideas of an initial origin, however, does lie within the scientific domain. The point I am making here is this: for theologians to raise again the prospects of *creatio ex nihilo* understood in terms of a beginning to time and space is to be consonant with discussions already taking place within scientific cosmology.[17]

Peters sees himself in sharp opposition to both Barbour and Gilkey over which side of the distinction posited by Gilkey should carry the weight of *ex nihilo*. Thus Peters warns us: "If we were to follow the path led by Barbour and Gilkey, we might end up making no definitive theological commitments whatsoever regarding whether the cosmos ever had an initial origin, or, if it did, just how God was involved in this origin."[18] The essential point here is that he

---

[15] Peacocke, "Theology and Science Today," in *Cosmos as Creation: Theology and Science in Consonance*, ed. Ted Peters (Nashville: Abingdon Press, 1989), 28-43. See especially 33-34.

[16] Peters, "On Creating the Cosmos," in *Physics, Philosophy and Theology: A Common Quest for Understanding*, ed. Robert John Russell, William R. Stoeger, and George V. Coyne (Vatican City State: Vatican Observatory, 1988), 273-296, esp. 291.

[17] Ibid., 288.

[18] Peters' criticism of Barbour may be based on an inaccurate assessment of Barbour's position. According to Peters, "what Barbour has done here is virtually

seems to agree with the distinction though he disagrees over which of its interpretations is correct.

2.1.4 *Ian Barbour: t=0 is Relatively Unimportant Theologically* Ian Barbour's position seems to lie somewhere between Peacocke and Peters, and shifts, if I am not mistaken, from being initially closer to the position defended by Peacocke towards the interpretation advocated by Peters.[19]

Barbour's pioneering text, *Issues in Science and Religion*,[20] was written in the mid 1960s, during the competition between Big Bang cosmology and the steady state theory of Hoyle, Bondi, and Gold. Here Barbour's theological assessment of cosmology emerges in several key passages. As Barbour points out in chapter 12, Big Bang cosmology had often been defended by theologians while naturalists opted for the steady state theory. "We would submit, however, that both theories are capable of either a naturalistic or a theistic interpretation." Both contain unexplained features and neither resolve the problem of time. "We will suggest that the Christian need not favor either theory, for the doctrine of creation is not really about temporal beginnings but about the basic relationship between the world and God."[21] He concludes that either theory is compatible with Christian theology, and only science can tell us which one is correct.

---

equate *ex nihilo* with initial beginning, discard the idea of initial beginning, and thereby discard *ex nihilo*." As we shall see below, however, the position Barbour adopted in *Issues in Science and Religion* (New York: Harper & Row, 1966) was that we can discard the historical interpretation of creation in favor of the ontological interpretation for two reasons both reminiscent of Gilkey: science cannot settle the matter and revelation doesn't contain factual data of this kind. Rather than rejecting the *ex nihilo* tradition, as Peters suggests he has, Barbour wants to keep it by reducing it to its ontological interpretation. Peters, on the other hand, is intent on keeping the historical interpretation alive within *ex nihilo*; this may account for his perceiving Barbour as rejecting *ex nihilo* when rejecting a historical beginning.

This certainly seems clearer in his more recent writings, where Barbour is inclined to keep something of the *ex nihilo* position. Whether he was so persuaded in 1966 remains an open question, and Peters may be correct about that part of Barbour's writings. My point here is that, assuming the distinction between ontological and historical interpretations for the sake of argument, Peters may be incorrect in his analysis that Barbour reduces ("equates") ex nihilo to t=0 and then discards the latter. Rather, Barbour may be seeking to keep them separate using the ontological/historical distinction, and then to discard the latter and keep the former.

[19] Here and elsewhere I am taking Peacocke and Peters to defend opposite sides of the distinction Gilkey sets up, and I am ignoring the historical sequence of the statement of their arguments. For instance, at this point in the paper I am letting Peacocke and Peters stand for opposite interpretations of the significance of the *ex nihilo* argument independent of when they wrote, and analyzing the movement of Barbour's thought between them from 1966 to 1990, even though Peacocke and Peters themselves wrote on the subject several times during this period.

[20] Barbour, *Issues in Science and Religion*.

[21] Ibid., 366-368.

Next Barbour turns to Neo-Orthodoxy and its interpretation of creation. After citing Emil Brunner[22] briefly, Barbour focuses his attention on Gilkey's *Maker of Heaven and Earth*, which he describes as "the best recent exposition of the doctrine by a Protestant theologian. . . ."[23] Barbour attributes to Gilkey the claim that "creation is not as such a historical account. . . . It is not really about temporal origins in the past, but about the basic relationship between God and the world in the present." Barbour admits that Neo-Orthodoxy leads to "a radical separation of scientific and religious questions" but he agrees with Gilkey who ". . . emphasized that creation is a relationship, not an event; the doctrine deals with ontological dependence, not temporal history."[24]

In the concluding section of chapter 12, Barbour again seems content to accept Gilkey's formulation of the distinction without further analysis. He states clearly that "we agree with neo-orthodoxy (Gilkey) that the doctrine of creation is not fundamentally a hypothesis about origins but an affirmation of our dependence on God. . . ."[25] Finally, his previous agreement with Gilkey is taken up into his general conclusions to the entire book, where in chapter 13 Barbour writes: "We may have to give up *creatio ex nihilo* as an initial act of absolute origination, but God's priority in status can be maintained apart from priority in time."[26]

When Barbour's first series of Gifford Lectures were given in 1989,[27] more than two decades had passed since the demise of the steady state model and the commensurate strengthening of support for Big Bang cosmology. In this changed scientific context, Barbour returns to the problem of the status of time and the meaning of *creatio ex nihilo*. What response should theologians *now* give to the singularity at $t=0$, given that this event is "inaccessible" to the scientist and thus represents "a kind of ultimate limit to scientific inquiry"?[28]

On the one hand, Barbour cites those like Pope Pius XII who found support for creation theology in Big Bang cosmology. On the other hand, he reports on those like Peacocke who claim that the topic is inconsequential to theology. Barbour's own position lies between these poles, although he again leans towards the position taken by Neo-Orthodoxy.

---

[22] Further evidence of the possibility that Barbour depended on Gilkey in this discussion is the fact that Brunner is cited by both he and Gilkey; in fact they cite the same text and chapter.

[23] Barbour, *Issues in Science and Religion*, 377.

[24] Ibid., 380.

[25] Ibid., 414.

[26] Ibid., 458. Relying on Jarislov Pelikan's arguments, Barbour is also critical of the *ex nihilo* tradition since it has historically suppressed the tradition of continuing creation. According to Pelikan, this has "made it difficult for us to interpret evolution as the means of creation." (Ibid., 384) Barbour urges that we merge continuing creation with providence, and he deploys a number of arguments in support of this move. These arguments are worth pursuing, but, however, not in the limited context of this paper.

[27] Barbour, *Religion in an Age of Science*, The Gifford Lectures 1989-1991, vol. 1 (San Francisco: HarperSanFrancisco, 1990).

[28] Ibid., 128ff.

He is cautious about "gaps"-type arguments in general, although he admits that the case of t=0 is different precisely because it is inaccessible in principle to science. He is also concerned that changes in cosmology could alter the entire situation. Hoyle had avoided the problem of a beginning with his steady state model, and other models, such as an oscillating universe, could still accomplish the same. Moreover, "it is equally difficult to imagine a beginning of time or an infinite span of time."

Still Barbour's underlying reason for caution towards t=0 is basically theological:

> I do not think that major theological issues are at stake (in the case of t=0), as has often been assumed. If a single, unique Big Bang continues to be the most convincing scientific theory, the theist can indeed see it as an instant of divine origination. But I will suggest that this is not the main concern expressed in the religious notion of creation.[29]

Barbour will again locate this concern more in terms of continuing creation than in terms of creation *ex nihilo*, though a discussion of this argument would take us beyond the scope of my paper.[30]

2.2 *Critique and Constructive Proposals*

2.2.1 *Specific Criticism of Peters and Barbour* First I want to offer some specific criticisms of the views of Peters and Barbour relating t=0. Then I want to suggest a general criticism of the views of all three—Peacocke, Peters, and Barbour—by turning to the underlying source of the problem, namely the strict dichotomy between ontological and historical/empirical origination advocated by Gilkey. Finally, I will propose a constructive approach to the problem which, I believe, circumvents these problems as a whole and opens the door to a detailed theological appraisal of divine action in terms of the doctrine of creation seen in the light of cosmology.

2.2.1.1 *Critique of Ted Peters* Unlike Peacocke and Barbour, who seems anxious to purge *ex nihilo* of any empirical content about t=0 (though as we shall argue, Barbour shifts his position somewhat with time), Peters wants to relate, to the point of equating, *ex nihilo* with t=0. The gain would ostensibly be to achieve consonance between theology and science, and thus to secure a starting point by which further discussions can be pursued.

---

[29] Ibid., 129.

[30] Barbour makes extensive use of the philosophical category of contingency to relate cosmology to *creatio ex nihilo* in the Gifford Lectures. This is an important point; however, contingency tends to remain a merely philosophical category if its meaning is restricted to the existence *per se* of the universe. If contingency is extended to include the finite age of the universe, then contact can be made with specific cosmologies such as Big Bang (which tends to *confirm ex nihilo*) or steady state (which tends to *disconfirm ex nihilo*). Thus finitude and not contingency plays the pivotal role in subjecting *ex nihilo* to the empirical test offered by cosmology.

My concern with Peters' proposal is drawn from three distinct areas: 1) Science: If (and it is inevitably a "when" as we shall see with Hawking's research) science changes its mind about t=0, the kind of consonance sought by Peters will be lost. 2) Philosophy: It is far from clear in principle whether the "event" designated by t=0 is open to scientific investigation. Certainly if there truly were an event which was both uncaused and yet the cause of future events, its scientific status would be highly ambiguous.[31] Indeed, as Bill Stoeger argues, although a point of *absolute* beginning might have occurred, cosmology in principle would *not* be able to discover it.[32] 3) Theology: The identification of *ex nihilo* with t=0 would seem too narrow. Peters clearly wants to avoid removing all empirical content from contingency. Still, there are many forms of contingency besides the finite age of the universe.

Moreover, in seeking to avoid an overly abstract rendition of contingency, Peters, as we have seen above, writes:

... just what does it mean for the world to owe its existence to God? One sensible answer is this: had God not acted to bring the spacetime world into existence, there would be only nothing.

Unfortunately, this comment actually fails to contain any empirical content, and is no less abstract than the strict dependence argument Peters wants to dismiss. Indeed, it could be taken as a definition of the claim of ontological origination!

2.2.1.2 *Critique of Ian Barbour* Barbour's views on t=0 seem to move between those of Peacocke and Peters. In the 1960s he was indifferent to the theological implications of cosmology. By the time of the Gifford Lectures in 1989, Barbour seemed more willing to concede that cosmology could play a modest role in the debate over *creatio ex nihilo*. His views there, however, still seem clearly stamped by the position Gilkey set out in *Maker of Heaven and Earth* more than three decades ago.

To see this, let us return to what Barbour says about *ex nihilo* in the passages we lifted up previously. In both *Issues in Science and Religion* and the Gifford Lectures, three critical points emerge: i) Barbour accepts the sharp ontological/historical distinction proposed by Gilkey; ii) he argues that the ontological interpretation carries the central meaning of *ex nihilo*; and iii) hence t=0, being empirical, plays no role in the *ex nihilo* interpretation. This position is reflected relatively succinctly in the following passage on the contingency of the universe in the Gifford Lectures:

---

[31] If we mean by events open to scientific investigation those which are both the effects of previous causes (which presumably it would not be) and which in turn are the cause of future effects (which presumably it would be), then an uncaused event would be a contradiction in terms. Note that one need not limit the form of causality assumed here to Laplacian determinism; statistical determinism would be equally acceptable. Moreover, quite different examples of statistical determinism are provided by statistical mechanics and quantum mechanics.

[32] For an excellent treatment of this and other philosophical aspects of cosmology see W. R. Stoeger, "Contemporary Cosmology and Its Implications for the Science-Religion Dialogue" in *Physics, Philosophy and Theology*, 219-247.

> ... I agree with the neo-orthodox authors who say that it is the sheer *existence* of the universe that is the datum of theology, and that the details of scientific cosmology are irrelevant here. The message of creation *ex nihilo* applies to the whole of the cosmos at every moment, regardless of questions about its beginning or its detailed structure and history. It is an ontological and not a historical assertion.[33]

It is only in the context of the Gifford Lectures, and not in *Issues in Science and Religion*, that Barbour turns to the possibility of a historical interpretation of *ex nihilo* and its relation to t=0. Should an initial singularity be supported scientifically, Barbour believes this *would* provide an "impressive example of dependence on God."[34] Still he reminds us that even an infinitely old universe would be contingent.

This inclusion of even the marginal significance of t=0 in his theology of creation represents something of a shift from Barbour's earlier position. Still, Barbour draws the distinction between ontological and historical origination very sharply and places essentially all the theological weight on the ontological interpretation, with only nodding attention to the possibility of the historical interpretation being relevant.

2.2.2 General Criticism: Gilkey's Dichotomy My main contention with Peacocke, Peters, and Barbour is that they seem to have accepted the sharp dichotomy Gilkey makes between ontological and historical/empirical origination. They differ only in how they decide which one is theologically important. Peacocke and, to a slightly lesser extent, Barbour, accept Gilkey's distinction and, with Gilkey, abandon the historical/empirical aspect. Peters accepts Gilkey's distinction but, *contra* Gilkey, reduces the ontological meaning of creation to its historical/empirical context since, without this, ontological dependence would supposedly be a mere abstraction.

Now I agree that the critical issue at stake in the *ex nihilo* tradition is ontological origination: the radical dependence of finite being on God as the absolute source of being. The question is whether ontological origination should be seen as *devoid* of historical/empirical meaning, as Gilkey argued. To understand Gilkey's reasons and then to evaluate them, we must go back to *Maker of Heaven and Earth*.

There, as I see it, Gilkey was faced with a fundamental problem which occurs throughout Neo-Orthodoxy and the ensuing literature in biblical theology: how to relate religious and empirical/scientific language. This problem may have been forced on him by the issue of t=0, but it is really ubiquitous to theology, as he so poignantly expressed in his now-famous article "Cosmology, Ontology, and the Travail of Biblical Language"[35] written just three years after *Maker of Heaven and Earth*. There Gilkey argues that Neo-Orthodoxy and biblical

---

[33] Barbour, *Religion in an Age of Science*, 144.
[34] Ibid., 129.
[35] Gilkey, "Cosmology, Ontology, and the Travail of Biblical Language," *The Journal of Religion* 41 (1961): 194-205.

theology consists in an unsuccessful attempt to combine biblical language and scientific cosmology. In the process it empties biblical language of its initial content, leaving it at best analogous, at worst equivocal. I would add to this that in doing so theological epistemology is isolated from secular knowledge, robbing it of explanatory, let alone predictive, power and confining it to the sphere of religious language.

Although he rejects one possible solution to the problem, that is, the attempt to incorporate facts about a beginning of the universe into the doctrine of creation, Gilkey in no way minimizes the theological importance of the question raised by t=0 nor does he dismiss t=0 as a minor topic in the pursuit of what is of "real" theological interest to the doctrine of creation, as so many theologians still do. Instead, even in *Maker of Heaven and Earth*, Gilkey sees the issue of t=0 as critically important, since it forces us to confront a *foundational* problem which governs and characterizes *every* major doctrine in Christian theology. "This same dilemma dogs the heels of every major theological idea. Every doctrine of Christian faith expresses the paradoxical relation between the transcendent God and the world of facts."[36] In the writings that followed this early work, we can watch Gilkey continue to wrestle with this problem of biblical language and secular thought.

Although I will be quite critical of Gilkey's resolution of the problem, we owe him a great deal for his lucid insistence on its importance as such. If Gilkey is correct, the epistemological problems surrounding t=0 are well worth *our* pursuing because they are *inherent* to the theological agenda as such, and not a minor issue best forgotten. Indeed, they raise the fundamental and unavoidable problem of the rationality of theology in an age of scientific epistemology.

The strategy Gilkey adopted to resolve the problem was to view religious language about historical/empirical origins as myth, thus insulating it from scientific discourse.[37] In specific, he argued that theological language about matters which might seem to be factual, like t=0, and thus open to scientific inquiry, is actually discourse at the level of myth having nothing in common with

---

[36] It is worth quoting Gilkey at length here, for this point is crucial: "If religious ideas stay wholly within the world of fact and experience, they lose that saving content which is the transcendent power and love of God. . . . If, on the other hand, they transcend the world entirely, and if they describe God in the purely negative and impersonal terms of speculative philosophy, that relatedness of God to the world of which the Gospel speaks is lost. Theological truth must maintain a dialectic or tension between God's transcendent eternity and the finite world of change and time, if it is to express the Christian Gospel of God's salvation. It must continually relate to the facts of our experience what transcends fact; for what theology seeks to formulate is the activity of God who transcends our experience. . . . One of the basic problems of theology, then, is to express the relation of eternity to time as Christianity understands it, without on the one hand competing with our scientific knowledge of the origins of the natural universe of space and time, and without on the other losing all positive relation to the world of actual experience." Gilkey, *Maker of Heaven and Earth*, 315-316.

[37] Ibid., 316 ff.

science. Once we recognize this we can avoid any potential conflict between theology and science, for when properly conceived they are seen to be completely separate worlds of discourse, and thus the paradox is resolved.

Does this strategy work? In my opinion it does not, since it marginalizes theology by insulating it from the cognitive claims and discoveries of secular inquiry. Moreover, I believe we now have an alternative. Perhaps the best way to deploy the alternative will be to examine the way in which Gilkey initially approaches the problem of relating theological doctrine and empirical facts.

Gilkey bases his approach on a premise which is located at the outset of his analysis and which governs his entire line of reasoning, including his resolution of the problem in terms of myth. The crucial premise consists in what I consider to be an unnecessarily rigid distinction between ontological and historical/empirical origination. Gilkey might appear to be appropriating the traditional distinction between ontological origination and historical/empirical origination as found in Aquinas. However, in a move which is crucial for all that follows, he then *recasts* the distinction into a *sharp dichotomy*. He then claims that, when properly conceived, theology is *only* concerned with one side of the dichotomy, namely ontological origination. From this it follows that language about historical/empirical origination has no valid cognitive role in theology. Where it does surface, it actually functions as myth. Since only the ontological claim is proper to theology, we should disregard the historical/empirical language and focus on the ontological meaning of creation. The result is that we avoid a conflict with science—and in the process, *any* cognitive encounter with science.

2.2.3 ***Overcoming Gilkey's Dichotomy: A Lakatosian Theological Research Program*** I believe that Gilkey's approach ought to be set aside. In my opinion, it is based on a unnecessary premise, namely that ontological origination and historical/empirical origination *must* be seen as strictly dichotomous. Instead I propose we embed historical/empirical language within the broader context of ontological origination, thus giving a factual basis to which the core philosophical/theological generalization about ontological origination can be related without literalization *or* equivocation.

To do so I first return to the distinction as it was traditionally given. Now it is my understanding that the distinction is more complex and flexible than Gilkey allows. I take Aquinas' strategy, for example, to have been something like the following. On the one hand, if science supports an eternally old universe, as Aristotle argued, one can still maintain that the universe is ontologically dependent in the philosophical sense by the mere fact of its existence *per se*. Moreover, there might be other factors which would flesh out the ontological dependence of an Aristotelian universe in empirical/historical terms, involving motion, causality, agency, purpose, and so on. On the other hand, if science supports a universe with a finite age, as the Big Bang suggests, this can count as empirical evidence in support of ontological origination, although other evidence might count against it, too. Ontological dependence is thus the *crucial*, but not the *exhaustive*, meaning of creation.[38]

Next I propose that we adopt a more complex strategy for relating ontological and historical origination. To do so I will draw on recent work in theological methodology, first anticipated by Barbour[39] and very recently developed in detail by both Nancey Murphy[40] and Philip Clayton.[41] These scholars appropriate current research in philosophy of science for the purposes of theological method, focusing specifically on the writings of Imre Lakatos.[42] Following their pioneering work, I propose we structure the doctrine of creation *ex nihilo* and its relation to data in cosmology in terms of a Lakatosian research program. This will include a central, or "core," hypothesis surrounded by a protective belt of auxiliary hypotheses which can be tested against relevant data, and a set of criteria agreed upon in advance by which to decide rationally between competing research programs. In this way evidence for empirical origination from contemporary science, such as the Big Bang offers in terms of t=0, could be related to a core theological hypothesis in such as way as to allow it to confirm ontological origination without the evidence being somehow directly identified with core hypothesis.

However, lest I appear to be setting up a "no lose" scenario, I hasten to add that this method is meant to cut both ways: evidence against empirical origination, such as offered in the 1950s and 1960s by Hoyle's steady state model or which might arise through new cosmological models of a universe with an infinite age, would count *against ex nihilo*.[43] Still ontological origination could never be absolutely disproven by *any* empirical evidence, since the sheer existence of the universe (let alone the existence of such "empirical evidence") is the foundational basis for the central philosophical argument for *ex nihilo*.

---

[38] One should also note that, even in the context of strictly philosophical contingency it is, after all, the *existence* of the universe which provides the datum on which the philosophical claim is warranted. Thus the claim is never *entirely* emptied of empirical context.

[39] Barbour, *Myths, Models and Paradigms: A Comparative Study in Science and Religion* (San Francisco: Harper & Row, 1976). See especially chs. 6 and 7.

[40] Nancey Murphy, *Theology in the Age of Scientific Reasoning* (Ithaca: Cornell University Press, 1990). See also her article in this volume.

[41] Philip Clayton, *Explanation from Physics to Theology: An Essay in Rationality and Religion* (New Haven: Yale University Press, 1989).

[42] See in particular his "Falsification and the Methodology of Scientific Research Programmes," in *The Methodology of Scientific Research Programmes: Philosophical Papers*, vol. 1, ed. John Worrall and Gregory Currie (Cambridge: Cambridge University Press, 1978), 8-101.

[43] Even the Big Bang model can be taken as offering disconfirming evidence if we can argue that its past can be characterized as infinite. As it turns out, using temperature to define the age of the universe makes even the Big Bang model infinitely old. Finally, quantum gravity can lead to proposals in quantum cosmology which either give the universe an infinite age or which, as in the Hartle/Hawking model, redefine the meaning of a finite age to be one having no beginning, as we shall see below.

I suggest we place at the core of the theological research program the hypothesis, *"creatio ex nihilo* means ontological origination." Next we deploy a series of auxiliary hypotheses which surround the core and relate it to relevant types of data including those from cosmology. I propose as one such auxiliary hypothesis the claim, "ontological origination entails finitude." I chose "finitude" because it seems a particularly fruitful concept for relating ontological origination and t=0. By finitude I mean the traditional Aristotelian concept of something with determinate status, measure or boundary, as opposed to the *apeiron*, that which is unbounded, unlimited, endlessly extensible.[44] We use this concept in an auxiliary hypothesis closest to the theological core theory. Thus on the one hand, finitude can function as an abstract philosophical concept located near the theological core where it will resist being reduced to a univocal, empirically testable meaning. Keeping a substantive distance between the theological core and the data from science will, in turn, at least partially address the concerns voiced by Gilkey, Peacocke, and Barbour.

Yet the concept of finitude need not be restricted to an abstract, philosophical context. Instead it can be developed in a variety of ways to make increasing contact with physics and cosmology, and thus with an empirical meaning of origination. To do so we construct a second auxiliary hypothesis, "finitude includes temporal finitude," that is, that which is bounded in time. From this we construct a third auxiliary hypothesis, "temporal finitude includes past temporal finitude," that is, the property of finite age. Now we are in a position to connect this series of hypotheses to Big Bang cosmology in which the data of astrophysics, the theory of general relativity, and other factors, assumptions, and simplifications, lead to the theoretical conclusion that the universe has a finite age and an initial singularity, t=0. In this way the concept of finitude can serve as a bridge between the core theory, ontological origination, and the data for theology, here seen in terms of the origin of the universe at t=0, thereby satisfying at least part of Peters' intent of making a connection (albeit *not* a direct one) between ontological and empirical origination.

Through this process I claim we can argue that the empirical origination described by t=0 in Big Bang cosmology tends to confirm what is entailed by theological core theory, *"creatio ex nihilo* means ontological origination." Moreover, to the extent that Big Bang cosmology continues to gain scientific support, the confirmation of *ex nihilo* is strengthened. Recent evidence from the COBE satellite showing the existence of minute structure in the microwave background constitute such additional support for Big Bang cosmology, and thus indirectly it constitutes additional support for *creatio ex nihilo*. I want to emphasize again, though, that this method allows for disconfirmation as well as confirmation. Ideally, one would compare several theological and/or non-

---

[44] I do not want to make this into a *definition* of finitude, since I want to argue that the distinction between the finite and the infinite need not include the concept of boundary. In recent mathematics, and especially since the work of Gregor Cantor, we have extended the notion of finitude to include unbounded finitude. We shall see the importance of this latter concept as it surfaces in the work of Hartle/Hawking below.

theological research programs, each of which attempt to relate *creatio ex nihilo* to cosmology in its own way, and assess which program is most progressive by the way it predicts novel facts and avoids *ad hoc* moves.[45] In this way, we can move theology out of its closed hermeneutical circle and allow it to make cognitive contact with empirical knowledge.[46]

Let us pause for a moment, and assess the situation. We have been focusing on the relationship between *creatio ex nihilo* and inflationary Big Bang cosmology. We know, however, that severe conceptual problems revolve around the status of t=0, problems which involve both technical issues in mathematical physics and in the philosophy of science. Moreover, Big Bang cosmology depends upon special relativity and *its* assumptions are "classical" (i.e., pre-quantum mechanical). Thus much attention is now being devoted to the goal of obtaining a quantum treatment of gravity. One of the goals being pursued via quantum gravity is to obtain a quantum cosmology which will be free of the initial singularity, t=0. As such a goal is approached, what will happen to the theological research program I have sketched above which relates? It is to this question that we now turn.

3 Quantum Cosmology in Theological Perspective

3.1 *Two Possible Routes to Quantum Cosmology: Quantum Fluctuations Versus the Hartle/Hawking Model*

There are a variety of approaches to quantum gravity and the cosmological models which emerge from it, that is, the models of quantum cosmology. In this paper I will focus on the Hartle/Hawking model, first by comparing it with an alternative (quantum fluctuations), and then through a detailed look at the model itself. Much of the discussion will draw from a recent paper by Chris Isham, "Creation as a Quantum Process,"[47] and from the Stoeger/Ellis and Isham papers in this volume.[48] At the outset I want to underscore the *speculative* nature of such

---

[45] For an excellent example of such a comparison between theological and non-theological research programs and cosmology, see the article by Murphy in this volume.

[46] In short, I believe it is better to be wrong than to be meaningless. In order to be wrong, a theory must be capable of being falsified. As long as it could be falsified, then if it isn't falsified but instead it received partial confirmation (since it can never be verified), it must bear some resemblance to the truth. This clearly should be a goal of theology, to include as part of the domain in which it tells the truth the world God so loved (John 3:16).

[47] Isham, "Creation as a Quantum Process," in *Physics, Philosophy and Theology*.

[48] In this paper I focus attention on Isham's analysis of the Hartle/Hawking model, drawing primarily from his article in *Physics, Philosophy and Theology*. In the future, I hope to expand my arguments to include his analysis of Vilenkin's model and other proposals in quantum cosmology.

proposals, a point that Isham makes forcefully in his paper. What I hope to show is their value for theology in such a way that it is relatively independent of their long-term scientific viability.

3.1.1 *Quantum Fluctuations* One attempt at quantum cosmology begins by supposing that what is actually "created" is matter alone, and not spacetime. More precisely, one assumes: i) the existence of a background spacetime (or some other manifold); ii) filled with the appropriate quantum fields in their ground (lowest energy) state; and iii) governed by appropriate laws of nature. Now the *material* universe arises as a spontaneous quantum fluctuation of the fields in the background spacetime.[49]

3.1.2 *Hartle/Hawking Model* The Hartle/Hawking model does not assume a background spacetime out of which the universe arises. It does, however, assume a set of three-dimensional spaces out of which spacetime can be constructed. It also presupposes the existence of the appropriate laws of physics, including quantum physics (in particular the Feynmann "path integral" approach) and general relativity, and it employs a mathematical trick to simplify calculations, namely the introduction of complex numbers.[50] We will look at the Hartle/Hawking model in a bit more detail below. For now the important points to note are that: i) It succeeds in describing a universe with a finite past but *no* initial singularity;[51] this accomplishes what previous theories had been unable to do, and changes the scientific mode of discussion about the origination of the universe. ii) Time arises phenomenologically in this model; it is not a given, external parameter which describes the evolution of the universe as in the FLRW model.

---

[49] To understand how this is possible, one needs to bear in mind two points. 1) Gravitational binding energy is negative, since gravity is an attractive force. Rest mass and kinetic energy are positive. It is likely that in our universe the amount of gravitational binding energy just equals the total rest mass and kinetic energy of the universe, so that the universe as a whole has no net energy. 2) Quantum mechanics predicts that the lifetime of a quantum system is inversely proportional to the uncertainty in its energy. The result of these two observations is that a quantum fluctuation of the background spacetime could produce a universe with a lifetime of billions of years, as our has, without violating conservation of energy.

[50] The use of complex numbers is commonplace in mathematical physics. Complex numbers consist of real and imaginary parts, the latter involving the square root of a negative number. They have nothing whatsoever to do with the meaning of "imaginary" in artistic, literary, or personal contexts.

[51] To be accurate, we cannot simply extrapolate backwards in ordinary time to a *unique*, finite past since time, in quantum gravity, is a highly ambiguous concept. Our ordinary concept of time assumes that the universe is a single four-dimensional spacetime manifold, and that time is unambiguously defined in terms of one of these dimensions. In the case of quantum gravity, however, one is faced with a set of such four-dimensional spacetime manifolds, all of which contribute to what we consider to be ordinary time.

3.2 *Exploring the Hartle/Hawking Model Further: The Roles and Status of Time and Space*

In the Hartle/Hawking model, time is treated as an "internal variable." A similar term is often used about general relativity, but in the quantum context the meaning is significantly different.

3.2.1 *General Relativity* For each solution to the field equations of general relativity, time and space form a *single* four-dimensional curved spacetime manifold. The spacetime manifold is *static*: it does not change in 'time' since it includes time as one of the four dimensions of spacetime.

The spacetime manifold can be sliced in a variety of ways. Each set of (spacelike) slices gives a specific "history" for the universe.[52] Each history depicts the universe as a sequence of slices or three-geometries[53] sewn together smoothly by an *external* parameter, time. Now each slice can be characterized in several ways: by the value of external time (i.e., through its relation to other slices) or by a property internal to the slice itself, such as the temperature of the matter field or the radius of curvature of the slice. When we use such properties as temperature or radius of curvature to serve as a temporal label, we call this label *internal* time.

There are an infinity of ways to slice spacetime, and hence an infinity of histories all explicitly different, yet all are consistent with a single spacetime. Hence one can say that general relativity provides both a *dynamical* (or *temporal*) view of the universe as a three-space evolving in time, of which there are an infinity of equivalent but distinguishable versions corresponding to different slicings, and a *static* (or *timeless*) view of the universe as a unique four-geometry (spacetime) which can be decomposed to yield an infinite set of different but complete dynamical histories. So time in general relativity can be an *external* parameter which sews together an infinite set of three-geometries or an *internal* parameter which functions as a property of a three-space, like temperature or curvature.[54]

3.2.2 *Hartle/Hawking Model* Returning to the Hartle/Hawking model, we find a further diminution in the role and status of time in relation to space. Here we start by imagining all possible curved three-spaces, c, with all possible ways in which matter can be distributed on each of them; together these form the "configuration space" for the theory.[55] Next we define a quantum mechanical state function Ψ (c,f) for each curved three-space, c, and for its associated matter fields, f. Ψ (c,f)

---

[52] Isham, "Creation as a Quantum Process," fig. 4, 390.

[53] For this paper the terms "three-space" and "three-geometry" are essentially interchangeable, and both refer to a three-dimensional space with an arbitrary curvature. Ordinary Euclidean three-space is an example of a three-space with zero curvature (i.e., a flat three-space).

[54] Isham, "Creation as a Quantum Process," 391.

[55] Ibid., 396.

is a measure of the probability of finding the particular curvature and matter fields, c and f.[56] Note that time has not entered in explicitly in the definitions so far. Instead it is treated as an *internal* property of c and f.

Now comes a crucial difference between quantum and classical gravity regarding time. In some cases, the three-spaces may be sewn together to form a continuous, four-dimensional spacetime. In these cases, the path in state space would be equivalent[57] to a single four-dimensional spacetime in general relativity. However in other cases, the non-zero three-spaces will *not* fit into, or be derivable from, a single four-dimensional spacetime.

One can think of the situation as follows: a single four-dimensional spacetime, as understood in classical physics, would be represented by a single, sharply delineated path between two different three-spaces $c(1)$ and $c(2)$.[58] In the quantum context, however, the path is "fuzzy" and diffuse. In effect, the classical picture of a well-defined spacetime arises as the limiting case of the quantum perspective, and the sharp trajectory of the classical spacetime is just that portion of the fuzzy pattern in quantum state space where the state function Ψ is maximal. Hence, as Isham concludes, "the concept 'spacetime' only has an unambiguous meaning within the framework of non-quantum physics, whereas the idea of three-dimensional 'space' can be applied to both the quantum and the classical theories."[59]

In other words, in the Hartle/Hawking model *time is less fundamental than space*. Time only has an unambiguous meaning for a *given* three-space, where it is best understood as internal time. It need not have an unambiguous meaning as the external parameter characterizing the evolution of a set of three-spaces, because there may be no single four-dimensional spacetime for which all these three-spaces are sections. Without such a four-dimensional spacetime, there is no context by which to view a series of three-spaces (each characterized by

---

[56] This interpretation of $\Psi(c,f)$ is only one of several possible ways of understanding the wave function. This fact is deeply connected with the problem of time in quantum gravity. I am indebted to Isham for this and other important comments on my treatment of quantum gravity. I refer the reader to the excellent but highly technical paper for further research: Isham, "Canonical Quantum Gravity and the Problem of Time," (Lectures presented at the NATO Advanced Study Institute "Recent Problems in Mathematical Physics," Salamanca, June 15-27, 1992).

[57] This must be understood modulo the very real problem of whether a realist interpretation of quantum mechanics in general can be sustained. To the extent that a realist interpretation is attributed to quantum physics, then the 'path' linking three-spaces would be interpretable as a four-dimensional spacetime. A somewhat related issue is the measurement or observer problem, i.e., to the problem of what it means for the universe as a quantum system to be 'observed' and thereby to be in a single state. How can an 'observer' be 'outside' the universe?

[58] See Isham, "Creation as a Quantum Process," fig. 4, 390. Strictly speaking, a single four-dimensional spacetime is represented by a whole collection of paths connecting $c(1)$ to $c(2)$, each path corresponding to a different foliation of the manifold whose boundaries are $c(1)$ and $c(2)$ (and hence to a different definition of time).

[59] Ibid., 397.

some value of internal time) as linked together and thus as "evolving." *In the Hartle/Hawking model, the static view surpasses the dynamic view.*

We also have the problem of the initial singularity, t=0, in the standard cosmology. To tackle this in the context of general relativity, Hartle/Hawking first represent time by an imaginary number. This changes the formalism in such a way that time is treated on strictly the same footing as space. One can solve Einstein's field equations for matter distributed in such a four-space and obtain solutions which are *not* singular![60]

Next Hartle/Hawking look for a *quantum* treatment using imaginary time and a state function for the universe. Their proposal is governed by the following stipulation: The spacetime must have only *one* boundary, not two as in standard Big Bang cosmology. In particular, it *cannot* have a singular initial boundary.[61] In the Hartle/Hawking model the universe has no "beginning" at the initial singular boundary, since there is *no* initial boundary at all! It simply has the "present" boundary (its current "size," to speak loosely) and a past history.[62] It is this feature of their model which led Hawking to refer to the boundary condition as the "no boundary" condition, and to the claim that, rather than being created, the universe "would just BE."[63]

The lack of an initial singularity, the 'fuzzy' path problem, the mathematical treatment of time as a complex variable, and the general difficulty of giving a realist interpretation of quantum physics all make a uniform global interpretation of time in the Hartle/Hawking model very difficult.

3.3 *Summary: Four Views of Time*

3.3.1 *General Relativity*
In general relativity one finds two interchangeable views of time: external and internal.

3.3.1.1 *External Time in General Relativity*
labels the sequence of three-spaces that form a given four-dimensional spacetime. The sense of their being sewn together as a process in time, three-space by three-space (like sewing beads on a string), offers a dynamic description of an evolving universe. The sense of their having been sewn together, that they now exist as a whole spacetime (like a necklace), leads to a static description.

---

[60] Ibid., fig. 6, 398.

[61] This is represented by the vertex of the cone in Ibid., fig. 5, 391.

[62] Again one should keep in mind that, strictly speaking, there are many space-times involved in the Hartle/Hawking model, not just one. Their approach requires integrating over all those four geometries which are consistent with the specified three-manifold.

[63] "The quantum theory of gravity has opened up a new possibility, in which there would be no boundary to space-time and so there would be no need to specify the behavior at the boundary. One could say: 'The boundary condition of the universe is that is has no boundary'. The universe would be completely self-contained and not affected by anything outside itself. It would neither be created nor destroyed. It would just BE." Hawking, *A Brief History of Time: From the Big Bang to Black Holes* (New York: Bantam Books, 1988), 136.

3.3.1.2 *Internal Time in General Relativity* is a function of a given three-space. It is transformable into some other characteristic of the three-space (such as its radius or the temperature of its matter field). It conveys a static view of the givenness of a specific geometry, losing the ordinary sense of time as passage.

However, in general relativity one can move freely between these two perspectives, since the only three-spaces admitted are those which can be sewn together (by the external time parameter) to form a unique four-dimensional spacetime. Conversely every four-dimensional spacetime can be decomposed in a variety of ways to form different sets of three-spaces. Hence dynamic (temporal) and static (spatial) views are equally fundamental.

3.3.2 *Quantum Gravity* In the *quantum treatment of gravity*, one expects time to have a less fundamental status than space in ways which simply do not occur in general relativity.[64] Internal time arises in quantum gravity, but in general external time does not. Moreover, the relation between quantum gravity and general relativity leads to two interpretations of the *status* of time in quantum cosmology: as a uniform characteristic of the entire universe viewed as a single domain or as a differing characteristic of the universe seen in terms of multiple domains (see 3.3.2.2 below).

3.3.2.1 *Internal Quantum Time* In quantum gravity, as in general relativity, we start with three-spaces which have a characteristic internal time (or, equivalently, some other parameter such as size or temperature). The additional problems that surface in the quantum context, however, are clearly in evidence when we turn to quantum cosmology. Here generally speaking, one is faced with numerous three-spaces which cannot be sewn together to form an overall four-dimensional spacetime. Thus the individual internal times of these three-spaces cannot be given an external time interpretation. In this sense internal time in quantum gravity loses even more of its temporal meaning than in general relativity.

However, in some specific quantum cosmological proposals, such as the Hartle/Hawking model, one can once again obtain an overall four-dimensional spacetime into which many of the relevant three-spaces can be sewn together.[65] In this sense the internal times of many of these three-spaces *can* be given a temporal interpretation via the context of the overall spacetime in ways similar to general relativity. Still, even here, many three-spaces must be included which cannot be joined together smoothly (i.e., the path of three-spaces is "fuzzy"). Thus some domains of the universe are more classical in appearance, others more quantum mechanical, even though all are generated from a strictly

---

[64] Note again that these remarks about quantum gravity and time reflect only a part of the overall problem of interpreting quantum gravity. Actually, the concept of time in quantum theory compared with the concept of time in general relativity is far more complicated than can be treated here. For a recent, technical treatment see Isham, "Canonical Quantum Gravity and the Problem of Time."

[65] Technically one must note that there are multiple four-geometries which in turn incorporate the sets of relevant three-geometries in the Hartle/Hawking *ansatz*.

quantum mechanical treatment of gravity. How should we view the overall temporality of such quantum cosmologies? We seem to be lead to two different interpretations of time in quantum cosmology, characterized by single or multiple domains.

3.3.2.2 *Quantum Time: Single or Multiple Domains?* There are at least two ways to view the relation between quantum gravity and general relativity, and each of them in turn has implications for the status of time in cosmology.

A) Single domain. Although all we have so far are tentative proposals for quantum gravity, in its eventual form as a full-blown theory it should be viewed as a *fundamental* theory replacing general relativity and thus describing the universe as a whole. This means that quantum gravity actually describes the entire universe, even the domain where its effects are negligible and where general relativity is a good approximation (i.e., the classical domain from the early universe to the present and into the far future).

Thus, even if the "timeless" character of quantum gravity is hidden to our macroscopic eyes, it must be taken as describing the way things really are even in our "temporal" region of the universe. Time as we know it is merely phenomenological; the entire universe is really only one timeless domain, all of it described by the three-spaces of quantum gravity. Hence the lack of a global external time parameter (i.e., the problem of not being able to sew together all relevant three-geometries into a single spacetime) means that the internal time of quantum geometry gives time even less status that does the internal time of general relativity.

B) Multiple domains. In a more *phenomenological* approach, quantum gravity is taken to be limited to the very early universe, describing it in quantum terms (as, for example, in the Hartle/Hawking proposal), whereas general relativity describes the universe from a fraction of a second to the present and into the far future using classical (i.e., non-quantum mechanical) terms. Hence time is domain-dependent, a real feature of the general relativistic domain but not of the quantum gravity domain.[66] We can conveniently name these the Einstein and the Hawking domains, respectively. Presumably it would be in the Einstein domain that we would also find the production of those characteristics, such as asymmetry and varying ontological status, needed for the "arrow of time," that is, the flow from past to present into future, but this issue takes us beyond the scope of this paper.

Thus time is neither globally present nor globally absent in this interpretation. Rather, external time is a phenomenon characterizing the Einstein domain of spacetime in which we live, far from the early, Hawking domain. No

---

[66] Note, however, that there is a serious debate regarding the status of time even in the classical domain, as discussed below and by Isham/Polkinghorne in this volume: should spacetime be considered a static four-geometry (a world of being) or a dynamic evolving three-geometry (a world of becoming)? The point here is that in the quantum domain, as opposed to the classical domain, *time is not even on a par with space*, rendering the static view of being *even more compelling* than the dynamic view of becoming.

longer a global feature of the model, external time is the local[67] result of the changing geometry of spacetime, that is, the change from being essentially a four-dimensional fuzzy space into being a four-dimensional well-defined spacetime.

In effect, as one thinks back from the present towards the very early universe (Hawking domain), the imaginary components of the time coordinate begin to dominate and the three-spaces become increasingly disjunct, making it increasingly hard to give a physical interpretation of time. Reversing directions and moving outward from the early universe towards the present, time as we know it seems to arise as spacetime becomes more clearly defined. In this sense, time as we know it, including time's arrow, is a phenomenon only manifested by a portion of the topology—even though in scope this domain extends from the present back to the first microseconds of the universe!

Finally, in the multiple domains interpretation one must deal with the transition between the timeless Hawking domain and the temporal Einstein domain. The key point here is that a more fundamental concept than time will be needed to describe the relation between the Hawking domain where quantum gravity predominates and the Einstein domain we inhabit. We can call the domain which connects them the transition domain.[68] Clearly in the previous, single domain interpretation, there is no need to discuss a transition domain since the entire universe is described in principle by quantum gravity as a fundamental theory.

Finally it should be noted that the distinction between single and multiple domains does not affect the validity of the claim that the universe has a finite past in the Hartle/Hawking model, only its meaning. It suggests how difficult it is to give a consistent interpretation of the concept of a "finite past" when dealing with quantum gravity.

3.4 *Relating Theology and Cosmology: Constructive Proposals*

3.4.1 *Theology as Offering Criteria in the Choice Between Quantum Fluctuations and the Hartle/Hawking Model*
How ought one to chose between these two approaches: quantum fluctuations versus the Hartle/Hawking model? The criteria are, in fact, largely philosophical in nature.[69]

---

[67] "Local" is meant here in a topological, not a geometrical, sense. The local region of spacetime that includes an external time includes all events from the first microsecond through the present into the far future.

[68] I resist the temptation to talk about the transition domain as the domain where time "emerges," even though this might be a natural way to think about it. Emergence seems to carry with it temporal overtones, and this is precisely what we need to avoid here. The 'emergence of time' is *not* a *temporal* process, i.e., it is *not* a process in time. Neither is it a sharp boundary from one domain to another, but a distributed process more like a gradual shift in character.

[69] As we know, philosophical and aesthetic commitments often serve as criteria of theory choice in science.

3.4.1.1 *Is "Creation in Time" Reasonable?* Isham illustrates this fact quite nicely when he makes his case against quantum fluctuations in favor of the Hartle/Hawking model: ". . . these (quantum fluctuation) theories are prone to predict, not a single creation/seed-point, but rather an infinite number of them."[70] The problem stems from the fact that the background spacetime is homogeneous and infinite. "There is simply no way of distinguishing any particular instant of time" at which the material universe would spontaneously appear. Hence an infinity of universes should be created throughout the background spacetime. This obviously leads to predictions in gross contradiction with our astronomical observations,[71] and theories like this have received little sustained attention.

Isham then suggests that this problem in physics was "pre-empted" by Augustine in his "refutation of this demiurgic concept of God." In the famous response to the question about what God was doing before God made the universe, Augustine argued that the question itself involves a false assumption about time. God did not create the world in *time*, but rather God created time *along with the world.*[72] As Isham remarks, "it is singularly striking that, sixteen centuries later, theoretical physicists have considered precisely the same subterfuge as a means of avoiding the question of the 'when' or 'before' of creation."

A position similar to Isham was developed by Paul Davies in *God and the New Physics.*[73] In a stimulating chapter entitled "Did God create the universe?" Davies writes:

> . . . as we have seen, modern cosmology suggests that the appearance of the universe involved the appearance of time itself. . . . [Thus] it is clearly meaningless to talk about God creating the universe in the usual causal sense, if that act of creation involves the creation of time itself. . . .
>
> This point seems to have been well appreciated by St. Augustine . . . who ridiculed the idea of God waiting for an infinite time and then deciding at some propitious moment to create a universe. . . . This

---

[70] Isham, "Creation as a Quantum Process," 387.

[71] In the quantum fluctuation model, the infinity of universes would all coexist, in a sense, within the same spacetime continuum. This means that we would see galaxies moving in all directions in the sky, depending on the location in spacetime of the 'creation/seed-point' from which they arose. Clearly we see nothing of the sort in the visible universe. This way of speaking about an infinity of universes is to be sharply distinguished from the concept of an infinity of universes that appears in "many-universe" theories arising in inflationary quantum cosmology or from many-worlds interpretations of quantum physics, and so on. In the latter cases, the universes are each more or less distinct spacetimes, somewhat like pseudopods arising from some common source and no longer in communication.

[72] As Isham points out, Philo of Alexandria took a similar position to that of Augustine.

[73] Paul Davies, *God and the New Physics* (New York: Simon & Schuster, 1983).

[insight] is a remarkable anticipation of modern scientific cosmology.[74]
...

Returning to Isham's argument, if we take an Augustinian perspective we should think of the motion of matter as coincidental with the origin of time. Hence time should be a derivative of something ontologically prior, namely a derivative of the properties of matter such as temperature or field configuration, and so on. As Isham points out, the problem of the beginning of time and the universe can be avoided in principle if a physical scheme can be found to implement the Augustinian insight, such that as we approach the initial point of the universe time itself loses its fundamental status. An instance of such a scheme is realized in the Hartle/Hawking model of the universe.

Wim Drees[75] has written extensively on the topic of the Big Bang. In discussing Isham's work, Drees points out how the Hartle/Hawking model avoids the problem of the beginning of time and recalls Isham's reference to Augustine and Philo. But Drees goes further than Isham here by suggesting an important difference between the concerns of Augustine and Hawking.

> Augustine probably understood the beginning of time with the creation as an event outside the scope of natural knowledge. But Hawking holds it to be only the beginning of a co-ordinate without reference to any special event, and without a breakdown of physical describability.[76]

I think this is a valuable clarification of the difference between Augustine and Hawking. According to the former, the beginning was an event which could not be discovered by reason alone. As Drees suggests, the Hartle/Hawking model challenges this point in that in their model there simply is no such event. It is not *hidden* from view; it *doesn't exist*.[77]

3.4.1.2 Is "Nothing" Creative? I would like to add a second argument from a theological perspective against the quantum fluctuation model and in favor of the Hartle/Hawking model. It is related to, but not reducible to, the argument proposed by Hawking and Isham which uses Augustine's claim about creation in time.

First we must make a *philosophical* distinction between the kinds of non-being used in Greek metaphysics. According to Paul Tillich, there were two

---

[74] Ibid., 38.

[75] Drees, *Beyond the Big Bang: Quantum Cosmology and God* (LaSalle, IL: Open Court, 1990).

[76] Ibid., 55-56.

[77] Drees also focuses on *creatio ex nihilo* in the context of the Hartle/Hawking cosmology. He begins with the common distinction between two meanings of *ex nihilo*: historical origination and ontological dependency. Drees claims that the former leads to deism, whereas theism, with its sense of the dependence of all moments on God, is more compatible with cosmology. He points out that the removal of the t=0 event does not undercut theism, only deism. ". . . theology is not necessarily tied to an absolute beginning, an edge to time. . . . Sagan's argument 'no edge, hence no God' is not decisive." (Ibid., 70-71.) Thus, Drees defends what I previously called the Peacocke/Barbour position, a position I criticized in some detail.

conceptions of non-being in Plato: *ouk on* and *me on*. The latter signified an undifferentiated form of potential reality out of which structures could spontaneously emerge. The former signified the utter lack of anything whatsoever, potential or actual. Tillich claimed that the classical formulation of *ex nihilo* meant creation out of *no thing*, not creation out of a reified "nothing" of any sort. Thus Christianity rejected the concept of *me on* and used *ouk on* instead to signify that there was nothing (i.e., there wasn't anything) apart from God before God created, that without God nothing could exist per se.[78]

Now consider the concept of a "filled background spacetime" used in the quantum fluctuation model in terms of these two Greek categories of nonbeing. The background spacetime includes the quantum fields (albeit in their quiescent state, i.e., vacuum state), the spacetime itself, and the laws of nature. It seems to me that the background spacetime would be a *meonic*, rather than an *oukonic*, form of non-being *on all three* counts.

Tillich then makes the *theological* claim that it is *ouk on*, or the non-being incapable of spontaneously creating the world, which is employed in the doctrine of creation *ex nihilo*, since God should not depend on anything (matter or form) existing prior to the act of *absolute* creation. It follows that acceptance of *creatio ex nihilo* would lead us to chose *against* the quantum fluctuation model.

Of course one should also ask whether the *ex nihilo* argument would *unilaterally support* the Hartle/Hawking model. This too would depend on the way in which our concept of non-being enters into *this* model. The issue has received some discussion already, notably by Wim Drees, John Barrow, Michael Heller,[79] and George Ellis.[80] Closely related arguments have surfaced before in discussions which preceded the work of Hartle/Hawking by, among others, Paul

---

[78] "*Ouk on* is the 'nothing' which has no relation at all to being; *me on* is the 'nothing' which has a dialectical relation to being. The Platonic school identified *me on* with that which does not yet have being but which can become being if it is united with essences or ideas . . . Christianity has rejected the concept of *me-ontic* matter on the basis of the doctrine of *creatio ex nihilo*. Matter is not a second principle in addition to God. The *nihil* out of which God creates is *ouk on*, the undialectical negation of being." Paul Tillich, *Systematic Theology*, vol. 1(New York: Harper & Row, 1951/1967), 188.

[79] Drees, *Beyond the Big Bang*; J. D. Barrow, *The World within the World* (Oxford: Clarendon Press, 1988); and Michael Heller, "Big Bang on Ultimate Questions," in *Origin and Early History of the Universe: Proceedings of the 26th Liege International Astrophysics Colloquium*, July 1-4, 1986, CointeOugree (Belgique). See in particluar Drees, *Beyond the Big Bang*, 72.

[80] In private correspondence Ellis points out that the Hartle/Hawking proposal includes pre-existent Hilbert spaces, quantum operators, Hamiltonians, and so on, "whose existence is if anything more mysterious than that of the universe itself. . . . In fact the argument (about creatio ex nihilo) supports the idea that the old-fashioned big bang should be correct—and thus that quantum gravity, if indeed necessary, does not change the nature of the big bang from that predicted by classical theory."

Davies and myself.[81] In the Hartle/Hawking case, the arguments tend to revolve around the idea that, although a previously existing spacetime of some sort is *not* proposed, a *pre-existing set of laws of nature* are presupposed to exist. But such laws would be included in what the traditional doctrine considers as falling within God's creation and not something external to, and co-eternal with, God. Hence even the Hartle/Hawking model does not assume the creation of the universe out of *oukonic* non-being.

Thus although the theological commitment to *ouk on* might weigh more heavily against the quantum fluctuation model than against the Hartle/Hawking model, it certainly does not support the latter unambiguously. Indeed it drives us to look for additional ways in which even the laws of nature may be created.[82]

3.4.2 Theological Appropriation of the "No-Boundary" Condition

The Hartle/Hawking proposal makes evident that a self-consistent scientific conception of the universe can be constructed in which the universe has a finite age but no "beginning," that is, the finite past is unbounded.[83] This may be the most important aspect of the Hartle/Hawking proposal for theologians to consider. Before turning to this discussion, however, we should consider what Hawking himself makes of the problem, since without doubt he speaks for much of secular society in finding reasons for rejecting the claim that God acts in the world.

3.4.2.1 *Hawking's Theological Conclusions and my Counterarguments*

3.4.2.1.1 *Hawking's Conclusions*

Hawking draws two conclusions from his cosmology. 1) Since there is no beginning to the universe, there is nothing left for God to do except choose the laws of nature. 2) By understanding the underlying laws of nature as chosen by God we can, in effect, read the mind of God. We will not analyze the second conclusion here, but turn instead to the first. How does Hawking warrant this conclusion?

---

[81] Davies, *God and the New Physics*, chap. 16, "Is the universe a 'free lunch'?" Here Davies asks: "Does such a (quantum model of the universe) have any need for God? . . . Physics can perhaps explain the content, origin and organization of the physical universe, but not the laws (or superlaw) of physics itself. Traditionally, God is credited with having invented the laws of nature and created things . . . on which those laws operate." 216-217; and Russell, "Cosmology, Creation, and Contingency," in *Cosmos as Creation*, especially 198-201.

[82] Would something still be left? Perhaps the laws of logic? It would be fascinating to explore *these* possibilities. We should keep in mind, however, the important argument raised by Stoeger in this volume against viewing the laws of nature as having an ontological standing of their own. Perhaps the laws merely describe nature, and do not act to prescribe the behavior of natural phenomena.

[83] I am using the term "unbounded" here as meaning "having no boundary." Mathematicians often use the term "unbounded" in a different sense. They might instead say that the Hartle/Hawking model involves four-manifolds that are past-bounded. Similar remarks apply to the use of the word 'open'.

Hawking's argument depends on a prior argument: that if God acts at all, God can *only* act at the beginning of the universe and not during the course of natural processes. This in turn rests on two assumptions: i) that the universe runs according to scientific laws which exclude God's action except by divine intervention; and ii) that God doesn't intervene in nature. Hawking makes these assumptions clear in several passages:

> With the success of scientific theories in describing events, most people have come to believe that God allows the universe to evolve according to a set of laws and does not intervene in the universe to break these laws.[84]

> These laws may have originally been decreed by God, but it appears that he has since left the universe to evolve according to them and does not now intervene in it.[85]

Let us inspect both i) and ii) in order to understand more fully the reasons for Hawking's position.

(i) Throughout the modern period, assumption i) was supported in large part by the *deterministic* nature of the reigning scientific paradigm: Newtonian mechanics. Deterministic laws, such as those of Newtonian mechanics, allow one, in principle, to predict the future exactly if all physical forces are known, and if the exact present position and momentum of all particles is given. Though in practice one could never hope to know the precise initial conditions or total operative forces, in principle one thought of the future as actually determined by these conditions and forces and not open to uncertainty—in the process, leaving the meaning of human free will very much in limbo, it should be noted. Classical physics thus led to a deterministic, clock-work interpretation of nature which dominated the seventeenth through nineteenth centuries.

(ii) Hawking also assumes that God does not intervene in the orderly workings of natural processes. This too is a modern assumption, shared by many theologians and virtually all scientists. The argument usually given since Hume is that it would be irrational, unaesthetic, or suggestive of weakness or shortsightedness for God to tinker with the very laws of nature set up by God in the first place and that the concept of miracle is in the end incoherent.

Thus, given a worldview shaped by Laplacian determinism and a rejection of the possibility of divine intervention, Hawking can draw his basic conclusion: God's only choices are i) to select the initial conditions at t=0 which govern the universe as a whole, making it the kind of universe in which life could evolve; and, of course, ii) to make such a universe actually exist at t=0.

---

[84] Hawking, *A Brief History of Time*, 140.
[85] Ibid., 122.

After that, God is through. But if the universe had no beginning, then Hawking draws his central theological conclusion: there is nothing *whatsoever* left for God to do—except to chose the laws of nature.

> So long as the universe had a beginning, we could suppose it had a creator. But if the universe is really completely self-contained, having no boundary or edge, it would have neither beginning nor end: it would simply be. What place, then, for a creator?[86]

3.4.2.1.2 *Counterarguments to Hawking* How might we respond to Hawking? First I want to challenge assumptions i) and ii) which are essential to his conclusion. In their stead I want to suggest how the "no boundary" feature of the Hartle/Hawking proposal *can* be important to theology in terms of a reinterpretation of finitude.

i) There are two important exceptions to the assumption that the universe runs according to deterministic laws. a) Quantum physics employs statistical laws which suggest that nature may be indeterministic, the future not entirely predictable from knowledge of the present. b) Chaos theory suggests that even classical physics may involve indeterminacy in a fundamental way, as indicated by the extreme sensitivity to initial conditions of many complex classical systems. Now both a) and b) are debatable, and thus, until their resolution, the question regarding the deterministic character of the laws of nature clearly remains open. Hawking even notes the possibility of a) but rejects it because of the second assumption about intervention:

> If one likes, one could ascribe (quantum) randomness to the intervention of God, but it would be a very strange kind of intervention: there is no evidence that it is directed toward any purpose.[87]

This leads us to the problem of intervention.

ii) There are a variety of strategies by which theologians and philosophers of religion address the question of intervention and divine action. For convenience these can be brought together under two types:[88]

(a) Many theologians have rejected intervention but have still found ways of speaking of God as acting in the world. For example, some have identified important historical or natural events as bearing supreme religious significance and thus as attributing them subjectively to the acts of God known by faith. Others have used a variety of agential models, including embodiment and non-embodiment models, to suggest how God can act objectively in the world without violating the laws of nature.

(b) Other theologians have seen 'intervention or nothing' as a false dichotomy. Often these writers draw on the open character of nature (e.g., quantum chance and/or chaos) to enhance the argument that there

---

[86] Ibid., 141.
[87] Ibid., 166.
[88] For references, see the footnotes in the "Introduction" to this volume.

is plenty of room within natural processes for God to act (the "intrinsic gaps" argument). Some have challenged the assumption that the deterministic character of the laws of nature requires God to intervene in order to act directly in specific events. Instead they argue that we can never know enough to justify excluding divine action. Others have suggested that the laws of nature are descriptive, not prescriptive, and thus God's action should not be thought of in terms of a "violation" of the laws in the first place. Finally, most agree that "intervention" is a misleading term since God is present to all events as their creator, whether or not a specific event is considered "special."

Thus the laws of nature may not imply a deterministic view of nature and even if it does God may be viewed as acting in specific events. Hence, Hawking's fundamental conclusion is undercut: even if the universe had no beginning, God is free to act throughout nature and history as creator and as redeemer.

3.4.2.2 *Finite Creation without a Beginning* How might theologians appropriate the possibility that the universe has a finite past but no beginning? And how might this fit into the framework we began to explore previously, that is, of a theological research program based on *creatio ex nihilo*?

My initial approach to this comes from my previous interpretation of the theological significance of $t=0$. Recall that, in my approach, $t=0$ confirms the claim that the universe is finite in the past, that is, that it has a finite age. With the Hartle/Hawking proposal, this claim is maintained in a certain sense, even though the meaning and status of time as such is radically reconceived, as we have already seen. With the Big Bang model, having a finite age was understood in terms of having had a beginning. Now with Hartle/Hawking we discover the fascinating claim that there can be a finite age without a beginning, that is, that the universe can be finite in time without being bounded in time, and that the meaning and status of temporality itself is radically altered.

What I believe this teaches us theologically is that we can and should distinguish between a first claim and a second, closely related but unnecessary, claim: i) that the universe, as God's creation, must have a finite past (i.e., that it has not existed forever); and ii) that in order to have a finite past, the universe must have had a beginning. Hartle/Hawking have shown us that the latter claim is not logically or mathematically necessary to the former claim. And it is the former claim, the finitude of the past, and not the latter claim, having a beginning point, which I propose we take to be in the last analysis the real theological importance of *creatio ex nihilo* seen in this empirical context.

And so, from my perspective, Hawking's work has the effect of disabusing us of an unnecessary adumbration to the central implication about the finitude of creation as it devolves out of the doctrine of *creatio ex nihilo*. Because of his insistence on the distinction between a finite past and a beginning of time, Hawking has, in effect, helped us claim that the universe is indeed a creation of God even if it had no beginning. For this, Hawking's work, even if it represents a *transitory stage* in the pursuit of a fully developed theory of quantum gravity, will have been enormously helpful to the task of Christian theology. While engaged in purely scientific research, Hawking's results have inadvertently illuminated a subtle distinction in the concept of finitude which is highly pertinent to Christian theology. For this we ought to be very grateful.

I believe this sort of interaction between theological and scientific insights signals the promise of a new, highly creative style for discussing theology and science. Previous work has often been limited to well-established conclusions in science in order not to base theological positions on changing scientific grounds. The present approach, however, is *indifferent* to the long-term status of the Hartle/Hawking model, since the survival of the Hartle/Hawking proposal is *irrelevant* to the validity of the mathematical concept of an unbounded finitude. It is precisely this fact which allows theologians to ponder the significance of something as *controversial* as the Hartle/Hawking proposal as they theologize about time and eternity.

We might go even further. The possibility in principle of an unbounded finite past suggests that theological issues about "the beginning" might best be excised from theological discussions about creation. Since Augustine we have known that it is preferable to think of the creation of time and not creation in (pre-existing) time. Nevertheless the theological and empirical status of the beginning of time has been a continuing problem for theologians down to the present, as Gilkey emphasizes in *Maker of Heaven and Earth*. With the concept of an unbounded finitude in mind, however, theologians might now simply leave aside questions about how to treat the "moment of creation" theologically, since it is seems to be an unnecessary element in the discussion of the "finite creation of the universe." We can think of the past universe as a set of events which have no past boundary, rather than a set of events with a boundary (t=0). The universe is in this sense a *finite creation with no beginning*. Every event is on a par with every other event; since time is unbounded, each event has temporal neighbors past and future (or at least in the early universe there is no 'first three-space'), and we need no longer focus on t=0 in order to retain a historical/empirical sense of creation *ex nihilo*.

Finally, we might frame this point in terms of the "Lakatosian theological research program" approach advocated above. The Hartle/Hawking claim that the universe has no initial singularity would count as evidence against the theological claim that the universe has a finite past, if finitude is taken in its usual sense of requiring a boundary. Hence we are led to frame an additional auxiliary hypothesis, namely that the claim of a finite past can include either a bounded or an unbounded past. This additional hypothesis allows us to include either the bounded finitude of Big Bang cosmology or the unbounded finitude of the Hartle/Hawking cosmology as confirming evidence of the core hypothesis, *"creatio ex nihilo* means ontological origination."

However, for this to avoid being an *ad hoc* move, we would want it to generate some novel predictions. To do so, we must begin with some minimal understanding of the concepts of "bounded" and "unbounded" drawn from set theory. By bounded we can think of a closed set containing all points on the x-axis from, and including, 0 and 1. By unbounded we mean an open set such as the closed set just described but with its boundary deleted, that is, the set $0<x<1$. We can use an open set to describe something of the qualities of a smooth transition from one domain to another in nature, where no sharp border occurs. With this in place we should now be in a position to speak about God's action as creator not only of the universe as such but of many of its theologically crucial features, such as the origin of life, mind, and spirit, without needing to look solely for a sharp discontinuity between domains of nature (e.g., the inanimate/animate distinction) that potentially signal divine agency.[89] We shall see how this might be the case in what follows.

3.4.3 *Creator of the Transition to Temporality* Previously I suggested that there are two ways to view the temporal properties of the universe from a quantum gravity/cosmology perspective: as a single domain characterized by quantum gravity as the fundamental theory, or as consisting in multiple domains, characterized by quantum and classical gravity (general relativity) and the transition between them. These alternative views lead to a reconceptualization of God's action as creator of the universe. In both cases we must recognize God's action as creator of the universe as a whole. However, in the case of multiple domains, we must discuss the additional issue of God's action as creator of its domains with their specific features and of the transition between domains.

3.4.3.1 *Single Domain* We start with God's action as creator of the universe as a whole, viewed as a single quantum domain. This claim requires little additional discussion beyond what we have already give it above. It leads once again to viewing *ex nihilo* in terms not only of the existence but also of the finitude of the universe as such.

3.4.3.2 *Multiple Domains* God's action as creator of multiple domains in the universe still entails the basic *ex nihilo* insight about ontological origination as well as the finitude of the temporal past. However, the possibility of viewing the universe in terms of multiple domains brings additional issues to bear on the topic of God's action as creator.

In the Einstein domain we encounter two standard alternatives in the debate over time. We can follow the "temporalists"[90] who accept time (and time's arrow) as real and adopt theological language about God's action *in* time. This leads naturally to *creatio continua* as an additional component of *creatio ex nihilo*, although the very existence of the universe always warrants an ontological type of *ex nihilo* argument. Alternatively we can follow the "atemporalists"[91] and develop an interpretation of God's action in atemporal categories.

Regarding the Hawking domain we must think about how to conceptualize God's relation to this "timeless" domain, since God is the creator here as well. Here one has a set of separate three-geometries rather than a

---

[89] This additionally implies something about the internal coherence of the doctrine of creation in light of quantum cosmology, namely that we could relate the *continua* and *ex nihilo* traditions of the doctrine in a more fundamental way. I will explore this suggestion in future work.

[90] Barbour, Peacocke, Lucas, Polkinghorne, and Ward are representatives of this position.

[91] Isham and Drees represent this position—as did Einstein.

continuous four-geometry. We might say that God relates directly to each three-geometry, supplying its existence *ex nihilo*.[92]

Finally we focus on God's relation as creator of the transition domain from Hawking to Einstein domains, that is, from "timeless" to "temporal." Here we see the glimmerings of a model, co-opted from theoretical physics, for the theological problem of understanding God's relation to the process by which time (and time's arrow) is created. Thus God acts in the world not only to create new structures in nature, as suggested by discussions of chaos theory, thermodynamics, and quantum physics; at a more fundamental level and in certain regions of the universe, God acts to create time (and time's arrow). Hence we should not just say with Augustine that God creates time along with the universe, and that God creates new phenomena in time. We should add a *third* claim: the universe is such that in certain broad transition domains within it *God creates the transition to time and time's arrow*.

This would pick up the sense of "novel prediction" suggested above in discussing creation in terms of a Lakatosian theological research program. There we added as an auxiliary hypothesis the claim that created finitude need not be bounded. By doing so it became possible to discuss God's creative action *ex nihilo* not only in terms of the ontological dependence of all things and of their historical origin in a unique beginning event, but in terms of the radically new arising smoothly within the processes of nature. Here we have evidence of such a phenomena, namely the transition domain from the Hawking to the Einstein domains (and with it the creation of the arrow of time in the Einstein domain as part of what God creates in certain portions of the universe). Thus the creative action of God would pertain not only to the creation of novelty within natural processes, it also involves the *creation of temporality as we know it*. The possibility of including the transition domain as *novel evidence* for the theological research program we are exploring in this paper through the auxiliary hypothesis about finitude marks this as a *progressive* theological research program over its competitors. Those writers dismissed such issues as t=0 or other empirical aspects of *creatio ex nihilo* as anomalous or irrelevant, or disconfirmed these issues by too strictly identifying *creatio ex nihilo* with t=0.

3.4.4 *Closing Reflection on Trinitarian Creation* Finally, my own view—which I can only point to here but not develop—is that we should think of the divine eternity as the source of the temporality of the universe. Here divine temporality is taken in the sense of authentic temporality, that is, divine eternity.[93] God acts eternally as God the creator to the universe. Hence the divine eternity must be understood in a sufficiently complex fashion as to at least allow for time so understood. The challenge then to theology is to think through the meaning of

---

[92] See Isham, "Creation as a Quantum Process," 404.

[93] I am indebted to the recent discussion of eternity stemming from the writings of Karl Barth and Karl Rahner and picked up in various ways by Ebehard Jungel, Wolfhart Pannenberg, Jürgen Moltmann, and Ted Peters. I hope to develop this approach in future writings.

divine eternity and divine action to embrace at least these meanings of creaturely time.

I would close by noting once again how closely tied are the debates over temporality in physics and cosmology, in philosophy of religion (God's action in the world), and in systematic theology (time and eternity). We need now to rethink the current models of divine agency in terms of our more complex understanding of temporality as suggested by quantum physics and quantum cosmology. This discussion in turn needs to be integrated into the systematic issues regarding time and eternity. Although highly tentative and difficult to access due to their mathematical abstraction, the scientific discussions of time ought to help theologians and philosophers of religion critically rethink their own presuppositions about time and eternity as they articulate in new ways God's action in the world.

Acknowledgments: I want to thank Ian Barbour, Wim Drees, George Ellis, Stephen Happel, Chris Isham, Nancey Murphy, Arthur Peacocke, Ted Peters, Bill Stoeger, Claude Welch, and John Wright for their help with this manuscript.

A CASE AGAINST TEMPORAL CRITICAL REALISM? CONSEQUENCES OF QUANTUM COSMOLOGY FOR THEOLOGY

Willem B. Drees

1 *Introduction*

This paper intends to explore the consequences of contemporary physical cosmology for theology. It will be argued that quantum cosmologies, and the underlying ideas about quantum gravity, pose major challenges to our understanding of time and knowledge. Though *explorative* in intent, the paper is *polemical* in structure. The most widely held positions in the contemporary Anglo-Saxon discussion of science-and-religion combine a "critical realistic" appreciation of science and religion with an evolutionary understanding of the world and temporality as an essential element in our understanding of God. This combined view I shall call *temporal critical realism*. The polemical thesis is that these positions are unable to cope with quantum cosmologies, and hence unable to fulfill their own promise of taking science seriously. They need to change either their ontology or their epistemology.

Central to the epistemology of critical realism is the idea that science aims at depicting the world. The approximations become ontologically more accurate as science develops better theories. The discussion will focus on varieties of critical realism which assume that the most well-established theories provide a view of the world which is reliable for metaphysical purposes, or which combine emphasis on the unity of the sciences with a hierarchical structuring of these sciences.[1] I will argue that such assumptions make it difficult

---

[1] There are various versions of "critical realism." J. Leplin lists in the "Introduction" to *Scientific Realism* (Berkeley: University of California Press, 1984) ten characteristic realist theses, about which actual realists happen to disagree. However, "what realists do share in common are the convictions that scientific change is, on balance, progressive and that science makes possible knowledge of the world beyond its accessible, empirical manifestations" (2). Among the defenders of critical realism in the context of science-and-religion, mostly in combination with an emphasis on temporality, are I. Barbour, *Myths, Models and Paradigms: A Comparative Study in Science and Religion* (San Francisco: Harper & Row, 1976); E. McMullin, "A Case for Scientific Realism," in *Scientific Realism*; A. R. Peacocke, emphasizing the hierarchical structuring, especially *Intimations of Reality* (Notre Dame: University of Notre Dame Press, 1984); T. Peters, "On Creating the Cosmos," in *Physics, Philosophy and Theology: A Common Quest for Understanding*, ed. Robert John Russell, William R. Stoeger, and George V. Coyne (Vatican City State: Vatican Observatory, 1988); J. Polkinghorne, *One World* (Princeton: Princeton University Press, 1986); J. M. Soskice, *Metaphor and Religious Language* (Oxford: Clarendon Press, 1985); and various

to avoid the challenges from quantum gravity and quantum cosmology (section 5).

Big Bang cosmology appears to be in line with temporal critical realism: the expanding universe seems to line up well with evolutionary worldviews developed in dialogue with biology. Various temporal critical realists have pleaded caution with respect to the beginning, claiming at most a form of consonance. However, they have often appealed to the Big Bang theory as describing a dynamical universe. But analysis of the limitations of the Big Bang theory shows that one should be equally cautious with statements regarding the fundamental nature of time in Big Bang cosmology (section 2). Quantum cosmology modifies the concept of time rather than extending time beyond the initial singularity of the Big Bang theory. Quantum gravity and quantum cosmology thus result in challenges for temporal critical realism (section 3).

An alternative to temporal critical realism might be a more positive appraisal of timelessness. This may give a Platonistic flavor to the theological schemes considered (section 4). If one is not prepared to grant timelessness such a place, one might consider deviating from realism with respect to the status of cosmology itself. However, if this is done in a rather *ad hoc* fashion, the claim that critical realists take science seriously, even though not literally, loses its force and thus its apologetic significance. Moreover, dismissing cosmology brings in other questions regarding the status and the unity of the sciences (section 5). Playing down cosmology might be acceptable if one were to give up on the unity of the sciences, seeing them as different partial constructs which need not fit into a neat whole. Such a theology would be less inclined towards ontological statements, as is illustrated by Mary Hesse's work (section 6).

One final introductory remark. Physicists, such as Chris Isham, engaged in quantum cosmology stress the tentative character of contemporary quantum cosmology, noting that there are in fact no real theories here. I agree that that is reason for caution, but I disagree when that is taken to free the philosopher and the theologian from reflecting upon these theories. The arguments below focus on general characteristics of quantum cosmology, rather than on specific schemes. These developments affect our understanding of time, primarily in a negative way, as research in quantum cosmology deviates for good reasons from our previous ideas about time. Research in this area thus invites us to rethink ideas based upon the concept of time as developed in relation to experience at our own, classical level of reality.

2 *Big Bang Cosmology and Theology*

Some authors, such as the astronomers Robert Jastrow and Gerald Schroeder, have claimed remarkable parallels between the creation narrative of Genesis and

---

process theologians in the Whiteheadian tradition. These authors may be taken to exemplify what has been referred to as "temporal critical realism" in the present article.

Big Bang cosmology.[2] Other authors have taken the discoveries to show the reality of a sudden beginning of the universe, which could serve as a premise in a cosmological argument for the existence of God.[3] Both arguments are wanting as they pay insufficient care to the nature of the scientific and religious statements at hand.[4] Temporal critical realists have pleaded for caution with respect to such claims. A major argument for them has been the possibility that "t=0" represents a kind of epistemological limit rather than an absolute beginning. The harmony they envisage is not related to "t=0," but rather to the basic role of time, that is, the dynamic rather than static nature of the universe (2.1). It will be argued that both the spacetime theory underlying Big Bang models (2.2) and the limitations of such models (2.3) make it possible to challenge not merely the absoluteness of the apparent beginning, but the dynamic nature of the universe as well.

2.1 *Critical Realists: Cautious Consonance, but Temporality*

Temporal critical realists have not been lured into arguments based on the initial singularity of the Big Bang theory. Even less have they been tempted by naive claims with respect to parallels between inflation—a brief period of extremely rapid expansion of the early universe—and "the wind of God moving over the waters" of the second verse of Genesis.[5] They are generally open to the possibility that Big Bang theory might describe only a cosmic epoch, perhaps in the context of an oscillating universe. *Creatio ex nihilo* is taken as an ontological claim of dependence, rather than as a historical claim about a beginning. For example, Ian Barbour writes:

> *The contingency of existence* corresponds to the central religious meaning of creation *ex nihilo*. On both sides the basic assertions can be detached from the assumption of an absolute beginning. On the scientific side, it now appears likely that the Big Bang was indeed an absolute beginning, a singular event, but we cannot rule out the possibility of a cyclic universe or infinite time.... With respect to the central meaning of creation *ex nihilo* (though not with respect to continuing creation) I agree with the neo-orthodox and existentialist authors who say that it is the sheer *existence* of the universe that is the

---

[2] R. Jastrow, *God and the Astronomers* (New York: Warner Books, 1980); G. L. Schroeder, *Genesis and the Big Bang: The Discovery of Harmony Between Modern Science and the Bible* (New York: Bantam Books, 1990).

[3] W. L. Craig, *The Kalām Cosmological Argument* (London: Macmillan, 1979).

[4] For more developed criticisms, see Drees, *Beyond the Big Bang*, especially ch. 1; and *idem*, "Potential Tensions Between Cosmology and Theology" in *Interpreting the Universe as Creation*, ed. V. Brümmer (Kampen, The Netherlands: Kok Pharos, 1991), 71-75.

[5] Schroeder, *Genesis and the Big Bang*, 93f. Schroeder also correlates the decoupling of matter and photons with God's injunction "Let there be light" (88). The six days of the biblical narrative and Big Bang theory deal with "identical realities . . . described in vastly different terms" (26).

datum of theology and that the details of scientific cosmology are irrelevant here.[6]

There is no support from cosmology for theology, or vice versa, though there may be something less stringent, "consonance." These authors generally consider as alternatives the steady state theory and an oscillating universe, as well as the possibility of redefining time, with the effect that the singularity is located in an infinitely distant past. And if a theory of quantum gravity could be found, "we would move the frontier of knowledge further back" (note the temporality of this expression).[7] The cautious attitude with respect to an appeal to Big Bang theory has to do with the possibility that time might be extended, rather than with the possibility that *the concept of time itself* might be of limited validity.

Whereas the apparent beginning of the universe is not taken as important,[8] authors such as Barbour, Peacocke, and Peters have claimed that Big Bang theory is significant in that it shows us the dynamical nature of the universe, the essential role of time. As Barbour expresses it:

> ... astrophysics adds its testimony to that of evolutionary biology and other fields of science. Time is irreversible and genuine novelty appears in cosmic history. It is a dynamic world with a long story of change and development. On the theological side, continuing creation expresses the theme of God's *immanence* and *participation* in the ongoing world.[9]

These authors all avoid simplistic identification of the initial singularity with the theological concept of creation, although they feel at home with, consonant with, the Big Bang cosmology. They intend to restrict themselves to the more general, apparently more safe, features of the model. The alternatives that are considered

---

[6] Barbour, "Creation and Cosmology," in *Cosmos as Creation: Theology and Science in Consonance*, ed. Ted Peters (Nashville: Abingdon Press, 1989), 141f. Similarly Peacocke, *Creation and the World of Science* (Oxford: Clarendon Press, 1979), 79; idem, "Theology and Science Today," in *Cosmos as Creation*, 33f; and McMullin, "How Should Cosmology Relate to Theology?" in *The Sciences and Theology in the Twentieth Century* (Notre Dame: University of Notre Dame Press, 1981), 38.

[7] Peters, "Cosmos as Creation," in *Cosmos as Creation*, 107.

[8] Except, to some extent, by Peters, who believes that there is "surprising and salutary consonance" between the theological idea of *creatio ex nihilo* "formulated in terms of an original beginning of time and space" and contemporary astrophysics (Peters, "On Creating the Cosmos," 276). However, he makes a distinction between the abstract *creatio ex nihilo* as ontological dependence and the assertion that there was a beginning as a concrete form of explicating *creatio ex nihilo*. "It is, of course, possible for a theologian to speak metaphysically about the utter dependence of the creation on its creator without reference to a temporal beginning" (Ibid., 274). He acknowledges that "oscillationism" is still an alternative type of theory and that the Big Bang (*i.c.* the Planck time) is "a methodological frontier and not the full ontological affirmation made by Christian theology" (Peters, "Cosmos as Creation," 107).

[9] Barbour, "Creation and Cosmology," 143.

are, basically, alternatives with an infinite past—a longer extension of the dimension time. This would *not* affect the claim that we live in a dynamic world.

There is no doubt that we have indeed come to realize that darkness at night implies that we are not living in a static universe (Olbers' paradox). Stars age. The average distances in the universe are increasing; we live in an "expanding universe." The abundances of chemical elements have been traced back to nuclear processes in stars or to processes during "the first three minutes." We do live in a dynamic universe, as seen from the inside.

But is this the most fundamental view of reality? Is it a fair "approximation" to "the way it really is," to use a critical realistic phrase? It is essential to be aware that this is a view "from inside." What is the significance of this perspective? It may not be the final perspective, "the view from nowhere," as Thomas Nagel puts it.[10] It is rather a view from "now-here." Are there any reasons to believe that a wider, more fundamental, perspective will seriously affect our understanding of the dynamical nature of the universe? First, we will consider the spacetime view of relativity theory and its implications for the emphasis on time (2.2). Second, we will consider some limits of the Big Bang theory and options for future theories (2.3).

2.2 *The Challenge of "Spacetime"*

One way to challenge the emphasis on time would be to invoke the special and general theories of relativity. In the special theory of relativity the notion of simultaneity as having a universal meaning with respect to a "now" is lost. This in turn raises serious issues for statements about God having time, being related in a special way to "the past" or acting as to influence "the future." "Past" and "future" can be used as concepts relative to a observer located at some position on a specific worldline in spacetime. The problem arises when a definite article is used, speaking about "the past" and "the future," as if these are global concepts. Thus, problems arise in theologies which insist that "God's future" is open, or make other claims which assume the existence of a universal notion of time.[11] As long as God lacks a specific location and state of motion, it is difficult to understand the meaning of God knowing "the past" or influencing "the future." At least three ways to get around the theological consequences of this loss of a single universal time have been proposed. One is to allow for the co-existence of more than one time in God. Another, defended by Polkinghorne, is to invoke God's omnipresence. A third option is to argue that there is a physical basis for a universal time by taking into account the cosmological background radiation. We will consider these options in turn:

---

[10] T. Nagel, *The View from Nowhere* (New York: Oxford University Press, 1986).

[11] The description of the problem as one with the definite article has been taken from an unpublished note from Isham; see also the article by Isham/Polkinghorne in this volume. The problem is discussed theologically, but not solved, in C. Hartshorne, *A Natural Theology for our Time* (La Salle, IL: Open Court, 1967), 93f.

(1) Some have argued that one might consider the co-existence of our time with other time series. However, if they are taken to be unrelated except for their co-existence in God,[12] "multiple time sequences" are of no help with the relativity problem. They refer to different spacetimes and therefore they don't solve the real problem of coordinating multiple observers existing in *one* spacetime. If one were to apply the idea of "multiple times" to the various times arising for different observers in one spacetime, these time sequences are strongly correlated. One might say that God is related to all those times. However, that would not introduce a universal notion of simultaneity which would allow for statements about God's relation to "the future" or "the present." To say that God is related to a multitude of times which are correlated as the various times of a single spacetime are, is equivalent to saying that God is related—in some non-temporal way—to the whole of spacetime.

(2) Polkinghorne suggests that omnipresence provides a way out. He argues that an omnipresent God is spatially coincident with every spacetime point, and thus has no need to use signaling to tell him what is happening and so he has instant access to every event as and when it occurs. That totality of experience is presumably the most important thing to be able to say about God's relation to world history.[13]

But Polkinghorne's description is ambivalent. "When it occurs" may be read as a reference to a hidden background of universal time, making possible a reading of "the totality of experience" as a three dimensional present. This seems to be Peacocke's reading when he states that "the future does not yet exist in any sense, not even for God," where "God is conceived as holding in being in physical time all-that-is at each instant and relating his own succession of divine states (the divine "temporality") to the succession of created instants."[14] Such a reading of Polkinghorne's solution is not in line with relativity theory, as it introduces a universal sense of now, correlated with that three-dimensional "totality of experience."

Another, perhaps better, reading of Polkinghorne's proposal takes it that "when it occurs" means that God has equal access to events at all spacetime points—whether deemed future, past, or present from any spacetime point. But then God's temporality is lost; the "totality of experience" covers four-dimensional spacetime as a whole.

(3) One might suggest that classical Big Bang cosmology solves some problems with respect to God's time, as there might well be a way to define a global time in an expanding universe. For example, one might use as the frame of reference that frame in which the background radiation is homogeneous.

---

[12] Suggested by K. Ward, *Rational Theology and the Creativity of God* (Oxford: Basil Blackwell, 1982), 166; discussed by B. Leftow, *Time and Eternity* (Ithaca: Cornell University Press, 1991), 29.

[13] Polkinghorne, *Science and Providence: God's Interaction with the World* (Boston: Shambhala Publications, 1989), 82; see also the opponent of the block universe in the article by Isham/Polkinghorne in this volume, especially his point (8).

[14] Peacocke, *Theology for a Scientific Age: Being and Becoming—Natural and Divine* (Oxford: Basil Blackwell, 1990), 131.

However, it is not clear that there is such a universal time outside the simplified homogeneity and isotropy of the Friedman-Robertson-Walker models. Besides, general relativity, on which the Big Bang theory is based, seems to make problems worse. Time may be a locally applicable concept, but still there may be no definition of time that covers the whole spacetime manifold.

Similar "block" views of time arise in other physical theories which deal with whole possible histories at once as trajectories in phase space. Temporal critical realists have not dealt satisfactorily with such "block" views. However, the following will not develop the consequences of classical spacetime views; rather we will ask whether the problem changes once we turn towards quantum cosmology. It might be the case that a notion of a single, flowing, universal time is possible, once general relativity as the framework for cosmology is succeeded by a quantum theory which integrates space and time as well as matter. Running ahead of the argument, the conclusion of the reflections on quantum cosmology will be that such a hope for a recovery of time will *not* be fulfilled. Rather, things will become worse. Before turning to quantum cosmology, however, it is important to see why such a further scientific development beyond the Big Bang theory is needed.

2.3 *Limits of the Big Bang Theory*

The Big Bang theory is the accepted theory about the evolution of the universe over billions of years. It relies upon two types of theories: general relativity describing spacetime and quantum theories describing matter. Both theories have their limitations; furthermore, they are hard to combine.

From our perspective, the theories of matter are the first to present problems near to t=0. Current theories about matter are valid only up to a finite temperature, and hence only valid *after* the first fraction of a second following the singularity at t=0, the initial moment entailed by general relativity. This implies an epistemological boundary to the domain where the Big Bang theory can be trusted. Further speculations have to deal with circumstances (temperatures and densities) for which the relevant particle physics is not yet well established.

Closer to the singularity comes a moment, presumably the "Planck time" (a number constructed from fundamental constants of quantum theory and gravity, about 10^{-43} seconds after the initial singularity), where general relativity must be replaced by a quantum theory of gravity. Such theories do not exist yet, but some current ideas will be discussed below (section 3). The meaningfulness of "time" will turn out to be uncertain. This is a troublesome conclusion; once "time" is no longer unambiguous, it becomes unclear what can be meant by "before" or by "the Planck time."

If one were to continue backwards in time, the initial singularity itself would be a third limit, where the theory of general relativity, the theory about spacetime, breaks down. However, as this limit lies beyond the Planck time, and thus in a realm where general relativity has to be abandoned anyhow, it is not clear in what sense this limit might be relevant. This cannot be decided *a priori*, without considering the actual theories of quantum gravity that have been proposed. Whereas the first and second limits are limits to our present knowledge, the third seems to be an edge, an ontological discontinuity—but it is hidden behind the other two.

Temporal critical realists seem to hold that the major uncertainty regards the third limit. Has there really been an absolute beginning, or is there continuity—for example in an oscillating universe, extending our "past time" indefinitely or even infinitely. That was also the issue between the Big Bang and steady state theories. If the alternative is seen this way, we might well trust the Big Bang theory in its proper domain, and leave the speculations of quantum cosmologists aside as not relevant to our understanding of the dynamical nature of our cosmic epoch. For instance, Peacocke discusses the Hartle-Hawking concept of imaginary time, but dismisses it as irrelevant, since:

> ... by the point at which biological organisms appeared on the Earth, the postulated imaginary component in Hartle and Hawking's physical time would have diminished to insignificance in their theory. So, with this cosmology, we are still free to employ the concept of the personal to interpret God's relation to the universe which goes on being created by God.[15]

If the problem is formulated as a problem about the absoluteness of the initial singularity, there is an implicit assumption of a fixed background time which could be extended beyond "t=0." The problems with such an approach are the problems facing creation *in* spacetime, as discussed by Isham in this volume and previously.

An alternative possibility, however, is that it is the conceptuality which is at stake, as a consequence of the breakdown of general relativity, and hence of its notions of time and space. In that case, we are not merely stumbling upon a possible ontological discontinuity, but rather the ontology—the basic conceptuality in terms of which we think of our world—has to change radically. As the successor to classical cosmology, quantum cosmology would imply a reinterpretation of the meaning of the Big Bang theory as well. Rather than considering creation *in* spacetime, this is the approach that considers creation *of* spacetime.

Changes in conceptuality have been typical of fundamental transitions in physics, such as those from classical physics to quantum physics and from Newtonian conceptions of space and time to those of the special and general theories of relativity. Knowledge was not merely extended to the very small or the very fast, but rather restructured. Thus from a realist perspective, new theories led to a reinterpretation of the world. In an instrumentalistic vein, focusing on predictive power, the old theory is a continuous limit of the new one; but conceptually or ontologically it is radically different. The empirical or observational consequences of previous theories, as far as they are corroborated by experiment, are reproduced by the new theory, though the new theory is cast in radically different conceptions. Such a transition is at stake with respect to quantum cosmology as well: it leads to a reinterpretation of our concepts regarding the world—especially the concept of "time." And that change is not

---

[15] Ibid., 133f.

restricted to cosmology, as the theory at stake is quantum gravity, intended to be a universal replacement of Newtonian and Einsteinian views of space and time. *If such a radical change in our ontological conceptuality is possible, due to the second limit of the Big Bang theory, the fundamental dynamical nature of the universe is open for reinterpretation, and not merely the absoluteness of the apparent beginning.*

3 Consequences of Quantum Cosmology and Quantum Gravity

Time in the context of relativistic spacetime theories is a phenomenological, "internal" construct. One might well see this as a modern day equivalent of Augustine's view of *creatio cum tempore*, time being part of the created order. The discovery of "internal" time, as characteristic of the theory of general relativity, has paved the way for a second discovery, the discovery of the limited applicability of the concept of time, as is typified in of quantum cosmologies and quantum gravity. We will consider these two developments in turn.

3.1 *External and Internal Time*

Isham, in his contribution for this volume, explains why origination of the material universe *in* a *fixed background spacetime* is problematic. One of the major problems is the problem of choice: "within an infinite, pre-existent, and homogeneous timeline, there is simply no way whereby the mathematics can select one particular time at which creation occurs."[16] Quantum theories which work with probabilities (e.g., instances per unit time) tend to introduce a plurality of origination points. This would lead to interacting "universes," contrary to the available empirical evidence.

Hence, physicists have turned to the development of theories which describe creation *of* time rather than creation *in* time. General relativity theory offers a fundamental hint in that direction. Whereas in a fixed background, time may be seen as external with respect to the system, the situation in general relativity is different. Here time is understood as an "internal" variable. For example, one might attempt to define time in relation to the average distance between "test-particles" such as galaxies, or one might use the temperature of the background radiation or features of other material phenomena. The evolution of properties of the universe in time is thus transferred to statements about the correlation between, for example, the temperature and other properties of the universe.

In traditional quantum theory, the fundamental equations (like the Schrödinger equation) describe the evolution of the wave function (or state vector) in time. This means that the properties of the system are given by a time dependent entity. "Thus 'time' is arguably part of the classical background which

---

[16] This objection to creation *in* time is not a new insight; for example, it was considered by Augustine (*Confessiones* 11, XII, 14) and, centuries earlier, by an Epicurean, as told by Cicero in his *De natura deorum* I, 9, 21.

plays such a crucial role in the 'Copenhagen' interpretation of the theory."[17] In some approaches to quantum gravity there might be a background structure which is sufficiently rich as to include some concept similar to classical time. However, those approaches have the same problem as described above for creation *in* a fixed background spacetime.

Quantum gravity and quantum cosmology have really taken a different approach.[18] The background structure is not a four-dimensional spacetime, but rather a three-dimensional space. On this three-dimensional model a wave function is defined which specifies curvature and matter. The "notion of time (and therefore spacetime) has to be *extracted* in some way from these variables."[19] Thus, the dynamical evolution might be recovered by defining a time variable on the basis of a suitable variable constructed either out of the curvature or out of the matter fields (hence either like average distance, or volume, or like background temperature). Though the external time has disappeared, the quantum scenario could be seen as open to two equivalent representations: an evolutionary one and a frozen one. However, the evolutionary representation is slightly odd, compared with evolutionary equations which arise in conventional quantum theory. This deviation is an advantage in the context of the program of quantum cosmology, the attempt to construct a genuine theory of origination of the universe. It leads us to the idea of "imaginary time."

3.2 *The Limited Applicability of Time*

Superspace is the collection of all possible configurations of curvature and matter on a given three-dimensional manifold. As Isham explains, a series of such possible configurations—a possible history of a universe—corresponds to a path in that superspace. However, quantum theory is such that it does not result in a single history, as if there were only a single path in superspace with non-zero probability. Just as in conventional quantum theory, each classical history will be slightly fuzzy, as slightly deviating histories have lower, but still non-zero, probability amplitudes. Besides, the wave function turns out to have a number of

---

[17] Isham, "Conceptual and Geometrical Problems in Quantum Gravity. Lectures presented at the Schladming Winter School," (Preprint Imperial/TP/90-91/14), 74.

[18] Beginning at least with B. S. DeWitt, "Quantum Theory of Gravity: I, II," *Physical Review* 160 (1967): 1113-1148; 162 (1967): 1195-1239.

[19] Isham, "Conceptual and Geometrical Problems in Quantum Gravity," 75. In his contribution to this volume, Isham discusses the problem of selecting a unique wave function for the universe. A genuine "creation" theory would have to predict all features of the universe, but assumptions are always fed into such theories. For instance, in the scheme he discusses here, part of the (contingent) background is the assumption of a three dimensional manifold of fixed topology. Issues of uniqueness and background might be very relevant entrances to philosophical and theological reflection, closely related to questions of contingency and necessity. However, these issues are beyond the scope of the present article, which focuses on time.

paths around which it is peaked, and thus to describe a whole set of (approximately) classical histories. The spacetime picture would not be unique, even if the wave function were unique.

Isham distinguishes two regions in superspace. In one region the spacetime picture corresponding to a path is fairly well in accord with the general relativistic view (and thus, aside of the spacetime problem discussed in 2.2, with our common sense understanding of time). But other regions of superspace do not lend themselves to such an interpretation. These regions are part of the theoretical structure and cannot be omitted, unless in an unsatisfactory *ad hoc* fashion. For trajectories in this realm of superspace, space and time do not correspond to the relativistic case (recognizable by the Lorentz distance formula $x^2 + y^2 + z^2 - c^2t^2$). For trajectories from this realm, time and space variables appear in the equivalent formula completely on a par, as there appears a plus sign rather than a minus sign in *all* four terms of the distance formula. It is in relation to this realm that some have spoken of "imaginary time." Taking time to be imaginary, one might retain the minus sign in the formula. However, it seems as accurate to say that another formula is applicable. The corresponding four-dimensional space is highly quantum mechanical. It is far more fuzzy than the spacetimes corresponding to paths in the other region of superspace, as the solutions peak less around certain paths in superspace in this second region. Isham calls this quantum mechanical four-dimensional realm "imaginary spacetime." Whether one wants to say that the concept of time has become meaningless or that time has become imaginary may be a matter of taste. However, it certainly is not the kind of time which allows for clear successions of events.

By way of summary, two features of the canonical approach to quantum gravity may be emphasized. First, the fundamental ontology (background structure) assumes a three-dimensional manifold, rather than a four-dimensional spacetime. The ontology contains as well a whole collection of possible configurations of geometrical and material configurations on this manifold: superspace. "Time" is a derivative notion, well defined only for certain subsets of, or certain paths in, this superspace.

Second, wave functions arising in this theory may be interpreted as describing a realm where the concept of time is meaningless and a realm where a relativistic concept of spacetime is meaningful. Even this more classical reality is, however, fuzzy for two reasons. A wavefunction corresponds to a plurality of paths describing pure classical space times, and each path has the fuzziness which is typical of conventional quantum theories.

It is important to note that the relevance of these ideas cannot be restricted to considerations regarding the quantum theory of the origination of the universe. Rather, it purports to be the quantum view of the universe or, even more significantly, the quantum theory of time (and space, but less so as some features of space are still assumed as part of the background structure). As the quantum theory of matter or radiation is different from classical theories of matter or radiation, so is the quantum theory of space and time different from classical theories of space and time.

3.3 Challenges for Temporal Critical Realism

"Temporal critical realism" was taken to combine an epistemological and an ontological position: science has to be taken seriously, but not literally, and reality is best described in dynamical terms, including its relation to God. Quantum cosmology and quantum gravity seem to challenge these positions. Let me begin with a remark about the epistemological issue.

Critical realists tend to see science as a more or less continuous series of successive approximations of increasing accuracy in depicting reality. However, the concept of approximation ties in with instrumentalistic approaches to the mathematics, whereas quantum cosmologists tend to be more Platonistic, for example, with respect to the reality of mathematics. If mathematics is seen as a tool, it may be more or less adequate in describing properties of entities. For example, in stating that an object has a mass of 4.3 kilograms, one means that it has that mass with the required precision, say, give or take at most 50 grams. Physical reality is modeled mathematically, but the model is considered to be an approximation. If, on the other hand, reality itself is assumed to be mathematical, one does not deal with approximations. For example, it is not clear in what sense the fundamental symmetry group underlying the particle world could be *approximately* group X—it is group X or it is not. This may need some qualifications in relation to spontaneous broken symmetries. However, it remains the case that the alternatives seem to be much more discrete than in an instrumentalistic approach to the mathematics. "One of the most fascinating features of mathematical structures as models of the world is their apparent ability to justify themselves. These structures are so strictly connected with each other that they seem to be necessary and to be in no way open to arbitrary, speculative alterations."[20]

In addition, the theories developed successively in cosmology, from Newton through Einstein to quantum cosmology, tend to be continuous with respect to the numerical outcome of certain calculations, but to differ radically in their fundamental conceptualities. Hence, it is hard to make clear in what sense these models depict reality approximately, and thus how they refer. Dismissing quantum cosmology for this reason as too speculative seems an unfair, *ad hoc*, move, which would be a betrayal of the intent to take science seriously.

The epistemological consequences from cosmology, as well as from other significant discontinuities in fundamental physics, may be at odds with "critical realism." However, the main targets of the defenders of "critical realism" seem to be sociological, psychological, and idealistic reductions of physical and religious reality to ideas produced by humans. The platonic realism under consideration here does not suffer from such a reductionism. It might therefore be possible for critical realists to change their epistemological position by accepting a kind of platonic realism (as quite a number of mathematicians tend to do). We will return to epistemological issues below (section 5).

---

[20] Michael Heller, "The Experience of Limits: New Physics and New Theology," quoted from the abstract in *Science and Religion*, ed. J. Fennema and I. Paul, (Dordrecht: Kluwer Academic Publishers, 1990), 207.

The temporal critical realists take time to be fundamental; reality is dynamic. This position has been developed for a variety of reasons; prominent among them is the desire to accommodate insights from the evolutionary sciences. Reality is, of course, dynamic and evolving, if considered on an intermediate scale from a point of view within an almost Newtonian epoch. However, questions arise already when one considers larger scales and has to take account of the conceptuality of general relativity. The dynamic picture may be extendable to the quantum level, the finer detail of photons and electrons. However, further down in scale, to the quantum gravity level, the conceptuality of dynamism breaks down. "Deep down" the ontology is different. It is not limited to quantum cosmology, as it has to do with quantum gravity a theory with universal applicability. And it is not just a detail at some irrelevant scale, because it affects, or should affect, the concepts of space and time as they are used at *all* levels. At the Newtonian level of description, space and time are taken to be universal, infinitely extendable continua. The special theory of relativity has raised problems with the universal simultaneity of time. General relativity calls into question the extendability of time, as singularities arise. And quantum gravity takes away the fundamental states of time. The still speculative ideas at the frontier of cosmological research, and even the standard theory of spacetime (General Relativity), thus suggest that the evolutionary presentation is one of limited validity, and not the most fundamental one. Hence, theological insights developed in the dialogue with the evolutionary understanding of the natural world are *not* directly extendable to the dialogue with cosmology.

It seems as if the temporal critical realists have not been considering such a shift in conceptuality as a reason for caution. They pleaded caution with respect to t=0 in Big Bang theory because of the possibility of an extension of past time. However, as Isham makes clear in his contribution, a cyclic view seems incompatible with the spacetime picture in contemporary theories of quantum cosmology and quantum gravity. The alternative is not an extension to earlier times, but a reinterpretation of time, and hence a reinterpretation of the meaning of the Big Bang model itself.

If the temporal perspective is considered to be essential to Christianity, there is a conflict with cosmology. However, it might be that it is possible to accept the cosmological view of time, embedding the common sense temporal view in a wider timeless view *sub specie aeternitatis*, provided a meaningful formulation for human responsibility in relation to human actions (within the spacetime framework) could be found. We thus will turn from theologies which emphasize temporality to theologies which grant timelessness a more prominent place.

4 *Platonizing Theology*

A metaphysical view of reality as timeless and as of a self-justifying mathematical nature seems at odds with the emphasis which many theologians place on contingency and time as major aspects of the Christian doctrine of creation.[21] A defensive approach would thus be to emphasize all the contingency that is left. However, might it not be that the contingency and temporality under consideration are not necessary to a proper view of God? One may be able to defend such a claim if one does not presuppose a voluntaristic understanding of God as creator, but rather supposes that God creates according to certain "internal necessities." For instance, God might not have been able to create something logically contradictory; the fundamental rules of logic would reflect God's rationality. Similarly, perhaps God could not have created something wicked or ugly, as goodness as well as aesthetic elements are intrinsic to the divine.[22] Michael Heller proposes to interpret the rationality displayed by the universe as an ultimate rationality, which is God's. Platonic and Neoplatonic philosophies, and theologies inspired by them, have always been inclined to regard the world as a reflection of "eternal objects" (for which one may read "mathematical objects") that dwell in God's mind. Thus, he points out that the metaphor of "God thinking the Universe" is well rooted in the history of theology.[23] Platonistic tendencies in cosmology may well be developed into a philosophy which extends the issues from mathematical intelligibility to wider concepts of rationality, and from there into values. Another further development of such a Platonistic philosophy might extend the discussion on mathematical intelligibility so as to introduce the notion of spirit or mind. "From the theological perspective, there is an intimate relationship between the spirit of rationality and the Christian idea of the Logos."[24]

In suggesting that quantum gravity might correlate well with a Platonizing theology, I do not intend to make a historical claim. The concept of "Platonism" certainly needs to be defined with greater precision. For the moment I intend only to draw attention to a number of different features which seem to apply to almost all fundamental scientific cosmologies, and which may transfer to the way ideas about God are formulated.

The following sections will give a limited survey of discussions regarding elements of such an understanding of God, considering divine eternity, divine action, and the way in which God may be conceived of as explanation of the universe. It is not an exhaustive presentation of such theologies, but an initial exploration of contemporary discussions. Is it possible to think theologically along such lines? Some hesitations will be expressed (4.4). Whereas this section explores the possibility of changing the ontological assumption underlying

---

[21] In addition to the temporal critical realists mentioned above, one could also refer to T.F. Torrance, *Divine and Contingent Order* (Oxford: Oxford University Press, 1981); W. Pannenberg, "The Doctrine of Creation and Modern Science," *Zygon: Journal of Religion & Science* 23 (1988): 3-21; and the discussion of his work by R. J. Russell, "Contingency in Physics and Cosmology: a Critique of the Theology of W. Pannenberg," *Zygon* 23 (1988): 23-43.

[22] H. G. Hubbeling, *Principles of the Philosophy of Religion* (Assen, The Netherlands: Van Gorcum, 1987), 148.

[23] Heller, "The Experience of Limits," 207.

[24] Heller, "Scientific Rationality and Christian Logos," in *Physics, Philosophy and Theology*, 141.

temporal critical realism, further sections will explore possible changes in epistemology.

4.1 *Divine Eternity*

> Recent philosophers and theologians tend to think that anything that could count as *God*—as the living, loving person whom the Old and New Testaments depict as in dialogue with the creatures of history—must be in time. Their message is that the deity of the atemporalists is too remote and impersonal to be God. Yet medieval philosophers and theologians tended to think that anything that could count as *God*—as the transcendent, perfect source of all that is other than Himself—could not be in time. The medievals would say that the deity of the temporalists is too small or too creaturelike to be God.[25]

"God is eternal" may be understood in two ways, either as everlastingness (through time) or as timelessness, without extension or location in time.[26] Early defenders of timelessness have been Augustine, Boethius, and Anselm. Brian Leftow has analyzed in detail their different ways of conceiving timelessness, arguing that divine timelessness is a consistent option. If one is able to think consistently about divine timelessness rather than everlastingness, one may ask whether that is an appropriate view of God. Opinions diverge on this issue.

Nelson Pike has analyzed the logical relations between the classical understanding of divine eternity as divine timelessness and other doctrines, such as immutability, omnipresence, and omniscience. Timelessness has consequences for the interpretation of these other attributes; consequences which he finds objectionable. Pike sees timelessness as a Platonic influence with hardly any scriptural basis, and points to the devastating consequences for other doctrines, ending with the question: "What reason is there for thinking that the doctrine of God's timelessness should have a place in a system of Christian theology?"[27]

The case for divine timelessness has been defended in the contemporary Anglo-Saxon philosophy of religion by Paul Helm and Leftow. Timelessness is not understood as a separate attribute, but rather as God's way of possessing certain attributes. For God's timelessness "justification can be found in the need to draw a proper distinction between the creator and the creature." Thus, "properties which the creator and his creatures have in common are distinguished by their mode of possession."[28] Though the biblical narratives describe God as speaking, and performing other temporal acts, Helm understands timelessness as offering a metaphysical underpinning for God's functioning as the biblical God. The question is not whether timelessness is a Greek notion or not, but "whether

---

[25] Leftow, *Time and Eternity*, 3.

[26] N. Pike, *God and Timelessness* (London: Routledge & Kegan Paul, 1970), ix.

[27] Ibid., 189f.

[28] P. Helm, *Eternal God* (Oxford: Clarendon Press, 1989), 17, 19; similarly Leftow, *Time and Eternity*, 66.

the thought that God is timeless is a necessary truth-condition of all else that Christians want to say of God."[29] It has consequences at the spiritual level: "The idea of God as timeless, as the changeless ground of all that changes, has profound implications for . . . the focusing of faith, hope, and love in what is unseen and eternal rather than what is visible and transient."[30] Leftow acknowledges that we tend to speak about God in temporal terms. However, existing timelessly may be treated as existing at some "date" called "eternity"—a date which is, however, not part of the time series, and hence does not stand in a relation of "before," "after," or "simultaneous."

The contemporary discussion regarding the viability of the concept of divine timelessness, framed in "possible worlds" semantics and the like, is subtle. It is beyond the scope of the present paper to take sides with respect to the outcome in the debate between philosophers such as Pike and John Lucas[31] on the one hand and Helm and Leftow on the other. Leaving the possible worlds of analytic philosophers of religion, we will now consider reasons provided by the given world—as described by quantum cosmology and quantum gravity—for taking seriously the option of timelessness. Against Pike's "I see no reason," there are three reasons arising from the encounter with cosmology as to why timelessness rather than everlastingness might have a place:

(1) Time is part of the created order. This is Augustine's view of *creatio cum tempore*, but it also seems a reasonable interpretation of most contemporary cosmologies, with their "internal" understanding of time (see above, 3.1). Hence, it is not meaningful to talk about God as if there was time before the creation—God as everlasting. Everlastingness would fit in the context of creation *in* time, rather than in theories which attempt to understand the creation *of* time.

(2) "Time" is not universally applicable in quantum cosmologies; classical spacetime is recovered only as an approximate, fuzzy notion and does not correspond to the whole of reality as described by the wavefunction, as it excludes, for example, the "imaginary time" realm (see above, 3.2). Hence, time is unlike traditional time at the most fundamental level of description, that of quantum gravity.

(3) The presence in physics of timeless descriptions, for example in terms of trajectories in phase space or of spacetimes, where the whole is a unit including all moments, suggests that it is possible to talk about the relation of God to this whole—and not only of the relation between God at one moment to the universe at that moment, differentiating moments in God.

---

[29] Helm, *Eternal God*, 22.
[30] Ibid., xiv.
[31] J. R. Lucas, *The Future: An Essay on God, Temporality, and Truth* (New York: Basil Blackwell, 1989).

I therefore maintain that it may be useful to attempt to understand, at least partly, God's transcendence with respect to spacetime as timelessness. This emphasizes God's unity with respect to the world. This leaves us with at least two possibilities.

If God is understood as *a being*—more or less the common sense theistic understanding, an assumption shared by Pike and Helm—there still might be an order, and perhaps even a flow, within God which could be labeled God's time. As my teacher in philosophy of religion, Huib Hubbeling, liked to ask: How could God otherwise enjoy music? If music is not enjoyable when all notes are played at the same moment, God's perfection, also with respect to aesthetical appreciation, requires that God has God's time. Karl Barth seems to have defended a similar distinction between ordinary time and God's time when he understood Jesus as the lord of time and distinguished between an uncreated time which is one of the perfections of the divine being and created time, with its succession of past, present, and future.[32] However, such a notion of "God's time," which would even be a universal time, is hard to fit in once time is thoroughly physicalized. One is not free to add one spatial dimension in contemporary superstring theories; adding another temporal dimension is at least as problematic.

An alternative would be to deny that God should be understood as a being, a single individual with attributes. God might, perhaps, be understood differently, say as "being itself," "the Good," or—as might perhaps be appropriate in the context of the natural sciences—"Intelligibility," hence more as an abstract entity.

4.2 *Divine Action*

Many theologians emphasize that time is a necessary component of a meaningful concept of divine action. There has been considerable discussion of divine action in recent philosophy of religion.[33] Some of these theologians and philosophers have defended the notion of specific acts of God in time. Others, such as Gordon Kaufman and Maurice Wiles, have opted for a more revisionistic position, seeing the whole universe as a single *master-act* of God. Leaving aside the many issues

---

[32] K. Barth, *Kirchliche Dogmatik* III/2 (Zürich: Theologischer Verlag, 1948), par. 47.

[33] See W. Alston, *Divine Nature and Human Action* (Ithaca: Cornell University Press, 1989), G. K. Kaufman, *God the Problem* (Cambridge: Harvard University Press, 1972); H. Kessler, "Der Begriff des Handeln Gottes: Überlegungen zu einer unverzichtbaren theologischen Kategorie," in *Kommunikation and Solidarität* (Freiburg, Schw./Münster: Exodus/édit. liberación, 1985); *Marburger Jahrbuch Theologie I*, hrsg. W. Härle, R. Preul (Marburg: Elwert, 1987; art. by R. Preul, W. Härle, H. Deuser, and C. Schwöbel); T. V. Morris, ed., *The Concept of God* (Oxford: Oxford University Press, 1987); *idem*, *Divine and Human Action* (Ithaca: Cornell University Press, 1988); T. F. Tracy, *God, Action, and Embodiment* (Grand Rapids: Eerdmans, 1984); K. Ward, *Divine Action* (London: Collins, 1990); and M. Wiles, *God's Action in the World* (London: SCM Press, 1986).

which arise from the dispute about science and specific acts, I will briefly summarize the "single act" position as it seems congenial to a cosmological point of view.

Kaufman regards activity proceeding from a single agent which is ordered toward a single end "one act," regardless of the complexity of the act or its end. Hence, "this whole complicated and intricate teleological movement of all nature and history should be regarded as a single all-encompassing act of God, providing the context and meaning of all that occurs."[34] Taking the whole as a single divine act might avoid problems linked with a more interventionistic account of divine acts in nature. God's master-act is understood to be the source of the overarching order itself. However, Kaufman insists on temporal order:

> It is meaningful to regard the fundamental structures of nature and history as grounded in an *act* (of God), however, only if we are able to see them as developing in time. An act is intrinsically temporal: it is the ordering of a succession of events towards an end. If we could not think of the universe as somehow developing in unidirectional fashion in and through temporal processes, it would be mere poetry to speak of God's act.[35]

Kaufman then continues with the claim that modern science, the Big Bang theory explicitly included, makes such an understanding of the universe possible, even though the teleological end is not well discernible to humans.

I have serious doubts about the possibility of defending a teleological end, in a temporal sense, to the universe, though some physicists have speculated about life in an indefinite future.[36] Besides, it is not clear whether Kaufman would be satisfied with the partial ordering of most spacetimes, the "block" view of relativity theories, or whether he insists on a flow of time and a universal notion of simultaneity. Such problems have already been raised above. Here, I want to pay some attention to the notion of temporality as used by Kaufman.

---

[34] Kaufman, *God the Problem*, 137.

[35] Ibid., 128.

[36] F. J. Dyson, "Time without End: Physics and Biology in an Open Universe," *Reviews of Modern Physics* 51 (1979): 447-460; idem, *Infinite in All Directions* (New York: Harper & Row, 1988); F. J. Tipler, "The Omega-Point Theory: A Model of an Evolving God," in *Physics, Philosophy and Theology*; idem, "The Omega Point as Eschaton: Answers to Pannenberg's Questions to Scientists," *Zygon* 24 (1989): 217-253; see also Drees, *Beyond the Big Bang*, 117-141.

The notion of temporality seems to be used at two levels. Kaufman ascribes temporality to God because he ascribes intentions to God. Time does function, for him, both within the universe, the created order, and beyond it—as a concept applicable to God (acts, intentions) as well. Is such an understanding of time as a universal background not challenged by the idea of the creation *of* time, which links time intimately with the whole created order rather than making it a universal category applicable both to God and to the created order? Would it not be possible to take more distance from the language of "acts," "causes," and the like in considering the relation between the physical universe and the divine? If the whole of spacetime is understood as a single act of God, could one not drop the notion of time (and causal action) at the meta-level? Might the concept of explanation in discussing the relation between God and the whole universe be more appropriate at that level than the concept of causation? This brings us to the next section.

4.3 *Divine Creation: God as Explanation of the Universe?*

> The only way of explaining the creation is to show that the creator had absolutely no job at all to do, and so might as well not have existed.[37]

Is there any need for introducing a creator of the universe? I will first point out some weaknesses in the claim that science provides a complete explanation of the universe. Next, I will consider two proposals for religious explanations of the universe (R. Swinburne, J. Leslie).

4.3.1 *Science as Explanation of the Universe?* Atkins, an eloquent defender of the view that science leaves nothing to be explained, puts great weight on reduction to simplicity. Beings such as elephants and humans arise through an evolutionary process given sufficient time and atoms; atoms arise given even more simple constituents. Perhaps the ultimate unit to be explained is, as Atkins suggests, only spacetime; particles being specific configurations, knots of spacetime points. The second major component in his argument is chance: through fluctuations, nothingness separates into +1 and -1. With such dualities, time and space come into existence. The +1 and -1 may merge again into nothingness. However, by chance a stable configuration may come into existence—for instance, our spacetime with three spatial dimensions and one temporal dimension.

Atkin's idea is based on a notion considered over a decade ago, "pregeometry," promoted by Wheeler. However, the fundamental issue has not changed significantly. For example, Hartle and Hawking wrote in their first article on the "no-boundary" cosmology that the wave function gives "the probability for the universe to appear from Nothing."[38] I would like to suggest that such claims face at least three kinds of problems.

---

[37] P. W. Atkins, *The Creation* (Oxford & San Francisco: Freeman, 1981), 17.
[38] Hartle and Hawking, "Wavefunction of the Universe," *Physical Review* D 28 (1983): 2960-2975; 2961.

(i) *Testability.* There is a plurality of fundamental research programs in cosmology. Experimental tests and observations may well be insufficient to decide among the more able contenders. Aesthetic judgments are, at least partly, decisive in opting for a specific scheme. However, what one considers elegant, another may reject.[39]

(ii) *Exhaustiveness.* Could a single and relatively simple complete theory be fair to the complexity of the world? Or, as Mary Hesse suggests, is it the case that for "the explanation of *everything* there must in a sense be a conservation of complexity, in other words a trade-off between the simplicity and unity of the theory, and the multiplicity of interpretations of a few general theoretical concepts into many particular objects, properties and relations."[40] We will return to this question, as it suggests a significant objection to a Platonizing theology related to cosmology.

(iii) *A vacuum is not "nothing."* The universe might be equivalent to a vacuum as far as conserved quantities go. Those conservation laws that are believed to be valid for the universe as a whole conserve a total quantity which may be zero. Take, for example, electric charge. Negative charges of electrons are matched by positive charges of protons. Atoms are electrically neutral. And so is, it seems, the observable universe. Even if negative and positive charges match, there still seems to be a lot of mass. The universe is, as far as mass is concerned, far from a vacuum: we encounter stars, planets, and people. However, in physics mass is not a fundamental concept; it is one of the positive forms of energy ($E = mc^2$). We need to take negative energy into account. It takes energy to launch a rocket; hence we say that the rocket has negative energy before being launched. In the universe the negative energy due to gravitational binding might equal the positive energy due to the mass-energy of the universe. Hence, the universe might well be equivalent to a vacuum, as far as energy is concerned. Similar arguments can be made about other properties: either they may total up to zero or they are not conserved. The universe may have arisen "out of nothing," at least without a source of materiality. The universe might be equivalent to a vacuum.[41]

The equivalence of the universe to "nothing" only holds net. It is like someone borrowing a million Dutch guilders and buying stock for that amount. That person would be as wealthy, fiscally speaking, as someone without any debts and without properties. However, the first would be of more significance on the financial market than the second. The first strategy also assumes more

---

[39] Stoeger, "Contemporary Cosmology and its Implications for the Science-Religion Dialogue," in *Physics, Philosophy and Theology*, 229; J. D. Barrow, *The World Within the World* (Oxford: Clarendon Press, 1988), 373; and Drees, *Beyond the Big Bang*, 66.

[40] Mary Hesse, "Physics, Philosophy, and Myth," in *Physics, Philosophy and Theology*, 197.

[41] See E. P. Tryon, "Is the Universe a Vacuum Fluctuation?" *Nature* 246 (1973; reprinted in *Physical Cosmology and Philosophy*, ed. Leslie [New York: Macmillan, 1990]): 396f. See also the discussion of creation in a background spacetime by Isham in this volume.

than the second: the financial system is taken for granted. Hence, as far as the conservation laws are concerned, the universe might come from a "vacuum," but such a vacuum is not nothing. The vacuum discussed here in the context of creation *in* time is a vacuum that behaves according to the (quantum) laws which allow for the fluctuations to happen—just as the apparent millionaire only can get started once there is a concept of money, of borrowing. Similar assumptions, though not about time, are in the background of the schemes regarding the creation *of* time, as Isham's contribution in this volume explicitly acknowledges. Thus, Atkins' account might still need some explanation for the laws or similar entities which govern the vacuum.

To conclude, perhaps scientific explanations may achieve a lot, but they do not explain without remainder. Could the remainder, such as the existence and the laws of the vacuum, be in need of a religious explanation, or at least support the plausibility of such a view? I will present two examples of the latter argument. The first is based on the theistic conception of a personal God as the preferred explanation for the universe. Another approach does not make the transition from a causal to a personal explanation, but rather from facts to values.

4.3.2 *Richard Swinburne: A Personal Explanation?* Assume that the most fundamental law, and its effectiveness, is scientifically (causally) inexplicable. In that case, Swinburne has argued, one has to face two possibilities: either that law is completely inexplicable or it has an explanation of another kind. Swinburne distinguishes between causal and personal explanations. A personal explanation should take its starting point from a person with intentions and certain capacities. These together determine the basic acts open to that person, such as raising one's hand. According to Swinburne a personal explanation cannot be reduced to a causal explanation. Even though physical concepts (such as muscle strength) are relevant to one's capacities, and brain states are linked to intentions, the correlations are not logically necessary.

Using this notion of "personal explanation" the most fundamental law of the universe might have such a personal explanation: that is the way God intended the universe to be. "The choice is between the universe as stopping-point and God as stopping-point."[42] According to Swinburne, a universe is much more complex than God. The supposition that there is a God is an extremely simple supposition. A God of infinite power, knowledge, and freedom is the simplest kind of person there could be, since the idea has no limitations in need of explanation. The universe, on the other hand, has a complexity, particularity, and finitude which cries out for explanation. Hence, the religious option is to be preferred over its alternative.

There is no explicit use of science in this argument. It might be rational and valid, but that is to be debated at the level of philosophical reasoning without

---

[42] Swinburne, *The Existence of God* (Oxford: Oxford University Press, 1978), 127.

support from science.[43] The scientific contribution lies in the description of the universe. However, if the choice between accepting the universe as a brute fact or as in need of an explanation of a different kind is justified by comparing the simplicity of the two hypotheses (as Swinburne does), it is a matter of the utmost importance to understand how complex or simple the two alternatives are. Many cosmologists believe that their theories are of an impressive simplicity and elegance in structure and assumptions, even if the mathematics are difficult. Whether this makes it more or less reasonable to regard the universe as a "creation" is not clear (why could one not believe that God made a universe with a simple structure?) but it does tend to undermine Swinburne's argument based on simplicity. And the more general idea of using a person as explanation for the universe introduces the problematic concept of a disembodied person and suggests another question: Why does that person exist? In addition, it seems to suffer from the problems surrounding the use of a single notion of "time" on two levels, as discussed above in relation to Kaufman.

4.3.3 *John Leslie: Creative Values?* Swinburne's approach does not offer an answer to the obvious question: Why does that person exist? If the person (God) explains the universe, who or what explains that person? A person is, according to our experience, an entity that can be or not be. However, values seem to be different. They seem to have something absolute about them. They might therefore be better candidates for a stopping-point in the quest for explanation than either causal or personal explanations. However, a value lacks effectiveness. Honesty may be a value, even though it is not realized automatically. In general, values do not bring about the corresponding states of affairs. Leslie, however, has defended the concept of creative values. He thereby places himself in a long philosophical tradition, which places the "Good" at the origin of all things. Plato seems to have held that knowledge and existence are both dependent upon the Good, the Good surpassing all existents in dignity and power.[44] Assuming Leslie's axiarchic principle that what is of value tends to come into existence, it may not be too difficult to argue for the necessity of consciousness, and hence for the necessity of characteristics like those our environment happens to have. Holding such a philosophical position, it is not surprising that Leslie has developed a strong interest in the argument from design in its contemporary cosmological form, based on the anthropic coincidences. Swinburne's position seems voluntaristic, the emphasis being on the will. Something would be good because God wills it. Leslie takes another stand: God may will something, if "will" is an adequate concept at all, because it is good.

The idea that values could be creative is highly speculative. Our experience is different: all too easily the good is neglected. Furthermore,

---

[43] For philosophical criticisms of Swinburne's argument, see J. L. Mackie, *The Miracle of Theism* (Oxford: Clarendon Press, 1982), 95-101; and J. Hick, *An Interpretation of Religion* (Basingstoke: Macmillan Press, 1989), 104-109.

[44] Plato, *Republic*, book VI (nr. 509). On creative values, see Leslie, *Value and Existence* (Oxford: Basil Blackwell, 1979). On the anthropic principles in this context, see Leslie, *Universes* (London: Routledge, 1990).

according to our experiences, values find their expression in human decisions. Thus, in his criticism of Leslie's position J. L. Mackie has stressed that the concept of "creative values" may well be a projection of our desire for things judged good on these things themselves, an objectifying of human desires and judgments.[45] Do values have a Platonic existence of their own, *a priori* of the things in which they are expressed? Or are dis-embodied values as problematic as dis-embodied persons? I wonder too whether such a view as Leslie's does sufficient justice to the problem of evil: the vulnerability of the good, the discrepancies between the real and the ideal.

Religious explanations of the universe, its existence and laws, seem to need assumptions about dis-embodied persons or values which are at least as problematic as the unexplained existence of the universe or its laws. Not being able to accept the finality of a scientific or a religious explanation, I think one does best in joining Charles Misner:

> To say that God created the Universe does not explain either God or the Universe, but it keeps our consciousness alive to mysteries of awesome majesty that we might otherwise ignore.[46]

4.4 *Some Theological Objections to Platonizing Theology*

Is a "Platonic" view of reality, which seems to correlate well with quantum cosmology and quantum gravity, a problem for Christian theology? Is there in this respect a genuine conflict between contemporary science and a Christian understanding of existence?

A Platonizing theology is certainly different in its understanding of God and the relations between God and the world from another view which would put more emphasis on temporality both in God and in the world. Some of those differences, with respect to divine eternity and divine action, have already been explored briefly above. There seem to be possibilities, however, for reconciling the changes in our understanding of time (and similar changes with respect to contingency, and the like) with a concept of God which is not totally discontinuous with the Christian tradition.

Temporal critical realists might object to the understanding of cosmology and its consequences for theology as presented here. As an epistemological position, it seems to be at odds with critical realism's view of the world as consisting of substantial entities and of science as continuously approximating the true ontology. Developments in cosmology exhibit significant discontinuities with respect to the ontologies suggested (see above, 2.3).

---

[45] Mackie, *The Miracle of Theism*, 230-239, especially 239; see for another friendly critic of Leslie's position J. J. C. Smart, *Our Place in the Universe* (Oxford: Basil Blackwell, 1990), 176, 180.

[46] C. W. Misner, "Cosmology and Theology," in *Cosmology, History and Theology*, ed. W. Yourgrau and A. D. Breck, (New York: Plenum Press, 1977), 95.

However important this objection may be, it concerns a philosophical issue, or perhaps even an empirical one,[47] rather than a religious issue.

There may be another objection, closer to the existential core of Christian belief. A Platonic view of reality seems to depict an abstract world such as it might exist in the mind of God or, perhaps, of a finite knowing subject, rather than the reality of matter and history. A Platonic view tends to emphasize the unity and coherence of the universe. Everything fits into an encompassing mathematical structure. Primacy of unity tends to go with a top-down approach. Reality, with its diversity, seems an illusion, since deep down there would be no diversity nor individuality. As far as there is diversity, it would not be good. The Good and the One are together on the divine side of being. In contrast, "nominalism" may be the symbol for the emphasis on diversity, and thus for a bottom-up approach. The diversity of things is the reality we encounter. Unity is our contribution in the process of description. Christian theology has its Platonizing trends, but it is also interested in particulars, especially in relation to our own being. The theologian Langdon Gilkey is thus wary of demands for total coherence:

> ... the incoherent and the paradoxical, the intellectually baffling and morally frustrating character of our experience, reflect not merely our lack of systematic thinking but also the real nature of creaturehood, especially "fallen creaturehood."[48]

Any comprehensive theological scheme is as much an attempt to think about diversity within unity as a complete scientific theory would be. They both remain open to a further consideration of their unity, and of the way that unity deals with the diversity. Gilkey's statement should not be misconstrued as to suggest that theology has no interest in the unity of an encompassing view. But Christian theologies should remain open to the diversity of experiences, even to the confusing and contradictory aspects of existence, such as evil. As I see it, the danger of too much emphasis on a timeless overall view of the universe might be that it enforces values which overemphasize unity and neglect diversity. Both unity and diversity should be part of a satisfactory view of the world. One might distinguish in the Christian tradition between strands that have more affinity with "mysticism," with a sense of unity or harmony with the divine, and strands that emphasize more the distance between the actual world or the actual behavior and the way it is intended by God, a "prophetic" stance. A Platonizing philosophy of nature might accommodate more easily a mystical strand in the Christian tradition than the prophetic strand, with its critical stance towards the existing order.

Is this an important conflict between a Platonizing interpretation of the universe and a Christian, existentially shaped, attitude in life? Is this conflict, if real, due to the limitations of such an understanding of the universe, or even a

---

[47] That critical realism might be empirically testable is suggested by. McMullin, "A Case for Scientific Realism," 29.

[48] L. Gilkey, *Maker of Heaven and Earth: The Christian Doctrine of Creation in the Light of Modern Knowledge* (Garden City: Doubleday, 1959), 37.

limitation to any understanding based on the natural sciences with their abstraction from particulars and from the present?

If the abstract, Platonizing character of cosmology is deemed a problem, one might opt for a "bottom-up" approach with respect to our knowledge, and thus make a turn to the subject who develops such Platonic views. However, before turning to more constructivistic approaches with respect to religion and science, I still need to rebut one possible objection against my argument so far: Why can temporal critical realists not dismiss quantum cosmology and quantum gravity as too speculative to be relevant?

5 *Why Critical Realists Cannot Dismiss Quantum Cosmology*

One might argue that a logical conflict may always be avoided. For instance, Philip Henry Gosse integrated evolution with belief in a recent creation by holding that the world has been created with all the evidence of a longer history—Adam with a navel and trees with rings.[49] Almost any conflict may be avoided by choosing appropriate additional hypotheses. However, I have not entered into this line of thought, as it tends to result in *ad hoc* solutions and escapes. And certainly, such *ad hoc* moves are not in line with the serious work done in science and religion.

An apparently more credible way for temporal critical realists to escape the challenges posed by quantum cosmology and quantum gravity would be to dismiss quantum cosmology as too speculative. Barbour, for example, has suggested that "we should consider only the broadest and most well-established features of the world disclosed by science, not its narrower or more speculative theories."[50] I will argue below that such a restriction to the scientific consensus is ineffective. This way of dismissing quantum gravity is *not* open to critical realists.

Another way to attempt to escape taking account of quantum cosmology would be to emphasize the intermediate, "human" or organic level of reality as the most relevant, and thus to defend passing over quantum cosmology, for example with respect to the non-applicability of time at "early times," such is the substance of Peacocke's remarks on the Hartle-Hawking imaginary time, referred to above (section 2.3). With respect to conventional quantum theory, Polkinghorne once played down its significance as it "only manifests its idiosyncratic character in processes of a smaller scale than normally concerns us."[51] Believing instead that quantum Russell humorously replied that "atoms may be small, but they're everywhere." He went on to show that quantum physics is relevant to all sorts of everyday phenomena, including visual

---

[49] P. H. Gosse, *Omphalos* (1857); see E. Gosse, *Father and Son* (London: Heineman, 1907; reprint, Harmondsworth: Penguin Classics, 1986), ch. 5.

[50] Barbour, "Creation and Cosmology," 143; similarly Peacocke, *Intimations of Reality*, 60f.

[51] Polkinghorne, "The Quantum World," 334.

perception, the stability of matter, and a host of other phenomena.[52] There is, of course, some relevance to Polkinghorne's position. One can still apply Newton's law of gravity for calculating orbits, and chemistry can be done without paying much attention to the quark structure of the atomic nuclei. I will nonetheless argue below (5.2) that in their quest for theologies with a credible ontology it is impossible to escape reflection on quantum gravity. This is because of the way the authors discussed here have emphasized the unity of the sciences and of reality: both are structured in hierarchical manners.

5.1 *Consensus and Speculation*

Restriction to the most soundly established features is typical of critical realism as defined by McMullin, for instance:
> ... the longterm success of a scientific theory gives reason to believe that something like the entities and structure postulated by the theory actually exists.[53]

The length of time during which the theory must be successful is unspecified, but the general requirement is nonetheless defended as reasonable. On this assumption, it seems reasonable to take the description of the universe as of a fraction of a second after the apparent "t=0" as reliable, since it has been accepted as successful for two decades—at least since the discovery of the cosmic background radiation in the mid-sixties. Future developments, as envisaged by the critical realists, may confirm a finite past or suggest an earlier cosmic epoch before the apparent "t=0."

The problem with this view is that it only considers one type of future development, an extension back in time. As argued above (2.3), the nature of the epistemological limit—the need to integrate quantum theories and spacetime theories around or before the Planck time—is such that it may well affect the entire interpretation of Big Bang theory. On the view of quantum gravity which we have followed so far, the quantum view of time turns out to be very different. Therefore, one also needs to reinterpret the concept of spacetime in Big Bang theory, and thus the part upon which there is consensus. Similarly, one could argue that the co-existence of a plurality of serious research programs which all accept the Big Bang theory in its "consensus" domain, shows that the Big Bang theory may well be open to a variety of future developments or interpretations, suggesting different ontologies.[54]

The variety of possible developments and interpretations of cosmology seems to offer temporal critical realists the possibility that there might be a cosmology which would fit their view of time, for example by assuming a Minkowski background spacetime. If a Minkowski background spacetime would

---

[52] Russell, "Quantum Physics in Philosophical and Theological Perspective," 369.

[53] McMullin, "A Case for Scientific Realism," 26.

[54] See the comparison of the approaches by Linde, Hawking, and Penrose in Drees, "Quantum Cosmologies and the 'Beginning'" *Zygon* 26 (1991): 373-396; and in *idem, Beyond the Big Bang*, 62-69.

not do, they could perhaps even opt for a cosmology with a universal notion of simultaneity so as to allow for a univocal definition of concepts such as "God's past" and "God's future." Among the serious contenders in cosmology the programs of Penrose and Vilenkin might be slightly more attractive to temporalists than Hawking's.[55] However, Penrose too has expressed the conviction that time and space may be notions of limited significance, and Vilenkin's approach was the one actually used by Isham in his contribution to set forth the features of quantum cosmology and quantum gravity, with "internal" time and "imaginary time."

If one allows theological or metaphysical preferences to be decisive in the choice of theories or in the interpretation of theories, some relief might be available for temporalists. However, a critical realist has to accept the fact that by the standards of critical realism certain possibilities are no longer open. One would have to accept two major transitions of "longterm success": the transition from Newtonian time to the conceptuality of special relativity, which led to the dismissal of absolute simultaneity, and the transition from the special to the general theory of relativity, which led to the transition from external to internal time. Hence, though some eclecticism might be permissible, one cannot—by the standard of longterm success—back away from quantum cosmology and gravity where it is in the process of developing such concepts.

In general, there seem to be four possibilities with respect to the way one deals with the consensus, and the lack of consensus, in science.

(i) *Eclecticism*: one takes whatever fits best. This attitude is present in much religious use of more speculative scientific statements, such as those of David Bohm or Wheeler. Taking one's selection from science in such a manner might be a contribution to the development of an intelligible and coherent view. However, it makes no contribution to the credibility of the position under consideration. It may even be to its disadvantage if the selection is made in a rather arbitrary manner.

(ii) *Cheap dismissal*: theologians need not pay attention to science, since the scientists are not themselves certain of their claims. Thus, theologians might claim to be free to hold whatever position they like. As long as the scientists, in this case the cosmologists, do not reach a consensus, anything goes. Such an approach would neglect the partial consensus among scientists. Though there is some variety of positions, many alternatives have been ruled out. Hence, to dismiss some science because of a lack of consensus is not really warranted.

(iii) *Cautiously wait and see* what will become the consensus. "Viable theologies need not, at the moment, even be consistent with such ideas, since these ideas are speculative at this time."[56] There is, I admit, no strict need for consistency with all speculative scientific theories. However, the scientific consensus is not that clear and safe either, as it is open to future developments

---

[55] Peters, "On Creating the Cosmos," 295; and Drees, *Beyond the Big Bang*, 68.

[56] H. L. Shipman, "The Creation of Order from Chaos," in *Creation and the End of Days*, ed. D. Novak and N. Samuelson (Lanham: University Press of America, 1986), 9.

which may cause significant reinterpretation. As Russell observed: "In active areas of scientific research, there are always numerous competing theories as well as competing interpretations of theories. If our strategy is to wait for agreement, I fear we will be limited to historical studies."[57] Moreover, he stresses that agreement is seldom unambiguous: what guarantees that it really has been reached? What about the eventual replacement of even "accepted" theories?

(iv) *Reverse eclecticism*: take the worst possible case. If one were able to show how it might be incorporated in a certain religious-metaphysical scheme, one would really have made progress. Taking science "where it hurts most" offers the greatest challenge, but also the greatest profit with respect to credibility.[58]

5.2 *The Unavoidability of Quantum Gravity*

Arthur Peacocke and other temporal critical realists have pleaded for a hierarchical view of reality, whereby higher levels of reality are constrained by lower levels but not determined by them, nor epistemologically fully reducible to them. A biological description of reality is not reducible to one in physical terms, though it cannot contradict the physical laws involved, such as the conservation of energy. Through this hierarchical structuring of the sciences, they are able to combine interest in the unity of the sciences ("One World") with an anti-reductionistic stance which, among other aims, intends to do justice to the distinct contributions offered by different disciplines. Roughly speaking, the hierarchy extends from the physical sciences through the life sciences up to psychology and cultural sciences, with theology occupying the uppermost level.

One may question the linear structuring of the sciences. It may be that some disciplines do not so much fit at one level as serve to integrate different levels. For example, one might defend the view that genetics in biology serves to integrate evolutionary biology, ecology, and ethology as higher levels with physiology, histology, and molecular biology at lower levels.[59] Similarly, one might claim that theology serves an integrating role rather than that it occupies the uppermost level. It might perhaps be more suitable to consider a network structure than a hierarchy. However, a network would do less justice to the

---

[57] Russell, "Quantum Physics in Philosophical and Theological Perspective," 370.

[58] L. Eaves, "Spirit, Method, and Content in Science and Religion: The Theological Perspective of a Geneticist," *Zygon* 24 (1989): 203. Eaves deals with the role of genetics, also in a response to the papal message of the 1987 conference: ". . . biologists need to be assured that their science is to be accorded the same sensitivity and respect that His Holiness' message has extended to physics" (L. Eaves, "Autonomy Is Not Enough," in *John Paul II on Science and Religion: Reflections on the New View from Rome*, ed. Robert J. Russell, William R. Stoeger, and George V. Coyne [Vatican City State: Vatican Observatory, 1990], 22). Similarly, I intend to warn against eclecticism with respect to the treatment of time in quantum cosmology.

[59] E. T. Juengst, "Response: Carving Nature at Its Joints," *Religion and Intellectual Life* 5 (Spring 1986): 70-78.

intuition that some sciences are more basic than others, though specific sciences could still be considered as constraints upon other sciences.

Let us assume the hierarchical structuring. A change in the concept of energy in physics would then imply changes in the way the metabolism of living organisms is understood. The discovery of the conservation of energy in the nineteenth century, as well as the proper definition of concepts such as free energy, has had its impact on higher level sciences, for instance those studying the metabolisms of organisms. If "energy" were discovered to be wrongly defined, biology would have to adapt. The constraints imply that changes in our understanding at lower levels of the hierarchy of the sciences have consequences for higher levels insofar as the higher levels use the same concepts and laws.

Now in the case of quantum gravity it is not the concept of energy which is at stake, but rather the concept of time. The changes initiated by the general theory of relativity have affected our understanding of the Newtonian theory. Not that it led to much change at the level of calculations done in a Newtonian framework, but it affected the assessment of the metaphysical adequacy of its view of space and time as absolutes. Similarly, quantum theories affected the assessment of the metaphysical adequacy of the billiard-ball view of material substance. The same should hold for a theory which affects our concepts of space and time: it should rank extremely low in the hierarchy of the sciences, due to the very general and basic concepts involved, and thus affect our view of all the other sciences. If, for example, time and space were shown to have a discrete rather than a continuous character, this would, in principle, affect our understanding of all the laws of physics—formulated as they are in terms of differential equations. If time were shown to be a derivative and not a fundamental concept, it would not be acceptable to treat time as a Newtonian, external absolute at higher levels—at least not for metaphysical purposes. Thus, one cannot dismiss quantum gravity in such a perspective as dealing with distances and durations which are too small to be relevant. The issue is not just quantum cosmology, but quantum gravity—the theory which would be the physical theory below the levels occupied by quantum theories and general relativity theory.

It would seem an unacceptable, *ad hoc* move for critical realists to dismiss quantum cosmology and quantum gravity just because its resultant view of time displeases them. And its understanding of the nature of time carries over to all levels of the sciences, including the life-sciences, the humanities and theology, due to the realist view which unites the sciences into a hierarchy, since quantum gravity would have to be located at the bottom, fundamental level.

An alternative might be to put less emphasis on the unity of the sciences. It might be that different sciences lead to different views of reality, but without allowing for the coherence and unity suggested by the hierarchical view. Rather, in the various sciences different views of reality are constructed, without claiming that they need to be ordered in such a hierarchical fashion. Thus, it might be possible that the concept of time used at different levels can vary. Such an attitude towards the sciences would, of course, be more modest in that it withdraws the metaphysical intention of achieving, or at least approximating, an encompassing view of all aspects of reality.

6 *Cosmology and Theology as Myths*

It is far beyond the scope of the present paper to attempt to do justice to the various ways in which less "realistic" approaches to theology have been developed. One might think of nominalistic strands in the history of theology, the various turns to existentialism, ethics, narrative, social struggle, and the rules of discourse in language games. Many of those less realistic approaches lack interest in the dialogue with the natural sciences. However, this is not necessarily the case. In order to show the existence, and to suggest the viability, of a third approach in science and religion, in addition to temporal and Platonic realism, some aspects of the work of Hesse on science and theology will be discussed.[60] The last section (6.2) will suggest some lines along which I think that one might develop such ideas into a substantial theological position.

6.1 *Mary Hesse: The Construction of Reality*

> The problem is essentially not one of scientific "realism," but of communicative strategy.[61]

Mary Hesse has defended a network model of science. Such a model tends to stress instrumental goals, valuing prediction, and control, "at the expense of realism, if realism is interpreted in terms of universalizable theoretical explanation."[62]

She has called her position on various occasions "realism," but qualifies that significantly. For example, what counts as the primary individuals is theory-relative; they may well be superseded in another theory. There is, taking her view, no reason to deny the existence of something formerly referred to as "phlogiston." However, its "what" is not decided thereby. "What those substances or those atoms actually *are*, is something whose description changes from theory to theory, and will never be finally settled as long as science continues to develop. Theories about essences are neither stable nor cumulative, and are therefore not part of the realistic aspects of science." Thus, theoretical descriptions asserting that space is Euclidean or that it has non-zero curvature are

---

[60] Another example would be the hermeneutical approach advocated in science-and-religion by a group of the University of Neuchatel (Switzerland). An extensive presentation of the theoretical ideas and various applications, for instance on Artificial Intelligence, came to my knowledge only after concluding this contribution. P. Bühler, P.-L. Dubied, C. Karakash, O. Schäfer-Guigner, and G. Theissen, *Science et foi font système: Une approche herméneutique* (Geneva: Labor et Fides, 1992).

[61] Hesse, "Retrospect," in *The Sciences and Theology in the Twentieth Century*, 287.

[62] Hesse, *The Structure of Scientific Inference* (London: Macmillan, 1974), 284. One could also consult her *Revolutions and Reconstructions in the Philosophy of Science* (Brighton: Harvester, 1980), especially 63-110. She acknowledges her debt to W. V. Quine for the term "network model."

unstable and non-cumulative. This does not exclude accumulations of approximate forms of law. "If such accumulation of approximations is thought insufficient for 'realism', then this account of science may be called instrumentalist, but there are other respects in which it is nearer to realism." It certainly is not the case that anything goes. Science is a learning process, with systematic self-correction. "Natural scientific inference has rational grounds, but these are essentially finite and local in application, and determined by empirical conditions of testability and self-correction."[63]

In the Gifford lectures delivered by Michael Arbib and Hesse, learning is discussed in a wider context. Taking up a suggestion from Piaget, the concept of a schema becomes central. Schemas may represent objects or actions, with perceptual schemas serving to supply the parameters that afford the action. In the line of Piaget's work on the development of schemas, one might think of them in a dynamic Kantian way; "the categories are no longer a priori, but change over time."[64] Schemas do not arise as ideas in isolation. They are not closed semantic nets, as dictionaries are—words explaining words. Interaction with the world, in perception and action, is central, and thus is embodiment. It is essential to link the development of knowledge structures in artificial intelligence with "being in the world."

In the context of the network model, Hesse has emphasized the limited scope of theories and the role of the pragmatic, instrumentalistic criterion. This has:

> . . . negative implications for the universal ontological and cosmological consequences that have sometimes been held to derive from natural science. There has been a constant tendency for the prestige of instrumental success to flow back into temporary ontologies and analogies, and to infect social and metaphysical thought about the nature and destiny of man and the universe.[65]

Thus, "no truths about the substance of nature which are relevant to metaphysics or theology can be logically derived from physics." "No substantial consequences about the world can be drawn from this game [science] except what were put into it."[66]

The use of science in theology might be apologetic, a matter of communication and status. "It would be a mistake now, as it was then, to build the details of such models of causality too firmly into our doctrine of God. They may provide useful analogies for apologetics and a useful liberation from too constrained a notion of God, but they are not essential to central theological beliefs, nor can they logically disprove such beliefs."[67] In the context of the

---

[63] Hesse, *Structure of Scientific Inference*, 299, 300, 302.

[64] M. A. Arbib and Hesse, *The Construction of Reality* (Cambridge: Cambridge University Press, 1986), 45.

[65] Hesse, *Structure of Scientific Inference*, 301.

[66] Hesse, "Physics, Philosophy and Myth," 189, 198; cf. earlier remarks regarding the technical role of the background in quantum cosmology (e.g., notes 17 and 19).

[67] Hesse, "Physics, Philosophy and Myth," 191.

present article, it may be of interest that she applies this also to considerations regarding the static or dynamic nature of reality, even paying some attention to quantum cosmology.[68]

In any case, it is unprofitable in an antimetaphysical age to seek to make the world safe for religion by metaphysics. Such a procedure is anachronistic and intellectually barren for believers and unbelievers alike. But there is no need for it. In relation to the Christian religion, at least, there are no intellectual foundations for belief except in the continuing tradition of practice, theology, and changing historical experience, which are all rooted in the Great Schema itself.[69]

Rather than looking for scientific contributions to a metaphysical theology of nature in the traditional sense, it may be more fruitful to regard science as consisting in "debates about an appropriate *language* for theology, and a source of appropriate *models*."[70] Hence, the issue is how theological concepts "may be expressed in a language accessible to those nurtured in the scientific framework." Science and theology "meet on the ground of different but comparable *social symbolisms* rather than of common subject matter or of method."[71]

6.2 *The Christophoric Circle*

Hesse focuses on epistemology, interpretations of physics and theology, rather than on interpretations of nature. However, some ontological consequences seem to follow. One is the emphasis, especially in the Gifford lectures, on "embodied existence," on the essential role of perception and action in relation to thinking. Another, related one, is the emphasis on the interaction if not interwovenness of truth and value, for example in her understanding of language. If embodiment, perception, and action, are taken to be central, the *imago Dei* notion cannot be focused exclusively on human reason or rationality, but should relate to action, as expressed in the phrase *imitatio Christi*. This may well be illustrated with the legend of Christopher.

Christopher—not yet bearing that name—was an impressive figure, strong as a bear. He wanted to serve no one but the highest king. He thus went to serve the greatest human king—until he discovered that the king was afraid of the devil. He then served the devil, but discovered that the devil avoided crossings. Thus, Christopher discovered that Christ must be greater than the devil. He then longed to serve Christ, but could not find him. Advised by a hermit, Christopher took on a humble task that fitted his capacities—helping people to cross a river. After many years a child called upon him to help him across. But the child turned out to be unexpectedly heavy. It was Christ, who carries the world—depicted in many images as a globe.

---

[68] Ibid., 200 n.13.
[69] Arbib and Hesse, *The Construction of Reality*, 243.
[70] Hesse, "Retrospect," 287.
[71] Ibid., and 282.

There is a circularity in the story which is not fully captured in the common depictions of it: Christopher carries the child while the child carries the world—on which Christopher should be understood to be standing. It seems to me an apt representation of a religious attitude which acknowledges the human, constructive side of faith: images of God are our schemes, they exist in us, individually and socially. But those human images are images that intend to express something that transcends us, both quantitatively (a persistent mystery beyond) and qualitatively (a greater and different love, a higher perfection than we will ever realize and an otherness which confronts us). The child that we carry may be the child that carries the world, including ourselves. As acting and thinking come together, the relation between values and facts—axiology and cosmology one might say—comes in sight. Some contemporary contributions to "science and religion" seek to relate religious thought to the contemporary world view suggested by the sciences. If that approach is taken in isolation, cosmology may, for example, by analogy and extrapolation, be related to order and design, to a positive view of God as the Maker of Heaven and Earth. The extrapolation might also, due to awareness of epistemological limitations on cosmology, be seen to suggest a mystery beyond knowledge.[72] Such ways of understanding God may be deemed pale and irrelevant to our individual and social existence.[73] An attractive feature of a more self-consciously limited view of scientific knowledge is that it leaves room for a more independently formulated understanding of God. But the advantage may also turn into a disadvantage: too much freedom, as mutual irrelevance, may lead to a loss of credibility.

One way to keep a proper distance from cosmology, of allowing for the correct amount of independence of cosmology from theology, is the turn to axiology, to the ethical and existential decisions that humans have to face, as the primary locus of theology. This may lead to questions about the nature of values in a world of facts. One might consider granting values a Platonic kind of existence, or point out how they function in the social world, arguing that science is unable to measure them as values. Elsewhere I have tried to define theology formally as the attempt to think the unity of cosmology and axiology, of "facts" and "values," whether as being in harmony or not.[74] I prefer to hold that values should interrupt and confront our behavior (facts) with something else, with what might be considered as God's intentions, if we are to do justice to the prophetic strands in the Christian tradition. This may be neglected when too much emphasis is placed on the coherence of the various sciences and theology, on consonance between our understanding of physical reality and of God, as seems to be the intent of many more realistic theologies, and even more so of so-called "new age" religious philosophies. A lack of interest in the manifold tensions typical of human existence, as seems to characterize most Platonizing theologies,

---

[72] For instance in M. K. Munitz, *Cosmic Understanding* (Princeton: Princeton University Press, 1986).

[73] Hesse, "Physics, Philosophy and Myth," 199.

[74] Drees, "Theologie en natuurwetenschap: onafhankelijkheid en samenhang," in Hans Küng *et al.*, *Godsdienst op een keerpunt* (Kampen, The Netherlands: Kok, 1990).

may as well endanger the possibility of a confrontive, interruptive style of ethical thinking. On the other hand, religious thinking which restricts itself to ethical issues may fall short in providing a basis for motivation and empowerment, even in the face of failures. As in the image of Christopher and the child, we have to act, to carry the child, but we do so on the assumption that we are carried by something—power, mystery, love?—far beyond us.

V
THEOLOGICAL IMPLICATIONS 2: THE LAWS OF NATURE

THE THEOLOGY OF THE ANTHROPIC PRINCIPLE

G. F. R. Ellis

1 *Introduction*

The anthropic principle is concerned with the issue of why the universe is of a nature that allows intelligent life to exist. There are three different contexts for the analysis of this principle: namely, on the basis of a religious understanding alone; on the basis of scientific understanding alone; and on the basis of a combination of religious and scientific understanding together. The first has been discussed down the ages, particularly in the form of the argument from design. The second has been the subject of much recent discussion.[1] The third has been formulated for example by Ernan McMullin[2] and Robert Russell,[3] but has not been developed as much as the others. It will be my aim to pursue the topic from this third stance, taking into account both modern scientific understanding and also a Christian view of the religious world order. The new aspect of the argument will be taking into account a more detailed view of the nature of Christianity than is customary in such arguments (where the aspect emphasized is often simply that of God as creator).

To keep the argument within bounds, I will make no effort to consider all possible interpretations of the Christian tradition, but will base my theological comments mainly on the first chapter of John's Gospel, and in particular on the analysis of that chapter given by William Temple in his superb book, *Readings in St. John's Gospel*.[4] The result is what may be termed old-fashioned theology; there is nothing new here in the theological description. Nevertheless by placing this in the context of present questioning of cosmology and the anthropic issue, the analysis may be useful as a contribution towards a synthesis, an integration of scientific and theological world views, as suggested by Ian Barbour.[5] This

---

[1] See J. Barrow and F. Tipler, *The Anthropic Cosmological Principle* (Oxford: Oxford University Press, 1988); J. Leslie, *Universes* (London: Routledge, 1989); and Y.V. Balashov, "Resource Letter AP-1: The Anthropic Principle," in *American Journal of Physics* 59 (1991): 1069-76.

[2] E. McMullin, "How Should Cosmology Relate to Theology?" in *The Sciences and Theology in the 20th Century*, ed. A. R. Peacocke (Notre Dame: University of Notre Dame Press, 1981).

[3] R. J. Russell, "Cosmology, Creation, and Contingency," in *Cosmos as Creation: Theology and Science in Consonance*, ed. T. Peters (Nashville: Abingdon Press, 1990).

[4] W. Temple, *Readings in St. John's Gospel* (London: Macmillan, 1961).

[5] I. Barbour, "Ways of Relating Science and Theology," in *Physics, Philosophy and Theology: A Common Quest for Understanding*, ed. Robert John

attempt is motivated by (1) recent developments in cosmology (the anthropic debate) which begin to touch on these boundaries, and (2) the wish to attain a unity of understanding, rather than a schizophrenic situation of two separated worlds of science and religion. Indeed the renewed moves to attain such a unified view may be regarded as contributing to one of the most important debates in the history of human thought, certainly worthy of much attention and careful analysis.

To put the whole into perspective, brief summaries are given first of the nature of argumentation involved, and of the standard scientific approaches to the anthropic idea. This provides the framework for the later sections, where the core of the argument is presented.

2 Modes of Knowing and Confirmation

Science and theology each provide a pattern of understanding.[6] Each uses a model of reality to provide explanation of aspects of the real world; the way they do so, by arguing deductively from the basic hypotheses made, is broadly similar in the two cases. However, each has a different area of interest, and employs different kinds of confirmation. Science has explanatory power, enabling understanding of the impersonal forces that control the natural world and, to some extent, of living systems and even of the mind. It can give very precise predictions of the behavior of matter, in the case of simple physical systems. Theology relates to the nature of the mysterious reality underlying both personal and community moral and religious lives. It does not make precise predictions, as sometimes are attainable in the natural sciences. It does, however, provide an understanding of our own humanity, its deepest needs and desires, of how we ought to live, and why we ought to live as we know we should. It also provides images of what is acceptable behavior and what is the good life.

The fundamental issue that arises in both theology and science is *the basis of knowledge*. This question occurs because of the problem of the contrast between surface appearances and underlying reality; more precisely, because of the hidden nature of the physical world on the one hand, and the hidden nature of God on the other. We cannot, for example, tell simply by looking around us the nature of the chemical elements, or of the four fundamental forces of physics; nor can we tell the nature of fundamental reality (the foundation of our being) or of the ethical life, in any easy manner. There is not even any obvious evidence of the existence (or non-existence) of God. Thus a major question is: "What are the methods we can use to discern the true situation?" This is relatively clear in science in general (although there are scientific domains where it is not so clear),

---

Russell, William R. Stoeger, and George V. Coyne (Vatican City State: Vatican Observatory, 1988).

[6] I am indebted to Nancey Murphy for helping formulate the following. See N. Murphy, *Theology in the Age of Scientific Reasoning* (Ithaca: Cornell University Press, 1990).

but is more problematic in the field of theology and ethics, where there is less agreement.

The basis of knowledge in science is formation of theories of behavior that are subject to confirmation by (a) observations and repeatable experiments, leading to predictions (based on an underlying theory) of the behavior of physical systems, which can be tested and confirmed—if the same conditions are set up again ("the systems are prepared identically"), the same behavior will follow. This enables us to establish general laws of behavior. In principle we can each have the personal experience that confirms scientific theory, by performing such experiments. However in practice we cannot each carry out all the experiments at the base of modern science, so in fact we also rely on (b) acceptance of scientific tradition and authority, in particular (c) as contained in recognized authoritative papers and textbooks. Nevertheless it is essential at the base of the subject that these sources are, at least in principle, subject to the test of experimental confirmation or denial, and their authority is limited by that feature. Furthermore we demand of a good theory (d) overall satisfactoriness as a system of explanation (does it "look right"), particularly through unification of previously disparate areas into a logically satisfactory scheme; however here problems arise in terms of the subjectivity of this criterion.

Theology (whose object is totally different from that of science) cannot be based on this strict kind of testing as it deals with more complex events, where it is impossible to set up the same situation again to test what will happen, or to obtain certainty of what will happen in any particular case. One can nevertheless try to detect common patterns underlying the experiences of many individuals and their understandings of ultimate reality. In the case of Christianity, the basis of knowledge is a combination of (a') personal experience and conviction of the presence and nature of God; (b') acceptance of tradition and authority of religious institutions such as the Church, and of those claiming to be religious leaders; and (c') belief in what we take to be God's word as revealed in the biblical record, and most specifically in the record of the life of Christ.

There are two problems with this approach. First, these sources of religious knowledge are often in disagreement with each other. There are competing revealed texts and authorities; how do I decide which is the true one? This must be by criteria outside the text itself. Again, how can I tell true visions from false ones? This must be through some independent faculty of evaluation.[7] Second, each of these kinds of source can be shown to have led to serious error in the past (indeed the historical record of those claiming to follow them is, in a large part, a very sorry story); so none of them is infallible. It is very hard to escape from delusion or self-serving in our interpretations. While science has developed tests that are quite effective in exposing false belief within its domain of competence, religion has not been so effective in this enterprise.

In the end perhaps the real tests of belief in the religious sphere are (d') overall satisfactoriness as a system of explanation of the world and of our lives (does it "look right"); and (e') the test "by their fruit ye shall know them," which

---

[7] See the section on *discernment* in the article by Murphy in this volume.

is the major way of confirming or denying what we believe in this sphere.[8] Together they may be summed up, "The test of the validity of the exercise of theological investigations will lie in its ability to discern pattern, to offer coherent understanding of human experience at its most profound."[9] Two issues arise here.

First, the result of the test of "good fruits" is far from controversial. A skeptic can claim that neither the Church as a whole (historically considered, or at present), nor most individuals in it, are examples making one want to follow their ways. Old examples are legion, including the inquisition, the crusades, the corruption that led to the Reformation (which itself soon led to vicious repression); in recent times we have the lynching parties of Southern Baptists, the Apartheid Church in South Africa, the killings in Ireland for which a religious base is claimed, and so on. In a less dramatic setting, it is possible that a statistical analysis of an average suburban congregation would show that the members are more loving and caring than their neighbors who do not belong to the Church, but that outcome is far from evident. One can perhaps claim it is the evidence of the few, those exceptional shining cases, that is the proof in the case of religion, indeed in the end this is probably a major source of belief. But then this is quite unlike science, where we may dispute odd experiments that contradict our theories, but it is the overwhelming statistical weight of evidence that carries the day, when evaluated in the light of the explanatory power of different theories. In the case of religion, we seem to believe despite the negative results of the test in a large number, possibly the majority, of cases. It seems to be the exceptional cases that provide the weight of evidence. A different kind of logic is at work in the evaluation.

Second, stating these criteria leaves unanswered the question of how we decide when a system of explanation is satisfactory or the outcome of some action is good; these judgments again have to based on some combination of personal conviction and belief in authority, once more resulting in a potential or actual relativization of standards. Thus, inevitably, there is an element of subjectivity in these matters, in the sense that our conclusions depend on our culture and education, as well as our particular historical circumstances and our maturity and level of understanding—that is to say, on our wisdom and personal insight, for these underlie our moral judgment.

This individuality is the basis of the wide divergence of opinions on religious matters. Yet the capacity of the human mind to critically discern the true and the good is mysterious but undeniable. Furthermore we need such standards and understanding in order to live meaningfully. Those who hold to a religious faith must believe that it is possible to discern the meaning of life and how we ought to live, however inadequately, and that over time humankind will move towards a truer understanding of religious as well as scientific issues—towards a consensus, if not on ultimate meaning, at least on codes of behavior and responsibility.

---

[8] In the case of science, this relates to the social impact and use of science, but not its truth value.

[9] J. Polkinghorne, *Science and Creation*, (Boston: Shambhala, 1988), 96.

Thus theology and science construct different symbolic universes, making truth claims supported in different ways, but with clear similarities, as emphasized by the above comparison. However it is fundamental that science rejects any revealed scripture as an ultimate source of knowledge. Furthermore, it must be emphasized that science is not just a social construct; it has hard predictive power. The impression of relativization derived from Kuhn's work is a partial truth; the answers obtained by science do indeed depend on the questions asked but, given that choice, there is often no ambiguity about the answer (at the classical level). What happens (given the initial conditions of the system and a description of its dynamical nature) is certain. For example, if I am hanging from a tree and then let go, I will fall; there is no uncertainty about this. Furthermore, in effect, we do not have a choice about some of the questions that are asked, for nature itself asks them; for example, the motion of planets in their orbits and the passing of time happen whatever we may wish or do, and quite independently of any social or cultural factors.

There is a large difference in the end between science and religion, as evidenced by different degrees of belief in their value and applicability. The scientist will ask: "What is an experimental test I can make that will support or deny the central theses 'God exists' or 'God is Love'?" It is precisely because there are no indisputable criteria for answering such questions conclusively that many scientists do not believe in religion. For some, the theological scheme is powerful as an integrating view and is supported by personal experience, but there are undoubtedly people for whom it is meaningless; and there is no indisputable evidence to make them change their stance. This underlies the unequal relationship of these disciplines in the modern world. However it is precisely because of the demand for solid confirmation that the domain of science is limited and omits many aspects of life of fundamental importance to humanity, particularly relating to value judgments and meaning. It is for this reason particularly that the thoughtful person must search for a broader world view than that provided by science alone.[10]

3 *The Issue of Ultimate Causation*

The religious and scientific approaches are both in principle relevant when we consider the nature of the universe itself and the underlying issue of ultimate causation. Here is one of the areas where both may make truth claims. Indeed some viewpoint on this issue is necessarily implied by scientific discussions of the origin of the universe, even if this is not made explicit.[11] However it should be noted that particular models of the nature of the physical universe (a "Big-Bang" origin, a steady state universe without temporal origin, a universe originating according to the Hartle/Hawking "no-boundary" ideas) do not

---

[10] This theme is developed in depth in G. F. R. Ellis, *Before The Beginning: Cosmology Explained* (London: Bowerdean Press, 1992).

[11] Ellis, "Major Themes in the Relation between Philosophy and Cosmology," *Memoirs of the Italian Astronomical Society* 62 (1991): 553-605.

necessarily affect the issue of ultimate causation; *a priori*, any of the possible ultimate causes could be compatible with any of these modes of realization of a physical universe.[12]

There appear to be five basic approaches to ultimate causation: random chance, high probability, necessity, universality, and design. We briefly consider these in turn.

3.1 *Random Chance*

Conditions in the universe just happened initially, and led to things being the way they are now, by pure chance. Probability does not apply. There is no further level of explanation that applies.

This is certainly logically possible, but not satisfying as an explanation except to a total reductionist (we obtain no unification of ideas or predictive power from this approach). Nevertheless some scientists implicitly or explicitly hold this view.

3.2 *High Probability*

Although the structure of the universe appears very improbable, for various physical reasons it is in fact highly probable (the chaotic cosmology idea). These arguments are only partially successful even in their own terms, for they run into problems if we consider the full space of possibilities (many discussions implicitly or explicitly restrict the allowed set of possibilities *a priori*); and we do not have a proper measure to apply to the phase space, enabling us to assess these probabilities. Furthermore, application of probability arguments to the universe itself is dubious because the universe is unique. Despite these problems, this approach has considerable support in the scientific community.

3.3 *Necessity*

Things have to be the way they are; there is no option. This can be taken in a strong form or a weak one. The strong form is the claim that *the features we see and the laws underlying them are demanded by the unity of the universe* (coherence and consistency require that things must be the way they are; the apparent alternatives are illusory). Thus it is the claim that only one kind of physics is self-consistent; all logically possible universes must obey the same physics. The weak form is that *only one kind of physics is consistent with the sort of world we actually see around us*. To really prove either would be a powerful argument, potentially leading to a self-consistent and complete scientific view; but we *can* imagine alternative universes! Why are they excluded? Hermann Bondi has emphasized that insofar as the unity view is valid, sufficient study of

---

[12] See Barbour, *Religion in an Age of Science*, The Gifford Lectures, 1989-1991, vol. 1 (San Francisco: HarperSanFrancisco, 1990); Peacocke, *Creation and the World of Science* (Oxford: Clarendon Press, 1979); and W. Drees, *Beyond the Big Bang: Quantum Cosmologies and God* (LaSalle, IL: Open Court, 1990).

any part of universe will reveal its whole structure, because of bonds of necessity.[13] However a partial counter-argument is provided by considering the locality of physics: we are able to predict what will happen in a laboratory without knowing the total state of distant regions of the universe. Furthermore we run here into the problem that neither the foundations of quantum physics nor of mathematics are on solid, consistent foundations. Until these issues are resolved, this line cannot be pursued to a successful conclusion.

3.4 *Universality*

This is the stand that "All that is possible, happens";[14] an ensemble of universes *is* realized in reality. There are three ways this has been pursued.

The view may be that this happens *in space* through random initial conditions,[15] as in chaotic inflation.[16] While this provides a legitimate framework for application of probability, from the viewpoint of ultimate explanation it does not really succeed, for there is still then one unique universe whose (random) initial conditions need explanation. Initial conditions *might* be globally statistically homogeneous, but there also *could* be global gradients in some physical quantities so that the universe is not statistically homogeneous. This is really a variant of the "high probability" idea mentioned above.

Alternatively, it could occur through the existence of the Everett-Wheeler *"many worlds" of quantum cosmology*,[17] where all possibilities occur through quantum branching. This view is controversial; it is accepted by some but not all quantum theorists. If we hold to it, we then have to explain the properties of the particular history we observe. (Why does our macroscopic universe develop to have high symmetries when almost all these branchings will not?)

Finally they could occur as *completely disconnected universes* as a sort of ultimate logical extension of the Feynman approach to quantum field theory; there really *is* an ensemble of universes.[18] There is then the problem of what

---

[13] H. Bondi, *Cosmology* (Cambridge: Cambridge University Press, 1960).

[14] Leslie, *Universes*, ch. 4; and D. W. Sciama, "The anthropic principle and the non-uniqueness of the universe," in *The Anthropic Principle*, ed. F. Bertola and V. Curi (Cambridge: Cambridge University Press, 1989).

[15] Ellis "The Homogeneity of the Universe," *General Relativity and Gravitation* 11 (1979): 281.

[16] A. Linde, "Particle Physics and Inflationary Cosmology," *Physics Today* 40 (1987): 61.

[17] See J. J. Halliwell, "Introductory Lectures on Quantum Cosmology," in *Quantum Cosmology and Baby Universes*, ed. S. Coleman, J. B. Hartle, T. Piran, and S. Weinberg (Singapore: World Scientific, 1991); and R. Penrose, *The Emperor's New Mind: Concerning Computers, Minds, and the Laws of Physics* (Oxford: Oxford University Press, 1989).

[18] Sciama, "The anthropic principle and the non-uniqueness of the universe."

determines what is possible? What about the laws of logic themselves? Are they inviolable in considering all possibilities?[19]

In all three approaches, on the one hand, major problems arise in relating this view to testability, and so we have to query the meaningfulness of the proposals as scientific explanations; on the other, in each case, in order to explain our actual observations, one has of necessity to introduce an (anthropic) selection element (see the following section), for most of the universe(s) will *not* look like an isotropic model; why do we live in a region that does? Furthermore they all contradict the Occam's razor approach to physics: they are all very uneconomical in their mode of explanation.[20] Nevertheless this approach has an internal logic of its own some find compelling.

3.5 Design

The symmetries and delicate balances we observe require an extraordinary coherence of conditions and cooperation of laws and effects, suggesting that in some sense they have been purposefully designed, that is, they give evidence of intention, realized both in the setting of the laws of physics and in the choice of boundary conditions for the universe.[21]

This is the basic theological view. To make sense of it, one must accept the idea of *transcendence*: that the designer exists in a totally different order of reality or being, not restrained within the bounds of the universe itself.[22] A scientific analogue of this idea is given by the now commonplace concept of the imbedding of the universe in a higher dimensional space, where ultimate reality (the higher dimensional space) is of a different order from the reality experienced by those restricted by their structure and sensory apparatus to the four-dimensional imbedded spacetime. On this kind of view the universe can be viewed from outside itself as a whole (incorporating all history: past, present, and future) and creation involves the causation of this whole.[23] The ultimate issue of existence of this higher reality remains unanswered (as it does in all approaches).

Scientifically, the problem is that this view leads to no cosmologically testable predictions (we have no solid argument, for example, why the existence of God should imply the nature of the chemical elements or features such as the density of matter in the universe), and so to major problems of confirmation; thus much scientific effort is aimed at using the other approaches to avoid this option.

---

[19] Leslie, *Universes*, 97-98.

[20] P. Davies, *The Mind of God: The Scientific Basis for a Rational World* (New York: Simon & Schuster, 1992).

[21] Barrow and Tipler, *Anthropic Cosmological Principle*; see also G. Gale, "A Revised Design: Teleology and Big Questions in Contemporary Cosmology," *Biology and Philosophy* 2 (1987): 475.

[22] See McMullin "Natural Science and Belief in a Creator: Historical Notes," in *Physics, Philosophy and Theology*.

[23] See ibid.; and Peters "On Creating the Cosmos," in *Physics, Philosophy and Theology*.

However it should be noted that untestability is one of the major features of the roots of cosmology; it seems unavoidable in *all* approaches.

The concept of design invoked here is more complex than one might at first suppose, on two counts.[24] First, as will be discussed later, the kind of design envisaged includes leaving large areas of reality "free from definite design," so that we might freely respond by adding or determining those free sectors through our own "designing" activity. Second, a simple idea of design does not apply, because for every species now in existence, thousands of species are extinct, representing evolutionary experiments and blind alleys leading "nowhere."[25] Thus natural selection gives no support for supposing a direct process of design. What we have to do is reconcile a lack of design at these levels with the "meta-design" that allows and even favors our existence in the long term. Presumably the resolution is similar to that of the problem of evil, briefly touched on later: given the laws of physics, chemistry, and biology that allow our existence, an unavoidable concomitant is such dead ends; they are the price one has to pay for the incredible results achieved.

3.6 *Multiple Causation*

In approaching these topics, it is important to note that there can be multiple layers of explanation and meaning imbedded in any particular situation.[26]

A classic example is the case of a girl who dies after being run over by car, where the question "Why?" has many answers: because her mother told her to go to the shop; because she was inattentive at the crossing; because the car was going too fast; because it had been raining; because of Newton's laws of motion; because her heart stopped beating; and so on. Similarly if we consider the word-processor I am using to write this article, it expresses many levels of organization and so embodies many layers of meaning.[27] It results from economic interchange between Japan and the West; it embodies design principles developed by Turing, Von Neuman, and many others; it functions through the principles of quantum physics applied to silicon chips developed through a process of discovery involving the development of the transistor and of VLSI circuits; it is being used as a tool in a debate over the fundamental principles underlying the existence of the universe, in the context of institutions whose purpose is to further fundamental knowledge.

In particular, we can have "laws of organization" as well as "fundamental laws," each expressing important aspects of reality. The organization in information-based feedback-control systems is quite unlike that in chaotic or fractal systems, statistically based systems, or those directly based

---

[24] I am indebted to Bill Stoeger for these remarks.

[25] E. Mayr, *Towards a New Philosophy of Biology* (Harvard: Harvard University Press, 1988).

[26] Davies, *The Cosmic Blueprint: New Discoveries in Nature's Ability to Order the Universe* (New York: Simon & Schuster, 1988); and Barbour, "Ways of Relating Science and Theology," in *Physics, Philosophy and Theology*.

[27] Davies, *The Cosmic Blueprint*.

on idealized or broken group symmetries such as crystals. Furthermore significant information-processing seems to always be based on open systems with irreversible processes taking place, unlike any classical or quantum reversible process. But it is such organization and processes that underlie living systems, where there are multiple layers of emergent organization superimposed on each other.[28] They are an expression of higher laws of organization and meaning at work that will only be possible for restricted sets of fundamental laws, which serve as a substratum or basis for such higher organization. Not all sets of fundamental laws can do this.

Realizing these levels of meaning and organization, we can avoid the mistake of assuming that the various ultimate causes can be assigned to only one particular pattern of immediate cause. For example a designer, if there be one, need not be restricted to either a single act of creation, or evolution, or multiple universes, or any other specific mechanism. The same applies to pure chance; it could also operate through any of these mechanisms. It is not impossible that more than one of the lines of thought I have above characterized as "fundamental approaches" could apply to the real universe.

4 *The Anthropic Issue*

The anthropic issue is presented in depth in the book by Barrow and Tipler[29] and in Leslie's book.[30] The questions arising are: "Why do we exist at this time and place in the universe? Why does the universe permit the evolution and existence of intelligent beings at *any* time or place?"

To understand the import of these questions, one must appreciate the incredible complexity of what has been achieved: the structure and function of a single cell is immensely complex.[31] However, a human grows to an interconnected set of 10^{13} cells, all working together as a single purposive and conscious organism in a hierarchically controlled way (the *organization* issue), able to function continuously all the time as the number of cells increases from 1 to 10^{13} (the issue of *growth*); all of this happening within an interacting set of organisms of similar levels of complexity within a hospitable environment (the *ecosystem* issue); this system itself developing from a single cell to the level of complexity we see around us today (the *evolution* issue); all the while remaining functional. And all of this is possible because of the nature of quantum mechanics (essentially the Schrödinger equation) and of the forces and particles described by physics (essentially the electromagnetic force acting on the proton and the electron, together with the strong force binding the protons and neutrons in the atomic nuclei), which together control the nature of chemistry and hence of biological activity. The nature of this achievement is truly awesome.

---

[28] See N. A. Campbell, *Biology* (Redwood City, CA: Benjamin Cummings, 1990).

[29] Barrow and Tipler, *The Anthropic Cosmological Principle*.

[30] Leslie, *Universes*.

[31] See Campbell, *Biology*.

Furthermore, one must emphasize the *stability* of the order embodied in living systems; the design each creature embodies is very stable to changes in the environment (within a restricted range; for example, handling a range of temperatures and foods), to a variety of assaults on the creature (e.g., the way antibodies repel invaders), and even to some physical damage (through the healing of wounds). This is the antithesis of chaos, and is completely unlike any non-living form of order (except that of cybernetic systems; for example, aircraft automatic pilots or robotic machine tools designed to function along the same lines and resulting in apparently purposive actions, but immensely simpler than any living system). Particular forms are created and maintained in a stable way despite perturbations of many kinds and great individual variety.

4.1 *The Scientific Approaches*

There seem to be essentially two viable scientific approaches: the weak anthropic principle (WAP), and the strong anthropic principle (SAP).

4.1.1 *The Weak Anthropic Principle* On this view, the question should be seen as relating to a selection effect: we can observe the universe only from places and times where intelligent life can exist (and can have evolved).[32] At first this seems merely a tautology, but it is more than that. It can help provide new insights into relations in physics and biology by asking: "How much variation in laws and initial conditions can there be, and still allow intelligent life to exist, including the unique features of the mind such as consciousness and self-consciousness?"

Thus, for example, if we consider laws of physics different from those we experience but something like them certain things follow. We then require the existence of heavy elements, sufficient time for evolution of advanced life forms to take place, regions that are neither too hot nor too cold, restricted values of fundamental constants that control chemistry and local physics, and so on.[33] Thus only particular laws of physics, and particular initial conditions in the universe, allow the existence of intelligent life. There are four interesting specific features that are of importance. First, as emphasized by Davies, the concept of locality is fundamental (local systems are able to function independently of the rest of the universe). In order that we can function as we do, the laws of physics must have this nature. Second, the existence of an arrow of time, and hence of laws like the second law of thermodynamics are probably necessary for consciousness. Third, physical conditions must be in a quasi-equilibrium state (or the delicate balances that allow our existence and evolution will not be fulfilled). Finally, the

---

[32] B. Carter, "The Anthropic Principle: Self-Selection as an Adjunct to Natural Selection," in *Cosmic Perspectives*, ed. C. V. Vishveshswara (Cambridge: Cambridge University Press, 1988).

[33] B. J. Carr and M. J. Rees, "The anthropic principle and the structure of the physical world," *Nature* 278 (1972): 605; and Barrow and Tipler, *The Anthropic Cosmological Principle*.

emergence of a classical era from a previous quantum domain is necessary for the existence of forms of life anything like the ones we know.[34]

Given such laws and initial conditions, the WAP can in principle help determine from what situations we can observe a universe, not only in spatially homogeneous universe models (where life can only exist at particular times) but also in the case of completely random initial conditions, as in chaotic inflation, or if there are multiple universes (life will in general only be able to occur in rather specific places; for example, in our solar system, life can only develop in a rather narrow range of distances from the Sun). In particular we may attempt to argue (with various degrees of success) that life can only evolve in regions in the universe that are like a spatially homogeneous and isotropic universe model, thus explaining why we see such a high degree of isotropy in the cosmic background radiation.

Overall, this viewpoint provides a partial but illuminating answer. Its use is necessary to explain why we observe a spatially homogeneous region in models like the chaotic inflationary universe; this particular argument has not yet been carried through in a thoroughly convincing manner, however. It gives an indication that too great a deviation from the aged, almost isotropic universe we see around us would prevent the existence of life, but does not very accurately predict that it *has* to be as isotropic as we see; it indicates that the constants of the laws of physics cannot deviate too much from their observed values and allow life to function, but does not fix their values precisely.

Causally speaking, it is a conservative approach; it does not throw light on *why* life exists. In particular, it does not help us understand why the laws of physics and actual initial conditions allow our existence unless we use a "universality" picture where all different laws occur as well as all initial conditions, so that life is bound to occur; it then helps characterize the conditions that will be observed by intelligent observers.

4.1.2 *The Strong Anthropic Principle* This is the much stronger statement: intelligent life *must* exist in the universe. It is a necessity. The strongest scientific basis for the claim is the argument that intelligent life is required for the consistency of quantum mechanics (which in some formulations depends in a crucial sense on the concept of an "observer").[35] However there is a considerable problem in showing this is indeed a requirement, in view of the divergent interpretations of quantum theory. This is compounded by a lack of testable consequences so far, so this claim is very controversial.

Also, this is again an intermediate answer: suppose we could confirm this hypothesis. Then we can ask, why is this (quantum mechanics) necessary? It takes us back to a chain of regression; it is not obviously necessary that quantum mechanics should be valid, unless we can conclusively argue that physics *must* include quantum features to be consistent; that is, by using a "necessity"

---

[34] See the article by C. J. Isham in this volume.

[35] Barrow and Tipler, *Anthropic Cosmological Principle*; and Isham, "Quantum Gravity," in *GR13: General Relativity and Gravitation*, ed. M. A. H. MacCallum (Cambridge: Cambridge University Press, 1987), 99.

argument, such as one based on the stability of matter. However this is clearly difficult because quantum theory itself neither has fully satisfactory foundations, nor is fully self-consistent in the sense of avoiding singularities. This quantum mechanical argument therefore does not command a large following, and so there seems no genuine and widely accepted scientific justification for the SAP.

4.2 *Ultimate Causes*

The viewpoint taken here is that life is the ground of consciousness, which is such a unique feature of the universe that its existence seems to require special explanation.[36] The failure of the SAP, scientifically considered, means that if we take seriously the issue of why life exists, and pursue the chain of causation to some ultimate cause (see the last section), the pure scientific approach fails to give a satisfactory answer.

What ultimate cause can lie behind the existence of life? "Pure chance" is very difficult to sustain, despite its logical unassailability, for it seems a totally inadequate explanation of the complexity we see (it must be emphasized that the order of complexity in life is totally different from that in, say, a crystal structure, a rock, a mountain range, or a fire). We cannot at present satisfactorily complete the argument of "necessity," or even of "high probability."

Some form of "universality" is a possible ultimate view, although one might maintain that this is really just a more sophisticated version of pure chance. It certainly suffers from severe problems of confirmation. Comparing the different possibilities, it is difficult to avoid the conclusion that the "design" concept is one of the most satisfactory overall approaches,[37] necessarily taking us outside the strictly scientific arena (which is not surprising: as the universe is unique—we cannot observe any other universes—it cannot be subject to confirmable scientific laws).

One should note here that Gale made the suggestion that "teleological reasoning does not require conscious purpose [because] goal-directed systems may be fully described in terms of negative and positive feedback mechanisms, using the logic of its tendencies and potentials."[38] This is correct as far as it goes, but avoids the full implications of the design issue by not asking the next question: "Why do the tendencies and potentials existing in the universe have the precise form that leads to this result?" This leads back to the issue of the nature of fundamental forces, and why they not only admit the existence of life but even (on this argument) *prefer* it. A random set of physical laws certainly would not fulfill this condition. We may take it as established that the detailed design of life as we know it has been achieved through evolution. The laws of physics and chemistry are such as to allow the functioning of living cells, individuals, and ecosystems of incredible complexity and variety, and it is this that has made

---

[36] Davies, *The Mind of God.*

[37] See W. Norris Clarke "Is a Natural Theology Still Possible Today," in *Physics, Philosophy and Theology*; Murphy in this volume; and Ellis, *Before the Beginning.*

[38] Gale, "A Revised Design."

evolution possible. What requires explanation is why the laws of physics are such as to allow this complex functioning to take place. On the design view, it is precisely in the choice and institution of particular physical laws, allowing evolutionary development, that the profound creative activity takes place.

As noted previously, the scientific method by itself avoids this approach because of its lack of scientific confirmation and strict predictive power. This argument can provide only an indication of the existence of a designer, not a proof. Nevertheless it does provide such an indication, through the evidence of the fine-tuning required to allow life to exist, given the basic nature of physical laws as we know them.[39] It carries substantial weight if one adopts a viewpoint broader than that which can be contained in a purely scientific analysis, and takes seriously the incredible nature of what is achieved by the universe, in allowing the existence of humankind.[40]

4.3 *The Need for a Combined Viewpoint*

In summary, the essential possibilities that arise in the anthropic issue are, (a) it can interpreted in terms of a selection principle, but then there must be an ensemble of universe states in which it can act; one needs to account for the existence of this ensemble of universes, and give some hint of how the proposal could be confirmed; and no ultimate explanation is considered. Alternatively, (b) it can plausibly be interpreted in fundamental terms as either due to pure chance, or else as purposeful design. In the former case we have a complete but unsatisfying explanation; in the latter case there is nothing more for science *per se* to say (from the scientific viewpoint, it will have occurred just by chance, for science itself does not have room for a designer).

If we look at the situation from a purely scientific basis, we end up without any solid resolution, basically because science attains reasonable certainty by limiting its considerations to restricted aspects of reality; even if it occasionally strays into the area, it is not designed to deal with ultimate causation. Thus something like a religious viewpoint is required to make progress, because religion is indeed concerned with ultimate issues. The argument that follows will develop the hints above, taking into account both scientific and religious viewpoints. It will be claimed that the anthropic question can be viewed in this way without there being an incompatibility with science, and that, indeed, a far more satisfactory overall view is attained than if we restrict our considerations to the purely scientific.

---

[39] Leslie, *Universes*, ch. 3.
[40] Ellis, "The Homogeneity of the Universe."

5 *The Christian Setting*

The Christian basis I will assume will now be summarized. It is based on the kinds of foundation mentioned in the second section; it is supported by texts from the Bible that are central to the New Testament message, and by a particular church tradition, but in the end its validation lies in the "rightness" of the view, plus the nature of its potential outcome (the goodness of what can be achieved if this view is taken seriously and put into practice).

5.1 *A Particular View*

The discussion is necessarily very brief, and so partial. It contains what I believe to be an essential core of the message of the New Testament, but omits features that some regard as important (for example it contains no stand on the nature of atonement, of the Trinity, or of resurrection). However those aspects are often divisive, as evidenced by the different views of them held by different sectors of the Christian tradition. The supposition here is that a detailed position on these issues, which may be important to individuals, is inessential in our search for a common core of understanding and belief.

This stand is based on the Quaker view that much of the usual theological doctrine, together with creeds, are unnecessary, and are often as misleading as they are helpful (as will necessarily be the case with any attempt to capture the nature of ultimate reality through language).[41] In particular, any representation or human concept of the nature of God will be partial and suggestive only, and misleading in some aspects. Personal experience and common action, in the attempt to make a reality of the gospel message, are more important than many words, and in a profound way form the basis of our understanding. The revelations given to us through the events and reflections recorded in the Bible should be interpreted through our own personal understanding, as mediated and broadened through our life in the community of the faithful, for no form of authority—Scripture, tradition, bishops, priests, politicians, or any dogma—can be taken as absolute; indeed any attempt to represent any of them as absolute or infallible is idolatry, an attempt to replace the living God by some usurper, an idol created by human minds.

In fact the ultimate task of each person is to exercise his or her own personal responsibility and understanding in response to the message of the Gospels in his or her own life situation; acceptance of any outward authority can be a form of denial of that responsibility. A right response is made possible because each person has an innate understanding, an inner light ("the Spirit," in more common parlance), that will guide one if one makes oneself open to it and listens to it. As *every* person has such a capacity, one has to take seriously (but not necessarily accept) what any other person says about fundamental issues, in particular those others who also claim a Christian commitment. That is, at all times one must adhere to what one understands and believes, acting as best one

---

[41] See Rufus F. Jones, *The Faith and Practice of the Quakers*, 4th ed. (London: Methuen & Co., 1930).

can in this light, yet also allowing the possibility that one might be wrong, and so remaining open to the insights that others may have to impart. Thus much of what is often taken as doctrine (which is communally agreed) can, on this view, be regarded as incidental; such beliefs can be understood as an individual affair, reflecting the position of each person as best helps his or her understandings. What works well for one person may not mean much to another (this is inevitable when using any imagery to represent ultimate reality). This does not matter, provided the individual holds to the central broad themes and is open to new understandings, continually seeking the truth. This approach is in good accord with the understanding of knowledge that is central to this paper (see section 2).[42]

5.2 *Attributes of God*

"Theology faces a particular difficulty in that the nature of its Object transcends us and our power to grasp him. We do not have the words and concepts to encapsulate God ... thus the language of theology is the language of symbol."[43] We must continually bear in mind the limitations of our language and our concepts when referring to God's transcendent reality. Accepting this, I propose to work from the following basic viewpoint:

(1) God is the creator and sustainer of the universe and of humankind, transcending the universe but immanent in it.
(2) God's nature embodies justice and holiness, but is also a personal God and loving God who cares for each creature (so the name "father" is indeed appropriate).
(3) God's nature is revealed most perfectly in the life and teachings of Jesus of Nazareth, as recorded in the New Testament of the Bible, who was sent by God to reveal the divine nature, summarized in "God is Love."
(4) God has an active presence in the world that still touches the lives of the faithful today.

These four statements have many deep implications. Characterization as "creator" implies the notion of design, implemented in terms of particular substance ("matter"); in relation to that material, God is necessarily omnipotent and transcendent. As mentioned before, one can use as a model of transcendence the idea of higher dimensions and imbedding, implying a transcendence of space

---

[42] The reader may or may not accept the particular standpoint just stated. However in what follows, only the main thesis presented below as fundamental to Christianity, is essential. One may wish to add further items over and above what is proposed here (e.g., specific theories of the Trinity or of atonement). Doing so will tend to restrict the message to some particular section of the Christian Church, which is what I have tried to avoid.

[43] Polkinghorne, *Science and Creation*, 94.

and time limitations, and so allowing for immanence (the meaningful presence of God at each time and in each event).

The characterization as "father" implies that God's nature is personal in some sense, with the fullness that this involves of feelings, will, desires. This is confirmed by the concept that humankind is made in the image of God, therefore expressing in the fullness of humanity some of God's essential nature. Thus each person can be called a son or daughter of God. It should be emphasized one could equally call God "mother," indicating the fullness of his/her nature; purely for convenience I will use the traditional language here.

For Christians, the main vehicle of revelation of the full nature of God is the historical life of Jesus. The precise nature of the relation between God and Jesus is impossible to characterize fully in human terms, and is in any case irrelevant; we simply need to know that he was fully human and also fully the son of God, sent to give us a revelation by demonstrating in his life God's nature (as far as it is possible for someone who is fully human to do so). This mysterious gift of his life was an act of supreme love by God (John 3:16).

The nature of the "active presence" in the world will be discussed in more detail later on; it is what would be referred to by many as "the Spirit." Its important practical implication is that there is in each person an "inner light" (John 1:9), giving inspiration and direction to those open to its guidance.

5.3 *Modes of Action*

Given the omnipotence of God, as creator of all that there is, the question arises as to how God makes use of this omnipotence. The answer given to us is: in a completely loving way. It is this that needs explication, and is clarified by the life of Jesus; indeed it is the central message of the New Testament. One characterization of the manner of God's action, is that *it always transcends the immediate problem by changing to a context of loving care, moving to a higher plane where love and forgiveness are the basic elements*. This change of view has the possibility of transforming the situation. It became explicit in the life of Jesus, as he examined his possible choices of action, and saw the transcending possibilities opened by the way of love. This is described by William Temple as follows:

> The Lord, at his Baptism, is conscious of the call to begin now the work of the Messiah. At once, therefore, he goes into the solitude to consider what manner of Messiah he shall be. The story of the temptations is, of course, a parable of his spiritual wrestlings, told by himself to his disciples. It represents the rejection, under three typical forms, of all existing concepts of the Messianic task, which was to inaugurate the Kingdom of God. Should he use the power with which he was endowed to satisfy the creature wants of himself and his human brethren, so fulfilling the hope of a "good time coming?" Should he be the Caesar-Christ, winning the kingdoms of the world and the glory of them by establishing an earthly monarchy? Should he provide irresistible evidence of his divine mission, appearing in the temple courts upborne by angels, so that doubt would be impossible?

> Every one of these conceptions contained truth. Yet if any or all of these are taken as fully representative of the Kingdom, they have one fatal defect. They all represent ways of securing the outward obedience of men apart from inner loyalty; they are ways of controlling conduct, but not ways of winning hearts and wills. He might bribe men by promise of good things; he might coerce men to obey by threat of penalty; he might offer irresistible proof; [but] all these rejected methods are essentially appeals to self-interest; and the kingdom of God, who is love, cannot be established that way.
>
> He has stripped the Messiahship bare, repudiating all existing conceptions of it. Only the essential task remains—to inaugurate the Kingdom of God. He starts his ministry, leading the life of perfect love, and teaching the precepts of perfect love. What we find is power in complete subordination to love; and that is something like a definition of the Kingdom of God. The new conception which takes the place of those rejected is that the Son of Man must suffer. For the manifestation of love, by which it wins its response, is always sacrifice. The principle of sacrifice is that we choose to do or suffer what apart from our love we should not choose to do or suffer. When love is returned, that sacrifice is the most joyful thing in the world. The progress of the Kingdom consists in the uprising within the hearts of men of a love and trust which answer to the Love which shines from the Cross and is, for this world, the Glory of God.[44]

The fundamental importance of this revelation is two-fold. First, it shows us the way we should act if we are to be true to a Christian calling. The choice for the Kingdom is the choice for generosity and the forgiving spirit of the Kingdom. This does not mean compromising truth; it does mean creating hope of reconciliation—all activities can be forgiven; anybody can be redeemed. Our acts and spirit of forgiveness should demonstrate this reality in loving sacrifice. This entails learning to give up that which we cling to, accepting the loss as the basis of greater good. All of this leads to a profound view of how to live at all levels of life. According to Robert Bellah:

> The deepest truth I have discovered is that if one accepts the loss, if one gives up clinging to what is irretrievably gone, then the nothing which is left is not barren but is enormously fruitful. Everything that one has lost comes flooding back out of the darkness, and one's relation to it is new—free and unclinging. But the richness of the nothing contains far more, it is the all-possible, it is the spring of freedom.[45]

The attempt to follow this way is incredibly difficult. It will be much easier if we are able to practice the presence of God, and particularly the awareness of the Light of Christ within every man and woman.[46] Indeed the two are inexorably

---

[44] Temple, *Readings in St. John's Gospel*, xxix-xxxii.

[45] R. N. Bellah, *Beyond Belief* (New York: Harper & Row, 1976).

[46] J. Pickvance, *George Fox on the Light of Christ Within* (New Foundation Publications No 3, 1984).

linked, for if we are aware of that presence and its loving nature, we will see the present problems in this profound context and their nature will be transformed for us (for example, making it impossible to kill or torture other people, and demanding that we try to help them in any way we can in their struggles in life). Such awareness will also give us support in following this way.

Second, this revelation shows the nature of God's action in the world. Given the established natural order, created and sustained by God, divine action in human life is through images of love and truth, not through any form of coercion. This mode of action is a voluntary choice on the part of the creator, made because it is the only way to attain the goal of eliciting a free response of love and sacrifice from free individuals. It implies total restraint in the use of his omnipotent power, for otherwise a free response to his actions is not possible. It is this mode of action that can guide us in the interpretation of the anthropic principle on the basis of a Christian understanding.

6 A Combined View

Our aim now is an analysis of the anthropic principle, in relation to design and the issues of omnipotence and transcendence. I give a central argument relating to the nature of design and omnipotence, followed by subsidiary arguments related to design and multiple universes, and to the nature of immanence and revelation. Finally the synthesis is supported by reference to the prologue of St. John's Gospel. The mode of argumentation here attempts to exploit the common features of theology and science, explored in the second section, thereby taking both into account.

6.1 Anthropic Design in a Christian setting

The key idea is that *the fundamental aim of loving action, characterized in the previous section, shapes the nature of creation, in particular setting its meaning and limitations.* Thus we take seriously and pursue the implications of the concept that the purpose of the universe *is* precisely to make this kind of sacrificial response possible. We do so understanding that the creator *could* have ordered things differently, but has voluntarily and specifically restricted the nature of creation to that required for this purpose.

In this spirit, we answer the issue of the existence of the universe and the nature of physical laws by the highly controversial but clear statement, *the universe exists in order that humankind (or at least ethically aware self-conscious beings) can exist.*[47] The further question of "Why?" is answered by the claim that this is done so that love may make itself manifest, an obviously and patently worthwhile purpose.[48] It is in such an approach to the anthropic question

---

[47] See e.g., Polkinghorne, *Science and Creation*, 23.

[48] However, it must be noted that this overall view does not give clarity on the extent to which long-range purpose is incorporated in the "laws of nature" at any given level. See Mayr, *Towards a New Philosophy of Biology*.

that the design concept attains its greatest strength—provided we consider the context as not merely related to physics and chemistry but to the full nature of our existence, with our fears and hopes, love and caring, value judgments, ethical choices and moral responsibility, whose reality I take to be at least as indubitable as any other area of experience.

From a materialistic viewpoint, this is a grossly anthropocentric view of the nature of the universe, for—materialistically considered—the human species is an insignificant feature in the vast realms of space, filled with vast numbers of galaxies each of which is made of vast numbers of stars. However the religious view will be that it is precisely the highest levels of order in the known universe (the extraordinary structure and function of the human body), making possible moral and religious understanding, that gives the whole its ultimate meaning and indeed its rationale for existence.

This does not necessarily mean that the whole enterprise is concentrated on humankind specifically (that is, on the species that has evolved on the particular planet Earth in our particular galaxy); but rather that it has its meaning through laws of physics and chemistry that will allow evolution of intelligent and responsive life in *many* places in the universe.[49] It is the total moral, ethical, and religious response of all these beings that reflects the underlying purpose. Thus the viewpoint is not anthropocentric in a Copernican sense of being limited to our particular existence alone, but incorporates that existence—certainly taken as being valuable and of significance—into a broader and more democratic view of the value of all intelligent life in the universe (which almost certainly exists in many places as well as the earth). This concept raises religious issues of significance, discussed later.

Given this basic aim, we shall examine its implications for the creation process. There are five major points.

6.2 *The Ordered Universe*

First, there is a need for the creation of a universe where ordered patterns of behavior exist for, without this, free will (if it can be attained) cannot function sensibly. If there were no rules or reliable patterns of behavior governing the activity of natural phenomena, it would not be possible to have a meaningful moral response to the happenings around one. Thus the material world, through which sentient beings are to be realized, needs to be governed by repeatable and understandable patterns of events.

---

[49] I. S. Shlovsky and C. Sagan, *Intelligent Life in the Universe* (New York: Dell, 1966); and Barrow and Tipler, *Anthropic Cosmological Principle*. If we accept the pessimistic assumptions of Barrow and Tipler, there is one advanced civilization in our galaxy, and so about 10^{11} in the observable region of the universe. I discount here the dogmatic but purely hand-waving argument of Mayr (Essay 4 in Mayr, *Towards a New Philosophy of Biology*), which centers on talk of probabilities but estimates none. If he did estimate them, his final answers would be wrong by fourteen orders of magnitude, for Mayr apparently believes there are 10^8 stars in the observable region of the universe (see page 67), as against the real figure of about 10^{22}.

One way of attaining this is through physical laws as we know them. A difficult question is whether this is the *only* way to attain such repeatable patterns of behavior; the answer is not clear. This is partly because we do not understand the underlying basis of physical laws, in the sense of knowing how the behavior they characterize is imbedded in matter (Is there in some sense a mathematical formulation of these laws embodied in reality? Is there some kind of template for each kind of particle, embodying its physical behavior but not described in a mathematical sense? Is the behavior the result of the creator simply imaging the desired results, and requiring the realized structure to conform?) However it seems likely that in whatever way they are realized in practice, they will indeed be experienced as "laws" underlying the regularity required (the issue of why they should be describable in mathematical terms is a deep one, but the answer is not essential to the further argument here).[50]

For want of a better way of understanding them, I will assume the desired regularities are indeed attained by setting particular laws of physics in action and then letting them run their course; more accurately, sustaining them so that they remain in action for the life of the universe. Without this, everything would be chaotic and formless. Thus we envisage the creator at all times maintaining the nature of the physical world so that a chosen set of laws of physics govern its evolution. It must be emphasized that once this choice has been made, providing it is adhered to (as will be assumed), then the action of laws will be seen by us as absolute and rigorously determining the behavior of matter. One can then act freely within the confines of the laws, but the laws themselves cannot be altered by any human action.

6.3 *The Anthropic Universe: Free Will*

However, we require much more than this; we require that these laws and regularities allow the existence of intelligent human beings who can sense and react in a conscious way, and who furthermore have effective free will. The word "effective" here means that whatever the underlying mechanisms governing human life, there must be a meaningful freedom of choice which can be exercised in a responsible way (for without this, the concept of ethics is meaningless).

We here touch on issues that science has not seriously begun to comprehend. We do not understand the nature of consciousness, nor whether the "free will" we experience is real (albeit constrained by many psychological and behavioral factors), or illusory. Some scientists maintain that it is indeed illusory, but in practice do not behave as if this were true; they are liable, for example, to talk about the difference between responsible and irresponsible behavior—a distinction which is meaningless if free will is not real.

If we consider, then, how to implement laws leading to beings with free will, we imply acceptance of the conclusions resulting from the anthropic discussion of the previous sections. The SAP would in effect be realized, but without the need for a basis in quantum theory; and necessarily all the restrictions

---

[50] See the article by Davies in this volume.

implied by the WAP as conditions for the existence of life (for example, restrictions on the nature of physical laws and limits on the value of the fine structure constant) must be fulfilled. The strictly scientific criteria of the WAP come into play, and give observable statements on the nature of the physical universe (for example, that there must be times and places where the background temperature is lower than 3000K, above which temperature a living being's body could not function, for all atoms would be ionized).

However, more is implied for we need to ensure the conditions required to attain free will (not normally emphasized in discussions of the anthropic principle). As just stated, we do not know what these conditions are, except of course that they are compatible with the laws of physics and chemistry as we experience them. Nevertheless it seems probable that fixed laws of behavior of matter, independent of interference by a creator or any other agency, is a requisite basis for existence of independent beings able to exercise free will, for they make possible meaningful, complex organized activity without outside interference (physical laws providing a determinate frame within which definite local causal relations are possible). Thus we envisage the creator choosing such a framework for the universe (thus giving up all the other possibilities allowed by the power available to him, such as the power to directly intervene in events by overruling the laws of physics from time to time). This voluntary restriction on the nature of creation makes possible the other major desired features, as we shall shortly see. From this viewpoint, fine-tuning is no longer regarded as evidence for a designer, but rather is seen as a consequence of the complexity of aim of a designer whose existence we are assuming; it is not plausible that this complexity of function can be achieved (within the context of established laws of behavior) without fine-tuning.

A fundamental question is: "Are the features of pain and evil implied in every universe that allows free will?" Almost certainly, the answer is Yes—because of the very nature of free will; for any restrictions on the natural order that prevented the selfish use of will which is the foundation of evil action would simultaneously destroy the possibility of free response and loving action, which is the aim of the whole. This is discussed further by Polkinghorne.[51]

6.4 *The Provident Universe*

Given the existence of creatures with free will, one can still imagine universes arranged so that this will is constrained in an essentially unfree way, contrary to the spirit set forth in the last section. In the temptations, Christ rejected the use of force to establish allegiance. He also rejected the strategy of making material comforts conditional on obedience. The same factors need to be built into the creation of the universe, for otherwise a free response would not be possible.

This is achieved by the impartial operation of the laws of physics, chemistry, and biology, offering to all persons alike the bounty of nature,

---

[51] Polkinghorne, *Science and Creation*. See also the response to Polkinghorne by Russell, "The Thermodynamics of 'Natural Evil,'" *CTNS Bulletin* 10:2 (Spring 1990): 20-25.

irrespective of their beliefs or moral condition. The major requirement here is that the laws of physics allow the growth of food for all humankind. (This is part of the basic anthropic presupposition: humankind could not evolve were this not so, so the very existence of humanity, in a world governed by physical laws, guarantees this requirement.) Rain falls alike on believer and unbeliever, and makes their existence possible (as opposed to a universe where, for example, rain only falls if you praise God—presumably a perfectly possible arrangement). This mode of operation of the physical world thus fulfills the condition of freeing people from a need for obedience to God in order to survive, and so makes a free and unconstrained response possible.

6.5 *The Hidden Nature*

A further requirement must be satisfied to enable the free response envisaged by Jesus in his response to the temptations: that the created world not be dominated by God himself, striding the world and demanding obedience on pain of punishment (as in the myths of some other religions), or alternatively dominated by explicit marks of his activity so that belief in his existence and nature would be forced on everyone—with a resulting demand on their behavior. This would be true, for example, if providence were arranged so that food was a direct result of prayer and sacrifice, rather than of biological processes as we know them; if God established a temple on earth, in which to hold court, surrounded by angels; or if there were some kind of explicit marks in the creation making clear that there *had* to be a creator.[52]

This issue of the "epistemic distance," the divine hiddenness that is necessary to preserve freedom, is discussed by John Hick.[53] While direct proof of God's existence and nature would not by itself remove free will, it would eliminate the necessity for moral discernment. Furthermore direct insight into the ultimate cost of disobedience would invite only a grudging allegiance to God's ways.

The requisite hiddenness of God is satisfied through the nature of creation as we see it (see section 2.1), governed by impartial physical laws, which nevertheless allow hints as to God's existence and true nature. Sufficient evidence is given for knowledge of God's existence and an outline of God's will, but this evidence is not overbearing; human sinfulness (self-interest) can make us see without seeing and hear without hearing. The ability to see the truth is dependent on readiness to listen and openness to the message (John 3:3).

6.6 *The Possibility of Revelation*

This leads to the final requirement: that despite the hidden nature of the underlying reality, it still be open to those who wish to do so to discern its nature (indeed, on the view taken here, it is the wish of God that they should do so) and

---

[52] E.g., Carl Sagan's novel *Contact* (New York: Simon & Schuster, 1985), where a message to humankind is hidden in the digits of the fundamental number π.

[53] J. Hick, *Evil and the God of Love* (New York: Harper and Row, 1966).

to receive encouragement to follow the true way. There is a two-fold requirement: first that it should be possible to make specific intimations of this reality available to those who are ready to receive them; and second that there should be available to all, as a basis for ethics, a mode of revelation of what is right and wrong, of what is good and bad.

Sociobiologists suggest there is a scientific basis for ethical behavior simply in terms of evolution and social interaction. While it is possible God could choose to work in this way, I believe that this kind of basis is inadequate to explain our full ethical understanding and strivings.[54] I suggest instead that there could be a basic ethical understanding given us that is not simply the result of blind evolution, but rather is purposefully provided as a concomitant of consciousness and free will. It is a weak shadow of the kind of religious vision that can also be given to individuals—and has been vouchsafed for by a great many—without violation of the laws of physics. Undoubtedly this is highly controversial, but it seems as possible as any other explanation, and cannot be simply dismissed until we have a full understanding of consciousness that provides an adequate alternative source for ethical understandings—something we are nowhere near having at present.

A possible physical basis for this will be discussed in the subsection on immanence and revelation, but that particular basis is not required to validate the point made here: that some provision must be made for understanding the full nature of reality, by those who wish to do so, in order to realize God's plan. Part of the Christian answer to the need for revelation is to claim that this is precisely what the life of Christ was about; this does indeed provide a revolutionary answer to the problem of revelation in the face of an ordered providence and scheme of nature based on physical law. However this does not fully resolve the issue at hand, for it has been emphasized before that part of the supposition is that Christ was fully human; how then did he obtain his vision of reality? The same issue arises in the case of the other prophets and saints, and indeed for the ordinary believer.

In whatever way it is done, the feature I will assume is that *there is indeed a channel for visions of ultimate reality, available to those open to them*; allowing the nature of that transcendent reality to partially shine through into the immanent reality of the world, making available to us new patterns of understanding, and providing encouragement and strength to follow these visions. This feature (corresponding to the Quaker vision of "the light within")[55] is clearly intimately related to the issues of transcendence and immanence; how we model these will determine in a large measure the sorts of revelation we can conceive of.

6.7 *The Physical Universe*

It should be noted that none of this is in contradiction to standard physical understanding of the universe, including the standard picture of the origin of the

---

[54] Ellis, "The Homogeneity of the Universe."
[55] Jones, *Faith and Practice*.

universe in a hot Big Bang (or its modifications as envisaged by quantum cosmology), except conceivably the last, to which we return shortly. Rather what we have is an extra layer of explanation offered for the physical world we see around us, not in any way contradictory to physical understanding, but rather providing a kind of rationale for the need for such a reality, that is a metaphysical explanation for the existence of physical laws.

Straightforward physicists can simply claim that there is no need for this extra layer of explanation in order to understand the physical world, and they will be right—provided we attempt to explain only physical reality, accepting without question the given nature of physical laws, and ignore all the issues raised by the existence of a moral and ethical order (and indeed also the aesthetic dimension to life encompassed in great art and music). When we try to make sense of these extra dimensions of existence, the simple physical explanation is woefully lacking; something like that offered here is much more profound and satisfying. The whole represents exactly what is described by Temple: "What we find is power in complete subordination to love"; for on this view, the possible exercise of creative power by the creator is voluntarily restricted to that which enables a universe where a free and loving response by humankind is possible, despite the costs and sacrifice entailed. It opens to us an awareness of a genuinely religious dimension of existence, one related to a sense of transcendence and holiness, that simply is not encompassed in a narrower world view, and for which there is considerable evidence in terms of many people's experience and understandings.[56]

7 *Multiple Universes*

My argument so far is that free and loving response to God imposes a set of requirements on the nature of the physical universe. The next issue raised by these speculations is, how did God arrive at the solution, the satisfactory design of the universe?

7.1 *The Nature of Design*

In effect, there are two possibilities. One is that the way any particular laws of physics would work and their outcome is obvious to God. In this case, God need only formulate them in the way God envisaged, and they would give the desired results (namely, admitting intelligent life and free will). In physicists' terms, *inter alia* the required value of the fine structure constant was obvious to God, or at least could be determined mentally by God without having to perform experiments.

The alternative is that the nature of the divine creative activity was not so totally omniscient, and God had to experiment to get the results right. Thus the issue is, does God have to "learn" through trying various alternatives, or could the result of design be immediately foreseen? Is the design process one of trying

---

[56] Ellis, "The Homogeneity of the Universe."

out mathematical formulations, trying out physical behavior, or simply imaging results and demanding that the needed structure conform?

The last option seems difficult to accept. The point is as follows: the basic argument for design is the incredible complexity and harmony that has been achieved whereby the fundamental laws of nature and the boundary conditions of the universe not merely admit but even in some sense prefer the existence of life and allow the incredible complexity of human beings. If we ourselves were to try to fashion laws of physics to allow life we would try and fail again and again and again for millennia. This should not be confused with the issue of evolution: it is indeed true that beings of the complexity we encounter today require millions of years of evolution for their formation, and this does appear to be the way God has allowed life to develop. The point is that this evolution takes place on the basis of given laws of physics and hence given laws of chemistry. If we alter those laws appreciably, life as we know it will not be possible; no evolutionary process whatever will take place, because not even a single cell, with all its incredible complexity, will function. What is at issue is the choice of these particular laws of physics, which allow evolution to take place. How could we succeed in finding the right set of laws with the right set of constants?

So the intriguing issue arises: is the God who we believe created the universe capable of getting all this right first time, or is God's nature such that even with all its capabilities God would need to try more than once?[57] This is partly related to whether the universe is *really* mathematical in nature or only incredibly well described by mathematics; can we imagine, for example, God actually trying a universe with a fundamental theory based on the group $O(5)$, and then trying $SU(10)$ if the previous attempt fails? Or can we envisage God calculating the effects of a string theory of unified forces so accurately that God could see that life would be possible, before ever trying the experiment? Could God perhaps simply visualize the result of particular possible laws of nature, without having to do the calculations? And would this include getting the initial conditions in the universe right first time, so that galaxies, stars, iron, oxygen, carbon, and so on, would form and so allow suitable habitats for humans to result? What may be particularly pertinent is that, although presumably God knows the properties of matter directly, "from the inside out," so to speak, and so can easily determine its behavior, the requirement of leaving the universe and ourselves free to respond and able to effect so much for ourselves, is a highly non-obvious achievement, and so could have required that God had to try more than once to achieve this effect. Of course if God has to try more than once to get the numbers right this could perhaps be taken to contradict traditional notions of God's omniscience, but it is fairly common today to qualify this attribute, as for example in process theology.

This discussion can be dismissed as hopelessly anthropomorphic; however it is difficult to avoid the question if the design we are supposing is indeed the result of a design *process* (that is, a series of concepts of what might exist that proceeds through various stages of detail and clarity from an initial

---

[57] See the novel *Starmaker* by Olaf Stapledon (New York: Dover, 1968).

broad concept to final detailed specification). The very concept of *design*, however, suggests this nature for the creative process.

There are two sorts of issues that arise in posing this question.

7.2 *Many Universes?*

The issue first is in relation to the many-universe idea mentioned in a previous section. If God were to experiment as suggested above, there would indeed be an ensemble of universes apart from the one we actually experience. However it would not be an ensemble of all possible universes, but rather a set of experimental universes converging towards the desired result (perhaps ultimately leading to a set of small variations around the final scheme?)

The particular interest then is the notion that we might not be the last such experiment; the creator could now be working out a better universe as the result of observing ours in operation. Thus we arrive at the Pangloss question: "Is this the best of all possible universes?" Could it be that at some later time[58] there will be an even better one? One might suggest, for example, that the cost of evil in the present dispensation is so high that there is a need for a search for another solution where that cost is less. We cannot answer decisively, but it is an intriguing concept. Murphy points out that this then raises the Pangloss question at the next level: "Is this the best of all possible sequences of universes?"

7.3 *Limits to Transcendence?*

The second issue relates to the nature of the transcendence and omnipotence of God. As mentioned before, there are models in physics of certain kinds of transcendence (e.g., where four-dimensional theories result from imbeddings in higher dimensional spaces); in these cases, the transcendent arena (the higher-dimensional space) is in turn subject to (higher-level) laws of logic and physics. The issue raised here regarding theology relates to views on the bounds of transcendence. Thus the kind of process of trial and error I have mentioned might perhaps be needed or implied by certain kinds of transcendence but not by others. The problem in discussing this is that we have to model our ideas on our own experience, and in particular the fact that we are subject to the ideas of learning on the one hand and of the passage of time on the other; if God is able to learn, or needs to learn, then God too must experience some kind of passage of time.[59] Is it conceivable God does not need to learn?

Related to this question is the fact that there are, as it were, different possible levels of omnipotence; for example, we may believe that although God has the power to create the entire universe, free will cannot be created without also enabling evil and suffering. If so, there are some kinds of requirements (laws of logic, perhaps) that transcend the omnipotence of God. Thus in terms of

---

[58] I apologize for using the notion of time here, which is of course quite inaccurate, but I cannot see how else to express the idea.

[59] See Peters' discussion of time in God's experience in this volume.

Russell's notion of a "ladder of contingency,"[60] we can consider various possible levels of limits of creative activity. An ultimate mystery (along with the issue of the existence of God) would be the origin of such restrictions or transcendent laws. Any attempt to understand this must of course be hampered by the fact that we do not have a satisfactory basis for understanding the laws of logic in our own universe (*vide* Gödel's theorem), nor whether they precede the existence of our universe or come into existence with it.[61]

8 *Immanence and Revelation*

In the picture presented here, we envisage the transcendent creator as being immanent, that is, present in every event. It is this immediate presence that maintains the behavior of matter (i.e., the laws of physics) in accord with the desired regularities, irrespective of our view of the physical origin of the universe. Thus the nature of creation is intimately entwined both with the continued existence of the matter in the universe, and with sustaining the specific laws of behavior of that matter. To fulfill the view given here, the nature of this immanent presence is voluntarily constrained to ensure the kind of creation discussed above.

8.1 *A Mode of Revelation?*

As has been emphasized, a requirement for completion of the envisaged aim for creation is that we are able to attain glimpses of the underlying purpose and reality of the universe through revelation, that is, by the availability of images of this reality that are somehow communicated to those ready to receive them. But how is this possible in the real physical world with its deterministic laws? We seem to have a contradiction with the suggested resolution of the creator to abide by the laws of physics.

There are three ways of evading this contradiction. The first is that in fact evolutionary processes plus social interaction enable the human brain to see the nature of ultimate reality without any further intervention. As indicated above, I believe this is an inadequate explanation, particularly as it denies the true nature of mystical experiences that have for centuries been a central feature of religious life, in that they have been perceived as a result of the activity of God.

A second possibility is that the creator, who ordinarily sustains the laws of physics, might decide not to do so or to alter them in certain cases; that is, divine intervention could take place by local and temporary alteration of the ordained physical order (the creator who fashioned these laws and maintains them in operation, could choose that they not be fulfilled when desired). I believe this would violate the whole spirit of the account of creation given above; this

---

[60] Russell, "Cosmology, Creation, and Contingency."
[61] See the articles by Stoeger and Davies in this volume.

kind of intervention would invalidate the concept of control of the universe in an impartial way by physical laws underlying the desired order.

The third possibility is that the creator and sustainer could directly provide images of the desired life and the nature of reality to humankind, without any violation of known physical laws. How is this possible? The point is simple but fundamental:[62] at the foundation of modern physics is quantum indeterminacy. *Quantum theory does not predict the result of any particular physical event or process, but only predicts the statistical behavior of matter.* For example, quantum theory is unable to predict when a particular radioactive nucleus will decay, or an atom will change from an excited state to a ground state. Quantum mechanics only deals in probabilities, never in certainties. There is thus no reason whatever that a creator who is maintaining and sustaining physical reality cannot determine when particular quantum events take place, without in any way violating the laws of physics. This could even be related to the collapse of the wave function, although this would only be one of several possibilities.

One can thus envisage the creator providing images as desired to individuals, or stimulating specific memories already existent by controlling the specific energy exchanges between particular excited states in the brain, without violating quantum mechanics in any way. If there are very few such events needed to provide a particular image, they will have a negligible effect on the overall statistics; if there were many, compensating exchanges would have to be arranged to keep the statistics within reasonable bounds, which is presumably possible. However it seems to me the former is a more probable situation rather than the latter; that is, it is likely that rather few individual quantum events can shape thoughts in the brain.

One must emphasize that this proposal is not the same as quoted by Peacocke:[63] "God would . . . make some microevent, subsequently amplified, to be other than it would have been if left to its own natural courses." The whole point of quantum uncertainty is that there is no such thing as a "natural course" that events would have taken. It is precisely such a deterministic outcome that physics has failed to find at the microscopic level—and quantum theory denies its existence. This denial is not a small side-effect of the theory; it is its very foundation. The possible relevance of quantum physics was explained in the 1950s by William Pollard,[64] and more recently by Barbour,[65] Russell[66] and others.

---

[62] See Barbour, *Religion in an Age of Science*; Russell, "Quantum Physics in Philosophical and Theological Perspective."

[63] Peacocke, *Theology for a Scientific Age: Being and Becoming—Natural and Divine* (Oxford: Basil Blackwell, 1990), 154.

[64] W. G. Pollard, *Science and Providence* (New York: Charles Scribner's Sons, 1958), ch. 4.

[65] Barbour, *Issues in Science and Religion*, ch. 4.

[66] Russell, "Quantum Physics in Philosophical and Theological Perspective," 343-374.

This proposal will undoubtedly be unpalatable to many physicists and, perhaps, philosophers. However if asked to disprove it, they will be unable to do so, for it does not in any way conflict with the foundations of modern physics. From the religious viewpoint it provides a possible basis for believing in revelation by God to individuals, without requiring a "miracle" in the sense of a violation of the natural order; the process is possible by acting within the freedom allowed by quantum theory, without contravention of physical laws as we know them.

Of course quantum theory is not the only source of unpredictability in physics; indeterminacy in physical reality is also due to the various forms of macroscopic indeterminacy which have been realized in recent years, especially in systems that are far from equilibrium, and may be related to the great flexibility that we have in imposing boundary conditions. Thus the proposal is that, in accord with suggestions by Polkinghorne and others, God acts within the flexibility of unpredictable situations—not in gaps or isolated intrusions, but in the process itself. However it is quite unclear to me how such features could lead to some ordered mechanism of revelation (specific images could, for example, have been embedded in the boundary conditions of the universe, but then we would have no guarantee they would arrive at the right place and the right time—indeed this is not possible if free will is real). It does not seem to provide an alternative mechanism that could achieve the kind of purpose envisaged here.

8.2 *The Modes of Action*

My proposal, then, would be that the voluntary restriction of the activity of God in this world is to three forms of action:

(1) Initial creation of the universe and setting of its initial conditions, thereafter strictly *maintaining the physical laws that make our ordered life possible*;

(2) The one unique event of *the manifestation of God's nature through the life of Christ*—"the coming of eternity into the midst of time, in the form of a living, visible human-divine Person, through whom all life at its highest levels is to be interpreted";[67]

(3) *Direct actions in terms of mental interventions; for example providing to men and women images of ultimate reality and of the good life*, as just discussed;

but not in any other way forcing or influencing their lives or intervening in natural processes. I believe this would be in accord with religious experience, and with the vision of sacrificial creation and action discussed in the previous section.

---

[67] Jones, *Faith and Practice*, 15.

If accepted, this proposal does exclude other forms of intervention or action as sometimes proposed, such as direct intervention by God to prevent some natural disaster in response to prayer, either by altering the outcome that would otherwise have followed from the laws of nature, or perhaps by intervention within the uncertainty allowed by those laws; thus it is narrower than many versions of the Christian tradition. It is indeed true that one could envisage quantum uncertainty allowing a channel for intervention that would, for example, through amplification allowed by equations evidencing "chaotic" behavior, be able to influence weather patterns and so allow a "steering" of daily physical events by influencing the behavior of inanimate objects. However to allow such interventions seems to greatly exacerbate the problem of evil (once one such intervention is allowed, why should they not occur all the time? If God can part the Red Sea, why does he allow my toothache? More pertinently, why does he permit the Holocaust?) The danger is arriving at a God who capriciously allows evil, or prevents it.[68]

For present purposes, I choose the simple and straightforward view proposed above, recognizing it is narrower than some would like. However, it does not exclude changes of bodily conditions (e.g., alleviation of illness) as a result of prayer or mental visions, through "top-down causation"—action of the brain on the human body.[69] We can easily see how changes of mental state can alter physical features of the body (increasing blood pressure or adrenaline levels, for instance) thereby changing the environment within which cells function and hence causing a change in their responses. The point is that this kind of view does indeed have practical implications (e.g., concerning the nature of prayer); it is probably broad enough to accommodate reasonable views and practice of a large part of the Christian tradition.

8.3 *The Multitude of Beings*

The issue that remains, and could be worrying[70] is the problem of how God can attend to the multitude of beings not merely on this world but on all the other worlds in existence in the universe (see the remarks above about existence of other beings). Furthermore there is the profound issue of whether each other world too would demand the life of a Christ, or whether this one alone of all the millions of worlds has had such a presence on it?[71] A question going back to Aquinas and his discussion of the plurality of incarnation.

The first issue is in a sense compounded by the question of how God can maintain every particle in the universe true to its essential being, that is, obeying the ordained laws of physics prescribed for it? Both are problems of

---

[68] Polkinghorne, *Science and Providence*; Russell, "The Thermodynamics of 'Natural Evil.'"

[69] Peacocke, *Creation and the World of Science*, 157-58.

[70] McMullin, "Natural Science and Belief."

[71] See E. A. Milne, *Modern Cosmology and the Christian Idea of God* (Oxford: Oxford University Press, 1952); E. L. Mascall, *Christian Theology and Natural Science* (Oxford: Clarendon Press, 1952).

numbers: God relating to vast numbers of individuals and enormously more elementary particles[72]—having decided on the laws of physics, how to ensure that every particle obeys them? We simply have to assume that the divine nature allows this, ensuring the maintenance of the nature of the particles and of the essential nature of human beings. It is essential to the overall picture here that God does have the capacity to be, in meaningful sense, creator and sustainer of each individual.

Perhaps God's immanence implies a kind of knowing so different from ours that these problems simply do not arise (a knowing of things by direct intuition rather than by external knowledge). Even so, there is a cosmologically interesting statement that seems a plausible consequence of the view presented here. The point is that while we can (perhaps with some difficulty) envisage the creator attending to each particle and each individual in the visible part of the universe, if we accept the standard universe models with flat or negative spatial curvature, there are not a finite number of particles in the universe, but an infinite number; and almost certainly, also an infinite number of individuals.[73] The same conclusion would hold in models such as the chaotic inflationary universe models that are statistically homogeneous in the large, but have infinite spatial sections.

It seems to me to stretch credulity too far that the creator could succeed in giving the desired attention to an infinite number of beings (one must note here that "infinite" is completely different from a very large number, no matter how large).[74] The point here is that a mind cannot really have a satisfactory grasp of items of information unless it integrates them one with another, for a mind must be unified.[75] Now it is always possible to integrate a finite number of items one with another, keeping track of the relationship between each item and all the others through some kind of indexing scheme, but when you have an infinity of items the problem explodes into logical nonsense—you get infinities of infinities of infinities. These disastrous infinities threaten to enter if God has to keep track not only of the items of knowledge, but also of God's own knowledge of the items of knowledge, and then of the relations between them also; which I would take as essential for an overall synthesis of the kind presented here to have meaning.

Thus I would suggest that to make the overall view coherent, we must assume *there is a finite number of particles, and of beings, in the universe.* This is achieved *if the universe has finite spatial sections*; and that in turn can be achieved in two ways: (a) in a universe with positive spatial curvature (and a high density of matter), implying, in agreement with other arguments,[76] that most of the matter in the universe is dark matter, hidden from our view; or (b) in a

---

[72] Remember that every cubic centimeter of water contains about 10^{24} molecules!

[73] Ellis and G. B. Brundrit, "Life in the Infinite Universe," *Quarterly Journal Royal Astronomical Society* 20 (1979): 37.

[74] Ibid.

[75] I thank John Leslie for the comment.

[76] See L. Krauss, *The Fifth Essence* (London: Basic Books, 1990).

universe with zero or negative spatial curvature, that has compact (finite) spatial sections as a result of an unusual topology.[77] The former could be confirmed if the current searches for dark matter were to succeed and show that the density is greater than the critical density; the latter could be observable if the real universe turns out to be a small universe, which we have seen round already several times. I suggest that if neither is true, and we conclude the universe has infinite spatial sections, then the case for a caring creator looking after each individual would be weakened—the infinite numbers that would then be involved would make this difficult to contemplate.

The second issue (does each separate world need a Christ?) raises deep divisions depending on one's view of the nature of Christ's activity. It seems to me to be democratically solved: yes, there will be a Christ on each world for otherwise we end up placing ourselves, theologically speaking, at the center of the universe—implying in some sense a lower value for the other beings, because of the more central position for ourselves (the scientific analogy would be a return to a pre-Copernican view of the universe). Opposition has been expressed, based on the view that Christ needs only be incarnate once in the entire universe to accomplish his task,[78] and on the profound claim by John Lucas that the cry "My God, my God, why hast thou forsaken me?" must surely occur only once in the universe. There is a strong case both ways. One thing is clear: if one concurs with the many-Christ view, it must surely immensely strengthen the claim regarding the finitude of the universe—then we have to countenance only a finite number of civilizations needing redemption. Surely an infinite number of Christ-figures must be too much, no matter how one envisages God. And if there is only one Christ-figure, we then proclaim ourselves preferred to an infinite number of other beings—a true reversion to a pre-Copernican view of us on this planet as having an enormously preferred position in the universe.[79]

In any event, this argument does underline that one of most important things that could ever happen to the human race would be to attain contact with other intelligent beings who have originated on other planets, and to see what their religious world view was. Thus the search by some scientists for signals from other intelligent beings ("SETI") is not only of enormous scientific interest, but could be of tremendous religious significance as well.

---

[77] See, e.g., section 7.6 of Ellis and R. M. Williams, *Flat and Curved Space Times* (Oxford: Oxford University Press, 1989).

[78] Milne, *Modern Cosmology*; Mascall, *Christian Theology*; and Drees, *Beyond the Big Bang*.

[79] If we suppose Christ is only needed because of our fall, the implication would be that we are the only fallen creatures amidst a vast universe of unfallen, again placing us in an extraordinary special position amongst all the beings in the universe, and consequently beneficiaries alone of the most specific display of God's love.

9 *The Prologue to St. John's Gospel*

The view put forward so far seems coherent as a combination of both religious and scientific concerns. As has been emphasized, while it is not demanded by the scientific worldview, it is consonant with it. The final question is what evidence there is to support it from the other side; specifically, how does it relate to the biblical view of creation. I refer here specifically to the Prologue to St. John's Gospel, which reflects an important New Testament position. I base my remarks largely on Temple's commentary;[80] uncited quotations will refer to this commentary. The Genesis picture (although certainly not to be taken as a scientific treatise) is of course consonant with the concept of God as creator of the universe, even reflecting in a broad way some of the modern scientific views of the nature of that creation through a "hot Big Bang," but does not convey the same depth of understanding of the implications of that creation as does the author of *John*.

> 1:1-2. In the beginning was the Word. And the Word was with God. The same was in the beginning with God.

"The expression used here means both 'in the beginning of history', and 'at the root of the universe'. What is said to exist is 'the Word' . . . [this is] the Word of the Lord by which the heavens were made, and the Rational Principle which gives unity and significance to all existing things." This emphasizes the transcendent nature of underlying reality: "the subject for which he is claiming attention is the ultimate and supreme principle of the universe."

> 1:3. Through its agency all things came to be, and apart from it hath not one thing come to be.

"The supreme principle of the universe is not only its bond of unity, but its ground of existence. . . . All things exist or come to be as a result of God's activity by self-expression." This is the statement of the ultimate basis of creation; the definitive selection from the various possible ultimate causes of the universe, of design and intent.

> 1:4-5. What came to be in it was Life, and the Life was the light of men, and the light shineth in the darkness, and the darkness did not absorb it.

"Within that supreme principle is, and always has been, Life. Life is not said to be a product of its agency in the world, but rather to be one of its own inherent characteristics." Thus we arrive at the Christian anthropic principle, a profound version of the SAP but on a completely different basis than the quantum basis proposed by some physicists: the creation *had* to have as its product, life, for that is the nature of the creator. However the Word is not fully displayed in human

---

[80] Temple, *Readings in St. John's Gospel*, 3-16.

life: "Only of one occurrence is it true to say that it took place not only through but *in* the Logos; that is the nativity of Jesus." Rather, in all periods, "the divine light shines through the darkness of the world, cleaving it, but neither dispelling it nor quenching it.... The darkness in no sense at all received the light; yet still the light shone undimmed." This is a necessary consequence of the mode of creation (with the freedom of action of the Creator voluntarily restricted), with the light shining to show the way, but those with free will able to accept it or reject it. This was the method of revelation chosen by God (necessary in view of his hidden nature):

> 1:9-13. There was the light, the true light which enlighteneth every man, coming into the world. In the world he was; and the world through his agency came into being; and the world did not recognise him. To his own home he came, and his own people did not receive him. But to as many as received him, to them he gave the right to become Children of God—to those that believe on the name of him who was born, not of blood, nor of the will of the flesh, nor of the will of a man, but of God.

"From the beginning the divine light has shone. Always it was coming into the world; always it enlightened every man in his reason and conscience." This is through some mode of direct revelation, perhaps the mechanism discussed above, through which the nature of the divine can be apprehended by each person who listens to the "still small voice" within. The specific revelation of Christ is needed to make the message completely clear and unambiguous. Awareness of the Light within each person is the basis of Christ-centered action in the world.[81]

"And now the great declaration is made. This Word, which is the controlling power of the whole universe, is no longer unknown or dimly apprehended. The Light which in some measure lightens every man has shone in its full splendour":

> 1:14. And the Word became flesh and tabernacled amongst us—and we beheld his glory, glory as of an only begotten from a father—full of grace and truth.

"The Incarnation was an act of sacrifice and humiliation—real, however voluntary. The sacrifice and the humiliation are the glory. If God is Love, His glory most of all shines forth in whatever most fully expresses love." It was this message that Christ had to learn for himself, but that in the end is rooted in the nature of creation.

> 1:17-18. The law was given through Moses; grace and truth came through Jesus Christ. God hath no man ever yet seen; God only begotten, who is in the bosom of the Father—he declared him.

---

[81] Pickvance, *George Fox on the Light of Christ Within.*

"If our hearts are open to the love of God made known in Christ Jesus, we no longer ask chiefly what has God commanded or what has he forbidden, but rather what will please our Father; and we do this not to gain any reward or to avoid punishment, but because our desire and joy is to please our Father. We are won to the new dispensation by the grace and truth in Jesus Christ. The appeal is not to our self interest, though He uses that in its place, but to the free assent of heart and mind. And this is made by the disclosure of the Divine nature." This is the ultimate vision, the intention of what will be achieved by the creation, made possible by the nature of that creation and by the specific revelation making the ultimate meaning of that creation known.

The Christian view is that it is this vision, in all its fullness, that shapes the nature of the cosmos. The preceding paragraphs have tried to show that this leads rather uniquely to a particular version of the anthropic principle, characterizing the shaping of the universe. "St. John is intensely and profoundly sacramental; he sees the spiritual in the material, the divine nature in the human nature which it uses as its vehicle." Thus the whole vision in no way contradicts the usual view of the physicist as to the operation and results of the laws of nature; rather, it supplies this view with a profound rationale based on a different order of reality, and is to some degree confirmed by its agreement with that view.

10 *Comments and Conclusion*

This paper has attempted to give an overall view of the anthropic principle from a Christian viewpoint, while taking full cognizance of modern scientific views. The aim has been to show that this is possible, and that indeed this combined view has the possibility of giving a much more profound basis for the anthropic principle than obtainable from a purely scientific view, with all the restrictions on mode of argumentation that that entails.

The attempt has been to show that the physical universe as we see it is a solution to the Christian anthropic problem: the demand for creation of a universe that allows the implementation of the full meaning of the Christian message. What one would like to do is either to show that this is the only solution, or else to find all solutions. Regrettably both are far beyond what we can achieve. The general line of argument tends to suggest that the solution we see around us is the only one, but this may well be simply because we are limited in our imagination to what we know, and are unable to comprehend some of the alternatives that could possibly work. Indeed while the range of universes compatible with life as we know it seems not to be very wide, nevertheless there is some freedom of choice: the designer apparently could make an arbitrary choice of the fine structure constant, for example, within a range that would allow life to exist. God might or might not be constrained by some particular process of quantum creation. One would like to investigate more closely the way this ultimate purpose is imbedded in and made manifest by the laws of physics as we know them.

The argument is bound by certain assumptions about the way a creator can operate; these may well be unreasonable limitations, but we *have* to assume

God's nature is like ours in some ways, for otherwise we cannot begin to imagine what God's conceptions could be like. As usual, the same kinds of questions arise at a higher level, in relation to the reality in which God (the creator) exists; we cannot answer issues at this level, having to take as given his existence and nature.

What proof can we give that this overall view is correct? Solid proof cannot be given, but neither can it be given for any other fundamental view of the nature of the universe. However the evidence one can put forward is stronger than that which can at present be adduced for some fashionable theories in cosmology such as the inflationary universe idea or the quantum creation of the universe. Nevertheless, ultimately, belief in this viewpoint is an issue of faith, of appreciating the greatness of the overall plan presented here and seeing its intrinsic rightness. The kind of evidence put forward has been of the sort discussed in the second section (particularly the appeal to overall coherence and satisfactoriness on the one hand, and to a revealed text—St. John's Gospel—on the other). There will be some for whom the whole line of argumentation is vexatious and unnecessary, some for whom it is misguided and basically wrong-headed, and others for whom it seems sublime and without doubt the ultimate truth. No argumentation along the lines used in physics or the other sciences can resolve this issue, for it is a different kind of argument.

Acknowledgments: I thank John de Gruchy, Robert Russell, John Leslie, W. P. Alston, Wim Drees, and M. Heller for helpful remarks, and particularly Bill Stoeger and Nancey Murphy for detailed comments, many of which have been incorporated in the paper.

EVIDENCE OF DESIGN IN THE FINE-TUNING OF THE UNIVERSE

Nancey Murphy

> Thus says the Lord, the Creator of the heavens,
>   he who is God,
> who made the earth and fashioned it and by himself fixed it
>   firmly,
> who created it not as a formless waste
>   but as a place to be lived in:
> I am the Lord, and there is none other (Is. 45:18).

1 *Introduction*

The purpose of this paper is to examine what has been called the fine-tuning of the universe. Put simply, it has become apparent over the past few decades that the possibility of any sort of life evolving in the universe is dependent on the values of a small set of basic numbers (or "natural constants"): the strengths of the four basic forces, masses and charges of certain subatomic particles, and a few others. If any of these numbers had differed even slightly from its present value (typically by one part per million, or less!) the universe would have evolved into "a formless waste" rather than "a place to be lived in."[1] Some scientists, philosophers, and theologians conclude that such dramatic "fine-tuning" cries out for explanation. One possible answer is to postulate a Master Tuner, in which case we have a new sort of design argument. What are we to make of such suggestions?

    I bring no scientific expertise to this question, but concentrate instead on philosophical and theological issues. In particular I shall discuss the form a design argument must take in light of recent developments in epistemology and philosophy of science. This is the main goal of the paper; however, I do include tentative claims about the acceptability of the design hypothesis relative to its current competitors. An adequate presentation of a design argument following the guidelines presented here, along with a thorough evaluation of the competition, would require a study of much greater length.

    In brief, I shall argue that the appropriate way to assess the value of fine-tuning as evidence for God's existence is to consider it in terms of the additional confirmation it provides for already-existing theological research

---

[1] Two prominent books are J. D. Barrow and F. Tipler, *The Anthropic Cosmological Principle* (Oxford: Oxford University Press, 1986); and J. Leslie, *Universes* (London: Routledge, 1989). See also the article by G. F. R. Ellis in this volume.

programs. We shall examine one such program below, and shall see that fine-tuning does indeed add to the empirical confirmation of that program.

2 The Nature of Knowledge

2.1 Humean Reasoning about Causes

David Hume is taken by many to have shown conclusively the folly of any argument from design. In Hume's day knowledge was divided into two categories: matters of fact and relations of ideas. Hume's goal was to show that knowledge of God could be had within neither category.

> If we take into our hand any volume—of divinity or school metaphysics, for instance—let us ask, Does it contain any abstract reasoning concerning quantity or number? No. Does it contain any experimental reasoning concerning matter of fact and existence? No. Commit it then to the flames, for it can contain nothing but sophistry and illusion.[2]

Our interest here is in the argument from design, which falls into the category of factual knowledge. In order to assess the possibilities of something like a design argument, using the fine-tuning of the universe as its grounds, we shall have to consider the changes that have occurred in our understanding of such knowledge since the days of Hume and William Paley. Two vastly important developments in epistemology separate us from Hume. One is the recognition of the role of hypothetico-deductive reasoning; the other is holism—the view that our beliefs face the tribunal of experience not singly but as a body.

2.2 Hypothetico-Deductive Reasoning

Hume's final criticism of the argument from design rests on a theory of reasoning about causes that requires the observation of regular conjunction. Consequently, however much the universe might look as though it were the product of intelligent design (although Hume has pointed out that it is not unambiguously so) we can make no inferences about its cause, since we have observed no previous instances of universes produced either by a designer or by anything else.

The recognition of the role of hypothetico-deductive reasoning (so named by Carl Hempel,[3] but described earlier by William Whewell and others) freed science from dependence on induction for knowledge of causes. "According to such a view, the essence of genuinely scientific reasoning about matters of fact is the framing of hypotheses not established by given empirical

---

[2] D. Hume, *An Enquiry Concerning Human Understanding*, last paragraph.
[3] See C. Hempel, *Aspects of Scientific Explanation* (New York: Free Press, 1965).

data but merely suggested by them."[4] Much of what Hempel has written has been effectively criticized—his requirement that the data be deducible from the hypothesis (generally one can only deduce approximations), and that the hypothesis be in the form of a universal generalization. This latter criticism is extremely important for our purposes, since the requirement of law-like generalizations would bar inferences to unique events, and we would have gained nothing beyond Hume's epistemology. An important contribution of Hempel's that did survive later criticism is the view that explanation and confirmation are correlative: a hypothesis is confirmed by a set of data insofar as it provides the best explanation of those data.

2.2.1 *Consequences for the Design Hypothesis* So while epistemology generally and philosophy of science in particular have moved beyond the views of the neopositivists in many respects, it has become a fixture of our thought about "matters of fact" to expand our knowledge by forming hypotheses (often about unobservable entities and unrepeatable events), whose acceptability is based on their explanatory power. This suggests that the best way to think about a design argument is to take the creation of the universe by God as one of a number of competing hypotheses to explain the peculiar facts classed together under the heading of fine-tuning. Competitors include theories attributing these facts finally to chance or to logical necessity, multiple universe hypotheses of various sorts, the weak and strong anthropic principles, and other theological or philosophical proposals such as John Leslie's Neoplatonism.[5]

---

[4] Paul Edwards, ed., *Encyclopedia of Philosophy*, vol. 4 (New York: Macmillan, 1967), s.v. "Induction," by Max Black.

[5] See Leslie, *Universes*.

2.3 Holism

The most significant recent advance in epistemology is the development of holist accounts of the structure and justification of knowledge. Hempel *et al.* recognized that an additional factor to be taken into account in accepting or rejecting a hypothesis is its fit with the rest of our theoretical network. In fact, we never test one theory at a time. While this fact was considered an annoying aside by neopositivists, who were interested primarily in the relation between theory and data, it has become central to more recent theories of knowledge. W.V.O. Quine has argued specifically for an understanding of knowledge wherein each belief is supported in the first instance by its interconnectedness with the rest of the web or network of beliefs. The web has at its periphery beliefs closely related to experience, but in the face of a perceived clash between belief and experience there are always a variety of ways to restore consistency.[6] Quine's holism in epistemology is reinforced by philosophers of science such as Thomas Kuhn, who argues that it is the paradigm as a whole that stands or falls as science progresses.[7]

2.3.1 *Imre Lakatos' Holist Theory of Science* While Kuhn's description of the make-up of scientific knowledge is colorful and widely known, I find Imre Lakatos' account of the structure of scientific webs (which he calls research programs) more intelligible and useful.[8] A research program has the following structure: It includes a core theory that unifies the program. The core is surrounded by a protective belt of auxiliary hypotheses. The auxiliary hypotheses both define and support the core theory, while assorted data support the auxiliary hypotheses in turn. (Since explanation and corroboration are symmetrical, the auxiliary hypotheses nearest the edges explain the data; higher level hypotheses—that is, theories nearer the center—explain lower-level theories, and the core theory is the ultimate explanatory principle.) Also included here are theories of instrumentation and initial conditions. The auxiliary hypotheses are referred to as a protective belt, since potentially falsifying data are accounted for by making changes here rather than in the core theory—called the "hard core" since it cannot be abandoned without rejecting the entire program.

---

[6] See Quine, "Two Dogmas of Empiricism," *Philosophical Review* 40 (1951): 20-43; and Quine and J. S. Ullian, *The Web of Belief* (New York: Random House, 1970).

[7] See T. Kuhn, *The Structure of Scientific Revolutions* (Chicago: University of Chicago Press, 1970).

[8] Lakatos' theory is developed in "Falsification and the Methodology of Scientific Research Programmes," in *Criticism and the Growth of Knowledge*, ed. Lakatos and A. Musgrave (Cambridge: Cambridge University Press, 1970; reprinted in *The Methodology of Scientific Research Programmes: Philosophical Papers*, vol. I, ed. J. Worrall and G. Currie [Cambridge: Cambridge University Press, 1978]), 91-196 (page references are to original). I have summarized and amplified his methodology, and responded to some of his critics, in *Theology in the Age of Scientific Reasoning* (Ithaca: Cornell University Press, 1990), ch. 3.

It is more accurate to say that a research program is a series of networks of theory, along with supporting data, wherein the core theory stays the same while the belt of auxiliary hypotheses changes over time. A mature program also involves what Lakatos calls a positive heuristic—a plan for systematic development of the program in order to take account of an increasingly broad array of data.

Let us represent a (highly simplified) temporal cross-section of a research program as in *Figure 1*, where HC stands for the core of the program, AH for auxiliary hypothesis, and D for datum. The arrows represent logical connections between the various parts of the structure. Most of them are bi-directional to reflect the hypothetico-deductive character of most scientific reasoning. That is, the data follow (quasi-deductively) from the auxiliary hypotheses and the auxiliary hypotheses from the core theory, but in virtue of that very fact they support the auxiliary hypotheses and the core theory in turn.

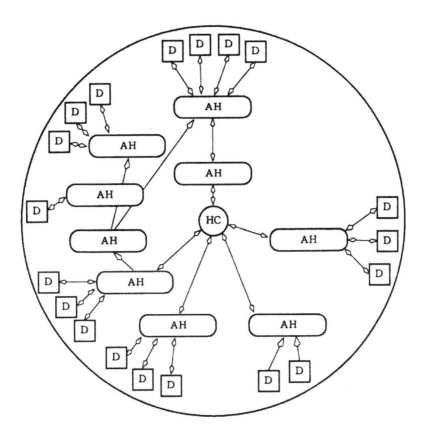

Figure 1: Temporal cross-section of a research program (highly simplified).

Any account of knowledge that places a high premium on coherence (as all holist theories do) will be open to two charges: (1) that the justification of any belief within the system will ultimately be circular, since it is a matter of each part fitting with the others; and (2) the charge of relativism—that is, that there is no way to judge between two competing and more or less equally coherent systems of belief. Lakatos made an important contribution to philosophy of science by proposing an external criterion for choosing among two or more competing research programs, each of which is fairly coherent. He recognized (as does Quine) that any belief system (research program) can be made consistent with conflicting data if one adds suitable auxiliary hypotheses. Sometimes these new hypotheses represent genuine improvements in our understanding of the subject in question; other times they are mere *ad hoc* devices to protect the theory from falsification. How is one to tell the difference? Lakatos proposed that a "progressive" move be defined as one where the new hypothesis not only accounts for the anomaly that led to its inclusion in the program, but also allows for the prediction and occasional corroboration of a "novel fact," that is, one not to be expected in light of the previous version of the research program.[9] When a research program makes such content-increasing moves it is said to be progressive, and is to be preferred to a "degenerating" program, where all or most of its auxiliary hypotheses are added in an *ad hoc* manner.

We can represent a progressive research program as in *Figure 2*, where successive versions of the two programs are "stacked" along a time dimension. For the progressive program the domain of facts (not the mere number of facts) increases over time. However, for a degenerating program the empirical content does not increase to keep pace with the increasing "density" of theoretical elaboration. Either the facts that are added to its confirmation are facts of the same type as before, or if facts from a new domain are added, the theories

---

[9] Lakatos changed his definition of 'novel fact'. In "Falsification and the Methodology of Scientific Research Programmes," he defined a fact as novel if it was improbable or impossible in light of previous knowledge. In "Why Did Copernicus's Programme Supersede Ptolemy's?" (with Eli Zahar, in *The Copernican Achievement*, ed. R. Westman [Los Angeles: University of California Press, 1978]) he described his earlier position as follows: "I originally defined a prediction as 'novel', 'stunning', or 'dramatic' if it was inconsistent with previous expectations, unchallenged background knowledge and, in particular, if the predicted fact was forbidden by the rival programme. The best novel facts were those which might never have been observed if not for the theory which anticipated it [*sic*]. My favourite examples of such predictions . . . were the return of Halley's comet, the discovery of Neptune, the Einsteinian bending of light rays. . . ." But by the time he co-authored "Why Did Copernicus's Programme Supersede Ptolemy's?" he was willing to include facts that were already known, but played no role in the formulation of that theory. I argue that while Lakatos' first formulation represents the ideal sort of confirmation for a research program, we have to recognize as well the value of "weakly novel facts." See my "Another Look at Novel Facts," *Studies in the History and Philosophy of Science* 20: 3 (1989): 385-88.

needed to account for them are *ad hoc*—"tacked on" rather than integrally connected with the rest of the program.

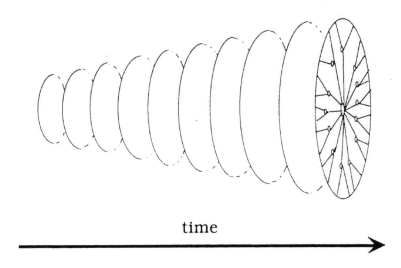

Figure 2: Successive versions of progressive and degenerating research programs.

Lakatos claimed not only to be proposing an intuitively plausible answer to the normative question of how scientists ought to decide which research program to work on, but also to have found the key to explain why scientists in the past abandoned one program in favor of another, and why they did so when they did. Subsequent studies in the history of science have gone some way toward establishing his historiographical claim in areas as diverse as economics, evolutionary biology, and high-energy physics.[10]

2.3.2 *Consequences for Thinking about Design* The obvious lesson to be learned from holist epistemology is that the only reasonable way to assess the claim that fine-tuning provides evidence for divine creation is to consider the design hypothesis not as a claim standing alone, and solely on the strength of the data provided by the fine-tuning, but rather as an integral part of a theoretical network with a variety of supporting data. In other words, the epistemology and

---

[10] For a partial list of references, see my *Theology in an Age of Scientific Reasoning*, 64.

philosophy of science considered so far call for the evaluation of a theological research program, one part of which is the claim that God created the universe with intelligent life as a prominent goal.

3 Theological Research Programs

The preceding discussion raises the question whether it makes sense to speak of a theological research program. I have argued elsewhere that it does, and have sketched out the theoretical organization of several programs, noting their core theories, typical auxiliary hypotheses, theories of interpretation (the theologian's equivalent of theories of instrumentation) and positive heuristics. For example, the core of Wolfhart Pannenberg's program is a statement to the effect that the God of Jesus Christ is the all-determining reality. His treatment of the theological loci or doctrines such as Christology, eschatology, and theological anthropology form clusters of auxiliary hypotheses. Data come from Scripture, history, and universal human experience.[11]

3.1 Data for Theology

Few objections will be raised to the suggestion that a theological program can be organized to mirror the structure of theoretical science. The difficult issue is what counts as data for theology.

3.1.1 Scripture Many theologians treat Scripture much the way scientists treat the results of experience or observation, so we might try replacing the boundary of the web in *Figure 1* (which represents experience) with the biblical texts. Then statements about the content (or meaning, or teachings) of the texts would become data.

This proposal has some drawbacks, however. The main problem is to answer the question why the texts should be relied upon to provide evidence for genuine knowledge of God, rather than evidence merely about Israel's and the early Church's beliefs about God. Now, one can include within the theoretical web a theory about the texts that certifies them in some way. The standard move would be a doctrine (theory) of revelation. Such a theory will function in the program much as a theory of instrumentation does in science. A theory of instrumentation explains why the readings obtained by the use of this device (e.g., a telescope) ought to yield reliable information about the entities being studied (e.g., planets). Similarly, a doctrine of revelation explains why the text ought to yield reliable information about God (and other things). But even so, the reasoning will appear viciously circular to anyone not already convinced of the evidential value of the texts.

A second drawback to this scheme is that it tends to suggest a "proof-text" approach to Scripture—that each or any bit of Scripture (sentence or pericope) can be understood and used to support a theological claim apart from

---

[11] See ibid., 176-178.

its historical context and its larger setting within the entire text or even the entire Bible.

A more suitable account of theological data can be devised by noting that theologians typically claim that their work is also subject to the constraints of contemporary experience. However, few (it seems to me) provide a workable account of how Scripture and experience are to be integrated in the theological task. Let me begin to sketch a proposal with another drawing (*Figure 3*).

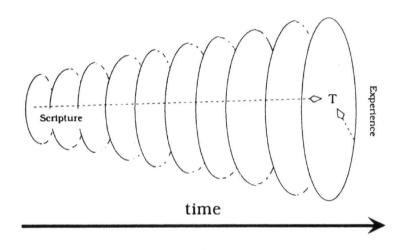

Figure 3: The 'shape' of a Christian theological research program integrating Scripture and experience.

This is intended to be a highly simplified drawing of the Christian tradition, beginning with the canonical text (which represents in itself a collection of traditions from Israel and the early Church, some of which are reinterpretations of earlier texts or oral traditions).[12] A more accurate sketch

---

[12] Keith Ward (in correspondence) suggests looking at Scripture as the result of a process "beginning with primary revelatory data—alleged, fallible revelations by God to humans in seven main forms: public events, inner experience, theoretical hypotheses and reflections, narratives, moral evaluations, writings coming out of liturgical practices, and those arising out of the social practices of the communities." The Scriptural texts are "a cumulative consequence of a process composed of such data and formed in a Canonical Matrix." This suggestion is not only a perceptive account of what we seem to be looking at when we examine the Bible, but has the further advantage of fitting nicely with the proposal being worked out in this section and the

would have to include branches representing the separation of the Christian church into relatively independent traditions: Roman Catholic, Protestant, Orthodox, Anabaptist. To see where specific theological research programs fit in (the "fine structure") we would have first to include a spaghetti-like tangle of separating and converging denominational mini-traditions.

Another aspect not reflected in the drawing is the differences among traditions and individual theologians in their use of the texts. There is no such thing as a naïve reading of the Bible—one entirely uninfluenced by intervening interpretations. However, some theologians seek as nearly as possible to rely on the biblical texts alone, while others intend to give great weight to intervening formulations: to creeds, the writings of the Reformers or canonized theologians, and others. These intentions form part of what corresponds in theology to the positive heuristic of a scientific research program—that is, the plan for the development of the program.

The point here is not to draw an accurate picture of Christian history, but rather to suggest that for each successive formulation of the tradition (i.e., for each successive version of Christian theology) there are constraints on any theological proposal (T) from two directions: from current experience (the circular boundary for each time-slice) and from the formulations of the past. The authority of Scripture, now, needs no special pleading (no doctrine of revelation). In fact, in Alasdair MacIntyre's words, a tradition is an ongoing argument about how to interpret and apply its classic texts.[13] Or, as David Kelsey maintains, it is analytic to say that the Christian Scriptures are authoritative for Christian theology.[14] What this authority amounts to, how it is brought to bear in authorizing particular theological proposals, will differ considerably from one theologian to another. For example, will it be a reconstructed history-behind-the-texts that carries authority; or a pattern of church life; or the texts' doctrinal implications? These factors are variable, but they are not beyond reasonable discussion and argument. Part of the task of a theological research program is to give an account of and rationale for taking these aspects of the text to be the relevant ones. These decisions constitute another aspect of the program's positive heuristic.

Now, is this so different from science? I believe that Lakatos and Quine would have to say no. Lakatos would emphasize the analytic point. Holding fast to some aspect of a research program is essential in order to be able to claim that one is still working on this program. For him it is the core theory that gives the program its identity (and we do indeed find such unifying theories in specific theological research programs). But to claim that one's research program is a part of the Christian tradition is to take the formative texts as something with which each new formulation must come to terms in one way or another.

---

following on the process of sorting and judging various kinds of 'religious experiences' as to whether or not they are indeed the product of contact with God.

[13] A. MacIntyre, *In Whose Justice? Which Rationality?* (Notre Dame: University of Notre Dame Press, 1988), 12.

[14] See D. Kelsey, *The Uses of Scripture in Recent Theology* (Philadelphia: Fortress Press, 1975), 99.

Quine emphasizes a pragmatic factor. The process of revising our web of beliefs in the face of recalcitrant experience always involves conservatism; it is simply too impractical to do otherwise.

So while it appears that theologians treat passages or aspects of Scripture the way scientists treat data, a closer analogy may be the role of classic texts (such as Newton's *Principia*) in defining a Kuhnian paradigm. Notice that this does not mean that theologians are forever tied to the Bible (any more than physicists are to Newton). The point is that if and when they reject it, they cease to be Christian theologians; they have *ipso facto* changed traditions.

3.3.2 *Experience* The next issue to be dealt with is the role of experience in theology. Here, I believe, theologians are much more vague than they are about the role of Scripture, and at least as diverse in their assessments of what aspects or kinds of experience are relevant. However, two general categories might be distinguished here; the difference is reflected in the nuances of the phrases: "religious experiences" versus "the religious dimension of universal human experience." I believe that the latter is too vague to be of use in searching for theological data and so I leave it to one side.

The outstanding problems in taking religious experiences as data are (1) their private nature and (2) the possibilities for counterfeit experiences—illusions, delusions, hallucinations. I have argued that a partial answer to these difficulties lies in the church's practice of communal discernment—a process whereby claims about putative religious experiences, revelations, divine guidance are brought to the community for assessment.[15] The exact criteria and the details of their application have varied from one community to another throughout the history of Christianity, but the general picture usually involves: (1) assessment of the coherence of the experience with accepted teaching; (2) fruits in the life of the individual and the relevant community, with particular emphasis on unity; and (3) independent confirmation from experiences of others involved. The New Testament itself calls for such testing of prophecy and teaching (e.g., 1 Jn 4:1; 1 Cor. 14:29); and in fact the very existence of the New Testament is due to the fact that these writings, when circulated among the early churches, were consistently judged to be in harmony with the teaching of Christ, either through the apostles or through the Spirit.

Faithful reflection of this sort within the context of prayer and with a genuine willingness to have one's mind changed would go a long way toward weeding out aberrant religious experiences. The results of the practice of discernment, I would argue, provide a sort of data that, while vastly different from the "hard" data of the natural sciences, still have some measure of objectivity in the sense of intersubjective agreement, reliability, and replicability under similar circumstances.[16]

---

[15] See my *Theology in an Age of Scientific Reasoning*, chap. 5.

[16] It remains to be seen exactly how reliable this practice can be under suitable circumstances. This, however, is an empirical question that I hope will be answered after a period of research.

How might such data have a bearing on theology? I offer one example. Two competing views of the "essence" of Christianity are found in Latin American liberation theology and in the writings of Mennonite historian and theologian John Howard Yoder. Liberation theology sees God as the God of the oppressed, whose work in history is the liberation of the economic underclass. Often proponents of this view endorse Christians' participation in armed revolution. In contrast, Yoder's God is best characterized by a phrase from a commentary on Revelation: "Suffering Love sits on the throne of the universe."[17] To participate in the acts of this God in history is to refuse to take up the world's weapons and to suffer for that refusal if one must.

Judgments made by communities faced with a live choice about how to support the oppressed in their midst can have a direct bearing, confirming one of these theological programs and disconfirming the other. If they judge that the Spirit of God is leading them to fight, the liberation program is confirmed. If instead they judge that they are being called to act in a nonviolent way, and are being prepared by God to suffer for it, this confirms Yoder's program.

4 *The Temple-Ellis Program*

I hope that the foregoing has established as well as is possible in a few pages the plausibility of thinking in terms of theological research programs. I turn now to a consideration of what I shall call the Temple-Ellis program. This discussion is based entirely on the paper by Ellis in this volume.[18] Ellis draws on the theological work of Anglican theologian William Temple to construct a theological-cosmological interpretation of fine-tuning.[19] I hope to show in passing that it does incorporate many of the features of a scientific research program.[20] However, my central argument will be that the additions Ellis makes to Temple's theology in order to account for the fine-tuning of the universe constitute a progressive move in Lakatos' sense. This is exactly the sort of result required by the epistemology discussed above if we are to claim support from the apparent fine-tuning of the universe for the designer hypothesis, treated as an auxiliary hypothesis incorporated into a theological research program.

I first quote Ellis' summary[21] of the main points of Temple's theology; this summary may be taken as the core of the program:

(1)　　God is the creator and sustainer of the universe and of humankind, transcending the universe but immanent in it.

---

[17] The phrase is W. T. Conner's, but it aptly summarizes the point of Yoder's commentary.

[18] See the article by Ellis in this volume.

[19] Ellis' sources are W. Temple, *Readings in St. John's Gospel* (London: Macmillan, 1961); and *idem, Nature, Man and God* (London: Macmillan, 1934).

[20] With more work it could be made to exhibit more such features, especially by attention to the theological data available for its support.

[21] See the article by Ellis in this volume, section 5.2.

(2) His nature embodies justice and holiness, but he is also a personal and loving God who cares for each of his creatures (so the name "Father" is indeed appropriate for him).

(3) His nature is revealed most perfectly in the life and teachings of Jesus of Nazareth, as recorded in the New Testament of the Bible, who was sent by God to reveal his nature, summarized in "God is Love."

(4) He has an active presence in the world that still touches the lives of the faithful today.

Next, Ellis notes several ways in which this core theory requires further elaboration. In Lakatosian terms, the implications of the core, (along with additional assumptions) constitute most of the program's auxiliary hypotheses.

- God as creator implies AH_1: that God is the designer of the material world.

- The personal character of God implies AH_2: that God manifests feelings, will, and desires.

- The fatherhood (motherhood) of God implies AH_3: that as sons and daughters we express some of God's essential nature.

- The active presence of God implies AH_4: the existence of what Christians refer to as the Holy Spirit.

- The claim that Jesus reveals the nature of God implies AH_5: that God's character can be further illuminated by examining the life and teachings of Jesus as reported in the New Testament. For example, we learn there that God's manner of action is unfailingly loving and forgiving.[22]

---

[22] Ellis notes that if Jesus, fully human as ourselves, was able to know God then there must be some channel for revelation open to him and perhaps to the rest of us as well. The relevant theory here has to do with the work of the Holy Spirit. But as Ellis rightly points out, we cannot take all such putative visions and revelations at face value. His suggestion that "good fruit" provides a criterion for testing putative revelations (Ellis, in this volume, section 2) is in close agreement with the view about the necessity for discernment expressed above.

So far we have had a glimpse of the theoretical structure of Temple's research program. Further along in Ellis' paper we see a sketch of a later version of the same program with new auxiliary hypotheses added in such a way as to be consistent with the core and with previously formulated auxiliary hypotheses, but also aimed at taking account of the new data from cosmology—the fine-tuning of the universe. Ellis' reasoning includes the following:

- From the core theory we know that Jesus reveals most perfectly the nature of God.

- From Scripture (e.g., the temptation story in Mt. 4:1-11 and Lk. 4:1-13) we know the character of Jesus; this includes his decision to allow a free response to his love rather than one compelled by wonders or power.

- From these two theories he derives a third: God wills that his creatures respond freely to his love (AH_6).

- From this latter conclusion and from added assumptions and observations about the requirements for free response (lawful regularities in the environment and intelligence—AH_7) Ellis derives conclusions about the kind of universe such a God could be expected to create; namely one that is law-governed on the natural level (AH_8) and finely-tuned to allow intelligent life to evolve (AH_9). What the fine-tuning amounts to will have to be spelled out with the help of an assortment of conclusions from the natural sciences, which in turn serve as data for this hypothesis.

Schematically the relevant parts of the program are represented in *Figure 4*. (See pages 412-413 above for the contents of the core.)

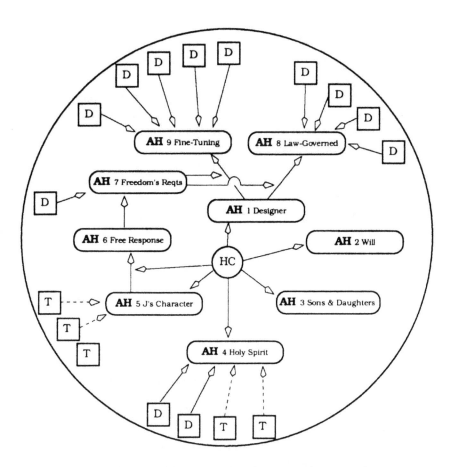

Figure 4: Schematic diagram of the Temple-Ellis research program. Auxiliary hypotheses 1-5 are Temple's; hypotheses 6-9 are Ellis' additions.

The main thing lacking from Ellis' program is attention to the data that might be provided to support it. A full discussion of this topic is beyond the scope of this paper, but I suggest the following as a few possibilities. The support for AH_5, regarding the character of Jesus, is from Scripture (the T's represent textual support; the dotted lines are intended to show that this is support from the past—that is, from the other "dimension," shown in *Figure 3*, to which theology must try to be faithful). AH_6, the claim that God wills free responses from his creatures has been understood by Anabaptists, Quakers, and others to imply that Christians must not force their wills on others, in religious matters or in any other. This further auxiliary hypothesis, not represented in the diagram, has in fact been confirmed by the process of discernment described above (among Quakers and Mennonites, and probably others), and it is open to replication now, as suggested by the imaginary example in section 3.3.2 (above). The doctrine of the Holy Spirit (AH_3) is probably the easiest of the purely theological hypotheses to confirm by means of current Christian experience, which agrees rather well with the relevant data from Scripture. There may be some observations supporting AH_7, regarding conditions for freedom, but I assume this hypothesis comes mainly from thought experiments.

Data for AH_8, the law-governed character of the natural world, are myriad; and those for AH_9, regarding fine-tuning, include both the conclusions regarding the conditions necessary for intelligent life to form and, at second remove, the data from which constants, particle masses, and so forth, are derived. To be more precise, the claim that the universe is fine-tuned is based on a number of sets of statements of the sort:

S_{1a}: Constant C must fall between x and y for life to evolve in the universe.

S_{1b}: C falls between x and y.

Or:

S_{2a}: The ratio of constants C_2 to C_3 must fall between y and z for life to evolve.

S_{2b}: The ratio of constants C_2 to C_3 falls between y and z.

Each of these statements is in turn based upon a variety of measurements, calculations, or both.

The auxiliary hypothesis—that the universe is fine-tuned for life—supports the designer hypothesis, and then indirectly all of the theological program. The immediate data for the fine-tuning are the pairs of statements as above, but it is obvious that the relation "evidence for" or "data for" is transitive: if A is evidence for B, and B is evidence for C, then A is evidence for C.[23]

---

[23] Wim Drees has pointed out an asymmetry in the hypothetico-deductive reasoning here. We argue from the fine-tuning to the existence of a designer because the designer hypothesis explains the fine-tuning. But the hypothesis of a designer only implies that the actual world fits within a band of life-allowing configurations, whether that band be wide or narrow. Thus, fine-tuning is irrelevant. However, Bayes's theorem shows that the amount of confirmation given to a hypothesis (H) by evidence (e) is greater if e itself has a low probability than if e has a high probability. What we mean by saying the universe appears to be fine-tuned is that the prior probability that the universe be suited for life is very, very low.

The drawback to Bayes's theorem in general is that there is usually no non-arbitrary way to ascertain prior probabilities, but we need not do so for present purposes. Our goal is only to show that the amount of confirmation provided by a particular piece of evidence varies with the prior improbability of the evidence.

Bayes's theorem states:

What I hope to have shown in this section is that theology can be regarded as a science (though one of a different level than cosmology—see below, section 5.1) and, with the help of Ellis' paper, that the fine-tuning of the cosmos fits in a perfectly natural way as an additional auxiliary hypothesis for this research program.[24] There are two reasons for making this point. The first is in response to changes in epistemology that suggest that no theory (and *a fortiori* no design argument) can be evaluated apart from the web of beliefs (network of

$$P(H/e) = \frac{P(H) \times P(e/H)}{[P(H) \times P(e/H)] + [P(-H) \times P(e/-H)]}$$

$P(H/e)$ is the probability of the hypothesis given the evidence.
$P(H)$ is the prior probability of H (i.e., before e is known).
$P(-H)$ is the prior probability of not H, i.e., $-H$.
$P(e/H)$ is the probability of the evidence given the hypothesis.

Since the point here is only to show the difference made to the Designer hypothesis by the fact that the conditions for life are highly improbable, we will simply assign a value to $P(H)$—let us say .2; $P(-H)$ is therefore .8, since the probability of the statement 'Either there is a designer or there is not' must equal 1.

Let us begin by assuming that a universe capable of supporting life is equiprobable with one that cannot. Then $P(e)$ is .5. But the probability that the universe will be suited for life if there is a designer—$P(e/H)$ is 1; that is, an omnipotent and omniscient designer would not have failed to meet these requirements.

One last assumption we have to make is the probability of e given not H. Let us assume for simplicity's sake that there are no other hypotheses to take into account, and that the $P(e/-H) = P(e) = .5$.

Thus:

$$P(H/e) = \frac{.2 \times 1}{(.2 \times 1) + (.8 \times .5)} = .333$$

Now, let us change $P(e)$ to .001 to show what happens as the probability of e goes down. (This is ridiculously high to represent the probability of the fine-tuning, but still makes the point.)

$$P(H/e) = \frac{.2 \times 1}{(.2 \times 1) + (.8 \times .001)} = .996$$

These rough calculations illustrate that the amount of confirmation provided by the evidence is proportionate to the prior improbability of the evidence, and account for our intuitive sense that the 'fineness' of the fine-tuning does make a difference to the posterior probability of the hypothesis.

[24] I want to make it clear that I am *using* Ellis' work to make a point that he himself did not intend; he did not set out to use the fine tuning as corroborative evidence for a theological research program.

theories) of which it is a part. The second reason is to begin to dispel a deep-seated bias against direct comparison of theological with scientific theories. We now turn to this second issue.

5 Competing Hypotheses

5.1 The Relations Between Theology and Science

Further pursuit of the viability of the design theory (over against its assorted competitors) as an explanation of fine-tuning requires some attention to the relations between theology and science. There are two opposing positions between which I intend to steer: One is the Fundamentalist or Creationist assumption that there is no difference in kind between theology and science. The other is the assumption more commonly found among heirs of the Liberal tradition that they are incomparably different. I want to maintain that they are different, but not incomparably so. The best approach here is to argue that theology is a science (or very science-like), but that it deals with reality at a higher level of complexity than do the other sciences—it takes its place at the top of the hierarchy of sciences.[25] A somewhat similar view is that of Wolfhart Pannenberg, who argues that theology is the science that provides the most all-encompassing context for the other sciences.[26]

Thus we need to replace Quine's two-dimensional model (a flat web of beliefs with different regions for different sciences) with a three-dimensional model—layers of webs where each layer represents tightly or loosely interconnected research programs making up the science of that level.[27] Each level is connected at points with the ones above and below, and also with more distant levels (See *Figure 5*).

When we come to labeling the sciences in the hierarchy, however, there is some ambiguity. The higher sciences study higher levels of complexity, but is the universe as a whole, which cosmologists study, more or less complex than a human society as investigated by the various human sciences?[28] Perhaps we need a branching hierarchy with three branches for the animate, inanimate, and the human systems or, more simply, two for the natural and human sciences. I suggest the arrangement in *Figure 6*, excluding intermediate sciences such as biochemistry.

---

[25] This is a view espoused in a number of Peacock's writings. See, for instance, *Theology for a Scientific Age: Being and Becoming—Natural and Divine* (Oxford: Basil Blackwell, 1990).

[26] See W. Pannenberg, *Theology and the Philosophy of Science* (Philadelphia: Westminster Press, 1976).

[27] Actually, we need a four-dimensional diagram in order to include the development of scientific and theological research programs in time.

[28] Or perhaps cosmology is the least complex because it reduces the features of the observable world to the simplest of laws.

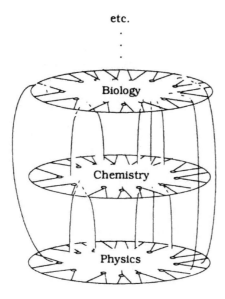

Figure 5: Hierarchical three-dimensional model with interconnection between the sciences.

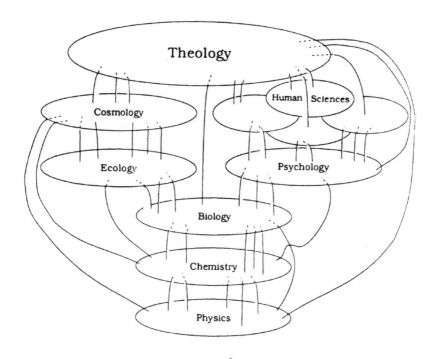

Figure 6: Branching hierarchy of sciences in relation to theology.

We know from looking at theories within science proper that there are connections between levels—both upward and downward. In fact, some sciences, such as genetics, cut through several levels and thereby tie adjacent and more distant levels together. Insofar as the levels in the hierarchy of science are interconnected they are mutually explanatory and mutually supportive. The chemistry of DNA helps explain (from below) how evolution works, and the existence of such an account adds credibility to evolutionary theory. Similarly, ecological factors such as ice ages help explain evolutionary changes from above.

The interconnection of levels not only works to support scientific theorizing, but constrains it as well. We usually think of this in terms of the requirement that laws and theories of higher levels conform to (i.e., avoid violation of) lower-level laws. However, the constraint goes both ways because the accounts given by lower levels have to permit (even if they cannot predict) the phenomena described by the levels above. So knowledge of the properties of atomic particles does not allow prediction of all of the characteristics and reactions of the chemical elements (and much less does it allow for prediction of the characteristics of entities at higher levels), but the laws of physics must be such that they do not rule out these higher-level phenomena. Another example: the laws of psychology should not be expected to predict the entire content or the specific character of religious experience, but must be such as not to rule out such experiences.[29] Similarly, Ellis argues that because theology assumes God attends to each individual in the universe, and because we cannot conceive of God attending to each individual in an infinite universe, this points to the conclusion that the universe must be finite (section 8.3); here the cosmological theory must not rule out the theological claim.

If we assume that Christian theology is the top level of the hierarchy of sciences, then it is not problematic to claim that insofar as the doctrine of creation explains the law-governed character of the natural world and the fine-tuning of the cosmos it is supported by those scientific findings. So the sketch in *Figure 4* of the Temple-Ellis program ought to be amended as in *Figure 7*, representing the claim that part of its support now comes from phenomena belonging to cosmology.

---

[29] Some might say that psychology does allow for the prediction of religious experience, but I believe this would be mistaken. Given our knowledge of religious experience, it might be said that psychology explains all or part of it after the fact. However, the reductionist claim to explain all of it seems simply false. While the imagined presence, say, of a Super-Father might be accounted for in Freudian terms, Freudian theory cannot account for what this Father seems to be saying or doing in the religious person's life except in the most general terms—meeting psychological needs. I cannot pursue this adequately here, but suffice it to say that even if most or all of religious experience could be accounted for *post hoc*, its particular forms and content could never have been predicted by psychological or sociological theories.

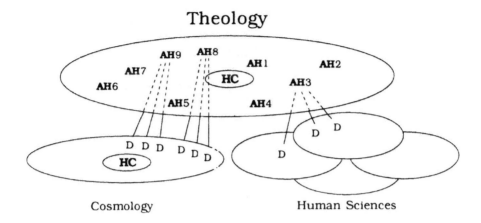

Figure 7: Support for cosmology and the human sciences of the Temple-Ellis program.

Notice, though, that to place Christian theology at the top of the hierarchy of the sciences is to beg the question we have set out to answer. Theology only deserves this position if in general its theories turn out to be the best explanations of the relevant phenomena, both from within its own realm and from within the natural and the human realms. Our inquiry is aimed at answering the limited question whether design is the best explanation for fine-tuning. To do so we need to look at the competition.

5.2 *Competitors*

I am in no position to evaluate the more scientific of the competitors for the design hypothesis. To some extent, only time will tell whether they are even capable of empirical support. Some remarks of a philosophical nature can be made, however. The competitors to be considered are:

(1) Pure Chance
(2) Mathematical (Logical) Necessity
(3) Many Universes
(4) Anthropic Principles
(5) John Leslie's Neoplatonic deity.

5.2.1 *Pure Chance* The claim that the universe just is the way it is and no explanation is possible can be taken in several ways, none of which falls into the category of natural science. It might be taken as a statement about the limits of knowledge—prediction that the null hypothesis will forever be our lot. This epistemological claim is falsifiable but not confirmable.

However, it may be better to take the chance hypothesis as a metaphysical claim that chance is in some way the ultimate principle behind reality. Standing alone, its only evidence is the fine-tuning itself. However, Peacocke and others have shown that chance can be meaningfully incorporated into a theistic worldview.[30] But then it is not a competitor of Christianity.

Another possibility is that the claim be incorporated into a Monod-style metaphysical system. Such a system would be supported whenever events appeared to be utterly random, but would have to explain or explain away all appearances of order and meaning. It is possible to do this, of course, but the question is whether it can be done in a coherent way, or would all the explanations be unrelated and *ad hoc*. My suspicion is that in the end it would turn out to be a "Chance of the Gaps" sort of argument.

5.2.2 *Mathematical Necessity* If it is someday shown that there is only one set of numbers that can be used to solve the equations comprising the basic laws of nature, this would in a sense provide a scientific explanation for the fine-tuning of the individual constants, masses, and so on. It provides no ultimate explanation, however, because we still can wonder at the coincidence that the only possible universe is also life-supporting, and at the fact that this one-and-only possibility is instantiated—the old question why there is something rather than nothing.

5.2.3 *Many Universes* A natural move to explain (explain away?) the wonder at finding the universe finely tuned for life is to propose that it is but one of vastly many universes—all different—and that we naturally find ourselves in the one (or one of the ones) that, by random variation, happened to be suitable for life. One or another of these many-universes explanations looks to be the most promising for straight-forward scientific status. It is not correct to claim, as some do, that such a hypothesis is untestable in principle since other universes are by definition not observable from this one; this would be to assume that we can only claim to have knowledge of what we can observe, and forgets the principle of hypothetico-deductive reasoning. The crucial issue has to do with whether any independent evidence for additional universes can be provided. At present, it seems that the standard many-universes hypotheses are *ad hoc*. They explain the appearance of fine-tuning but play no further role in our system of knowledge—

---

[30] See I. Barbour, *Religion in an Age of Science*, The Gifford Lectures, 1989-1991, vol. 1 (San Francisco: HarperSanFrancisco, 1990); J. Bartholomew, *God of Chance* (London: SCM Press, 1984); Peacocke, *Creation and the World of Science* (Oxford: Clarendon Press, 1979); J. Polkinghorne, *Science and Providence: God's Interaction with the World* (Boston: New Science Library, 1989); W. Pollard, *Chance and Providence* (New York: Charles Scribner's Sons, 1958).

they have no other connections within the web of beliefs. In this regard (other things being equal) they are at a disadvantage when compared with the design hypothesis. However, certain forms of quantum cosmology, drawing on quantum gravity, lead to a many-universe hypothesis. These quantum many-universe hypotheses are more robust than the standard version in that they arise out of a more general theoretical structure with additional empirical consequences. On the other hand, they are more speculative, since they reflect the overall tentativeness of all current proposals concerning quantum gravity. However, those based on an interpretation of quantum theory, including quantum creation of the universe (QCU) proposals, might avoid being *ad hoc* and deserve further study.

The design hypothesis has other advantages as well. Supposing that suitable scientific confirmation is provided for one or another of the many-universes hypotheses, the further question can still be raised: "Why are there all these universes?" The design hypothesis explains both the fine-tuning and the existence of the universe(s). In fact, the many-universes hypothesis is not incompatible with design. For example, theologies such as Augustine's include the principle of plenitude: God's creative purpose is in the first instance to create as much variety as possible; human life is important to God mainly because it fills one prominent rung on the ladder of being.[31] This theory of God's motivation for creation leads to the prediction that if other universes are possible, they will be created. Confirmation of a many-universes theory would provide confirmation for such a view of God's purposes or values in creating.

However, a more biblically-based version of Christian theology sees the creation of human beings with their capacity for relation to God as an extremely high priority, and suggests that much of the rest of God's creative action is ordered to that end. Confirmation that there is but one universe and that it is fine-tuned for intelligent life would confirm the more human-centered understanding of creation.

5.2.4 *Anthropic Principles* The various anthropic principles sometimes appear to be proposed to explain the fine-tuning of the universe, for example:

> The modern form of the Weak Anthropic Principle arose from attempts to relate the existence of invariant aspects of the Universe's structure to those conditions necessary to generate 'observers'. Our existence imposes a stringent selection effect upon the type of Universe we could ever expect to observe and document. Many observations of the natural world, although remarkable *a priori*, can be seen in this light as inevitable consequences of our own existence.
>
> Cosmological interest in such a perspective arose from attempts to explain the ubiquitous presence of large dimensionless ratios in combinations of micro and macrophysical parameters. . . . In this chapter we shall describe some of the background to these and other cosmological "coincidences" and show how, in the period 1957-1961,

---

[31] John Hick, *Evil and the God of Love*, 2nd ed. (London: Macmillan, 1977), 83.

they led to Dicke's proposal of an anthropomorphic mode of explanation.[32]

However, it seems that neither the weak anthropic principle (WAP) nor the strong anthropic principle (SAP) could possibly serve this function. The WAP has been formulated as follows:

> The observed values of all physical and cosmological quantities are not equally probable but they take on values restricted by the requirement that there exist sites where carbon-based life can evolve and by the requirement that the Universe be old enough for it to have already done so.[33]

This principle "explains" fine-tuning only in a Pickwickian sense. The need is not to explain why the universe is fine-tuned given the presence of life, but rather, to explain why it is fine-tuned in the first place. The relation between fine-tuning and life is actually the reverse; fine-tuning is a partial explanation for the existence of life. To turn it around and say that the presence of life somehow explains the fine-tuning is to reason backwards.[34]

The SAP has been formulated as follows:

> The Universe must have those properties which allow life to develop within it at some stage in its history.[35]

Here again we have an "explanation" of fine-tuning only in a very odd sense unless a further explanation is added to it: "Why does the universe have properties suited for the development of life?" Answer: Because it must have those properties in order to support life. But the next question is: "Why must it support life?" So far two answers have been given. One is based on the necessity for consciousness (an observer) to make sense of quantum theory, but the interpretation of quantum theory on which this is based is controversial and apparently untestable.[36] The other answer is that there must be life in the universe because God has ordained it, but now the SAP collapses into the designer hypothesis.

5.2.5 *Neoplatonism* John Leslie argues that the fine-tuning of the universe supports the hypothesis of the existence of God at least equally as well as it does the many-universes hypothesis. However he identifies his concept of God as Neoplatonic. By this he means that God is not a person but rather a metaphysical principle, which he expresses as follows: "Neoplatonism is the view that some ethical needs are themselves creatively effective, unaided by any mechanism."[37] This principle of the creative efficacy of ethical requirements is his explanation for the fine-tuning of the universe, and also for its existence.

He states further:

---

[32] Barrow and Tipler, *The Anthropic Cosmological Principle*, 219.
[33] Ibid., 16.
[34] Notice that I am not claiming that the WAP is empty or useless; only that it cannot properly be used as a complete explanation of the fine-tuning.
[35] Barrow and Tipler, *The Anthropic Cosmological Principle*, 21.
[36] However, see the article by A. Grib in this volume.
[37] Leslie, *Universes*, 171.

Whether this view is right is something which cannot be settled by conceptual analysis. The matter is not a logical, conceptual matter. But the actual existence of a universe (and of one, what's more, which seems fine tuned for Life) can suggest its rightness.[38]

So Leslie's proposal is a metaphysical one, a philosophical concept of God which we can imagine putting in place of Christian theology at the top of our hierarchy of sciences. However, he seems to be arguing for this thesis solely on the basis of the support it gets from fine-tuning and from the fact that it is less objectionable than some other hypotheses (including that of a personal designer-God) and no more objectionable than the many-universes hypothesis. He argues that it is preferable to the hypothesis of a personal designer because he sees no adequate answer to the further question of why this designer exists. But one might in turn ask him what accounts for the efficacy of his Neoplatonic principle; or why the idea of such a principle argues for its existence (cf. the usual criticism of Anselm's ontological argument).

The important issue, as emphasized above, is whether any of the competing hypotheses have confirmation independent of the fine-tuning, or whether instead they are merely *ad hoc* provisions to explain the fine-tuning alone. Leslie's bare metaphysical principle seems to be untestable in principle, just as is the pure chance hypothesis. However, Neoplatonism is a varied tradition and a more robust account of God and of God's relation to the universe might be drawn from it. It might turn out that a contemporary Augustinian-Neoplatonic research program would be more consistent with the many universes hypothesis than alternative versions of Christian theology, again because of the principle of plenitude.

5.3 *The Christian Design Hypothesis*

Our all-too-brief survey of the competition has served to show that the available explanations of fine-tuning are of varied types, most of which are non-scientific or meta-scientific or metaphysical. In our culture there is a strong tendency to value science and discount theology; so, for many, in a competition between a scientific and a theological hypothesis, the theological will be rejected out of hand. Therefore it is helpful to note that most of the design hypothesis' competitors are not scientific either.

If we can get beyond biases against theology in general and beyond the "two worlds view," that automatically discounts God as an explanation for anything scientific, then it appears that the design hypothesis has a great deal to say for it. Its advantage over all of the others considered here is the large amount of independent confirmation available to it. Even if there were no other evidence for it from the side of the natural sciences, it can be argued that there is a great deal from the side of the human sciences and from data peculiar to theology itself (historical events, religious experience, etc.).

But can the "confirmation" of a theological theory be compared to that of a scientific theory? If one of the many-universes hypotheses were confirmed

---

[38] Ibid.

scientifically would that not settle the matter? Others have pointed out that there are similarities and differences in the kinds of support available for theology and for science. In the natural sciences we have relatively clear-cut, "objective," "hard" data. In contrast, in the human sciences (including psychology) we have to do more to manufacture reliable and valid data. Theology's data, not surprisingly, are more like the latter than the former.

In science there are a number of criteria for evaluating a theory relative to its data. We look for close agreement between predictions and actual measurements (empirical fit), fruitfulness in leading to new discoveries, broad scope and unifying power, and simplicity or elegance. Lakatos' criterion, described above, incorporates all of these criteria to some extent. He has noted that what is wanted in developing a theory is that it not become a jumble of *ad hoc* additions as it is adjusted to take account of new data; this relates to the criteria of simplicity and coherence (unifying power). Empirical fit and fruitfulness are both covered by Lakatos' requirement for the prediction and corroboration of new, previously unexpected facts.

As noted above (see note 9), there has been a great deal of controversy over the definition of novel facts. At one point Lakatos amended his account of novel facts to include already-known facts that are shown to be relevant to the theory in ways previously unexpected. For example, the explanation of Mercury's perihelion have dramatic corroboration to Einstein's theory—even though as a low-level empirical proposition it had been known for almost a hundred years—because its exact solution was "an unexpected present," an unintended by-product of Einstein's program.[39]

I have argued elsewhere that theological research programs can be confirmed in the manner advocated by Lakatos.[40] If this is indeed possible, then what we would like to see here is novel confirmation of a theological program resulting from the incorporation of the fine-tuning hypothesis. Again Ellis has done the theological work. Certain well-established facts about the natural world that were irrelevant to the earlier version of the Temple-Ellis research program (i.e., Temple's own theology) have become relevant as an unintended by-product of the auxiliary hypotheses regarding the requirements for free will that Ellis added in order to take account of fine-tuning. These facts include all of the observations that show the natural world to be law-governed. Ellis did not set out to explain this feature of the universe, but in the process of considering the necessary conditions for free and intelligent life he added an auxiliary hypothesis (AH_8) explaining why, in the plan of God, this needed to be so. These facts are "weakly novel" since they were already known.

Note that this addition also provides a tie to the part of Christian theology known as theodicy—answers to the problem of evil. Polkinghorne has argued that we need to add a "free-process" defense, analogous to a free-will defense, in order to take account of natural evil.[41] That is, much of what we call

---

[39] Lakatos and Zahar, "Why Copernicus's Programme Superseded Ptolemy's."

[40] See Murphy, *Theology in the Age of Scientific Reasoning*.

[41] See, for example, his article in this volume.

natural evil (hurricanes, earthquakes, plagues, famines) are direct results of the operation of the laws of nature. Ellis' arguments regarding the necessity of a law-governed universe in order to make human free choice meaningful answers the inevitable question: "Why does God allow the universe to run according to these laws rather than intervene at many points to save us from their harmful effects?"

There is room for further theoretical development at this point. One of the laws of nature is the second law of thermodynamics. Both Ellis and Robert Russell have suggested that the existence of an arrow of time is a necessary prerequisite for consciousness, and that the arrow of time is in turn dependent upon the second law.[42] Now, the second law itself is responsible for much of what we call natural evil (hunger, fatigue, death), and these natural conditions provide powerful motivation for moral evil. It is interesting to note that Augustine's embellishments of the story of the Fall involve describing Eve's and Adam's former existence in the Garden of Eden as "on-entropic."

> In Paradise, then, man . . . lived without any want, and had it in his power to live eternally. He had food that he might not hunger, drink that he might not thirst, the tree of life that old age might not waste him. There was in his body no corruption, no seed of corruption, which could produce in him any unpleasant sensation. He feared no inward disease, no outward accident. . . . Body and spirit worked harmoniously together, and the commandment was kept without labour. No languor make their leisure wearisome; no sleepiness interrupted their desire to labour.[43]

So this amplification of the program (now the Temple-Ellis-Russell program) goes some distance toward explaining moral evil as well as natural evil.

To sum up: it is clear that the addition of the fine-tuning hypothesis to the Temple-Ellis-Russell program is, in Lakatos' terms, progressive rather than *ad hoc*. It allows previously unrelated facts to be related to the program, and opens the way for a better explanation of the anomalous existence of evil—one that is integrally related to the rest of the program.

6 Conclusion

This paper has been an attempt to answer the question: "What can we make of the fine-tuning of the universe epistemologically and theologically." I began with the suggestion that fine-tuning provides grounds for a new design argument. To consider this possibility, though, it was necessary to be explicit about how views of arguments based on matters of fact have changed since the heyday of natural theology in the seventeenth and eighteenth centuries. I claimed that any design argument based on fine-tuning would have to be hypothetico-deductive in form.

---

[42] See the article by Ellis in this volume. See also, R. J. Russell, "Entropy and Evil," *Zygon: Journal of Religion & Science* 19:4 (1984): 449-468; and *idem*, "The Thermodynamics of 'Natural Evil'," *CTNS Bulletin* 10:2 (Spring 1990).

[43] Augustine, *The City of God*, xiv. 26.

That is, the hypothesis that God created the universe with the intention that it produce intelligent life, and tuned it accordingly, would be shown to be acceptable if it turned out to be the best available explanation for the facts in question.

However, the relative merits of competing hypotheses can only be determined by seeing how coherently each fits into a network of other well-supported theories (Quinian holism), and also, according to Lakatos, by seeing whether or not it allows for the prediction of novel facts. I noted on the one hand that most of the design hypothesis' competitors had no network of theories to cohere with, and some seemed incapable of confirmation by means of any facts whatsoever. The design hypothesis, on the other hand, has been shown to fit progressively into a Christian theological research program, increasing the overall coherence of the program and (most important) tying it to previously unrelated facts.

In the midst of this argument I made several moves intended to dispel worries that a theological explanation of scientific findings is in principle unacceptable. I provided a brief account of how theology can be seen to fit Lakatos' description of a scientific research program. I also used and elaborated the suggestion made by Peacocke that we view theology as the topmost science in the hierarchy of sciences. If we regularly explain phenomena described in a lower-level science by means of concepts or theories from above, then the use of a theological theory to explain findings in cosmology ought not present any special problems.

However, to assume that Christian theology is the proper set of beliefs to top off our three-dimensional web is to beg the question somewhat. So I offer my arguments in favor of design as explanation for fine-tuning as one small step toward the justification of that position for Christian theology.

Acknowledgments: I would like to thank all of the conference participants who commented on an earlier draft of this paper, and especially George Ellis, Robert Russell, and Keith Ward for helpful criticisms of this version.

THE LAWS OF NATURE AND THE LAWS OF PHYSICS

John Polkinghorne

1 *Introduction*

When we think of the laws of nature, it is the laws of physics which come naturally to one's mind: Newton's law of gravity, or spontaneously broken gauge field theories, or even (if one is somewhat grandiose and speculative) a grand unified theory of everything.[1] Of course, historically these laws have been subject to some revision (as when Newton gave way to Einstein), but I hold to the critical realist view that this process can be interpreted verisimilitudinously, as the tightening grasp of an actual reality. I have defended that stance elsewhere, in the course of an account of the recent history of elementary particle physics[2] and I assume it in what follows.

A critical realist will want to take with great seriousness what fundamental physics has established about the nature of the world, and even listen with a respectful attention to speculative discourse concerning regimes (such as quantum cosmology) where imagination has outstripped knowledge. Yet that should not lead us to deny the authority and significance of straightforward macroscopic experience. I wish to assert a metaphysical egalitarianism in which the commonplace is accorded at least equal status with theories of the very large or very small. I am confirmed in this stance by the consideration that far and away the most interesting, puzzling, and surely significant thing about the physical world is that it has proved to be the home of self-conscious beings.

2 *Emergence*

To equate the laws of physics with the laws of nature would be to adopt a reductionist stance. I do not oppose a *constitutive* reductionism, for I believe that if you take me apart, all you will find is a collection of quarks and gluons and electrons. I accept that humans are psychosomatic unities and I do not anticipate that a separable spiritual component (soul or entelechy) is part of our make-up. Hence my approach to the perplexity of mind and matter will be to seek to

---

[1] See P. Davies, *Superforce* (London: Heinemann, 1984).

[2] J.C. Polkinghorne, *Rochester Roundabout* (Harlow: Longman, 1989), ch. 21. For a general discussion of the problems see: *Scientific Realism*, ed. J. Leplin (Berkeley: University of California Press, 1984).

embrace a dual-aspect monism. Like many others,[3] I oppose a *conceptual* reductionism (which would assert that biology and anthropology are merely elaborate corollaries to physics). If one accordingly adopts the admirable stance of asserting that wholes are more than mere sums of their parts, and combines this with the concept of a relatively orderly world, one is implicitly suggesting that there are holistic as well as constituent laws of nature. The structured richness of reality then results from an interplay between these two kinds of laws, and there is the problem how we should best think about their mutual interrelation.

The usual strategy has been to talk in terms of *upward emergence*. Increasing complexity provides an environment of relationships within which novelty comes to be. It is a strategy at once plausible yet opaque. The difficulty is to see how complexity gives rise to radical novelty rather than to mere proliferation of entities at the same level of possibility. Of course, without several atoms we would not have the notion of inter-atomic forces, but if those forces are calculable from the Schrödinger equation with electromagnetic forces inserted into it, what we encounter seems just additional complication, rather than anything really new. (Quantum statistics modifies the way the additions are made but it does not remove the constituent-based character, as a consideration of atomic theory illustrates.) The wetness of water is a concept which only makes sense when vastly many H_2O molecules are involved. Yet, because it is concerned with transactions of energy it is perfectly conceivable that it arises simply from the multiplicity of interactions of those molecules, unmodified in their essential nature by the context in which they occur.

Yet not all emergences have this character of conceivable simple continuity with the underlying processes. While I believe that "more is different," I do not think that this is always attributable simply to the fact that the computing time needed to make predictions rises exponentially with increasing degrees of freedom. The emergences which are truly interesting and truly puzzling are those where there seems to be a qualitative change—the coming-to-be of life or consciousness. (Even within physics itself, there are unresolved problems about how the irreversibility of thermodynamics emerges from the reversibility of the underlying dynamics; about how reliable classical measuring apparatus arises from its fitful quantum constituents.)

I do not think that we will be able to accommodate the appearance of the truly novel without being willing to go beyond a conceptual anti-reductionism to some form of what one might call "*process anti-reductionism*."[4] There is a residual reductive notion inherent in the upward approach alone, for its bottom-up argument is almost bound to imply that a certain fundamental status

---

[3] E.g., I. Barbour, *Religion in an Age of Science*, The Gifford Lectures, 1989-1991, vol. 1 (San Francisco: HarperSanFrancisco, 1990), ch. 6; A. R. Peacocke, *Creation and the World of Science* (Oxford: Clarendon Press, 1979), ch. 4; *idem*, *God and the New Biology* (London: Dent, 1986), chs. 1 and 2; and Polkinghorne, *One World: The Interaction of Science and Theology* (London: SPCK, 1986), ch. 6.

[4] See Peacocke, *God and the New Biology*, ch. 1, for a discussion of various forms of anti-reductionism.

should still be accorded to the laws of physics, since they are assigned so foundational a role. Yet the idea of ontological emergence must surely imply that the description in terms of parts is not exhaustive and that the total context has some genuine influence on the behavior of those parts. I want, therefore, to explore a more even-handed approach, which will also consider the possibility of *downward emergence*, in which the laws of physics are but an asymptotic approximation to a more subtle (and more supple) whole. Elsewhere I have written:

> It is possible that emergence is, in fact, a two-way process; that it would be conceptually valid and valuable to attempt to transverse the ladder of complexity in both directions, not only relating the higher to the lower but also the lower to the higher. Such a proposal goes somewhat beyond the mere acknowledgment of level autonomy, for it suggests a degree of reciprocity of understanding between levels.[5]

An approach of this kind may seem vulnerable to being characterized by the pejorative word "vitalism." It is always necessary when being threatened by refutation by slogan to inquire wherein lies the condemnatory character of the description being applied. Vitalism is an implausible concept if it is taken to imply that the addition to matter of a "magic ingredient" is necessary before it could become living or conscious. I have already forsworn such a view by frankly adopting a constituent reductionism. What is proposed here is not vitalism in that sense, but what might be called contextualism, the view that the behavior of parts is not independent of the nature of the whole that they constitute. This attitude is not contradicted by the impressive successes of molecular genetics, obtained by extrapolating physics and chemistry into the discussion of aspects of living beings, because the influence of context may be expected to vary with circumstance. There will doubtless be processes which are quasi-mechanical (that is, understandable in unmodified constituent terms) and these will always be the first to be understood in a discipline because they are the most readily accessible. To use parabolic language drawn from physics, we understand clocks before we understand clouds, and we should not conflate or confuse the two.

The endeavor I am advocating needs to be controlled by certain broad principles:

(1) *Coherence*. It will have to explain how the laws of physics, as we have come to know them, are related to this putative wider view. Their striking rational beauty will have to be perceived as in part the simplicity which results from some extreme regime in which process takes a particularly tractable and understandable form. The obvious sense in which one would expect this downward emergence of physical law to take place is the approximation in which entities are treated as isolable from the totality of their environment. As we shall see, this is only a good approximation for special kinds of physical system. In particular, it is the approximation within which the constituent investigations of elementary particle physics take place.

---

[5] Polkinghorne, *Reason and Reality: The Relationship between Science and Theology* (London: SPCK, 1991), 35.

(2) *Historic Continuity*. Acceptance of a constituent reductionism and of the insights of evolutionary history, embeds humanity within the physical world. Our understanding of that world must then be consistent with our being among its inhabitants. I take it, without further discussion (though I am aware that there has been much argument on this point!), that human beings have fundamental experiences of agency and choice which will have to be accommodated within our account of reality. Rationality would disappear with the denial of human freedom, for discourse would become the mere mouthings of automata (including, of course, the discourse denying human freedom). A world whose future was formed entirely by the unfolding of energetic causation and mechanical determinism would not, in my view, be open to human agency. Nor do I believe that a world whose openness to the future arose simply from the quantum ricketiness of its constituent parts would be able to satisfy that criterion either. To suppose so would seem to confuse randomness with freedom. Our ultimate aim must be to understand how a hot quark soup has given rise, after some fifteen billion years, to people; how intentional causality is compatible with physical causality.

(3) *Realism*. Underlying the whole venture is the realist conviction that we are able to gain actual knowledge of reality by scientific methods. I believe that what we can know and what is the case are intimately related. I possess a T-shirt with the stirring motto, "Epistemology models Ontology." Unless we believe ourselves to be lost in a Kantian fog, or reduced to an instrumentalist scrabble to get things done without knowing why we are able to do so, we must suppose that acquired knowledge is a guide to the way things are and resist attempts to drive a wedge between the two. The critical realist asserts this to the maximum degree consistent with careful evaluation. It is a strategy for the pursuit of understanding which has much to commend it. One can see how natural that is for a scientist by recalling the early history of quantum theory. When Heisenberg wrote his celebrated paper on the uncertainty principle he was concerned with analyzing what could be measured. His work was epistemological in character. Yet it was not very long before almost all physicists gave it an ontological interpretation. They see the Heisenberg principle as an expression of indeterminacy in nature, not merely of ignorance about it.

3 *Insights from Chaos Theory*

Even with these principles in mind, the task of accommodating intentional agency within a scientific world-view is a bold and speculative one. Ultimately it would require that one had the solution to the mind-brain problem within one's grasp. No such grandiose claim is being made here. Yet I think that a suggestive way of beginning to think about these issues has been provided by the modern development of the so-called dynamic theory of chaos.[6] For several years I have

---

[6] See J. Gleick, *Chaos: Making a New Science* (London: Heinemann, 1988); and I. Stewart, *Does God Play Dice?* (Oxford: Blackwell, 1989).

been concerned with what broader lessons we might learn from its insights[7] and Arthur Peacocke has recently been looking in a similar direction.[8] In this article it is only possible to indicate the general character of the discussion generated by this approach.

The theory of chaos tells us the surprising fact that those "tame" systems, from whose behavior we thought we had learnt about classical dynamics—such as the steady harmonic oscillator or the ceaselessly revolving single planet—are quite exceptional. I call them "tame" because small uncertainties in our knowledge, or small disturbances in circumstance, produce only correspondingly limited consequences for their behavior. To all intents and purposes we can know how they will behave. Yet most of the physical world, even when we describe it in the apparently dependable terms of classical Newtonian dynamics, is not like that. In Karl Popper's famous phrase, there are many more clouds than clocks around. Once systems attain even a modest degree of complexity they can become capable of an exquisite degree of sensitivity to circumstance which makes them intrinsically unpredictable. Their behavior exhibits apparent haphazardness, though not to an unrestricted degree—which is why "the theory of chaos" is really such an unfortunately inept name. Rather one sees a kind of structured randomness in the apparently haphazard exploration of a restricted domain of possibility (as with a "strange attractor").

The surprising feature of chaos in classical dynamics is that it presents us with this apparently random behavior arising from solutions to deterministic equations. It is an oxymoronic sort of subject. The critical question is: "Which shall we take the more seriously, the randomness or the determinism?" Let me put it this way. The most obvious thing to say about chaotic systems is that they are intrinsically unpredictable. Their exquisite sensitivity means that we can never know enough to be able to predict with any long-term reliability how they will behave. Unpredictability is an epistemological statement about what we can know. In my realist speculation I shall want to go beyond that to make an ontological guess about what is actually the case for the physical world. In other words, the epistemological limitations of classical physics act as the spur to seek a new kind of physics to which the old is an approximation and in which the former's unpredictability is re-expressed as an ontological openness. I want to say that the physical world is open in its process, that the future is not just a tautologous spelling-out of what was already implicit in the past, but there is genuine novelty, genuine becoming, in the history of the universe.

It is important to explain clearly what one means by the concept of an open future. Of course, I am not supposing that the future is some kind of lottery,

---

[7] Polkinghorne, *Science and Creation: The Search for Understanding* (Boston: Shambhala, 1988), chs. 3 and 5; *idem*, *Science and Providence: God's Interaction with the World* (Boston: Shambhala, 1989), ch. 2; and *idem*, *Reason and Reality*, ch. 3.

[8] Peacocke, *Theology for a Scientific Age: Being and Becoming—Natural and Divine* (Oxford: Basil Blackwell, 1990); and *idem*, "Natural Being and Becoming—The Chrysalis of the Human," in *Individuality and Cooperative Action*, ed. J. E. Earley (Washington, D.C.: Georgetown University Press, 1991).

that there is a whimsical randomness in the form it will take. There are causative principles which bring it about, but these are not exhaustively described by the presently known laws of physics, since these are construed as regulating constituent behavior only. The dead hand of the Laplacian calculator is relaxed and there is scope for forms of causality other than the energetic transactions of current physical theory. As we shall see there is room for the operation of holistic organizing principles (presently unknown to us, but in principle open to scientific discernment), for human intentionality, and for divine providential interaction. The character of such influence is perhaps best conceived as "active information," the creation of novel forms carried by a flexible material substrate. "Information" is not being used to mean a body of facts but a principle of organization, concerned with patterned structure rather than the transaction of energy. (For those acquainted with David Bohm's version of quantum theory,[9] the guidance of the pilot wave would be an example of active information impinging on the energetic motions of the particles, encoding details of the whole environment and producing a holistic constraint on particle behavior.) One could conceive that generalized forms of "active information" could correspond to actions manifesting freedom and that even lower-order forms (organizing principles) would be able to bring about previously unrealized possibilities of intrinsic novelty (e.g., life) and hence the phenomenon of becoming, which seems such an obvious feature of both human life and cosmic history. Other aspects of becoming could be modeled by the self-organizing behavior of systems far from equilibrium, where small triggers can produce remarkable macroscopic order.[10] However, the phenomena induced in this way, though possessing dynamical characteristics (Bénard instability; the Belousov-Zhabotinski reaction), do not seem to me to display the same degree of openness that chaotic systems manifest in their temporal development.

Thus I am encouraged to take the logically possible, but not logically forced, step of giving a primacy to behavior over equations, and so of interpreting deterministic chaos as pointing to an actual physical world of subtle and supple character whose laws are yet to be discovered and to which our presently known constituent laws are but asymptotic approximations. One is further encouraged in this speculative task by the hope that a world whose process is open to the future would be able to accommodate basic human experience within its description. On this view the deterministic equations from which our mathematical exploration began would be regarded as approximations in *an emergent downward direction* from this more subtle and supple reality. The approximation would arise as one made the simplifying assumption of treating the system in question as if it were isolable from the rest of what is going on. (We know, of course, that the exquisite sensitivity of chaotic systems, their

---

[9] D. Bohm, *Wholeness and the Implicate Order* (London: Routledge & Kegan Paul, 1980).

[10] See Peacocke, *Creation and the World of Science*, ch. 3; I. Prigogine and I. Stengers, *Order out of Chaos: Man's New Dialogue with Nature* (New York: Bantam Books, 1988).

vulnerability to the slightest external trigger, means that they are never truly isolable.)

You might think that this particular knot might more readily be cut by invoking quantum theory. After all, these exquisitely sensitive systems soon depend for the form of their future behavior on details more fine than Heisenberg will permit. Would it not be better to justify the openness of the future by employing this sensitive enmeshment of the everyday world with the indeterminate quantum world? I can see some attraction in the suggestion, but I hesitate for several reasons. One is the gut feeling I have already expressed that everyday openness should not have to depend on goings-on in the microworld; randomness is not freedom. Even if that were mistaken, there are other grounds for caution. One such centers on the unresolved interpretative problems of quantum theory,[11] particularly the measurement problem, whose perplexities arise precisely from our unsureness of how to treat the interaction between the macroscopic and the microscopic. Yet another reason lies in the difficulties that have been found in exhibiting chaotic behavior in relation to the Schrödinger equation.[12] I cannot but think that there must be at least some quantum analogue of classical chaos, even if it takes a somewhat different form (fuzzy fractals?), but it would be necessary to see this settled before one ventured very far along the route of invoking a quantum basis for openness.

So far I have argued that chaos theory presents us with the possibility of a metaphysically attractive option of openness, a causal grid from below which delineates an envelope of possibility (it is not the case that *anything* can happen, but many things can), within which there remains room for maneuver. How that maneuver is executed will depend upon other organizing principles active in the situation, viewed holistically. A chaotic system faces a future of labyrinthine possibilities, which it will thread its way through according to the indiscernible effects of infinitesimal triggers, nudging it this way or that. In the extrapolation I am making, which sees chaos theory as actually an approximation to a more supple reality, these triggers of vanishingly small energy input become non-energetic items of information input ("this way," "that way") as proliferating possibilities are negotiated. The way the envelope of possibility is actually traversed will depend upon *downward causation* by such information input, for whose operation it affords the necessary room for maneuver. Downward causality through information input could operate in various ways. There could be holistic organizing principles, inherent in nature and bringing about increasing complexification, which could provide an explanation of the existence of the "optimistic arrow of time,"[13] evident in cosmic evolution. If these principles act in repeatable circumstances, there is no reason why they should not be scientifically determinable, though presently they are unknown. This identification would correspond to the establishment of a new class of holistic laws of nature.

---

[11] Polkinghorne, *The Quantum World* (Harlow: Longman, 1984), ch. 6.

[12] See J. Ford, "What is chaos, that we should be mindful of it?" in *The New Physics*, ed. P. Davies (Cambridge: Cambridge University Press, 1989).

[13] Davies, *The Cosmic Blueprint* (London: Heinemann, 1987).

4 Mind and Brain

It is also possible that the ideas being discussed might afford some very modest help with the classical problem of the relationship of mind and brain. Having said that, I am mindful of the warning uttered by Thomas Nagel that those who today venture to speak about such matters are indulging in "pre-Socratic flailings around."[14] Nevertheless, we have to do the best we can, and I think there is at least a hopeful direction in which to wave our arms.

I have suggested previously that we might try to consider a complementary metaphysics of mind/matter,[15] what philosophers would call a dual-aspect monism: there is one stuff, but encountered in contrasting regimes it gives rise to what we call the material or the mental. The aim is to treat these two polar extremes in as even-handed a way as possible. Each pole would have its own anchorage in the appropriate dimension of reality; we participate in both a physical world and a noetic world.

The adjective "complementary" meant to invoke a celebrated aspect of quantum theory, to which Niels Bohr drew particular attention. If such invocation is to go beyond a mere slogan, it must indicate how the complementary poles are related to each other. In suggesting how this might be, I wish to use quantum theory as an analogical guide. One of its celebrated dualities is that of wave and particle. Quantum field theory tells us how the trick is done. A wavelike state is one with an *indefinite* number of particles in it (a possibility quite foreign to classical physics but permitted in quantum physics because of the latter's ability to mix together (superpose) states which classically would be immiscible). The key to complementarity always seems to lie in some dimension of indefiniteness. This suggests that mind/matter might be reconciled by their being different poles of the world's stuff in greater or lesser states of flexible organization; matter is the emergent downward side, mind the emergent upward side, where the arrow points in the direction of an intrinsic openness. Chaotic dynamics would represent the first primitive stirrings of openness as one mounted the ladder of complexity leading from matter to mind. The notion of causality through active information would be the first primitive glimmer of an understanding of how the mental decision of my will to lift my arm brings about the physical act of its raising. The attraction of complementarity, when it is properly invoked, is that it reconciles descriptions which otherwise might seem to be framed in terms of mutually exclusive categories (wave/particle; matter/mind). Of course there must also be some general area of discourse within which both terms fall. In the case of waves and particles, it is physics; in the case of mind and matter, it is ontology.[16]

---

[14] T. Nagel, *The View from Nowhere* (Oxford: Oxford University Press, 1986)

[15] Polkinghorne, *Science and Creation*, ch. 5.

[16] See the discussion of Polkinghorne, *Reason and Reality*, 25-28.

5 Divine Action

Having embarked upon my speculative metaphysical journey, I had better pursue it to the end. Agency through the holistic operation of information within the intrinsic flexibility of complex physical systems might also be a way of understanding God's action in creation.

Theologians have often, and rightly, wished to speak of God's continuing interaction with the world. (Only if we can do so will the fashionable notion of *creatio continua* amount to more than a pious gloss on deism.) Some have denied that we can say anything meaningful about its mode of operation.[17] I find this too intellectually despairing an attitude to take. Others (the process theologians) have wanted to use language of "lure" or persuasion, based on a panpsychic view of reality.[18] I find this latter view totally implausible. Other writers use words like "influence," "guidance," "improvisation," without indicating clearly what they could actually mean. A common thread in much of this discourse is the notion that in some way the analogical clue to how to speak about divine agency is provided by an appeal to the example of human agency. That is a strategy which I too would wish to adopt. The picture I have presented is of a world in which our willed action occurs through information input into flexible open process. That process is located in our bodies (presumably principally in our brains) but there is also open process elsewhere in the universe. I believe that God's continuing interaction with creation will be in the form of information input into the flexibility of cosmic history. This is a theme which I have tried to pursue systematically in *Science and Providence*. Let me summarize some of the points which I think are relevant to present consideration:

(1) Though pictured as acting through open process, God does not soak up all its freedom. It is an important insight of theology that in the act of creation, freedom is given to the whole cosmos to be and to make itself. I have made this the basis of a free-process defense in relation to physical evil (disease and disaster), which parallels the free-will defense in relation to moral evil (the erroneous and sinful choices of humankind).[19] God neither wills the act of a murderer nor the incidence of a cancer, but allows both to happen in a world which God has endowed with the ability to be itself. Of course, it is less clear what the moral value is of allowing a tectonic plate to slip and kill fifty thousand people in the Lisbon earthquake of 1755. Yet we are intimately connected with the physical world from which we have emerged and I think it probable that only a universe to which the free-process defense could apply would be one in which there could be people to whom the free-will defense could apply. The balance between the room for maneuver which God has reserved and that which God has given away is, of course, a delicate issue. It is also an issue familiar to theology, for it is the problem of grace and free-will, written cosmically large.

---

[17] A. Farrer, *Faith and Speculation* (London: A & C Black, 1967).

[18] See J. B. Cobb and D. R. Griffin, *Process Theology: An Introductory Exposition* (Philadelphia: Westminster, 1976). Griffin prefers the word panexperiential to panpsychic.

[19] Polkinghorne, *Science and Providence*, ch. 5.

(2) God's action within the cloudiness of unpredictable open process will always be hidden; it cannot be demonstrated by experiment, though it may be discerned by faith. Its nature also limits what it is sensible to pray for. Long ago, the Alexandrian theologian Origen recognized that one should not pray for the cool of spring in the heat of summer. Quite so: the succession of the seasons is a clockwork part of the physical world, whose regular pattern reflects the faithfulness of the creator. Divine action is self-consistent and one aspect will not overrule another.

(3) I have been criticized by some for what they believe is a return to the discredited notion of a "God of the gaps." My answer would be that what was discreditable about that illusory deity was that the gaps were epistemic, and thus extrinsic to nature, mere patches of current scientific ignorance. As they disappeared with the advance of knowledge, the "god" associated with them faded away as well. No one need regret this passing, for the true God is related to the whole of creation, not just the puzzling bits of it. Yet if the physical world is really open, and top-down intentional causality operates within it, there must be intrinsic "gaps" ("an envelope of possibility") in the bottom-up account of nature to make room for intentional causality, whether I have identified their nature correctly or not. We are unashamedly "people of the gaps" in this intrinsic sense and there is nothing unfitting in a "God of the gaps" in this sense either. Of course, the suggestions of this paper are crude guesses. What else can we manage today but pre-Socratic flailings around? I think we have to take the risk of such exploration and not rest content with a discussion in such soft-focus that it never begins to engage our intuitions about God's action with our knowledge of physical process.

(4) More subtly and disturbingly, one might fear that I have enmeshed God too closely with physical process, making the elementary and disastrous theological blunder of treating God as a cause among other causes, a mere invisible agent in cosmic process. I do not want to turn God into a demiurge and I do not believe that I have done so. God's interaction is not energetic but informational. I believe my account is a kind of demythologization of what is unclearly articulated by those who use words like "lure" or "influence" or "guidance."

(5) The picture of physical process encouraged by the metaphysical exploration of chaos theory is one of a world of true becoming in the sense I have explained. I think that this has important implications for God's relation to time.[20] The picture of classical theism is that the eternal God sees all that happens in time at once. God does not foreknow the future but simply knows it. It was a picture similar to that of classical relativity, with its spacetime diagrams presented for inspection as frozen chunks of history. That is all very well for a deterministic world, which is really a world of being and not becoming. Chaos theory's intimation of openness changes all that. It seems to me that the eternal viewpoint may no longer be a coherent possibility, for the future is not up there waiting for us to arrive; we help to make it as we go along.[21]

---

[20] Ibid., ch. 7.

[21] See the article by Isham/Polkinghorne in this volume.

This has two consequences for our view of God. First, God will have an intimate connection with the reality of time. In fact, without that connection, it would mean that fundamentally (that is, in God's view) time was less than truly real. God's temporal involvement will scarcely come as a surprise to those acquainted with the biblical God of Abraham, Isaac, and Jacob, involved in the history of Israel. Of course the divine nature must also have an eternal aspect, for God is not in thrall to the flux of becoming, only in intimate and interacting relationship with it. One of the emphases of much twentieth century theology has been to seek to give an account of such a dipolar (time/eternity) theism. Here I gladly acknowledge a helpful aspect of process thought.

The second consequence is more contentious, but it seems to me to be necessary if we take the reality of time seriously. God does not know the future. That is no imperfection in the divine nature, for the future is not yet there to be known. Of course God is ready for the future—God will not be caught out but is, in fact, exceptionally well-prepared for it—but even God does not know beforehand what the outcome of a free process or a free action will be.

At first sight, this is a rather unpalatable conclusion for believers. We have become used to the notion that God's act of creation involves a *kenosis* (emptying) of divine omnipotence, which allows something other than God to exist, endowed with genuine freedom. I am suggesting that we need to go further and recognize that the act of creating the other in its freedom involves also a *kenosis* of the divine omniscience. God continues to know all that can be known, possessing what philosophers call a current omniscience, but God does not possess an absolute omniscience, for God allows the future to be truly open. I do not think this negates the Christian hope of ultimate eschatological fulfillment. God may be held to bring about such determinate purpose even if it is by way of contingent paths.[22]

6 Conclusion

It is time to draw these speculations to a close. Science discerns a wonderful order in the world. We live in a regularly lawful cosmos. Yet that order is not so tight that it rigidifies into a mechanical determinism. We do not need the insights of chaotic dynamics to tell us that, for I believe that we know as surely as we know anything that we are not automata. If the universe does not look like a machine, neither, overall, does it look like an organism. Whitehead's panpsychism is as implausible in its way as was eighteenth century necessitarianism in its way. A just view of physical reality will require something subtle in its combination of reliability and openness that neither of these two simplified pictures could provide. The structured randomness of chaotic dynamics is the beginning of a more hopeful approach to the nature of physical

---

[22] See D. Bartholomew, *God of Chance* (London: SCM Press, 1984).

reality. One may expect that the laws of nature will give room for both energetic and informational causality and the laws of physics will find an honored place in that wider scheme, as asymptotically emergent-downward approximations to that greater truth.

LIST OF CONTRIBUTORS

William Alston, Professor of Philosophy Emeritus, Syracuse University, Syracuse, New York, USA.

George Coyne, Director, Specola Vaticana, Vatican City State, Europe.

P. C. W. Davies, Professor of Mathematical Physics, The University of Adelaide, Adelaide, South Australia, Australia

Willem B. Drees, Center for the Study of Science Society and Religion (Bezinningscentrum), Free University, Amsterdam, The Netherlands.

George F. R. Ellis, Professor of Applied Mathematics, University of Cape Town, Rondebosch, South Africa.

Andrej A. Grib, Professor of Mathematics, Leningrad, Russia.

Stephen Happel, Associate Professor of Religion and Culture, Department of Religion and Religious Education, The Catholic University of America, Washington DC, USA.

Michael Heller, Professor of Philosophy, Pontifical Academy of Theology, Cracaw, Poland.

Chris J. Isham, Professor of Theoretical Physics, The Blackett Laboratory, Imperial College, London, England.

J. R. Lucas, Fellow of Merton College, Oxford, England.

Nancey Murphy, Associate Professor of Christian Philosophy, Fuller Theological Seminary, Pasadena, California, USA.

Ted Peters, Professor of Systematic Theology, Pacific Lutheran Theological Seminary and The Graduate Theological Union, Berkeley, California, USA.

John Polkinghorne, President, Queens' College, Cambridge, England.

Robert John Russell, Professor of Theology and Science In Residence, Graduate Theological Union, Founder and Director, The Center for Theology and the Natural Sciences, Berkeley, California, USA.

William R. Stoeger, SJ, Staff Astrophysicist and Adjunct Associate Professor, Vatican Observatory Research Group, Steward Observatory, University of Arizona, Tucson, Arizona, USA.

Keith Ward, Regius Professor of Divinity, University of Oxford, Oxford, England.

INDEX

abduction, 207
absolute dependence, 6
absolute space, 5, 212
absolute time, 5, 96, 201, 212, 237, 238, 239
abundances of helium, deuterium, and lithium, 48
act of God, 6, 343, 344, 345, *see also* divine action
active information, 27, 434, 436
ad hoc, 22, 23, 30, 193, 194, 199, 206, 307, 322, 328, 337, 338, 351, 355, 406, 407, 422, 425, 426, 427
Adams, R. M., 195
aesthetic criterion (value), 255
algorithmic compressibility, 213
Alston, William P., 8, 16, 17, 28, 185, 207, 209, 343, 399
analogy, 7, 8, 9, 66, 114, 116, 121, 124, 155, 161, 244, 252, 254, 270, 284, 287, 359, 395, 411
Anselm, 119, 341, 425
anthropic principle, 23, 24, 30, 162, 175, 179, 230, 348, 363, 369, 373, 381, 384, 396, 398, 403, 423, 424, *see also* Christian anthropic principle, strong anthropic principle
apeiron, 306
Appleyard, Bryan, 151
Aquinas, Thomas, 4, 21, 248, 249, 250, 255, 292, 293, 295, 304, 393
Arbib, Michael A., 216, 357, 358

argument for the existence of God, 25, 329, *see also* design argument, Humean ontological argument, ontological argument
Aristotle, 97, 235, 240, 304
arrow of time, 21, 26, 63, 113, 142, 202, 272, 274, 283, 324, 373, 427, 435
Ashbery, John, 105
astronomy, theoretical, 1
astrophysics, 42, 106, 111, 154, 157, 306, 330
atemporality, 111, 122, 268
Athanasius, 286
atheism, 5
Atkins, P. W., 345, 347
Augustine, 4, 20, 59, 101, 111, 115, 235, 245, 247, 249, 267, 271, 275, 281, 315, 316, 322, 324, 335, 341, 342, 423, 427
auxiliary hypotheses, 22, 25, 293, 305, 306, 404, 405, 406, 408, 413, 414, 426
axiological explanation, models, 19

Bachelard, Gaston, 108
Barbour, Ian G., 5, 7, 9, 20, 21, 23, 52, 98, 101, 107, 108, 111, 152, 154, 189, 270, 292, 293, 296, 297, 298, 299, 300, 301, 302, 305, 306, 323, 325, 327, 329, 330, 351, 363, 368, 371, 391, 422, 430

Barrow, John D., 12, 50, 112, 114, 132, 150, 160, 162, 317, 346, 363, 370, 372, 373, 374, 382, 401, 424
Barth, Karl, 6, 18, 235, 285, 287, 324, 343
Bartholomew, D. J., 422, 439
Bartley, III, W. W., 121
basic acts, 347
Bauer, Edmond, 15, 166, 169, 170, 171, 173, 177
beginning of time, 51, 53, 249, 300, 316, 321, 322, 330
Bell, J. S., 15, 65, 86, 171, 172, 176, 181, 182
Bellah, R. N., 380
Bénard instability, 434
bending of light, 40, 406
Benjamin, Andrew E., 108, 166, 372
Bergson, Henri, 96, 180
Bertola, F., 369
Bible, 4, 6, 7, 165, 235, 236, 285, 329, 377, 378, 409, 410, 411, 413
bifurcation, 75, 96, 272
Big Bang, 10, 12, 17, 19, 21, 23, 27, 29, 42, 46, 47, 48, 49, 52, 53, 55, 58, 60, 61, 62, 70, 80, 98, 99, 100, 101, 105, 111, 112, 118, 122, 125, 165, 166, 173, 192, 199, 236, 240, 241, 242, 243, 244, 245, 246, 248, 264, 273, 275, 276, 277, 278, 283, 286, 288, 291, 292, 293, 294, 297, 298, 299, 300, 305, 306, 307, 311, 316, 317, 321, 322, 328, 329, 330, 331, 332, 333, 334, 335, 339, 344, 346, 352, 368, 387, 395, 396, *see also* inflation
biological sciences, 2, 115, 219

biology, evolutionary, 19, 296, 354, 407, *see also* evolution, biological
Black, Joseph, 212
black holes, 40, 41, 50, 95, 198
block universe, 13, 17, 19, 28, 96, 139, 141, 142, 143, 144, 145, 146, 147, 201, 332
Boethius, 20, 235, 264, 267, 271, 341
Bohm, David, 353, 434
Bohr, Niels, 15, 167, 168, 436
Bondi, H., 298, 368, 369
Boolean logic, 15, 167, 178, 183
Borgmann, Albert, 231
boundary conditions, 24, 54, 55, 56, 126, 153, 175, 217, 242, 243, 244, 261, 276, 277, 278, 370, 388, 392
boundary of superspace, 70, 71, 74
Braaten, Carl, 285, 287, 288
Brahman, 271
bridge rules, 94, 95
Brockman, John, 216
Brümmer, Vincent, 51, 329
Brundrit, G. B., 394
Brunner, Emil, 299
Buckley, Michael, 5
Bühler, P., 356
Bulgakov, Sergius, 179, 184
Burtt, E. A., 5, 151

CAP, *see* Christian anthropic principle
Calvin, John, 5
Campbell, N. A., 372
Cantor, G. N., 108, 306
Carr, B. J., 373
Carter, B., 373
causal structure, 84, 139, 140, 204
causality, 6, 7, 64, 67, 81, 83, 84, 124, 172, 179, 189, 191, 241,

252, 253, 259, 301, 304, 357, 432, 434, 435, 436, 438, 440
chaos, 3, 5, 9, 13, 14, 27, 30, 59, 112, 113, 152, 159, 189, 221, 272, 273, 275, 324, 373, 432, 433, 434, 435, 438
chaotic cosmology, 24, 368
Christian anthropic principle (CAP), 24, 25, 30, 396, *see also* anthropic principle
Christie, John R. R., 108
Christology, 21, 134, 286, 408
church, 2, 26, 377, 410, 411
Cicero, 335
Clarke, W. Norris, 245, 375
classical cosmology, 35, 43, 179, 334
classical electromagnetism, 142
classical mechanics, 85, 94, 95, 142, *see also* Newtonian mechanics
classical theism, 7, 438
Clayton, Philip, 293, 305
closed universe, 5, 44, 106, 273
Cobb, Jr., John B., 7, 437
COBE, 48
Coleman, S., 108, 369
collapse of the wave packet, 15, 166
Collingwood, R. G., 241
communal discernment, 26, 411
complementarity, 126, 166, 168, 436
complex numbers, 74, 308
complexity, 3, 5, 10, 14, 17, 24, 86, 150, 152, 153, 156, 158, 159, 160, 273, 274, 344, 346, 347, 372, 375, 384, 388, 418, 430, 431, 433, 436
comprehensibility of the world, 93, 102, 103
computability, 15, 163
concurrence, 5, 191
configuration space, 309

confirmation, 30, 40, 47, 226, 282, 289, 307, 364, 365, 367, 370, 375, 376, 402, 403, 406, 411, 416, 417, 423, 425, 426, 428
consciousness, 6, 14, 15, 24, 26, 79, 120, 128, 146, 147, 152, 156, 158, 159, 160, 162, 163, 164, 166, 167, 169, 170, 171, 177, 180, 188, 235, 236, 251, 253, 260, 266, 267, 268, 274, 277, 279, 348, 349, 373, 375, 383, 386, 424, 427, 430
consonance, 52, 98, 100, 101, 162, 265, 284, 289, 297, 301, 328, 330, 359
contingency, 19, 174, 229, 247, 251, 282, 296, 300, 301, 305, 329, 336, 339, 349, 390
continuing creation, 299, 300, 329, 330, *see also creatio continua*
coordinate transformation, 142
Copenhagen interpretation, 166, 167, 172, 173, 177
core theory, 25, 306, 404, 405, 410, 413, 414
Cornell, J., 8, 50, 185, 186, 194, 195, 250, 305, 332, 343, 364, 404
cosmic background radiation, 13, 143, 352, 374
cosmological principle, 43
cosmology, 1, 3, 6, 7, 10, 11, 12, 13, 14, 15, 17, 18, 19, 20, 21, 22, 23, 24, 27, 29, 30, 35, 39, 42, 43, 44, 45, 48, 49, 50, 53, 55, 56, 57, 60, 61, 65, 66, 76, 78, 79, 80, 81, 83, 86, 97, 99, 100, 103, 107, 125, 145, 165, 166, 167, 173, 175, 178, 179, 180, 185, 192, 198, 199, 203, 204, 205, 206, 208, 217, 224, 235, 240, 243, 247, 260, 261, 264, 272, 274, 275, 277, 283,

286, 288, 291, 292, 293, 294, 295, 296, 297, 298, 299, 300, 301, 302, 303, 305, 306, 307, 308, 311, 312, 313, 315, 316, 318, 322, 323, 325, 327, 328, 329, 330, 332, 333, 334, 336, 338, 339, 340, 342, 345, 346, 349, 351, 352, 353, 355, 357, 358, 359, 363, 368, 369, 371, 387, 399, 414, 417, 418, 420, 421, 423, 428, 429, *see also* Big Bang, steady state theory, inflation, quantum cosmology

Coste, Difier, 110

Coveney, Peter, 113

Coyne, George V., 2, 42, 52, 93, 96, 111, 180, 189, 279, 297, 327, 354, 364

Craig, W. L., 329

creatio continua, 267, 275, 296, 323, 437, *see also* continuous creation

creatio cum tempore, 335, 342, *see also* creation in time

creatio ex nihilo, 22, 59, 275, 278, 283, 293, 297, 299, 300, 301, 306, 307, 316, 317, 321, 322, 323, 324, 330

creation, 4, 8, 9, 10, 16, 19, 21, 22, 24, 25, 27, 29, 47, 51, 52, 54, 57, 59, 93, 99, 100, 101, 103, 105, 109, 111, 117, 134, 143, 155, 165, 166, 173, 174, 175, 184, 186, 191, 192, 193, 194, 195, 196, 197, 240, 245, 247, 248, 249, 250, 251, 253, 259, 260, 263, 264, 265, 266, 267, 269, 271, 272, 275, 278, 281, 282, 283, 284, 286, 287, 288, 289, 292, 293, 294, 295, 296, 297, 298, 299, 300, 302, 303, 304, 305, 315, 316, 317, 318, 321, 322, 323, 324, 328, 329, 330, 334, 335, 336, 340, 342, 345, 346, 347, 348, 351, 370, 372, 381, 382, 384, 385, 390, 392, 396, 397, 398, 399, 403, 407, 420, 423, 434, 437, 438, 439, *includes doctrine of creation*

creation in time, 335, 342, 347, *see also creatio cum tempore*

creation of time, 59, 315, 322, 335, 342, 345, 347

creation tradition, 21, 292, 294

creative emergence, 247, 253, 259

creative values, 348, 349

creativity, 19, 118, 129, 149, 253, 254, 255, 256, 257, 258, 259, 260, 275

creator, 3, 4, 19, 22, 24, 25, 51, 52, 94, 174, 248, 249, 254, 263, 277, 286, 289, 291, 292, 294, 320, 321, 323, 324, 330, 340, 341, 345, 363, 378, 379, 381, 383, 384, 385, 387, 389, 390, 391, 394, 395, 396, 398, 412, 413, 438

critical realism, 13, 23, 26, 27, 30, 214, 327, 328, 338, 341, 349, 350, 352, 353, *see also* realism, temporal critical realism

Crutchfield, James P., 189

Cupitt, Don, 105, 130

Curi, V., 369

Currie, Gregory, 305, 404

curvature, 38, 39, 42, 60, 61, 62, 66, 67, 70, 72, 76, 78, 214, 240, 309, 310, 336, 356, 394

D'Espagnat, Bernard, 227, 228

Dalihard, J., 172

data for theology, 306, 408

Davies, P. C. W., 12, 14, 15, 22, 28, 103, 110, 112, 113, 114, 132, 135, 149, 152, 156, 159, 164, 184, 213, 216, 242, 254, 274, 281, 293, 315, 318, 370, 371, 373, 375, 383, 390, 429, 435
de Gruchy, John, 399
de Molina, Luis, 194
Dear, Peter, 108
Deason, Gary B., 150
deceleration parameter q_0, 44
deism, 5, 18, 20, 271, 278, 291, 316, 437
depth, 154, 189, 218, 274, 367, 372, 396
Derrida, Jacques, 12, 123, 128, 129, 131
design argument, 15, 24, 25, 401, 402, 403, 417, 427, *see also* arguments for the existence of God
determinism, 8, 16, 18, 54, 83, 85, 142, 147, 187, 188, 190, 191, 193, 194, 198, 230, 247, 258, 301, 319, 432, 433, 439
deterministic laws, 8, 16, 320, 390, *see also* laws of nature
Deus ex machina, 240
Deuser, H., 343
Deutsch, David, 82, 163
DeWitt, Bryce S., 71, 173, 174, 175, 177, 336
dialectic, 13, 114, 134, 275, 280, 283, 303
differentiable manifold, 139, 147
directionality, 12, 110, 114, 122, 123, 158
discernment, 26, 365, 385, 411, 413, 415, 434
divine action, 1, 3, 4, 5, 7, 9, 10, 16, 17, 18, 22, 23, 24, 27, 28, 29, 30, 134, 144, 185, 186, 187, 188, 189, 190, 191, 192, 194, 197, 198, 199, 200, 204, 205, 206, 207, 260, 261, 274, 292, 294, 300, 320, 321, 325, 340, 343, 381, *see also* act of God
divine concurrence, 191
divine contingency, 19
divine creativity, 258
divine eternity, 20, 28, 265, 267, 268, 275, 279, 288, 324, 340, 341, *see also* eternity
divine foreknowledge, 204
divine freedom, 120
divine immanence, 6, *see also* immanence
divine necessity, 19
divine preservation, 5
divine sovereignty, 191
divine temporality, 3, 18, 28, 250, 271, 286, 324
divine transcendence, 275, *see also* transcendence
downward causation, 264, 281, 282, 284, 435
downward emergence, 26, 431
Dray, W. H., 241
Drees, Willem B., 22, 23, 29, 100, 101, 141, 205, 273, 293, 316, 317, 323, 325, 327, 329, 344, 346, 352, 353, 359, 368, 395, 399, 416
Dreyfus, H. L., 111
Dubied, P.-L., 356
dynamic theory of chaos, 432
dynamical systems, 95, 211
dynamics, 13, 21, 43, 141, 142, 211, 217, 221, 286, 430, 433, 436, 439
Dyson, Freeman J., 40, 344

Earley, J. E., 433
early universe, 11, 35, 52, 53, 173, 175, 244, 313, 314, 322, 329

Eaves, Lindon, 354
Eddington, Sir Arthur, 40, 166
edge of spacetime, 165
eigenvalue, 85
eigenvector, 85, 87
Einstein, Albert, 2, 35, 38, 39, 40, 43, 54, 56, 60, 62, 72, 74, 88, 94, 96, 98, 102, 103, 149, 165, 168, 169, 171, 181, 190, 227, 238, 269, 311, 313, 314, 323, 324, 338, 426, 429
Einstein field equations, 39, 40, 56
emergence, 11, 14, 26, 113, 122, 124, 125, 126, 127, 150, 152, 153, 156, 162, 164, 205, 223, 224, 247, 253, 258, 259, 260, 275, 281, 314, 374, 430, 431, *see also* creative emergence, downward emergence, upward emergence
emergence of mind, 14, 156, *see also* mind
emmanationism, 289
empirical realism, 227
empiricism, 115
epistemic distance, 385
epistemology, 26, 30, 143, 144, 292, 303, 327, 341, 358, 401, 402, 403, 404, 407, 412, 417
equation of state, 43
equations of motion, 54, 64
eschatology, 20, 107, 179, 408, *includes doctrine of eschatology*
essential singularity, 41, 291
eternal objects, 340
eternity, 19, 20, 27, 28, 29, 180, 181, 184, 235, 237, 249, 263, 264, 265, 267, 268, 269, 271, 275, 276, 277, 278, 279, 280, 281, 282, 283, 284, 285, 286, 288, 294, 303, 322, 324, 325, 340, 341, 342, 392, 439, *see also* divine eternity
event horizon, 40, 41
evil, 5, 8, 19, 25, 26, 27, 30, 117, 127, 184, 254, 256, 349, 350, 371, 384, 389, 393, 426, 427, 437
evolution, biological, 3, 155, 159, *see also biology, evolutionary*
Exodus, 4, 198, 343
explanation, 12, 14, 15, 19, 23, 25, 95, 110, 117, 124, 125, 130, 149, 164, 187, 199, 207, 208, 218, 222, 223, 224, 225, 226, 227, 240, 241, 242, 243, 247, 253, 256, 258, 259, 260, 278, 281, 340, 345, 346, 347, 348, 349, 356, 364, 365, 366, 368, 369, 370, 371, 375, 376, 386, 387, 390, 401, 403, 404, 418, 421, 422, 424, 425, 426, 427, 428, 435
external parameter, 309, 310
external properties, 82, 83
external temporal relations, 19, 249
external time, 145, 309, 312, 313, 314, 336

Farley, Edward, 109, 266
Farmer, J. D., 189
Farrer, Austin, 7, 437
Feddoso, Alfred J., 194
Feigenbaum, Mitchell, 159
Fennema, J., 338
filled background spacetime, 317
finitude, 21, 22, 263, 279, 286, 293, 294, 300, 306, 320, 321, 322, 323, 324, 347, 395, *see also* universe, finite age of
finitude, unbounded, 22, 294, 306, 322

Finkelstein, David, 167
first cause, 4, 240, 243
first moment, 19, 22, 193, 248, 275, 294, 295
Fischler, W., 175
flat universe, 44
Flew, Anthony, 294
Ford, J., 152, 435
Fraser, J. T., 108, 111, 112, 126
free will, 19, 24, 26, 147, 192, 194, 195, 197, 204, 236, 319, 382, 383, 384, 385, 386, 387, 389, 392, 397, 426, *see also* libertarian free will
fundamental laws, 29, 219, 372, 388

Gadamer, H. G., 106, 117, 121
Gale, Richard, 256, 370, 375
Galilean transformations, 237
Galileo, 2, 35, 149, 162
Geach, P. T., 254
general relativity, 10, 11, 13, 22, 23, 29, 35, 38, 39, 40, 41, 42, 43, 44, 47, 52, 53, 54, 55, 56, 59, 60, 62, 63, 64, 66, 67, 68, 73, 74, 77, 79, 97, 141, 142, 146, 165, 203, 209, 210, 276, 291, 294, 306, 308, 309, 310, 311, 312, 313, 323, 333, 334, 335, 339
Gerhart, Mary, 116
Gilkey, Langdon, 4, 6, 21, 292, 294, 295, 296, 297, 298, 299, 300, 301, 302, 303, 304, 306, 322, 350
Gleick, J., 272, 432
God, 3, 4, 5, 6, 7, 8, 9, 13, 14, 16, 17, 18, 19, 20, 21, 22, 23, 24, 25, 26, 27, 28, 29, 30, 52, 53, 54, 59, 93, 98, 99, 100, 101, 103, 107, 109, 112, 113, 114, 118, 119, 120, 132, 133, 134, 135, 139, 143, 144, 146, 150, 152, 155, 160, 161, 162, 164, 165, 167, 180, 184, 185, 186, 187, 188, 190, 191, 192, 193, 194, 195, 196, 197, 198, 199, 204, 205, 206, 207, 208, 209, 222, 230, 231, 235, 236, 237, 238, 239, 240, 243, 244, 245, 247, 248, 249, 250, 251, 252, 254, 256, 257, 258, 259, 260, 263, 264, 265, 266, 267, 268, 269, 270, 271, 272, 273, 274, 275, 276, 277, 278, 279, 280, 281, 282, 283, 284, 285, 286, 287, 288, 289, 291, 292, 294, 295, 296, 297, 298, 299, 301, 302, 303, 307, 315, 316, 317, 318, 319, 320, 321, 323, 324, 325, 327, 329, 330, 331, 332, 334, 338, 340, 341, 342, 343, 344, 345, 347, 348, 349, 350, 353, 357, 359, 363, 364, 365, 367, 368, 370, 375, 377, 378, 379, 380, 381, 385, 386, 387, 388, 389, 390, 391, 392, 393, 394, 395, 396, 397, 398, 399, 401, 402, 403, 408, 409, 412, 413, 414, 415, 420, 422, 423, 424, 425, 426, 427, 428, 430, 432, 433, 437, 438, 439
God:world::mind:body, 8
God's action, 3, 4, 5, 8, 14, 16, 22, 27, 28, 29, 30, 134, 161, 207, 209, 230, 231, 236, 263, 264, 274, 284, 291, 294, 319, 321, 323, 325, 379, 381, 437, 438, *see also* divine action
God's future, 266, 331, 353
God's new creation, 266
God's past, 266, 353
God of the gaps, 27, 93, 240, 438, *see also* divine action
God, the creator, 248, *see also* creation

Gosse, P. H., 351
grace, 107, 192, 397, 398, 437
gravitational fields, 35, 38, 40, 41, 42, 58, 210
gravitational redshift, 40
Gregory of Nyssa, 20, 267, 285
Grib, Andrej, 15, 28, 65, 79, 84, 165, 173, 177, 424
Griffin, David R., 7, 437
Grünbaum, Adolf, 99, 100, 101

Habermas, Jürgen, 105
Halliwell, J. J., 369
Happel, Stephen, 12, 27, 105, 120, 325
Härle, W., 343
Harrison, E. R., 50
Hartle, J. B., 11, 19, 22, 29, 68, 72, 74, 75, 101, 174, 175, 241, 242, 243, 292, 293, 305, 307, 308, 309, 310, 311, 312, 313, 314, 315, 316, 317, 318, 320, 321, 322, 334, 345, 351, 367, 369
Hartle/Hawking, 22, 29, 292, 293, 305, 307, 308, 309, 310, 311, 312, 313, 314, 315, 316, 317, 318, 320, 321, 322, 367
Hasker, William, 195
Hawking, Stephen, 11, 12, 19, 20, 22, 29, 53, 68, 72, 74, 75, 96, 101, 112, 113, 132, 174, 175, 193, 241, 242, 243, 264, 276, 277, 278, 279, 283, 286, 291, 292, 293, 301, 305, 307, 308, 309, 310, 311, 312, 313, 314, 315, 316, 317, 318, 319, 320, 321, 322, 323, 324, 334, 345, 351, 352, 353, 367
Hebblethwaite, Brian, 9, 187
Hegel, G. W. F., 116, 129, 134

Heidegger, Martin, 82, 117, 128, 129, 132, 133
Heisenberg, Werner, 80, 168, 239, 276, 432, 435
Heller, Michael, 2, 11, 27, 93, 96, 97, 184, 317, 338, 340, 399
Helm, Paul, 268, 269, 341, 342, 343
Hempel, Carl G., 25, 240, 402, 404
Henderson, Edward, 9, 187
hermeneutics, 108, 109, 114, 132
Hesse, Mary, 12, 114, 115, 116, 117, 121, 124, 132, 215, 216, 227, 328, 346, 356, 357, 358, 359
Hick, John, 348, 385, 423
hierarchy of the sciences, 355, 421
Highfield, Roger, 113
Hilbert space, 168, 176, 317
historical/empirical origination, 21, 292, 293, 294, 300, 302, 304
history of religion, 1
history of science, 1, 94, 407
Hodges, A. P., 163
Hodgson, P. E., 237
holism, 160, 281, 282, 283, 284, 402, 404, 428
horizon, 35, 40, 41, 49, 121, 190, 274, 291
horizon problem, 35, 49, 291
Hoyle, Fred, 298, 300, 305
Hubbeling, H. G., 340, 343
human agency, 5, 30, 432, 437
human freedom, 5, 16, 17, 28, 118, 119, 120, 124, 126, 132, 139, 184, 185, 187, 191, 192, 193, 198, 199, 200, 203, 206, 254, 282, 432
human intentionality, 27, 434
human sciences, 109, 132, 418, 421, 425, 426

Hume, David, 25, 261, 319, 402, 403
Humean ontological argument, 257, *see also* argument for the existence of God

idealism, 15
Ignatius of Loyola, 115
imaginary time, 75, 76, 244, 276, 277, 286, 311, 334, 337, 342, 351, 353
imago Dei, 358
imitatio Christi, 358
immanence, 6, 8, 134, 330, 379, 381, 386, 394
incompatibilist, 191
inertia, 5, 153
inertial frames, 36, 238, 269, 270, 271
inference, 6, 208, 225, 228, 357
infinite, 5, 19, 25, 29, 41, 44, 46, 58, 59, 62, 65, 68, 70, 71, 72, 84, 113, 114, 141, 161, 165, 177, 192, 193, 194, 198, 199, 212, 219, 223, 228, 240, 242, 248, 251, 256, 268, 283, 300, 305, 306, 309, 315, 329, 331, 335, 347, 394, 395, 420
inflation, 17, 25, 49, 50, 217, 276, 291, 329, 369, 374, *see also* Big Bang
initial singularity, 10, 20, 42, 46, 47, 54, 74, 277, 286, 291, 292, 302, 306, 307, 308, 311, 322, 328, 329, 330, 333, 334
instrumentation, 25, 26, 404, 408
intelligibility, 125, 126, 127, 128, 132, 149, 162, 253, 254, 255, 257, 258, 260, 270, 340
internal property, 63, 64, 310
internal time, 63, 67, 74, 75, 76, 309, 310, 312, 313, 353

interpersonal analogy, 9, *see also* analogy
intervention, 25, 126, 186, 188, 189, 193, 198, 206, 207, 288, 319, 320, 390, 393
intrinsic gaps, 27, 321, *see also* God of the gaps
intrinsic indeterminacy, 144
invariance, 218, 241
invariant interval, 37, 39
Isham, C. J., 11, 13, 20, 22, 27, 28, 29, 51, 59, 66, 68, 97, 101, 103, 111, 113, 115, 139, 141, 145, 164, 166, 184, 201, 203, 205, 210, 250, 255, 278, 293, 307, 308, 309, 310, 312, 313, 315, 316, 323, 324, 325, 328, 331, 332, 334, 335, 336, 337, 339, 346, 347, 353, 374, 438
Israel, W., 4, 96, 205, 408, 409, 439

Jantzen, Grace, 8, 9
Jastrow, Robert, 98, 99, 328, 329
Jeanrond, Werner G., 109
Jenson, Robert, 272, 284, 285, 286, 287, 288
Jesus, 4, 24, 119, 236, 250, 263, 266, 272, 283, 286, 288, 343, 378, 379, 385, 397, 398, 408, 413, 414, 415
John Paul II, 1, 2, 354
Juengst, Eric T., 354
Jung, Carl, 78
Jüngel, Eberhard, 287

Kant, Immanuel, 6, 7, 59, 115, 122, 123, 124
Karakash, C., 356
Kauffman, Stuart, 159
Kaufman, Gordon, 8, 159, 188, 343, 344, 345, 348

kenosis, 9, 28, 443
Kessler, H., 344
kinematics, 13, 141
Kingdom of God, 381, 382
Klebanov, I., 176
Kolb, E. W., 49
Krauss, L., 397
Kretzmann, Norman, 20, 206, 268, 269, 270
Kuhn, Thomas, 406

Lakatos, Imre, 22, 26, 293, 305, 406, 407, 408, 409, 410, 413, 415, 430, 431
Lakatosian theological research program, 323
Laplace's calculator, 142
lawlike behavior, 9
Lawrence, N., 108
laws of biology, 27
laws of nature, 1, 3, 9, 10, 13, 14, 15, 16, 17, 18, 20, 24, 25, 27, 28, 29, 112, 124, 125, 144, 153, 155, 156, 159, 161, 162, 164, 185, 193, 199, 206, 207, 208, 209, 218, 219, 220, 222, 223, 226, 227, 228, 229, 230, 231, 259, 261, 274, 277, 308, 318, 319, 320, 321, 384, 390, 391, 395, 401, 426, 431, 433, 440, 444, *see also* deterministic laws, physical laws
laws of organization, 374
laws of physics, 11, 15, 24, 25, 27, 29, 36, 47, 146, 147, 152, 153, 156, 158, 159, 161, 162, 163, 240, 263, 277, 308, 355, 372, 373, 375, 376, 378, 384, 385, 386, 387, 388, 390, 392, 393, 396, 423, 433, 435, 438, 444

Layzer, David, 113, 114, 128, 133
Leftow, Brian, 332, 341, 342
Leplin, Jarvett, 327, 433
Leslie, John, 167, 199, 345, 347, 348, 349, 365, 371, 372, 374, 378, 397, 403, 405, 425, 428, 429
Lewis, C. S., 190
libertarian free will, 195, 204, *see also* free will
light bending, 40
light cone, 13, 39, 181, 183, 272
limits of rationality, 100
limits of science, 15, 217
Linde, A., 353, 371
Lipton, Peter, 240
Logos, 340, 399
Lonergan, Bernard, 7, 12, 115, 117, 120, 121, 122, 123, 125, 126, 127, 128, 129, 132, 133, 134, 135
Lorentz transformations, 141
Lucas, John, 19, 29, 81, 142, 145, 146, 235, 236, 237, 266, 270, 285, 342, 398
Ludwig, G., 83, 106, 216
lure, 254, 441, 442

MacCallum, M. A. H., 376
Mach, Ernest, 96
Mackie, J. L., 348, 349
Macquarrie, John, 188
manifold, 39, 98, 102, 126, 139, 141, 145, 147, 203, 206, 217, 228, 308, 309, 311, 333, 336, 337, *see also* spacetime
many worlds interpretation, 80
Margenau, Henri, 81
Mascall, Eric L., 7, 396, 398
mathematical beauty, 255

mathematical objects, 340
mathematical physics, 72, 110, 307, 309
mathematics, 12, 15, 29, 59, 63, 97, 106, 108, 110, 111, 112, 118, 122, 123, 133, 147, 150, 153, 154, 162, 163, 164, 222, 278, 306, 335, 338, 348, 371, 390
Maxwell's equations, 212
Mayr, Ernst, 219, 373, 384
McFague, Sallie, 8, 9, 109, 117, 119, 120
McMullin, Ernan, 52, 59, 95, 99, 103, 231, 232, 327, 330, 350, 352, 365, 372, 396
me on, 60, 187, 317
measurement apparatus, 15
measurement problem, 11, 439
Melchin, Kenneth R., 126, 128
Merchant, Carolyn, 151
Messiah, 289, 381
metabolism, 355
metaphor, 12, 28, 98, 106, 107, 108, 111, 112, 115, 116, 117, 118, 119, 120, 121, 122, 123, 124, 125, 126, 129, 132, 133, 135, 136, 161, 188, 222, 340
metaphor, interactive theory of, 116, 118
metaphysics, 7, 14, 15, 27, 129, 130, 164, 285, 317, 358, 404, 440
methodology, 3, 84, 101, 240, 305, 406
metric tensor, 39, 41, 44
middle knowledge, 17, 194, 195, 197, 198, 207
Milne, E. A., 396, 398
mind, 3, 8, 10, 11, 14, 15, 27, 28, 29, 53, 54, 56, 125, 139, 153, 366, 368, 375, 380, 397, 400, 414, 433, 436, 440, *see also* emergence of mind
mind of God, 53, 155, 319, 350
Minkowski spacetime, 42, 179, 183, 242, 243, 244
miracle, 6, 156, 261, 320, 394
Misner, C. W., 349
Moltmann, Jürgen, 21, 287, 288
Moses, 193, 194, 400
Munitz, M. K., 359
Murphy, Nancey, 3, 22, 25, 26, 30, 102, 104, 121, 152, 165, 189, 222, 232, 293, 305, 307, 366, 367, 377, 391, 403, 430
mystical communion, 13, 134
mysticism, 134, 135, 136, 351

Nagel, Thomas, 331, 440
naive realism, 16, 168, 173, 214
narrative form, 110
natural causal determinism, 193, 194, 198
natural kinds, 215, 216, 227
necessity, 15, 18, 19, 24, 124, 143, 160, 207, 208, 210, 218, 219, 223, 224, 226, 228, 229, 230, 231, 235, 243, 247, 251, 255, 257, 259, 260, 267, 336, 349, 370, 371, 372, 376, 377, 388, 405, 417, 428, 431
Needham, Joseph, 150
New Testament, 4, 24, 341, 377, 378, 379, 396, 411, 413
Newton, Sir Isaac, 2, 5, 35, 38, 40, 41, 54, 55, 64, 74, 83, 94, 96, 149, 150, 153, 155, 157, 190, 210, 215, 236, 237, 338, 352, 371, 411, 429
Newtonian mechanics, 5, 84, 96, 190, 209, 213, 237, 319, *see also* classical mechanics

Nietzsche, F., 128, 129
nominalism, 350
nomological, 19, 202, 240, 241, 242, 247, 254
non-being, 317
Norris, Chistopher, 128, 375
Novak, D., 353
novel fact, 22, 25, 26, 294, 307, 406, 426, 428
novelty, 7, 26, 27, 152, 259, 324, 430, 433, 434

O'Keefe, John A., 99
objectivity, 105, 172, 196, 227, 411
observable, 42, 43, 49, 81, 82, 83, 84, 85, 86, 87, 169, 173, 174, 175, 176, 177, 207, 214, 220, 226, 259, 346, 382, 384, 395, 418, 422
observer, 11, 15, 24, 36, 39, 41, 65, 79, 83, 84, 86, 88, 116, 121, 124, 143, 166, 167, 169, 170, 171, 172, 173, 174, 175, 176, 177, 178, 179, 181, 182, 183, 202, 238, 269, 270, 271, 288, 310, 331, 374, 424
Olbers' paradox, 331
omnipotence, 19, 24, 144, 191, 247, 250, 252, 257, 267, 379, 381, 389, 439
omnipresence, 13, 143, 250, 331, 332, 341
omniscience, 27, 144, 195, 239, 251, 257, 267, 270, 341, 388, 439
ontological argument, 257, 425, *see also* argument for the existence of God
ontological dependence, 237, 294, 295, 299, 302, 304, 324, 330

ontological origination, 21, 292, 293, 294, 301, 302, 304, 305, 306, 322, 323
ontological reductionism, 18, 207, 230
open future, 204, 250, 252, 433
open universe, 44
openness, 9, 118, 139, 142, 143, 146, 147, 152, 156, 160, 266, 275, 385, 432, 433, 434, 435, 436, 438, 439
Oppenheim, P., 240
order parameter, 211
origin of life, 156, 323
origin of structure, 10, 49
origin of the laws of nature, 209, 230
oscillating universe, 17, 273, 300, 329, 330, 334
ouk on, 59, 317, 318
owing its existence to God, 275, 296

Packard, N. H., 189
Page, Don. H., 113
Pagels, Heinz, 153
panentheism, 8
Pangloss question, 389
Panikker, Raimundo, 133
Pannenberg, Wolfhart, 5, 20, 62, 264, 274, 279, 280, 281, 282, 287, 324, 340, 344, 408, 418
pantheism, 282
paradigm shift, 210
Park, D., 108
past temporal finitude, 21, 22, 293, 306
Paul, Ian, 1, 2, 12, 14, 59, 103, 106, 108, 110, 112, 115, 117, 156, 188, 213, 254, 268, 274, 281, 293, 315, 317, 338, 341, 354, 403, 434

Pauli exclusion principle, 154
Peacocke, A. R., 5, 7, 8, 9, 21, 23, 103, 152, 159, 189, 230, 252, 267, 274, 275, 279, 282, 292, 296, 297, 298, 299, 300, 301, 302, 306, 323, 327, 330, 332, 334, 351, 354, 363, 368, 391, 393, 422, 428, 430, 433, 434
Penrose, Roger, 62, 79, 96, 102, 111, 113, 114, 115, 171, 173, 291, 352, 353, 369
people of the gaps, 438
perfect fluid, 43
perichoresis, 263, 289
personal agency, 7, 9
Peters, Ted, 4, 20, 21, 29, 52, 263, 275, 292, 296, 297, 298, 300, 301, 302, 306, 324, 327, 330, 353, 363, 370, 389
phase transition, 49, 211, 213
phenomenology, 108, 117
Philo of Alexandria, 315
philosophical theology, 1, 198
philosophy, 1, 3, 5, 7, 17, 19, 25, 82, 88, 100, 103, 105, 106, 108, 115, 120, 124, 129, 132, 149, 170, 200, 208, 209, 216, 285, 295, 303, 305, 307, 325, 340, 341, 343, 350, 401, 403, 406, 408
philosophy, modern, 5, 100, 124
philosophy of religion, 1, 325, 341, 343
philosophy of science, 1, 19, 100, 105, 305, 307, 401, 403, 406, 408
physical cosmology, 17, 100, 206, 264, 272, 277, 284, 288, 327, *see also* cosmology
physical laws, 16, 24, 47, 50, 83, 87, 88, 155, 189, 190, 191, 197, 206, 207, 225, 354, 375, 376, 381, 383, 384, 385, 387, 391, 392, *see also* laws of nature
physical sciences, 207, 211, 219, 243
physics, 1, 3, 9, 10, 11, 12, 13, 14, 15, 18, 19, 20, 23, 24, 25, 26, 27, 29, 30, 35, 36, 37, 38, 42, 43, 47, 53, 54, 56, 58, 59, 60, 62, 64, 65, 66, 67, 68, 71, 78, 80, 81, 82, 83, 84, 86, 87, 93, 94, 95, 99, 101, 106, 109, 110, 111, 114, 119, 122, 123, 126, 139, 141, 142, 143, 144, 145, 147, 152, 153, 154, 155, 156, 157, 159, 161, 162, 163, 165, 166, 167, 178, 180, 188, 190, 202, 205, 206, 207, 208, 211, 212, 217, 221, 225, 226, 230, 237, 240, 247, 251, 263, 275, 276, 289, 294, 296, 306, 307, 308, 310, 311, 315, 318, 319, 320, 324, 325, 333, 338, 342, 346, 351, 354, 355, 357, 358, 364, 368, 370, 371, 372, 373, 374, 375, 382, 383, 384, 386, 387, 388, 389, 390, 391, 392, 393, 398, 399, 407, 420, 429, 430, 431, 433, 434, 436, 440
Pickvance, J., 380, 397
Pike, Nelson, 341, 342, 343
Piran, T., 369
Planck length L_p, 55
Planck time t_p, 173
Plantinga, Alvin, 252
Plato, 78, 129, 146, 155, 279, 317, 348
Platonism, 235, 340
Plotinus, 279, 280, 281
Podolsky, B., 88, 168, 181
Polchinski, J., 175
Polkinghorne, John, 9, 13, 23, 26, 27, 28, 29, 30, 97, 135, 139, 143, 146, 189, 201, 250, 254,

258, 271, 287, 313, 323, 327, 331, 332, 351, 366, 378, 381, 384, 392, 393, 422, 426, 429, 430, 431, 433, 435, 436, 437, 438
Pollard, William, 189, 391, 422
Popper, Karl, 80, 121, 433
positive heuristic, 25, 405, 408, 410
Poulet, George, 108
practical reason, 6, 123
precession of the perihelion, 40
Preul, R., 343
Price, James Robertson, 42, 133
Prigogine, Ilya, 110, 113, 127, 132, 135, 142, 145, 156, 272, 274, 275, 434
principle of causality, 83, 84
probabilistic perspective, 11
probability, 12, 24, 58, 59, 65, 66, 69, 71, 78, 79, 80, 85, 86, 87, 123, 125, 126, 127, 128, 158, 169, 170, 174, 178, 183, 188, 189, 225, 226, 260, 310, 336, 345, 368, 369, 375, 416, 417
probability amplitude, 336
problem of evil, 8, 25, 254, 349, 371, 393, 426, *see also* theodicy
prolepsis, 283
prologue of St. John's Gospel, 381
prophecy, 12, 133, 134, 411
providence, 5, 30, 191, 299, 385, 386
proximate origins, 230

quantum cosmology, 1, 3, 10, 11, 14, 15, 17, 18, 22, 23, 27, 29, 35, 55, 56, 57, 65, 66, 76, 78, 79, 80, 81, 145, 166, 167, 173, 175, 178, 180, 203, 217, 235, 243, 247, 286, 292, 293, 294, 305, 307, 308, 312, 313, 315, 323, 325, 328, 333, 334, 336, 338, 339, 342, 349, 351, 353, 355, 357, 358, 369, 387, 423, 429
quantum electrodynamics, 142
quantum field theory, 172, 219, 227, 369
quantum fluctuations, 255, 315
quantum gravity, 11, 22, 23, 27, 29, 42, 55, 56, 66, 67, 68, 69, 71, 72, 74, 77, 78, 95, 145, 167, 242, 264, 292, 294, 305, 307, 308, 310, 312, 313, 314, 323, 327, 328, 330, 333, 335, 336, 337, 338, 339, 340, 342, 349, 351, 352, 353, 355, 423
quantum indeterminacy, 25, 30, 197, 241, 391, *see also* intrinsic indeterminacy, uncertainty principle
quantum logic, 167, 172, 173, 176, 177, 182
quantum physics, 3, 9, 11, 13, 14, 15, 78, 81, 82, 178, 278, 294, 308, 310, 311, 315, 324, 325, 351, 369, 371, 391, 436
quantum theories of origination, 59, 63, 64
quarks, 47, 95, 212, 223, 429
Quine, W. V. O., 25, 356, 404, 406, 410, 411, 418

Rahner, Karl, 7, 21, 287, 324
Rahner's Rule, 21, 287
Raine, D. J., 97
real and abstract predicates, 215
real necessities, 230
real possibilities, 230
realism, 12, 14, 15, 18, 23, 24, 27, 84, 85, 117, 125, 145, 168, 173, 186, 208, 214, 216, 227, 327,

328, 338, 339, 341, 350, 352, 353, 356, 357, *see also* critical realism, empirical realism, naive realism, temporal critical realism
redeemer, 3, 4, 263, 291, 322
redshift, 40
reduction of the state vector, 87, 89, 145, *see also* collapse of the wave packet
reductionism, 14, 18, 24, 27, 207, 209, 230, 282, 339, 433, 434, 435, 436
Rees, Martin J., 375
reference frames, 35, 36, 38, 141, 146, 270
referentiality, 107, 112, 118, 122, 123, 129
Reichenbach, Hans, 113, 114, 115, 133
relativism, 26, 151, 408
relativity, theory of, 60, 83, 94, 96, 139, 141, 145, 200, 235, 269, 270, 331, 339, 353, 355
religious discourse, 12, 16, 107, 186
retroduction, 207, 225
revelation, 7, 16, 25, 26, 181, 248, 252, 254, 260, 261, 269, 295, 298, 381, 382, 383, 388, 389, 393, 394, 399, 400, 411, 413, 416, *includes doctrine of revelation*
Ricci curvature, 39
Ricoeur, Paul, 12, 13, 108, 112, 115, 116, 117, 118, 119, 120, 121, 122, 123, 125, 133
Roger, G., 63, 81, 103, 112, 113, 172, 291
Rolston, III, Holmes, 20, 271, 272
Rosen, N., 168, 169, 182
Rowe, William, 199
Russell, Allan, 117

Squires, E., 179
Stapledon, Olaf, 390
state of a system, 67, 84, 89
state vector, 87, 89, 145, 336
statistical determinism, 301
statistical ensemble, 81
steady state theory, 291, 298, 330
Stengers, Isabelle, 111, 113, 114, 128, 133, 136, 156, 273, 274, 438
Stewart, Ian, 152, 437
Stoeger, William R., 2, 3, 10, 18, 28, 29, 35, 42, 52, 93, 111, 181, 189, 207, 210, 279, 291, 297, 301, 308, 319, 327, 346, 354, 366, 373, 392
strong anthropic principle, 375, 405, 428
structured randomness, 27, 437, 444
Stump, Eleonore, 20, 268, 269, 270
sub specie aeternitatis, 340
suffering, 21, 254, 267, 287, 392
sufficient causality, 259
sunyata, 271
superlaw, 318
superspace, 69, 71, 72, 73, 74, 75, 76, 77, 337
superstrings, 67, 213, 217
surface of last scattering, 46, 47
surplus meaning, 116, 118
Susskind, L., 176
Swinburne, Richard, 245, 260, 345, 347, 348
symbol, 128, 166, 350, 380
symbolic logic, 118, 121

$t=0$, 10, 11, 21, 22, 23, 42, 46, 47, 273, 277, 279, 291, 292, 293, 296, 297, 298, 300, 301, 302,

303, 305, 306, 307, 311, 316, 319, 321, 322, 324, 329, 333, 334, 339, 352
Tao, 271
teleological narratives, 132, 134
teleology, 123, 130, 134, 160, 161, 162
telos, 123, 128
Temple, William, 24, 26, 30, 271, 363, 379, 380, 387, 396, 412, 414, 415, 420, 421, 426, 427
temporal critical realism, 23, 327, 328, 341
temporal finitude, 21, 22, 263, 293, 306
temporal narrative, 12
temporality, 3, 12, 13, 14, 16, 18, 20, 21, 22, 23, 28, 29, 106, 107, 110, 113, 114, 122, 123, 124, 126, 128, 132, 133, 134, 142, 143, 160, 200, 235, 237, 238, 245, 250, 251, 253, 264, 265, 266, 267, 271, 272, 279, 283, 284, 286, 287, 292, 293, 313, 321, 324, 325, 327, 330, 332, 339, 340, 344, 345, 349
testability, 357, 370
Theissen, Gerd, 356
theodicy, 5, 26, 426
theological research program, 3, 26, 30, 306, 307, 321, 322, 324, 402, 408, 409, 410, 412, 417, 418, 426, 428
theology, 1, 3, 5, 6, 7, 10, 11, 12, 14, 15, 17, 18, 19, 20, 21, 23, 24, 26, 28, 29, 30, 93, 97, 99, 102, 103, 106, 107, 119, 120, 122, 128, 132, 133, 139, 144, 150, 165, 186, 188, 198, 235, 243, 248, 263, 270, 274, 278, 283, 289, 291, 292, 293, 294, 295, 296, 297, 298, 299, 302, 303, 304, 306, 307, 308, 316, 320, 322, 325, 327, 328, 330, 340, 341, 346, 349, 350, 354, 355, 356, 357, 358, 359, 363, 364, 367, 378, 381, 388, 389, 408, 410, 411, 412, 415, 417, 418, 419, 420, 421, 423, 425, 426, 427, 428, 437, 439
theoretical knowledge, 153, 157
theory of everything (TOE), 9, 229, 240, 429
thermal equilibrium, 110
thermodynamics, 26, 264, 272, 273, 274, 324, 373, 427, 430
Theunissen, Michael, 132
Thomas, Owen C., 7, 9, 10
Tillich, Paul, 59, 188, 317
time, 3, 5, 10, 11, 12, 13, 14, 15, 17, 18, 19, 20, 21, 22, 23, 24, 26, 27, 28, 29, 35, 37, 39, 40, 42, 43, 44, 45, 46, 47, 49, 51, 52, 53, 54, 55, 56, 57, 58, 59, 60, 61, 62, 63, 64, 65, 66, 67, 69, 73, 74, 75, 76, 77, 78, 80, 81, 83, 84, 85, 86, 88, 94, 96, 97, 99, 101, 102, 106, 107, 108, 110, 111, 113, 114, 115, 120, 121, 122, 123, 125, 126, 128, 131, 132, 133, 135, 139, 141, 142, 143, 144, 145, 146, 147, 150, 153, 155, 158, 160, 165, 168, 171, 173, 174, 177, 178, 179, 180, 183, 184, 187, 189, 192, 195, 199, 200, 201, 202, 203, 204, 205, 206, 210, 212, 213, 214, 216, 235, 236, 237, 238, 239, 240, 241, 242, 243, 244, 245, 247, 248, 249, 250, 251, 252, 253, 259, 263, 264, 265, 266, 267, 268, 269, 270, 271, 272, 273, 274, 275, 276, 277, 278, 279, 281, 282, 283, 284, 285, 286, 287, 288, 289, 291, 294, 295, 296, 297, 298, 299, 300, 301, 303, 306, 308, 309, 310, 311, 312, 313,

314, 315, 316, 321, 322, 323, 324, 325, 327, 328, 329, 330, 331, 332, 333, 334, 335, 336, 337, 339, 341, 342, 343, 344, 345, 347, 348, 349, 351, 352, 353, 354, 355, 357, 366, 367, 372, 373, 379, 384, 388, 389, 392, 393, 404, 405, 406, 410, 418, 421, 427, 430, 435, 438, 439
time asymmetry, 12, 110, 113, 122
time, created, 237, 315, 343
time, flowing, 13, 97
timeless, 13, 18, 19, 20, 22, 23, 28, 101, 144, 146, 160, 184, 203, 206, 236, 247, 249, 250, 263, 271, 274, 277, 278, 284, 289, 294, 309, 313, 314, 324, 339, 342, 350
Tipler, Frank J., 50, 62, 112, 114, 132, 160, 162, 180, 344, 363, 370, 372, 373, 374, 382, 401, 424
Todes, S. J., 111
Toffoli, Tommaso, 150
Torrance, T. F., 162, 235, 340
Tracy, David, 9, 107, 119, 128, 133, 343
transcendence, 8, 24, 128, 207, 264, 275, 282, 283, 343, 370, 378, 381, 386, 387, 389
transitional character, 22
Trinity, 21, 29, 105, 133, 134, 143, 245, 263, 264, 267, 269, 272, 284, 285, 286, 287, 288, 289, 377, 378
Trinity, economic, 29, 134, 263, 264, 287, 288, 289
Trinity, immanent, 21, 29, 133, 263, 264, 287, 288, 289
trope, 115, 130, 131
Tryon, E. P., 58, 59, 346
Turner, M. S., 49

ultimate origin, 18, 149, 191, 207, 230
uncaused event, 301
uncertainty principle, 239, 244, 276, 432, *see also* quantum indeterminacy
uncreated time, 247, 343
unified field theory, 35
universal simultaneity, 13, 23, 339
universe, finite age of, 300, 301
unpredictability, intrinsic, 27, 142, *see also* ontological indeterminacy
upward emergence, 26, 430

Van Fraassen, Bas C., 215, 228, 229
vector space, 65, 84
very early universe, 11, 35, 53, 173, 244, 313, 314
Vilenkin, Alex, 11, 19, 55, 68, 72, 74, 77, 174, 241, 242, 244, 353
vitalism, 431
Von Neumann, J., 166, 167, 168, 169, 170, 171, 172, 173, 177

Ward, Keith, 1, 19, 28, 29, 59, 109, 247, 252, 323, 332, 343, 409, 428
weak anthropic principle, 373, 424
Welch, Claude, 6
Wheeler, John A., 163, 166, 167, 172, 173, 174, 175, 177, 345, 353, 369
White, Hayden, 130
Wigner, Eugene, 15, 102, 162, 166, 167, 170, 171, 177, 178
Wiles, Maurice, 8, 188, 283, 343
Williams, R. M., 50, 395

Wittgenstein, L., 82, 105, 106, 111, 216
Wood, David, 129, 131
Woolf, Harry, 163
Word, 396, 397
Worrall, John, 305, 404
Wright, John, 106

Yourgrau, W., 349

Zapatrin, R. R., 177
Zurek, Wojciech H., 213
Zycinski, Joseph, 2, 96